Studies in Computational Intelligence

Volume 778

Series editor

Janusz Kacprzyk, Polish Academy of Sciences, Warsaw, Poland
e-mail: kacprzyk@ibspan.waw.pl

The series "Studies in Computational Intelligence" (SCI) publishes new developments and advances in the various areas of computational intelligence—quickly and with a high quality. The intent is to cover the theory, applications, and design methods of computational intelligence, as embedded in the fields of engineering, computer science, physics and life sciences, as well as the methodologies behind them. The series contains monographs, lecture notes and edited volumes in computational intelligence spanning the areas of neural networks, connectionist systems, genetic algorithms, evolutionary computation, artificial intelligence, cellular automata, self-organizing systems, soft computing, fuzzy systems, and hybrid intelligent systems. Of particular value to both the contributors and the readership are the short publication timeframe and the worldwide distribution, which enable both wide and rapid dissemination of research output.

More information about this series at http://www.springer.com/series/7092

Anis Koubaa

Editor

Robot Operating System (ROS)

The Complete Reference (Volume 3)

 Springer

Editor
Anis Koubaa
College of Computer Science
   and Information Systems
Prince Sultan University
Riyadh
Saudi Arabia

and

CISTER Research Unit
Porto
Portugal

and

Gaitech Robotics
Shanghai, Beijing
China

ISSN 1860-949X   ISSN 1860-9503   (electronic)
Studies in Computational Intelligence
ISBN 978-3-319-91589-0   ISBN 978-3-319-91590-6   (eBook)
https://doi.org/10.1007/978-3-319-91590-6

Library of Congress Control Number: 2018940893

Printed on acid-free paper

This Springer imprint is published by the registered company Springer International Publishing AG
part of Springer Nature
The registered company address is: Gewerbestrasse 11, 6330 Cham, Switzerland

Acknowledgements to Reviewers

The Editor would like to thank the following reviewers for their great contributions in the review process of the book by providing a quality feedback to authors.

Reviewers

Anis Koubâa, Prince Sultan University, Saudi Arabia / CISTER Research Unit, Portugal

Abdulla Al-Kaff, Universidad Carlos III de Madrid

David Portugal, Ingeniarius, Ltd.

Maram Alajlan, King Saud University

Joao Fabro, UTFPR - Federal University of Technology-Parana

Valerio De Carolis, Heriot-Watt University

Ricardo Julio, UNIFEI

Andre Oliveira, UTFPR

Juan Jesús Roldán Gómez, Universidad Politécnica de Madrid

Kostas Alexis, University of Nevada, Reno

Christoph Rösmann, Institute of Control Theory and Systems Engineering, TU Dortmund University

Guilherme Sousa Bastos, UNIFEI

Walter Fetter Lages, Universidade Federal do Rio Grande do Sul

Lennart Kryza, TU Berlin

Vladimir Ivan, The University of Edinburgh

Elena Peña-Tapia, Universidad Politécnica de Madrid

L. V. R. Arruda, UTFPR

Michael Hutchinson, Loughborough University

Maximilian Krämer, Technische Universität Dortmund

Gonçalo Martins, University of Coimbra

Viswanath Bellam, Robotics Research Industry

Ali Bin Wahid, Shanghai Gentech Scientific Instruments

Paulo Drews Jr., FURG

Franz Albers, TU Dortmund

Christos Papachristos, UNR

João Santos, FARO
Moritz Luetkemoeller, TU Dortmund
Christos Papachristos, UNR
Alvaro Cantieri, IFPR

Acknowledgments

The Editor would like to thank the Robotics and Internet of Things (RIoT) Lab of the College of Computer and Information Sciences of Prince Sultan University for their support to this work.

Furthermore, the Editor thanks Gaitech Robotics in China for their support.

Contents

Part III Navigation, Motion Planning and Control

Part IV Contributed ROS Packages

Part V Interfaces for Interaction with Robots

Part I
Multi-robot Systems

A ROS-Based Framework for Simulation and Benchmarking of Multi-robot Patrolling Algorithms

David Portugal, Luca Iocchi and Alessandro Farinelli

Abstract Experiments with teams of mobile robots in the physical world often represent a challenging task due to the complexity involved. One has to make sure that the robot hardware configuration, the software integration and the interaction with the environment is thoroughly tested so that the deployment of robot teams runs smoothly. This usually requires long preparation time for experiments and takes the focus away from what is essential, i.e. the cooperative task performed by the robots. In this work, we present *patrolling_sim*, a ROS-based framework for simulation and benchmarking of multi-robot patrolling algorithms. Making use of Stage, a multi-robot simulator, we provide tools for running, comparing, analyzing and integrating new algorithms for multi-robot patrolling. With this framework, roboticists can primarily focus on the specific challenges within robotic collaborative missions, run exhaustive tests in different scenarios and with different team sizes in a fairly realistic environment, and ultimately execute quicker experiments in the real world by mimicking the setting up of simulated experiments.

Keywords Multi-robot systems · Multi-robot patrol · Simulation
Benchmarking · ROS package

D. Portugal (✉)
Ingeniarius, Ltd, R. Coronel Veiga Simão, Edifício CTCV, 3ºPiso,
3025-307 Coimbra, Portugal
e-mail: davidbsp@ingeniarius.pt

L. Iocchi
The Department of Computer, Control, and Management Engineering,
Sapienza University of Rome, Via Ariosto 25, Rome 00185, Italy
e-mail: iocchi@diag.uniroma1.it

A. Farinelli
The Department of Computer Science, University of Verona,
Strada Le Grazie 15, Ca' Vignal 2, Verona, Italy
e-mail: alessandro.farinelli@univr.it

© Springer International Publishing AG, part of Springer Nature 2019
A. Koubaa (ed.), *Robot Operating System (ROS)*, Studies in Computational
Intelligence 778, https://doi.org/10.1007/978-3-319-91590-6_1

3

1 Introduction

The field of Robotics has witnessed significant advances recently, and the generalized use of a common middleware for robotic applications, the Robot Operating System (ROS) [1], has contributed to this phenomenon. Nowadays, researchers do not reinvent the wheel when developing robotic applications, since they often benefit and build upon the community contributions (algorithms, tools, drivers, etc.) in fundamental tasks such as interfacing with sensors, debugging, localization, etc. This also led roboticists to increasingly make their code available as open source, allowing the community to improve and leverage the existing functionality, thus fostering innovation in the field.

Multi-robot systems (MRS) are a research area within Robotics, in which a set of robots operate in a shared environment in order to accomplish a given task. The applications of MRS are vast and have been documented previously in the literature [2]. In fact, when they are efficiently deployed, MRS have several advantages over single robot solutions, such as: distributed control, increased autonomy, ability to communicate, greater fault-tolerance, redundancy, assistance by teammates when needed, space distribution, performing different tasks in parallel, and quicker mission accomplishment [3]. In general, MRS have the potential to increase the robustness and reliability of the robotic solution, remaining functional with some degree of performance degradation in case a member of team fails. However, one of the main challenges in such systems is to coordinate multiple robots so as to execute collective complex tasks efficiently, while maximizing group performance under a variety of conditions and optimizing the available resources. Thus, a coordination mechanism is necessary to select actions, solve conflicts, and establish relationships between robots so they can effectively fulfill the mission [4].

In this work, we present a ROS-based framework for simulation and benchmarking of MRS. In particular, we focus on multi-robot patrolling (MRP) as a case study. In MRP, a set of mobile robots should coordinate their movements so as to patrol an environment. This is a widely studied and challenging problem for MRS coordination with a wide range of practical application scenarios (see Sect. 2).

In more detail, by making use of Stage [5], a scalable and fairly realistic multi-robot simulator, we provide tools for running, comparing, analyzing and integrating new algorithms for the coordination of multiple robots performing patrolling missions. Our main goal is to relief researchers from the effort of setting up complex MRS experiments, shifting the focus to the coordination mechanism between robots, enabling exhaustive tests in different scenarios and with different team sizes, and bridging the gap between simulations and real world experiments.

In the next section, we provide the motivation and background behind this work, and in Sect. 3, the proposed framework for MRP simulation and benchmarking named *patrolling_sim* is described in detail. In Sect. 4, we discuss challenges, benefits of using the framework and lessons learned, and finally the chapter ends with conclusions and future work.

2 Background

Multi-robot systems and related subjects, such as design [6], communication [7], and path-finding [8] gained increased attention during the 80s. Still, early work on inspection robots [9], navigation of patrol robots [10], and robot security guards [11] focused exclusively on single robot solutions. In the end of the 80s and beginning of the 90s, the first physical multi-robot systems have been documented in pioneering research works with small populations of robots by researchers from Japan and the USA [12–15]. During the 90s, a significant boost in work on MRS has been noticeable, with a growing involvement of European researchers. In this decade, robotics competitions, especially RoboCup [16] played an important role to foster MRS research.

Patrol is generally defined as "the activity of going around or through an area at regular intervals for security purposes" [17]. For MRS, this is a somehow complex mission, requiring an arbitrary number of mobile robots to coordinate their decision-making with the ultimate goal of achieving optimal group performance by visiting all point in the environment, which require surveillance. It also aims at monitoring, protecting and supervising environments, obtaining information, searching for objects and detecting anomalies in order to guard the grounds from intrusion. Hence, a wide range of applications are possible, as exemplified in Table 1.

Employing teams of robots for active surveillance tasks has several advantages over, for instance, a camera-based passive surveillance system [18]. Robots are mobile and have the ability to travel in the field, collect environmental samples, act or trigger remote alarm systems and inspect places that can be hard for static

Table 1 Examples of applications of multi-robot patrol

Area of application	Example
Rescue operations	Monitoring trapped or unconscious
Military operations	Mine clearing
Surveillance and security	Clearing a building
Supervision of hazardous environments	Patrolling toxic environments
Safety	Preventive patrol for gas leak detection
Environmental monitoring	Sensing humidity and temperature levels inside a facility
Planetary exploration	Collecting samples
Cooperative cleaning	Household vacuum and pool cleaning
Areas with restricted access	Sewerage inspection
Vehicle routing	Transportation of elderly people
Industrial plants	Stock storage
Computer systems	War-game simulations

cameras to capture. These capabilities are highly beneficial to safeguard human lives and provide a great amount of flexibility to the deployed system [19].

Early work on applications using teams of mobile robots in surveillance contexts addressed essentially cooperative sweeping, coverage, and multi-robot coordination [20–24]. The MRP as we know it, started to receive more attention following the pioneer work of Machado et al. [25], where the environment was abstracted using a topological representation, i.e., a patrol graph, which connected the key points in the area that needed regular visits by agents. The authors proposed and compared several patrolling architectures, using different agent behaviors, different communication assumptions and decision-making methods. Criteria based on the average and maximum idleness of the vertices were proposed to evaluate the performance of each technique using a simplistic JAVA-based patrolling simulator [26]. However, conclusions were solely drawn on two scenarios, and unweighted edges were used, meaning that agents always take the same time to travel from one vertex to another, independently of the distance between them.

Since then, several different MRP strategies with increasingly less relaxed assumptions have been presented, based on a wide variety of concepts, such as: simple reactive architectures [27], game theory [28–31], task allocation [32, 33], market-based coordination [34–37], graph theory [38–42], gaussian processes theory [43, 44], Markov decision processes [45, 46], evolutionary algorithms [47], artificial forces [48], reinforcement learning [49, 50], swarm intelligence [51–53], linear programming modeling [54], bayesian heuristics [55, 56] and others with sub-optimal results, leading to several detailed dissertations and thesis on the subject [57–66]. Lately, an effort for real world validation of MRP systems has been evident [67–70].

There are slight variations to the MRP. The idleness concept, i.e. the time that a point in the environment spends without being visited by any robot, has been broadly used in the literature as a basic performance metric, while other authors proposed alternatives such as the frequency of visits to important locations [71, 72]. Additionally, different coordination methods for the team of agents have been studied, such as centralized deterministic [73] and distributed probabilistic methods [74].

Important theoretical contributions on the area patrolling problem have also been presented previously [75–79], and it has been showed that the multi-robot patrolling problem is NP-Hard, and it can also be solved optimally for the single robot situation by finding a TSP tour in the graph that describes the environment to patrol (cf. Fig. 1).

In this paper, we propose a patrolling framework focusing on intelligent strategies for coordination of the team in order to effectively visit all the surveillance points that need vigilance inside a target area. Thus, we do not address other variants of the the problem, like border/perimeter patrol [80, 81], and adversarial patrol [28, 82]. Building teams of robots has costs, and producing robots in large quantities may be expensive. Moreover, doing experiments with physical robotic teams is a long-term effort, which requires an integrated implementation, and it lacks the possibility to easily repeat an experiment with different approaches. Thus, choosing a useful, flexible, scalable and realistic multi-robot simulator is a key task [83]. Simulations are repeatable, faster to deploy, and can be fully automatic, enabling the comparison of different algorithms with different setups (e.g., robots types, fleet sizes,

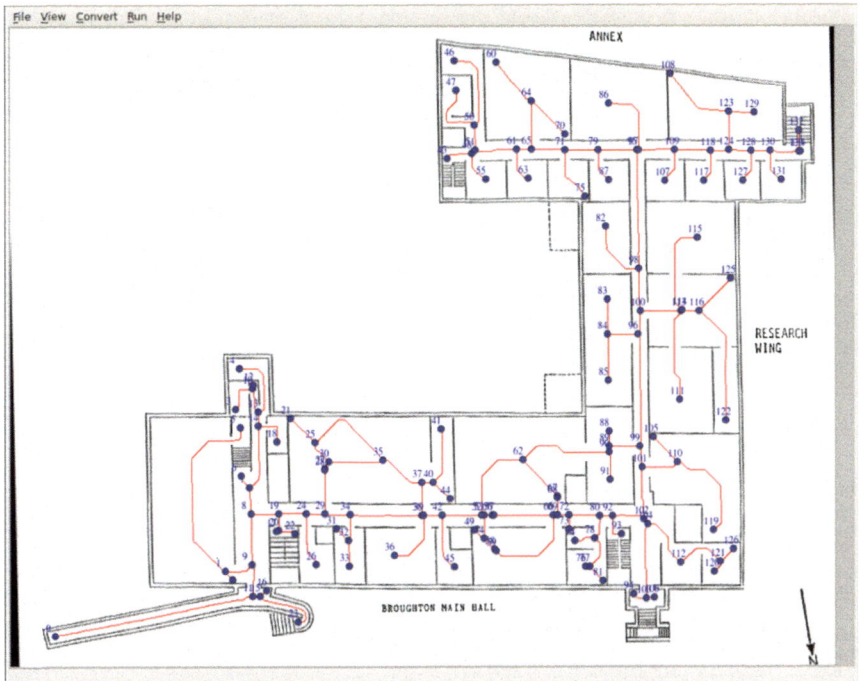

Fig. 1 A patrol graph displayed on top of a metric map to be used in multi-robot patrolling tasks. The blue dots represent the vertices of the graph that must be visited, while the red arcs represent the edges that connect pairs of vertices

environments), and simulation testbeds for MRS are nowadays crucial to rapidly reproduce experiments [84]. While ROS and Stage provide the key building blocks to develop realistic simulations of robotics systems, there is no ready-to-use framework that allows researchers to run experiments testing and validating multi-robot coordination strategies.

Against this background we present *patrolling_sim*,[1] a ROS-based framework for simulation and benchmarking of multi-robot patrolling algorithms, which has been developed and used by the authors in previous works [33, 85]. The *patrolling_sim* framework allows to run exhaustive tests in different scenarios and with different team sizes in fairly realistic environments, and ultimately to execute quicker experiments in the real world by mimicking the setting up of simulated experiments. In the next section, we describe such a framework in more details.

---

[1]http://wiki.ros.org/patrolling_sim.

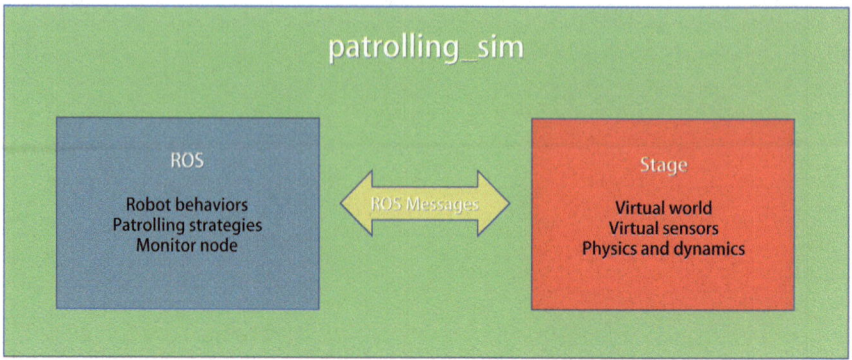

Fig. 2 High-level overview of the patrolling simulation framework

3 Patrolling Simulation Framework

Work on the *patrolling_sim* began in 2010 with the need to compare distinct multi-robot patrolling strategies [86] using a simulation environment and different team sizes. At the time, ROS CTurtle, the second official release of ROS, was used, and 5 patrolling strategies were implemented and integrated: Consciencious Reactive (CR) [25], Heuristic Conscientious Reactive (HCR) [57], Heuristic Pathfinder Conscientious Cognitive (HPCC) [57], Cyclic Algorithm for Generic Graphs (CGG) [38] and the Multilevel Subgraph Patrolling (MSP) Algorithm [38]. Over the years, several utilities, features and algorithms were progressively added, and the framework has been migrated to recent ROS distributions, being currently supported in ROS Kinetic Kame. Figure 2 illustrates the main components of *patrolling_sim*. In the next subsections, we take a deeper look into these components: ROS and Stage, and we overview and highlight some of the key design choices and features available in *patrolling_sim*.

3.1 Robot Operating System (ROS)

Despite the existence of many different robotic frameworks that were developed in the last decades, the Robot Operating System (ROS) has already become the most trending and popular robotic framework, being used worldwide due to a series of features that it encompasses, and being the closest one to become the standard that the robotics community urgently needed. The required effort to develop any robotic application can be daunting, as it must contain a deep layered structure, starting from driver-level software and continuing up through perception, abstract reasoning and beyond. Robotic software architectures must also support large-scale software integration efforts. Therefore, usually roboticists end up spending excessive time with engineering solutions for their particular hardware setup [87]. In the past, many robotic researchers solved some of those issues by presenting a wide variety of frameworks to manage complexity and facilitate rapid prototyping of software for experiments, thus resulting in the many robotic software systems currently used in academia and industry, like YARP [88], Orocos [89], CARMEN [90] or Microsoft Robotics Studio [91], among others. Those frameworks were designed in response to perceived weaknesses of available middlewares, or to place emphasis on aspects which were seen as most important in the design process. ROS is the product of trade-offs and prioritizations made during this process [1].

The major goals of ROS are hardware abstraction, low-level device control, implementation of commonly-used functionalities, message-passing between processes and package management. ROS promotes code reuse with different hardware by providing a large amount of libraries available for the community, like laser-based 2D SLAM [92], 3D point cloud based object recognition [93], among others, as well as tools for 3D visualization (*rviz*), recording experiments and playing back data offline (*rosbag*), and much more.

Regular updates and broad community support enable the users to obtain, build, write, test and run ROS code, even across multiple computers, given its ability to run distributedly in many processors. Additionally, since it is highly flexible, with a simple and intuitive architecture, ROS allows reusing code from numerous other open-source projects such as several Player robot drivers, the Stage 2D and Gazebo 3D simulation environments, Orocos, mostly for industrial robots and machine control, vision algorithms from the Open Source Computer Vision (OpenCV) library [94], etc. As a result, integrating robots and sensors in ROS is highly beneficial.

Due to its peer-to-peer, modular, tools-based, free and open-source nature, ROS helps software developers in creating robotic applications in a quick and easy way. These applications can be programmed in C++, Python, LISP or Java, making ROS a language-independent framework. Furthermore, ROS places virtually all complexity in libraries, only creating small executables, i.e. *nodes*, which expose library functionalities to ROS. *Nodes* communicate by publishing or subscribing to messages at a given *topic*. The *topic* is a message bus, typically named so that it is easy to identify the content of the *message*. Hence, a *node* that requires a certain kind of data, subscribes to the appropriate *topic*. There may be multiple concurrent publishers and subscribers for a single *topic*, and a single *node* may publish and/or subscribe to multiple *topics*. The idea is to decouple the production of information from its consumption.

Beyond the easiness of using the available tools, ROS also provides seamless integration of new sensors without the need for hardware expertise. As a result, the overall time spent in developing software is greatly reduced due to code reuse and hardware abstraction, when using available ROS drivers to interface with the hardware (Fig. 3).

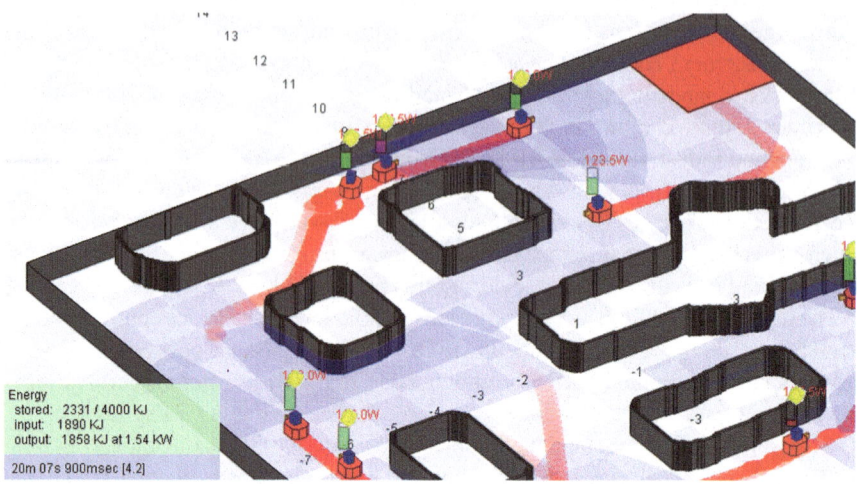

Fig. 3 Example of a simulation in Stage. Extracted from http://playerstage.sourceforge.net/?
src=stage

3.2 Stage Multi-robot Simulator

The scalability of multi-robot simulators has always been a known issue. 3D simulators like Gazebo [95], MORSE [96], and V-Rep [97] normally fail to keep up the frame rate and the *simulated versus real* time ratio with teams of low number of mobile robots, such as 5 or 6, with advanced navigation and perception capabilities in modern day computers. Clearly, in order to be able to simulate at least a dozen mobile robots under the abovementioned conditions, a more lightweight simulator is necessary. Stage [98] is a C++ software library designed to support research into multi-agent autonomous systems. Stage simulates not only a population of mobile robots, but also sensors and objects in a two-dimensional (2D) bitmapped environment. It is a 2D dynamic physics simulator with some three-dimensional (3D) extensions, thus commonly being described as a 2.5D (two-and-a-half dimensional) simulator. Its graphical interface is designed using OpenGL, which takes advantage of graphics processor (GPU) hardware, being fast, easy to use, and having wide availability.

Stage was originally developed as the simulation back-end for the Player/Stage system [5]. Player clients developed using Stage usually work with little or no modification on real robots and vice-versa. Thus, Stage allows rapid prototyping of controllers destined for real robots. This is a powerful argument to support the real world validity of Stage-only experiments and a major advantage of using a well-known simulator. Stage also allows experiments with realistic robot devices that one may not happen to have. Various sensors and actuator models are provided, including range-finders (sonars, laser scanners, infrared sensors), vision (color blob detection), 3D depth-map camera, odometry (with drift error model), and differential steer robot base. Stage is relatively easy to use, it is realistic for many purposes, yielding a useful balance between fidelity and abstraction that is different from many alternative simulators. It runs on Linux and other Unix-like platforms, including Mac OS X, which is convenient for most roboticists, and it supports multiple robots sharing a world. Moreover, Stage is also free and open-source, has an active community of users and developers worldwide, and has reached a well-known status of being a robust simulation platform.

Stage is made available for ROS, through the *stageros* node from the *stage_ros* package,[2] which wraps the Stage multi-robot simulator. Using standard ROS topics, *stageros* provides odometry data from each virtual robot and scans from the corresponding laser model. Additionally, a ROS node may interact with Stage by sending velocity commands to differentially drive the virtual robot.

3.3   Installation and Initializing Experiments

At the time of writing, the patrolling simulation framework supports the latest ROS Long-Term Support (LTS) release: ROS Kinetic Kame. Assuming one is running Ubuntu Linux OS, the installation steps are quite simple as seen below:

1. Install ROS Kinetic Kame, following the instructions at:
   http://wiki.ros.org/kinetic/Installation/Ubuntu
2. Install needed dependencies, by typing in the terminal:
   ```
   $ sudo apt install ros-kinetic-move-base ros-kinetic-
   amcl ros-kinetic-map-server
   ```
3. Setup your ROS catkin workspace, by typing in the terminal:
   ```
   $ mkdir -p ~/catkin_ws/src
   $ cd ~/catkin_ws
   $ catkin_make
   $ source devel/setup.bash
   ```
4. Add the following two lines to your bash configuration file (at /home/$USER/.bashrc):
   ```
   source ~/catkin_ws/devel/setup.bash
   export ROS_WORKSPACE=~/catkin_ws
   ```
5. Download and compile *patrolling_sim*:
   ```
   $ roscd; cd src
   $ git clone https://github.com/davidbsp/patrolling_sim
   $ roscd; catkin_make
   ```

After successfully downloading and compiling the *patrolling_sim* framework, one can easily initiate and configure multi-robot patrolling experiments by running the *start_experiment.py* script as seen in Fig. 4:
```
$ rosrun patrolling_sim start_experiment.py
```

---

[2]http://wiki.ros.org/stage_ros.

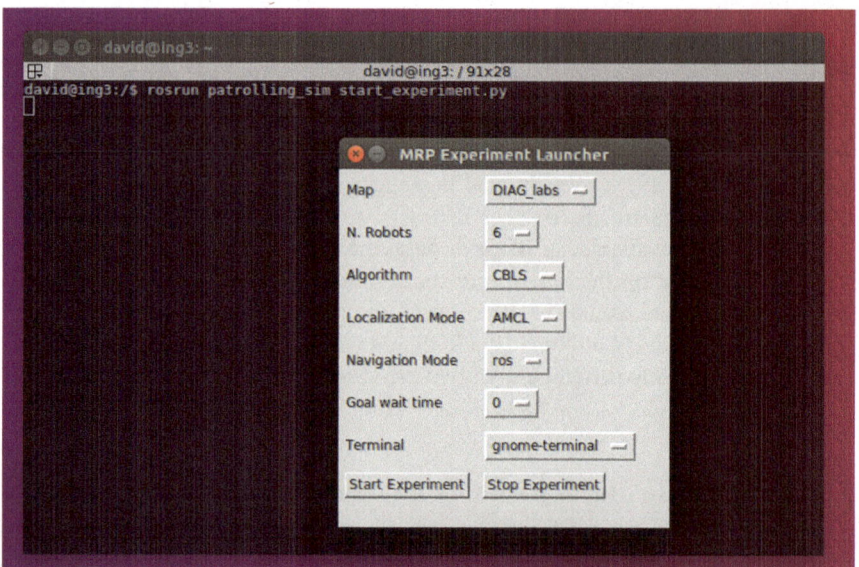

Fig. 4 User configuration interface

The script triggers a user configuration interface that has been implemented using the *TkInter* GUI Programming toolkit for Python [99]. This enables easy configuration of simulated patrolling missions using ROS and Stage. Namely, the configuration interface allows users to choose between different environment maps, robot team sizes, patrolling algorithms, localization modes, navigation modes, waiting times when reaching patrolling goals, and even different types of terminals. Due to the expandability and flexibility of the patrolling framework, the user can easily add additional maps, and patrolling algorithms beyond those referred in Sect. 3.

Currently, two localization modes are supported: Adaptive Monte Carlo Localization (AMCL) and fake localization. AMCL is a probabilistic global localization algorithm [100], which uses a particle filter to track the pose of a robot, by fusing laser scan matching with a source of odometry in order to provide the estimate of the robot's pose with respect to a known map reference frame. Fake localization is a much more lightweight localization node for simulations that provides the same interface to the robots as AMCL, and simply forwards perfect localization information reported by the Stage simulator with negligible computation cost.

Furthermore, the robots leverage from autonomous navigation by following two possible approaches: ROS navigation or *spqrel_navigation*. On one hand, ROS navigation [101] is available out of the box in ROS via the navigation metapackage. This way, given any physically reachable goal, the robot should be able to autonomously navigate to that goal, avoiding collisions with obstacles on the way by following a series of steps. The navigation system at the high level is fairly simple: it takes in a navigation goal, data from sensors, and localization information, and outputs velocity commands that are sent to the mobile robot base via the *move_base* node. Autonomous navigation in *patrolling_sim* is achieved with a known *a priori* map. Therefore, the robot will follow informed plans considering distant obstacles. The navigation algorithm includes several interesting features. For instance, Random Sample Consensus (RANSAC) is applied to filter out Light Detection And Ranging (LIDAR) readings that are invalid due to hardware limitations in the real world, such as false positives generated by veiling effects. Also, a planar costmap, which is initialized with the static map, is used to represent obstacle data and the most recent sensor data, in order to maintain an updated view of the robots local and global environment. Inflation is performed in 2D to propagate costs from obstacles out to a specified radius in order to conservatively avoid collisions. The global planner uses an A^* algorithm that plans in configuration space computed during obstacle inflation in the costmap, not taking into account the dynamics or the kinematics of the robot, which are considered instead by the local planner, which generates velocity commands for the robot, safely moving it towards a goal. The planner cost function combines distance to obstacles, distance to the path produced by the global planner, and the speed at which the robot travels. Finally, a few recovery behaviors can be performed, e.g. due to entrapment. The robot will perform increasingly aggressive behaviors to clear out space around it, and check if the goal is feasible.

On the other hand, *spqrel_navigation*[3] [102] is a lightweight alternative for ROS navigation, which includes two ROS nodes: *srrg_localizer2d*, a lightweight variant for the AMCL node, and *spqrel_planner*, a lightweight variant for the *move_base* node. They have the same interfaces as AMCL and *move_base*, so they can be used in their replacement with minimal effort. Also *spqrel_navigation* is open source and it has been created with the goal to run on systems with limited computational resources, thus it is very suitable for multi-robot simulations on a single machine or for low-cost multi-robot systems. At the high-level, the *spqrel_navigation* package has the same interfaces of ROS navigation and therefore it can be easily used as a replacement for it. However, a significant decrease in computation load when compared to ROS navigation can be expected.

After choosing the desired configuration, and pressing the "start experiment" button, the script will dynamically trigger ROS *launch* files, which will start the necessary ROS *nodes* and *parameters* to accommodate the configurations chosen (e.g. setting the initial position for localization estimation and the actual robots' position in the stage simulator). Moreover, the script will start each different robotic agent with navigation, localization, and communication capabilities in ROS. This is illustrated in Fig. 5.

In addition, one can run a set of experiments using the `run_exp.sh` bash script. After the time defined in the TIMEOUT variable, the command terminates and more instances can be repeated for performing multiple batch experiments. The script template runs a command-line version of the `start_experiments.py` script as many times as intended.

---

[3]https://github.com/LCAS/spqrel_navigation/wiki.

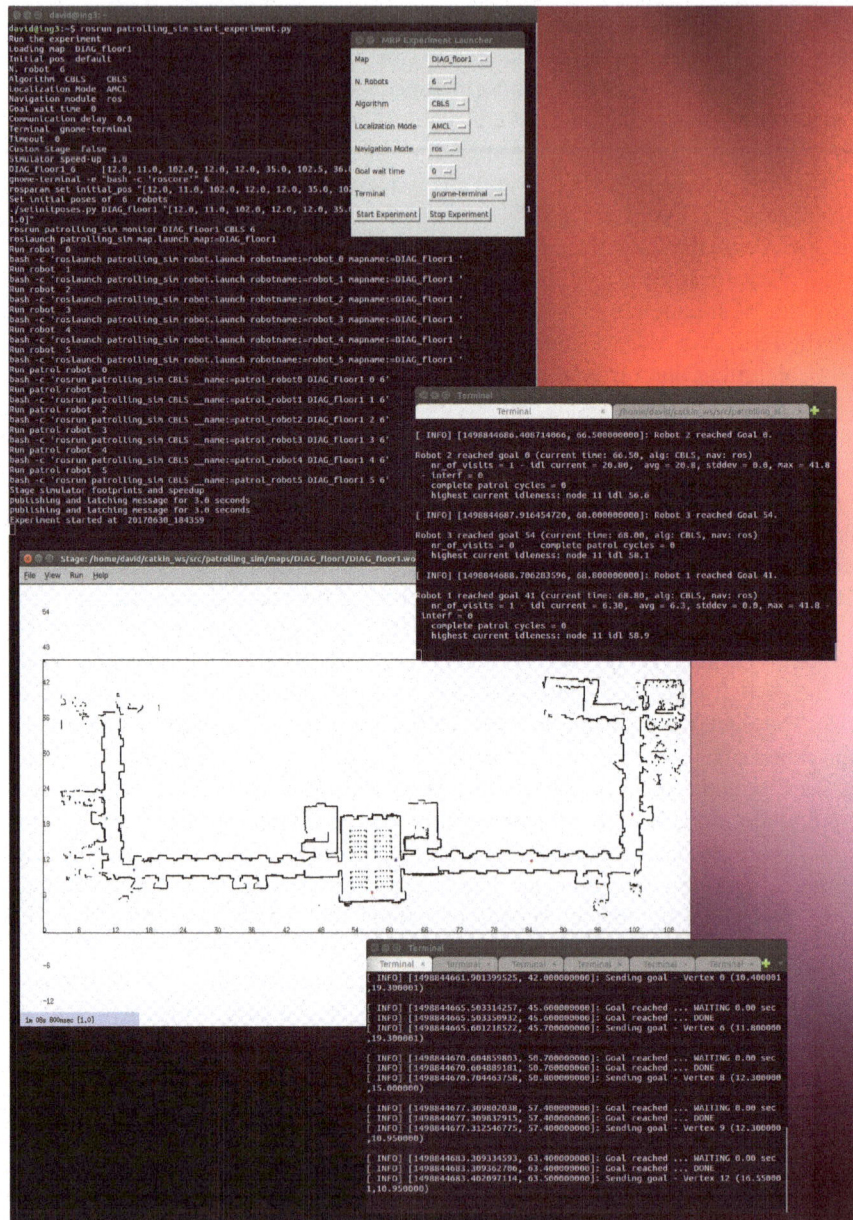

Fig. 5 An experiment running after the initial configuration

3.4 Patrol Agent and Additional Strategies

The patrolling behavior of each robot depends exclusively on the MRP strategy chosen. Typically, the mobile robots within a team follow a similar behavior, only changing their initial conditions, such as their ID and starting position in the environment. During the mission, robots are either given or compute their own waypoints, i.e. vertices of the graph that they should visit, and they continuously coordinate with teammates (for instance, by keeping distances between them, explicitly communicating, etc.) to collectively perform the patrolling task.

Considering the above description of a typical case, it becomes clear that there are several common behaviors within distinct patrolling strategies. Having this in mind, we have provided a general *PatrolAgent* foundation, which can be used for the implementation of robot behaviors. More specifically, *PatrolAgent* represents a base class with general behaviors, which can be extended in derived classes for each specific MRP algorithm that inherit its members and retain its characteristics, and in which they can add their own members.

The common properties of the *PatrolAgent* class include essentially the initialization of agents (assigning the robot ID, extracting relevant map and graph information, initializing control variables, starting positions, idleness tables, ROS publishers and subscribers, etc.); routines for announcing when the robot is ready to start the patrolling mission; actions to perform when the robot moves to a position in the environment, when it arrives there, in case of inter-robot interference and when the simulation finishes; routines for updating parameters based on events, for exchanging poses with other robots, and for saving and sending the robot's own pose.

This way, the inclusion of additional MRP strategies in the *patrolling_sim* framework becomes straightforward and is highly encouraged. One simply needs to create a derived class that inherits all the accessible members of *PatrolAgent*, and modify or add new functions and members to implement the required behaviors of the coordination strategy proposed. This flexibility to add MRP algorithms has resulted in a current total of 11 distinct approaches (cf. Table 2): the 5 original strategies referred in Sect. 3, and 6 additional strategies developed along the years, namely: Random Patrolling (RAND), Greedy Bayesian Strategy (GBS) [103], State Exchange Bayesian Strategy (SEBS) [55], Concurrent Bayesian Learning Strategy (CBLS) [69], Dynamic Task Assignment Greedy (DTAG) [33], and Dynamic Task Assignment based on sequential single item auctions (DTAP) [33].

3.5 Automatically Extracting Results for Analysis

The patrolling framework proposed is based on distributed communication, following a publish/subscribe mechanism, due to its built-in integration in ROS. In the beginning of each test, a specific ROS node is responsible for advertising the start of the mission when all robots are ready, and collecting results during the

Table 2 Overview of MRP strategies in *patrolling_sim*

MRP strategy	Short description
Conscientious reactive (CR)	Robots move locally to the neighbor vertex with higher idleness
Heuristic conscientious reactive (HCR)	Robots decide the neighbor locally, based on idleness and distance
Heuristic pathfinder conscientious cognitive (HPCC)	Robots decide the next vertex globally on the graph, based on idleness, distance, and the vertices in-between
Cyclic algorithm for generic graphs (CGG)	All robots follow the same global route, which visits all vertices in the graph
Multilevel subgraph patrolling (MSP)	Each robot patrols its own region of the graph, using a cyclic strategy in each subgraph
Random patrolling (RAND)	Robots decide randomly the next vertex
Greedy bayesian strategy (GBS)	Robots use local Bayesian decision to maximize their own gain
State exchange bayesian strategy (SEBS)	Similar to GBS, but considers their teammates in the decision to avoid interference
Concurrent bayesian learning strategy (CBLS)	Robots concurrently decide and adapt their moves, according to the system and teammates state, using a reward-based learning technique
Dynamic task sssignment greedy (DTAG)	Robots negotiate greedily the next patrol vertex to visit
Dynamic task assignment based on sequential single item auctions (DTAP)	Robots negotiate all vertices of the graph to build a partition of locations to visit

experiments. This *monitor* node is merely an observer, which analyzes the exchange of communication between robots in the network, and does not provide feedback to them whatsoever. The key objective is to collect experimental results independently, as seen in Fig. 6, which in turn allows benchmarking different MRP algorithms under the same test conditions.

During the patrolling missions, the *monitor* node (cf. Fig. 7) logs several relevant parameters, such as the current idleness of vertices in the graph and corresponding histograms, the average and standard deviation of the idleness of the vertices along time, the total and average number of visits per vertex, the number of complete patrols, the number and rate of inter-robot interference occurrences, the maximum and minimum idleness between all vertices, and the overall average, median and standard deviation of the graph idleness. All these data are saved on files in different formats for later statistical analysis. Some examples of performance metrics and results obtained with *patrolling_sim* are illustrated in [33].

Furthermore, the monitor node controls the patrol termination condition, which can be defined when reaching a given number of patrol cycles (typically a minimum number of visits to all vertices in the graph), as a time window, or any other measurable condition; thus announcing the end of the mission, and stopping the simulation.

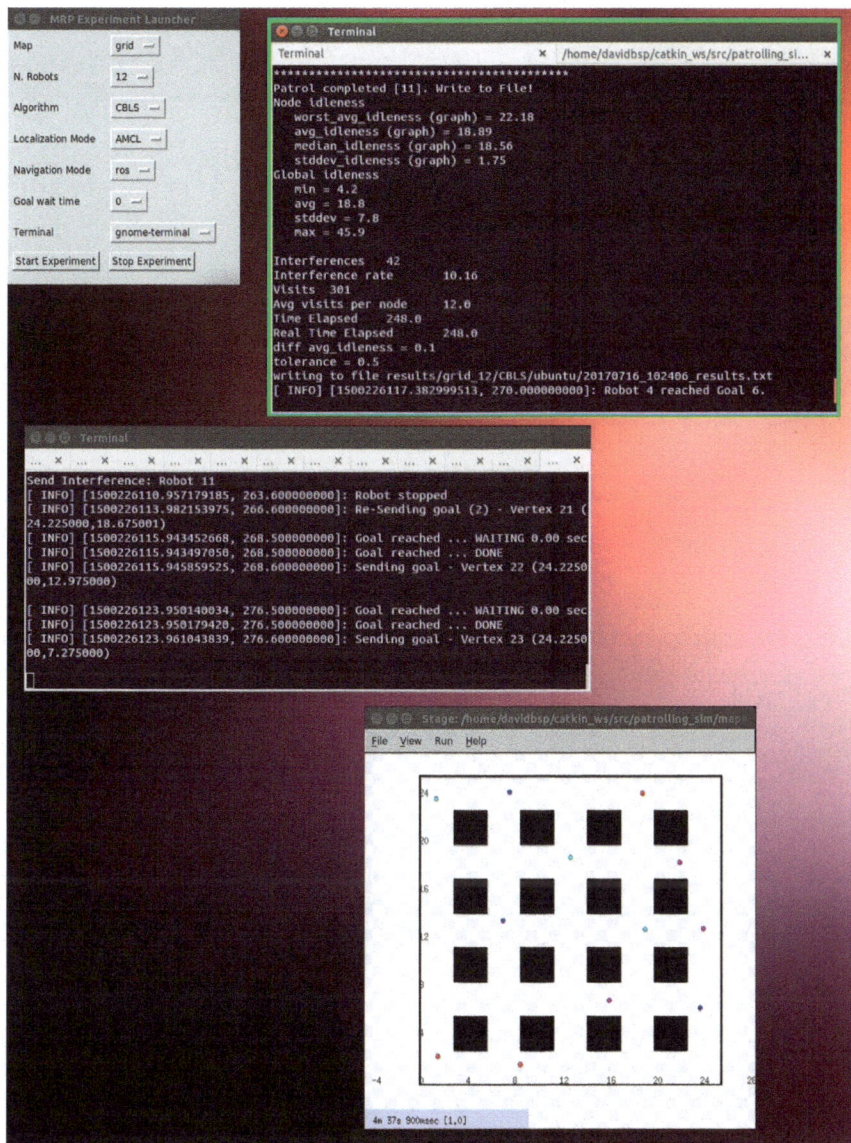

Fig. 6 The monitor node (highlighted in green) announcing the 11th patrol cycle in an experiment with 12 robots in a grid shaped map

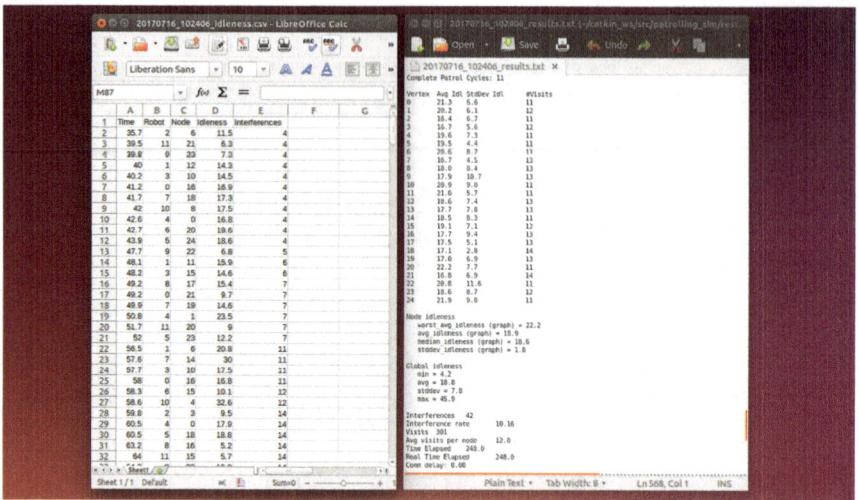

Fig. 7 Log files written by the monitor node, resulting from the experiment of Fig. 6

3.6 Transition to Other Simulators and Real World Experiments

Despite having been developed for use with the Stage multi-robot simulator, *patrolling_sim* can easily be tested with other simulators with minor modifications. To that end, one only needs to launch the framework without resorting to the User Configuration Interface, and replace the ROS launch file that starts the stage simulator with a launch file to start the alternative simulator instead. By having the simulator (environment, robots, sensors, ROS topics, tf frames, etc.) configured similarly to Stage, all the ROS nodes in the system will be able to communicate flawlessly, and simulations will run without issues. In fact, in [83], a multi-robot team on patrol employing the *patrolling_sim* framework, was used as a case study for comparing the Morse and the Gazebo 3D simulators. The quantitative analysis focused on CPU and GPU consumption, thus assessing the scalability of both simulators, and their ability to simulate a limited number of robots. This shows the flexibility and ease of use with other simulators. The patrolling framework has also been tested with the V-REP simulator, according to [104].

In addition to this, *patrolling_sim* can also be exploited for use with teams of physical mobile robots. Some research groups have tested patrolling strategies based on the proposed framework over the past few years. For instance, in [33, 34, 70], experiments with a team of three Turtlebot robots have been described in office-like and corridor-like environments. In [69], experiments with up to six Pioneer-3DX robots have been conducted in a real world large scale infrastructure, and in [68] this number was raised to a total of 8 Turtlebot robots in the experiments reported.

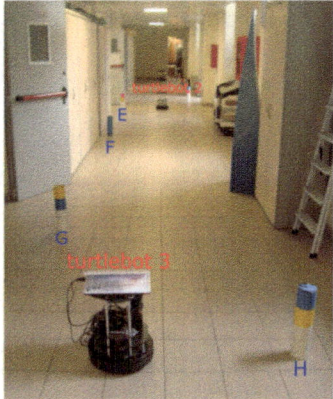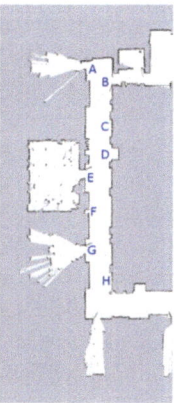

Fig. 8 Example of a real world experiment with 3 turtlebots. Extracted from [33]

According to [33], "*the tests with real robots have been performed by using the same implementation of the algorithms described [...]. ROS infrastructure indeed allows for an easy porting from a Stage-based simulated application to a real one. In particular, Turtlebots in the real environment and robots in the Stage simulator use the same map of the environment, the same configuration of parameters for localization and navigation, and the same implementations of the MRP algorithms*". This is illustrated in Fig. 8.

By testing the execution of the developed algorithms with real robots, the portability of the software to a real environment becomes evident. However, besides the complexity involved in setting up teams of mobile robots for patrolling tasks, these experiments present an additional challenge due to the intrinsic characteristics of ROS, which is typically used for centralized applications, e.g. in single robots or architectures with a common point for processing the information. According to [105], in MRS setups, topics and parameters are often complex and may result in duplicities, high computing costs, large demand for communications (specially over Wi-Fi), delay in the processes and other problems related to system handling by an overloaded single ROS master. Therefore, to avoid this situation, robots in a multi-robot ROS architecture commonly run a dedicated master node.

For the aforementioned physical experiments for MRP, roboticists have used several different solutions to enable the communication between robots running dedicated ROS masters. A few works use external tools, such as the lightweight communications and marshalling (LCM) [106], a library independent of ROS that is used to exchange information between robots in [70], or simply UDP broadcast, as in [68]. Moreover, a set of *multimaster* solutions have been integrated as ROS packages. For instance, in [103] the *wifi_comm*,[4] a multi-robot communication and discovery package was used. This was proposed in [107], being based on *for-*

---

[4]http://wiki.ros.org/wifi_comm.

eign_relay[5] to register topics on foreign ROS masters, and the Optimized Link State Routing (OLSR) protocol to monitor robots connecting and disconnecting from the network, and allowing the deployment of mobile ad-hoc networks. Another solution used for multi-robot communication in [33] is *tcp_interface*,[6] which provides a ROS node that allows easy translation from ROS messages to strings sent over TCP.

Recently, a very promising solution named *multimaster_fkie*[7] [108] has been employed in the hybrid simulated and physical robot experiments reported in [83]. This package offers a set of nodes to establish and manage a multimaster network, requiring no (or minimal) configuration, and all changes in the ROS system are automatically detected and synchronized. From the developer point of view, no specific routines are necessary, which shows the flexibility of the solution, relying only on a simple configuration of the shared topics, nodes and services between different masters. For this reason, the *multimaster_fkie* package will be put in use in the STOP R&D Project[8] [109], which aims at deploying a commercial security system of distributed and cooperative robots on patrol by 2020.

Supporting communication between different ROS master nodes, allow for the exchange of messages between different robots without the need of a server, and supporting the malfunction of any part of the system without compromising the integrity of the whole system, since there is no central point of failure.

4 Discussion

In this work, we take a step towards providing a standard benchmarking framework for running and comparing different patrolling strategies with teams of mobile robots. Due to the complexity of the MRP problem, and the absence of a superior coordination strategy for any environment with any number of robots, we believe that providing a common simulation testbed will allow important advancements in this field of research. Therefore, we welcome the integration of additional strategies by the MRS community.

The simulation of new MRP strategies allows to preliminarily validate them, while enhancing the coordination of robots, the decision-making abilities, and correcting small bugs before moving on to real world experiments. On one hand, real world experiments include noisy sensor readings, localization issues and even robot failures, which may not be precisely modeled in simulation experiments. On the other hand, they may benefit from significant code reuse, tools for analysis, debugging, etc., and the features provided by *patrolling_sim* and ROS itself.

In summary, the patrolling framework proposed is based on the current standard for robotics software – ROS, allowing the easy utilization and transition of experi-

---

[5]http://wiki.ros.org/foreign_relay.

[6]https://github.com/gennari/tcp_interface.

[7]http://wiki.ros.org/multimaster_fkie.

[8]http://stop.ingeniarius.pt/.

ments to the real world, as well as other simulators beyond Stage, e.g. MORSE or Gazebo. It provides various useful features, such as a graphical user interface for parameterizing the simulations or data logs for performance analysis, and provides a balanced trade-off between realism versus computation load, by making use of the 2.5 D Stage simulator.

Despite the focus on MRP, we hope that this framework can be useful for many other MRS applications due to the common intersections between these, e.g. setting up of MRS teams in simulated environments, launching navigation and localization nodes on several robots in parallel, creating a common structure to deal with the simulation of teams of mobile robots, allowing the communication between multiple robots within a team, etc.

5 Conclusion

In this chapter, we have proposed *patrolling_sim*, a ROS-based framework for simulation and benchmarking of MRP algorithms, which has been used in recent years to study the patrolling problem. The framework proposed enables researchers to run, compare, analyze and integrate new algorithms in commonly adopted simulation testbeds. Thus, it places the focus on the coordination between multi-robot teams, and facilitates the preparation of MRS experiments in the physical world with ROS, by mimicking the setting up of simulated experiments and reusing the source code.

Beyond the inclusion of more algorithms and simulated environments in the framework, in the future we would like to add more features, including the full integration of an automatic method to extract patrol graphs from occupancy grids and select initial robot positions for the robotic team, as well as support for running different simulators directly from the configuration GUI.

Acknowledgements This work was supported by the Seguranças robóTicos coOPerativos (STOP) research project (ref. CENTRO-01-0247-FEDER-017562), co-funded by the "Agência Nacional de Inovação" within the Portugal2020 programme.

References

1. M. Quigley, K. Conley, B. Gerkey, J. Faust, T. Foote, J. Leibs, E. Berger, R. Wheeler, A. Ng, ROS: an open-source robot operating system. ICRA Workshop Open Source Softw. 3(2), 00 (2009)
2. T. Arai, E. Pagello, L.E. Parker, Advances in multi-robot systems. IEEE Trans. Robot. Autom. **18**(5), 655–661 (2002)
3. R. Rocha, Building Volumetric Maps with Cooperative Mobile Robots and Useful Information Sharing: a Distributed Control Approach based on Entropy. Ph.D. thesis, Faculty of Engineering of University of Porto, Portugal, 2006
4. A. Farinelli, L. Iocchi, D. Nardi, Multirobot systems: a classification focused on coordination. IEEE Trans. Syst. Man Cybern. Part B: Cybern. **34**(5) (2004)

5. B. Gerkey, R. Vaughan, A. Howard, The player/stage project: tools for multi-robot and distributed sensor systems, in *Proceedings of the IEEE International Conference on Advanced Robotics (ICAR 2003)* (Coimbra, Portugal, June 30–July 3 2003), pp. 317–323

6. E. Freund, On the design of multi-robot systems, in *Proceedings of the 1984 IEEE International Conference on Robotics and Automation (ICRA 1984)*, vol. 1 (IEEE, 1984), pp. 477–490

7. K.G. Shin, M.E. Epstein, Communication primitives for a distributed multi-robot system, IN *Proceedings of the 1985 IEEE International Conference on Robotics and Automation (ICRA 1985)*, vol. 2 (IEEE, 1985), pp. 910–917

8. E. Freund, H. Hoyer, Pathfinding in multi-robot systems: soution and applications, in *Proceedings of the 1986 IEEE International Conference on Robotics and Automation (ICRA 1986)*, vol. 3 (IEEE, 1986), pp. 103–111

9. K. Takehara, Nuclear power plant facility inspection robot. Adv. Robot. **3**(4), 321–331 (1989)

10. S. Xie, T.W. Calvert, B.K. Bhattacharya, Planning viewpoints and the navigation route of a patrol robot in a known 2-D encironment, in *Cambridge Symposium on Intelligent Robotics Systems. International Society for Optics and Photonics, SPIE*, vol. 727 (1987), pp. 206–212

11. T. Kajiwara, J. Yamaguchi, J. Kanemoto, S. Yuta, A security guard robot which patrols using map information, in *Proceedings of the IEEE/RSJ International Workshop on Intelligent Robots and Systems (IROS 1989)* (Tsukuba, Japan, 4–6 Sept 1989)

12. S. Premvuti, S. Yuta, Y. Ebihara, Radio communication network on autonomous mobile robots for cooperative motions, in *Proceedings of 14th IEEE Annual Conference of the Industrial Electronics Society (IECON'88)* (Singapore, 25–27 Oct 1988), pp. 32-37

13. F.R. Noreils, Integrating multirobot coordination in a mobile-robot control system, in *Proceedings of the IEEE International Workshop on Intelligent Robots and Systems (IROS 1990)*, Towards a New Frontier of Applications (IEEE, 1993), pp. 43–49

14. A. Matsumoto, H. Asama, Y. Ishida, K. Ozaki, I. Endo, Communication in the autonomous and decentralized robot system ACTRESS, in *Proceedings IEEE International Workshop on Intelligent Robots and Systems (IROS 1990)*, Towards a New Frontier of Applications (IEEE, 1990), pp. 835–840

15. M. Matarić, Minimizing complexity in controlling a mobile robot population, in *Proceedings of the IEEE International Conference on Robotics and Automation (ICRA 1992)* (Nice, France, 1992), pp. 830–835

16. L. Iocchi, D. Nardi, M. Salerno, Reactivity and deliberation: a survey on multi-robot systems, in *Workshop on Balancing Reactivity and Social Deliberation in Multi-Agent Systems*, Lecture Notes in Computer Science, vol. 2103 (Springer, Berlin Heidelberg, 2001), pp. 9–32

17. Webster's Online Dictionary (2017), http://www.webster-dictionary.org

18. C. King, M. Valera, R. Grech, J. R. Mullen, P. Remagnino, L. Iocchi, L. Marchetti, D. Nardi, D. Monekosso, M. Nicolescu, Multi-robot and multi-camera patrolling, in *Handbook on Soft Computing for Video Surveillance* (CRC Press, 2012), pp. 255–286

19. D. Portugal, Effective Cooperation and Scalability in Mobile Robot Teams for Automatic Patrolling of Infrastructures. Ph.D. thesis, Faculty of Science and Technology, University of Coimbra, Portugal, 2013

20. F.R. Noreils, Multi-robot coordination for battlefield strategies, in *Proceedings of the 1992 IEEE/RSJ International Conference on Intelligent Robots and Systems (IROS 1992)*, vol. 3 (Raleigh, North Carolina, USA, 7–10 July 1992), pp. 1777–1784

21. D. Kurabayashi, J. Ota, T. Arai, E. Yoshida, Cooperative sweeping by multiple mobile robots, in *Proceedings of the 1996 IEEE International Conference on Robotics and Automation (ICRA 1996)*, vol. 2 (Minneapolis, Minesota, USA, 22–28 April 1996), pp. 1744–1749

22. L.E. Parker, B.A. Emmons, Cooperative multi-robot observation of multiple moving targets, in *Proceedings of the 1997 IEEE International Conference on Robotics and Automation (ICRA 1997)*, vol. 3 (Albuquerque, New Mexico, USA, 25–26 April 1997)

23. J. Feddema, C. Lewis, P. Klarer, Control of multiple robotic sentry vehicles, in *AeroSense'99, Proceedings of the SPIE*, Unmanned Ground Vehicle Technology, vol. 3693 (Orlando, Florida, USA, 7–8 April 1999), pp. 212–223,

24. I.A. Wagner, M. Lindenbaum, A.M. Bruckstein, Distributed covering by ant-robots using evaporating traces. IEEE Trans. Robot. Autom.n **15**(5), 918–933 (1999)
25. A. Machado, G. Ramalho, J. Zucker, A. Drogoul, Multi-agent patrolling: an empirical analysis of alternative architectures, in *Multi-Agent-Based Simulation II*, Lecture Notes in Computer Science, vol. 2581 (Springer, Berlin, 2003), pp. 155–170
26. D. Moreira, G. Ramalho, P. Tedesco, SimPatrol - towards the establishment of multi-agent patrolling as a benchmark for multi-agent systems, in *Proceedings of the 1st International Conference on Agents and Artificial Intelligence (ICAART 2009)* (Porto, Portugal), pp. 570–575
27. A. Almeida, G. Ramalho, H. Santana, P. Tedesco, T. Menezes, V. Corruble, Y. Chaveleyre, Recent advances on multi-agent patrolling, in *Advances in Artificial Intelligence (SBIA 2004)*, Lecture Notes in Computer Science, vol. 3171 (Springer, Berlin, 2004), pp. 474–483
28. N. Basilico, N. Gatti, T. Rossi, S. Ceppi, F. Amigoni, Extending algorithms for mobile robot patrolling in the presence of adversaries to more realistic settings, in *Proceedings of the International Conference on Intelligent Agent Technology (IAT09)* (Milan, Italy, 2009), pp. 557–564
29. J. Pita M. Tambe, C. Kiekintveld, S. Cullen, E. Steigerwald, GUARDS-innovative application of game theory for national airport security, in *Proceedings of the 22nd International Joint Conference on Artificial Intelligence (IJCAI11)*, vol 3 (Spain, Barcelona, 2011), pp 2710–2715
30. E. Hernández, A. Barrientos, J. del Cerro, Selective smooth fictitious play: an approach based on game theory for patrolling infrastructures with a multi-robot system. Expert Syst. Appl. **41**(6), 2897–2913 (2014). Elsevier
31. P. de Souza, C. Chanel, S. Givigi, A game theoretical formulation of a decentralized cooperative multi-agent surveillance mission, in *4th Workshop on Distributed and Multi-Agent Planning (DMAP)* (London, UK, 2016)
32. F. Sempé, A. Drogoul, Adaptive patrol for a group of robots, in *Proceedings of the International Conference on Robots and Systems (IROS 2003)* (Las Vegas, USA, 2003)
33. A. Farinelli, L. Iocchi, D. Nardi, Distributed on-line dynamic task assignment for multi-robot patrolling. Auton. Robot. J. **41**(6), 1321–1345 (2017). Springer
34. C. Pippin, H. Christensen, L. Weiss, Performance based task assignment in multi-robot patrolling, in *Proceedings of the ACM Symposium on Applied Computing (SAC 2013)* (Coimbra, Portugal, 18–22 Mar 2013)
35. K. Hwang, J. Lin, H. Huang, Cooperative patrol planning of multi-robot systems by a competitive auction system, in *Proceedings of the ICROS-SICE International Joint Conference* (Fukuoka, Japan, 18–21 Aug 2009)
36. C. Poulet, V. Corruble, A. Seghrouchni, Working as a team: using social criteria in the timed patrolling problem, in *Proceedings of the 24th IEEE International Conference on Tools with Artificial Intelligence (ICTAI 2012)* (Athens, Greece, 7–9 Nov 2012)
37. A. Sugiyama, T. Sugawara, Improvement of robustness to environmental changes by autonomous divisional cooperation in multi-agent cooperative patrol problem, in *Advances in Practical Applications of Cyber-Physical Multi-Agent Systems, 15th International Conference PAAMS 2017*, Lecture Notes in Artificial Intelligence, vol. 10349 (Springer, Berlin, 2017), pp. 259–271
38. D. Portugal, R. Rocha, MSP algorithm: multi-robot patrolling based on territory allocation using balanced graph partitioning, in *Proceedings of 25th ACM Symposium on Applied Computing (SAC 2010)*, Special Track on Intelligent Robotic Systems (Sierre, Switzerland, 22–26 Mar 2010), pp. 1271–1276
39. T. Sak, J. Wainer, S. Goldenstein, Probabilistic multiagent patrolling, in *Brazilian Symposium on Artificial Intelligence (SBIA 2008)* (Salvador, Brazil, 26–30 Oct 2008)
40. R. Stranders, E.M. de Coteb, A. Rogers, N.R. Jennings, Near-optimal continuous patrolling with teams of mobile information gathering agents, in *Artificial Intelligence* (Elsevier, 2012)
41. P. Fazli, A. Davoodi, A.K. Mackworth, Multi-robot repeated area coverage. Auton. Robot. **34**(4), 251–276 (2013)

42. A. Koubâa, O. Cheikhrouhou, H. Bennaceur, M. Sritim, Y. Javed, A. Ammar, Move and improve: a market-based mechanism for the multiple depot multiple travelling salesmen problem. J. Intell. Robot. Syst. **85**(2), 307330 (2017)
43. A. Marino, L. Parker, G. Antonelli, F. Caccavale, Behavioral control for multi-robot perimeter patrol: a finite state automata approach, in *Proceedings of the IEEE International Conference on Robotics and Automation (ICRA 2009)* (Kobe, Japan, 2009), pp. 831–836
44. A. Marino, G. Antonelli, A.P. Aguiar, A. Pascoal, A new approach to multi-robot harbour patrolling: theory and experiments, in *Proceedings of the 2012 IEEE/RSJ International Conference on Intelligent Robots and Systems (IROS 2012)* (Vilamoura, Portugal, 7–12 Oct 2012)
45. J. Marier, C. Besse, B. Chaib-draa, Solving the continuous time multiagent patrol problem, in *Proceedings of the International Conference on Robotics and Automation (ICRA 2010)* (Anchorage, Alaska, USA, 2010)
46. X. Chen, T.S. Yum, Patrol districting and routing with security level functions, in *Proceedings of the International Conference on Systems, Man and Cybernetics (SMC2010)* (Istanbul, Turkey, Oct 2010), pp. 3555–3562,
47. O. Aguirre, H. Taboada, An evolutionary game theory approach for intelligent patrolling. Procedia Comput. Sci. Part II **12**, 140–145 (2012)
48. P. Sampaio, G. Ramalho, P. Tedesco, The gravitational strategy for the timed patrolling, in *Proceedings of the International Conference on Tools with Artificial Intelligence (ICTAI10)* (Arras, France, 27–29 Oct 2010)
49. Y. Ishiwaka, T. Sato, Y. Kakazu, An approach to the pursuit problem on a heterogeneous multiagent system using reinforcement learning. Robot. Auton. Syst. (RAS) **43**(4) (2003)
50. H. Santana, G. Ramalho, V. Corruble, B. Ratitch, Multi-agent patrolling with reinforcement learning, in *Proceedings of the International Conference on Autonomous Agents and Multi-agent Systems*, vol. 3 (New York, 2004)
51. V. Yanovski, I.A. Wagner, A.M. Bruckstein, A distributed ant algorithm for efficiently patrolling a network. Algorithmica **37**, 3765–186 (2003)
52. H. Chu, A. Glad, O. Simonin, F. Sempé, A. Drogoul, F. Charpillet, Swarm approaches for the patrolling problem, information propagation vs. pheromone evaporation, in *Proceedings of the 19th IEEE International Conference on Tools with Artificial Intelligence (ICTAI 2007)*, vol. 1 (IEEE, 2007), pp. 442–449
53. H. Calvo, S. Godoy-Calderon, M.A. Moreno-Armendáriz, V.M. Marínez-Hernández, Patrolling routes optimization using ant colonies, in *Pattern Recognition, 7th Mexican Conference (MCPR 2015)*, Lecture Notes in Computer Science, vol. 9116 (Springer, Berlin, 2015), pp. 302312
54. B.B. Keskin, S. Li, D. Steil, S. Spiller, Analysis of an integrated maximum covering and patrol routing problem. Transp. Res. Part E: Logist. Transp. **48**, 215–232 (2012). Elsevier
55. D. Portugal, R.P. Rocha, Scalable, fault-tolerant and distributed multi-robot patrol in real world environments, in *Proceedings of the 2013 IEEE/RSJ International Conference on Intelligent Robots and Systems* (Tokyo, Japan, 3–7 Nov IROS 2013)
56. H. Chen, T. Cheng, S. Wise, Developing an online cooperative police patrol routing strategy. Comput. Environ. Urban Syst. **62**, 19–29 (2017). Elsevier
57. A. Almeida, Patrulhamento Multiagente em Grafos com Pesos. M.Sc. thesis, Centro de Informtica, Univ. Federal de Pernambuco, Recife, Brazil, Oct 2003 (In Portuguese)
58. D. Moreira, SimPatrol: Um simulador de sistemas multiagentes para o patrulhamento. M.Sc. thesis, Centro de Informática, Univ. Federal de Pernambuco, Recife, Brazil, Sept 2008 (In Portuguese)
59. D. Portugal, RoboCops: A Study of Coordination Algorithms for Autonomous Mobile Robots in Patrolling Missions, Master of Science Dissertation, Faculty of Science and Technology, University of Coimbra, Portugal, Sept 2009
60. A. Franchi, Decentralized Methods for Cooperative Task Execution in Multi-robot Systems. Ph.D. thesis, Department of Computer and System Science, Sapienza University of Rome, Italy, Dec 2009

61. Y. Elmaliach, Multi-Robot Frequency-Based Patrolling. Ph.D. thesis, Department of Computer Science, Bar-Ilan University, Ramat Gan, Israel, Jan 2009
62. N. Agmon, Multi-Robot Patrolling and Other Multi-Robot Cooperative Tasks: An Algorithmic Approach. Ph.D. thesis, Department of Computer Science, Bar-Ilan University, Ramat Gan, Israel, Feb 2009
63. F. Pasqualetti, Secure Control Systems: A Control-Theoretic Approach to Cyber-Physical Security, Ph.D. thesis, Department of Mechanical Engineering, University of California, Santa Barbara, USA, Sept 2012
64. P. Fazli, On Multi-Robot Area and Boundary Coverage, Ph.D. thesis, Department of Computer Science, University of British Columbia, Vancouver, Canada, Aug 2013
65. C.E. Pippin, Trust and Reputation for Formation and Evolution of Multi-Robot Teams. Ph.D. thesis, Georgia Institute of Technology College of Computing, Atlanta, Georgia, USA, Dec 2013
66. E.H. Serrato, Cooperative Multi-Robot Patrolling: A study of distributed approaches based on mathematical models of game theory to protect infrastructures. Ph.D. thesis, Universidade Politécnica de Madrid, Escuela Técnica Superior de Ingenieros Industriales, Madrid, Spain, Dec 2014
67. L. Iocchi, L. Marchetti, D. Nardi, Multi-Robot Patrolling with Coordinated Behaviours in Realistic Environments, in *Proceedings of the International Conference on Intelligent Robots and Systems (IROS 2011)* (San Francisco, CA, USA, 25-30 Sept 2011), pp. 2796–2801
68. C. Pippin, H. Christensen, Trust modeling in multi-robot patrolling, in *Proceedings of the 2014 IEEE International Conference on Robotics and Automation (ICRA 2014)* (Hong Kong, China, 2014), pp. 59–66
69. D. Portugal, R.P. Rocha, Cooperative multi-robot patrol with bayesian learning. Auton. Robot. J. **40**(5), 929–953 (2016). Springer
70. C. Yan, T. Zhang, Multi-robot patrol: a distributed algorithm based on expected idleness. Int. J. Adv. Robot. Syst. 1–12 (2016). SAGE
71. M. Baglietto, G. Cannata, F. Capezio, A. Sgorbissa, Multi-robot uniform frequency coverage of significant locations in the environment, in *Distributed Autonomous Robotic Systems*, vol. 8 (Springer, Berlin, 2009)
72. Y. Elmaliach, N. Agmon, G. Kaminka, Multi-robot area patrol under frequency constraints, in *Proceedings of the 2007 IEEE International Conference on Robotics and Automation (ICRA 2007)* (Rome, Italy, 10–14 April 2007), pp. 385–390
73. F. Pasqualetti, J. Durham, F. Bullo, Cooperative patrolling via weighted tours: performance analysis and distributed algorithms. IEEE Trans. Robot. **28**(5), 1181–1188 (2012)
74. D. Portugal, R.P. Rocha, Cooperative multi-robot patrol in an indoor infrastructure, in *Human Behavior Understanding in Networked Sensing, Theory and Applications of Networks of Sensors* (Springer International Publishing, 2014), pp. 339–358
75. Y. Chevaleyre, Theoretical analysis of the multi-agent patrolling problem, in *Proceedings of the 2004 International Conference on Agent Intelligent Technologies (IAT 2004)* (Beijing, China, 20–24 Sept 2004), pp. 302–308
76. F. Pasqualetti, A. Franchi, F. Bullo, On cooperative patrolling: optimal trajectories, complexity analysis and approximation algorithms. IEEE Trans. Robot. **28**(3), 592–606 (2012)
77. S. Smith, D. Rus, Multi-robot monitoring in dynamic environments with guaranteed currency of observations, in *Proceedings of the 49th IEEE Conference on Decision and Control* (Atlanta, Georgia, USA, 2010), pp. 514–521
78. S. Ruan, C. Meirina, F. Yu, K.R. Pattipati, R.L. Popp, Patrolling in a stochastic environment, in *Proceedings of the 10th International Command and Control Research and Technology Symposium (ICCRTS)* (McLean, Virginia, USA, 13–16 June 2005)
79. D. Portugal, C. Pippin, R.P. Rocha, H. Christensen, Finding optimal routes for multi-robot patrolling in generic graphs, in *Proceedings of the 2014 IEEE/RSJ International Conference on Intelligent Robots and Systems (IROS 2014)* (Chicago, USA, 14–18 Sept 2014)
80. Y. Elmaliach, A. Shiloni, G.A. Kaminka, A realistic model of frequency-based multi-robot polyline patrolling, in *Proceedings of the 7th international joint conference on autonomous agents and multiagent systems (AAMAS 2008)*, vol. 1 (2008), pp. 63–70

81. A. Marino, L.E. Parker, G. Antonelli, F. Caccavale, A decentralized architecture for multi-robot systems based on the null-space-behavioral control with application to multi-robot border patrolling. J. Intell. Robot. Syst. **71**, 423–444 (2013)
82. N. Agmon, D. Urieli, P. Stone, Multiagent patrol generalized to complex environmental conditions, in *Proceedings of the 25th Conference on Artificial Intelligence (AAAI 2011)* (San Francisco, CA, 711 Aug 2011)
83. F. M. Noori, D. Portugal, R.P. Rocha, M.S. Couceiro, On 3D simulators for multi-robot systems in ROS: MORSE or Gazebo?, in *Proceedings of the 15th IEEE International Symposium on Safety, Security, and Rescue Robotics (SSRR 2017)* (Shanghai, China, 11–13 Oct 2017)
84. Z. Yan, L. Fabresse, J. Laval, N. Bouragadi, Building a ROS-based testbed for realistic multi-robot simulation: taking the exploration as an example. Robotics **6**(3), 1–21 (2017)
85. D. Portugal, R.P. Rocha, Multi-robot patrolling algorithms: examining performance and scalability. Adv. Robot. J. Spec. Issue Saf. Secur. Rescue Robot. **27**(5), 325–336 (2013). Taylor and Francis
86. D. Portugal, R.P. Rocha, On the performance and scalability of multi-robot patrolling algorithms, in *Proceedings of the 2011 IEEE International Symposium on Safety, Security, and Rescue Robotics (SSRR 2011)* (Kyoto, Japan, 1–5 Nov 2011), pp. 50–55
87. A. Araújo, D. Portugal, M.S. Couceiro, R.P. Rocha, Integrating Arduino-based Educational Mobile Robots in ROS. J. Intell. Robot. Syst. (JINT) Spec. Issue Auton. Robot. Syst. **77**(2), 281–298 (2015). Springer
88. G. Metta, P. Fitzpatrick, L. Natale, Yarp: yet another robot platform. Int. J. Adv. Robot. Syst. (IJARS) **3**(1), 43–48 (2006)
89. H. Bruyninckx, Open robot control software: the OROCOS project, in *Proceedings of the IEEE International Conference on Robotics and Automation (ICRA 2001)*, vol. 3 (Seoul, Korea Rep., 21–26 May 2001), pp. 2523–2528
90. M. Montemerlo, N. Roy, S. Thrun, Perspectives on standardization in mobile robot programming: the carneggie mellon navigation (CARMEN) toolkit, in *Proceedings of the 2003 IEEE/RSJ International Conference on Intelligent Robots and Systems (IROS2003)* (Las Vegas, Nevada, Oct 2003)
91. J. Jackson, Microsoft robotics studio: a technical introduction. IEEE Robot. Autom. Mag. **14**(4), 82–87 (2007)
92. G. Grisetti, C. Stachniss, W. Burgard, Improved techniques for grid mapping with rao-blackwellized particle filters. IEEE Trans. Robot. **23**(1), 34–46 (2006)
93. R. Rusu, S. Cousins, 3D is here: point cloud library (PCL), in *Proceeding of the IEEE International Conference on Robotics and Automation (ICRA 2011)* (Shanghai, China, 9–13 May 2011)
94. G. Bradski, A. Kaehler, *Learning OpenCV: Computer Vision with the OpenCV Library* (OReilly Media, 2008)
95. N. Koenig, A. Howard, Design and use paradigms for gazebo, an open-source multi-robot simulator, in *Proceedings of the 2004 IEEE/RSJ International Conference on Intelligent Robots and Systems (IROS 2004)*, vol. 3 (Sendai, Japan, Sept 28–Oct 2 2004), pp. 2149–2154
96. G. Echeverria, N. Lassabe, A. Degroote, S. Lemaignan, Modular open robots simulation engine: Morse, in *Proceedings of the 2011 IEEE International Conference on Robotics and Automation (ICRA)* (Shanghai, China, 9–13 May 2011), pp. 46–51
97. M. Freese, S. Singh, F. Ozaki, N. Matsuhira, N., Virtual robot experimentation platform v-rep: a versatile 3d robot simulator, in *The IEEE International Conference on Simulation, Modeling, and Programming for Autonomous Robots (SIMPAR 2010)* (Darmstadt, Germany, Springer, 15,18 Nov 2010), pp. 51–62
98. R. Vaughan, Massively multi-robot simulation in stage. J. Swarm Intell. **2**(2–4), 189–208 (2008). Springer
99. M.J. Conway, Python: a GUI development tool. Interact. Mag. **2**(2), 23–28 (1995)
100. S. Thrun, D. Fox, W. Burgard, F. Dellaert, Robust monte carlo localization for mobile robots. Artif. Intell. (AI) **128**(12), 99–141 (2000)

101. E. Marder-Eppstein, E. Berger, T. Foote, B. Gerkey, K. Konolige, The office marathon: Robust navigation in an indoor office environment, in *Proceedings of the 2010 IEEE International Conference on Robotics and Automation (ICRA 2010)* (Anchorage, AK, USA, May 2010), pp. 300–307
102. M.T. Lazaro, G. Grisetti, L. Iocchi, J.P. Fentanes, M. Hanheide, A lightweight navigation system for mobile robots, in *Proceedings of the Third Iberian Robotics Conference (ROBOT 2017)* (Sevilla, Spain, 22–24 Nov 2017)
103. D. Portugal, R.P. Rocha, Distributed multi-robot patrol: a scalable and fault-tolerant framework. Robot. Auton. Syst. **61**(12), 1572–1587 (2013). Elsevier
104. L. Freda, M. Gianni, F. Pirri, Deliverable 4.3: communication and knowledge flow gluing the multi-robot collaborative framework, in *TRADR: Long-Term Human-Robot Teaming for Disaster Response (EU FP7 ICT Project #609763)* (2016), http://www.tradr-project.eu/wp-content/uploads/dr.4.3.main_public.pdf
105. M. Garzón, J. Valente, J. Roldán, D. Garzón-Ramos, J. de León, A. Barrientos & J. del Cerro, Using ROS in multi-robot systems: experiences and lessons learned from real-world field tests, in *Robot Operating System (ROS) - The Complete Reference (vol. 2), Studies in Computational Intelligence*, vol. 707 (Springer, Berlin, 2017)
106. A. Huang, E. Olson, D.C. Moore DC, LCM: lightweight communications and marshalling, in *Proceedings of the 2010 IEEE/RSJ International Conference on Intelligent Robots and Systems (IROS 2010)* (Taipei, Taiwan, Oct 1822, 2010), pp. 4057–4062
107. G. Cabrita, P. Sousa, L. Marques, A. de Almeida, Infrastructure monitoring with multi-robot teams, in *Proceedings of the 2010 IEEE/RSJ International Conference on Intelligent Robots and Systems (IROS 2010), Workshop on Robotics for Environmental Monitoring* (Taipei, Taiwan, 18–22 Oct 2010)
108. A. Tiderko, F. Hoeller, T. Röhling, The ROS multimaster extension for simplified deployment of multi-robot systems, in *Robot Operating System (ROS) - The Complete Reference (Vol. 1), Studies in Computational Intelligence*, vol. 625 (Springer, Berlin, 2016), pp. 629–650
109. D. Portugal, S. Pereira, M. S. Couceiro, The role of security in human-robot shared environments: a case study in ROS-based surveillance robots, in *Proceedings of the 26th IEEE International Symposium on Robot and Human Interactive Communication (RO-MAN 2017)* (Lisbon, Portugal, Aug 28–Sept 1 2017)

David Portugal completed his Ph.D. degree on Robotics and Multi-Agent Systems at the University of Coimbra in Portugal, in March 2014. His main areas of expertise are cooperative robotics, multi-agent systems, simultaneous localization and mapping, field robotics, human-robot interaction, sensor fusion, metaheuristics, and graph theory. He is currently working as a Senior Researcher at Ingeniarius Ltd. (Portugal), where he has been involved in the STOP R&D technology transfer project on multi-robot patrolling. He has been involved in several local and EU-funded research projects in Robotics and Ambient Assisted Living, such as CHOPIN, TIRAMISU, Social-Robot, CogniWin and GrowMeUp. He has co-authored over 55 research articles included in international journals, conferences and scientific books.

Luca Iocchi is an Associate Professor at Sapienza University of Rome, Italy. His research activity is focused on methodological, theoretical and practical aspects of artificial intelligence, with applications related to cognitive mobile robots and computer vision systems operating in real environments. His main research interests include cognitive robotics, action planning, multi-robot coordination, robot perception, robot learning, sensor data fusion. He is the author of more than 150 referred papers in journals and conferences in artificial intelligence and robotics, member of the program committee of several related conferences, guest editor for journal special issues and reviewer for many journals in the field. He has coordinated national and international projects and, in particular, he has supervised the development of (teams of) mobile robots and vision systems with cognitive capabilities for applications in dynamic environments, such as RoboCup soccer, RoboCup rescue, RoboCup@Home, multi-robot surveillance, and automatic video-surveillance.

Alessandro Farinelli is an Associate Professor at University of Verona, Department of Computer Science, since December 2014. His research interests comprise theoretical and practical issues related to the development of Artificial Intelligent Systems applied to robotics. In particular, he focuses on coordination, decentralised optimisation and information integration for Multi-Agent and Multi-Robot systems, control and evaluation of autonomous mobile robots. He was the principal investigator for several national and international research projects in the broad area of Artificial Intelligence for robotic systems. He co-authored more than 80 peer-reviewed scientific contributions in top international journals (such AIJ and JAAMAS) and conferences (such as IJCAI, AAMAS, and AAAI).

Multi-robot Systems, Virtual Reality and ROS: Developing a New Generation of Operator Interfaces

Juan Jesús Roldán, Elena Peña-Tapia, David Garzón-Ramos,
Jorge de León, Mario Garzón, Jaime del Cerro
and Antonio Barrientos

Abstract This chapter describes a series of works developed in order to integrate ROS-based robots with Unity-based virtual reality interfaces. The main goal of this integration is to develop immersive monitoring and commanding interfaces, able to improve the operator's situational awareness without increasing its workload. In order to achieve this, the available technologies and resources are analyzed and multiple ROS packages and Unity assets are applied, such as $multimaster\_fkie$, $rosbridge\_suite$, RosBridgeLib and SteamVR. Moreover, three applications are presented: an interface for monitoring a fleet of drones, another interface for commanding a robot manipulator and an integration of multiple ground and aerial robots. Finally, some experiences and lessons learned, useful for future developments, are reported.

Keywords Multi-robot systems · Virtual reality · Operator interfaces · Immersive teleoperation

J. J. Roldán (✉) · E. Peña-Tapia · J. de León · M. Garzón
J. del Cerro · A. Barrientos
Centro de Automática y Robótica (UPM-CSIC), Universidad Politécnica
de Madrid, Calle José Gutiérrez Abascal, 2, 28006 Madrid, Spain
e-mail: jj.roldan@upm.es

E. Peña-Tapia
e-mail: elena.ptapia@alumnos.upm.es

J. de León
e-mail: jorge.deleon@upm.es

M. Garzón
e-mail: ma.garzon@upm.es

J. del Cerro
e-mail: j.cerro@upm.es

A. Barrientos
e-mail: antonio.barrientos@upm.es

D. Garzón-Ramos
IRIDIA, Université Libre de Bruxelles (ULB),
CP194/6, Av. F. Roosevelt 50, B-1050 Brussels, Belgium
e-mail: dgarzonr@ulb.ac.be

© Springer International Publishing AG, part of Springer Nature 2019
A. Koubaa (ed.), *Robot Operating System (ROS)*, Studies in Computational
Intelligence 778, https://doi.org/10.1007/978-3-319-91590-6_2

29

(a)   **(b)**

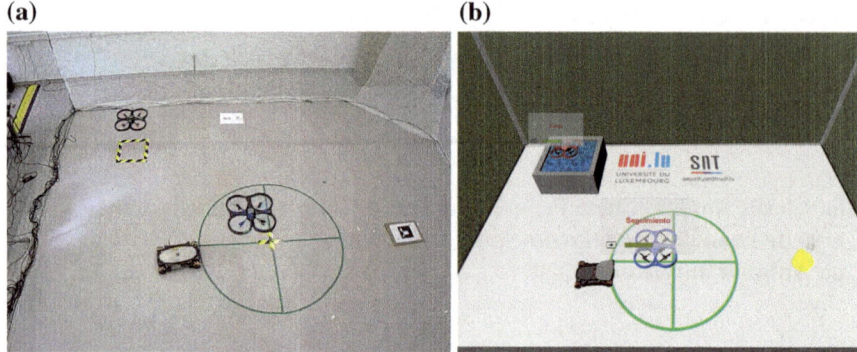

Fig. 1 Aerial robots during the training of a mission of fire search and extinguishing: the aim of this work is to connect **a** real world to **b** virtual world

1 Introduction

Nowadays, Multi-Robot Systems (MRS) are being applied to multiple areas, such as security [1], environmental monitoring [2] and search and rescue [3]. By using MRS it is possible to improve resource allocation, time cost, robustness and many other aspects of a real world mission. Nevertheless, due to workload, situational awareness and other factors, the management of these systems is a challenge for operators. As a result of this, most of the proposals for future interfaces involve multi-modal interactions [4], immersive technologies (e.g. virtual, augmented and mixed reality) [5] and adaptive displays [6].

The work presented in this chapter pretends to establish a connection between the real world, where robot fleets are performing their tasks in certain environments, and the virtual world, where the operators are discovering information about the mission, making decisions and generating commands for the robots (as shown in Fig. 1).

During the last decade, Robot Operating System (ROS) has become one of the most practical and popular frameworks for developing robot software. It is used in a very diverse range of applications, from industrial manipulators to service or mobile robots, in both low and high-level tasks, using single or multi-robot systems. However, the operator interfaces currently provided by the ROS community are still limited. Moreover, the development of immersive interfaces directly under ROS, remains a fairly complex task for the time being.

On the other hand, the world of Virtual Reality (VR), which also has been developed during the last decade, takes advantage of powerful and flexible tools, such as the *Unity*[1] and *Unreal*[2] game engines. These tools allow to develop a wide range of realistic scenarios, intelligent characters and objects with realistic dynamics and

---

[1]https://unity3d.com/.

[2]https://www.unrealengine.com.

kinematics. However, these tools have not been adapted to robotic systems and, sometimes, the work must start from scratch.

The connection between both worlds requires some ROS packages and Unity assets, which are described in detail later in this chapter. The main ROS packages are $multimaster\_fkie$[3] and $rosbridge\_suite$,[4] whereas the fundamental Unity assets are $RosBridgeLib$[5] and $SteamVR$.[6] This chapter describes the layout developed for the integration and it presents tests in both simulated and real scenarios.

The simulated scenarios sought to integrate both ground and aerial robots in the virtual reality environment. For this purpose, the $Gazebo$[7] simulator was used, which includes models of different robots and their respective plugins for controllers, sensors, actuators... The main outcome of this experiments was the integration of the main types of ROS messages, used by the robots to transfer their information to $Unity$, as well as the transformation of inputs in this game engine to commands for controlling the robots via ROS messages.

There were also several tests using real scenarios. These experiments allowed to test general performance in the tasks of controlling and monitoring of real robots within representations of real scenarios in a virtual reality environment. In order to achieve this, a robotic arm was modeled in virtual reality, whereas methods for displaying information and generating commands were developed and validated.

The rest of the chapter is organized as follows:

- Section 2 summarizes the relevant state of art for this work, including multi-robot missions, human factor challenges and operator interfaces.
- Section 3 addresses some guidelines to design and develop an immersive interface, discussing the requirements and resources for this kind of interfaces.
- Section 4 presents the architecture developed in this work, which allows to connect multiple real and simulated robots to conventional and immersive interfaces, as well as the ROS packages and $Unity$ assets implemented in this architecture, explaining how to configure and launch them for any given mission.
- Section 5 collects some applications with MRS, VR and ROS. These applications include a monitoring interface for drones, a commanding interface for a robotic arm, and a complete and flexible interface for MRS.
- Finally, Sect. 6 summarizes the main results and conclusions of these applications, as well as the learned lessons and future challenges.

---

[3] http://wiki.ros.org/multimaster_fkie.

[4] http://wiki.ros.org/rosbridge_suite.

[5] https://github.com/MathiasCiarlo/ROSBridgeLib.

[6] https://steamcommunity.com/steamvr.

[7] http://gazebosim.org/.

2 State of Art

This section collects the main contributions of the literature related to MRS, VR and ROS. There are multiple topics that converge on this application, such as the multi-robot missions, robot fleets, operators, human factor issues, human-robot interaction, interfaces, etc. Therefore, the section is organized in multiple subsections to cover these topics:

- Section 2.1 collects a set of recent multi-robot missions, analyzes them in terms of robots, operators and scenarios, and classifies the different types of MRS.
- Section 2.2 addresses the main human factor challenges that are posed by these missions, considering their problems, methods for detection and potential solutions.
- Finally, Sect. 2.3 analyzes the problems related to the conventional interfaces and reports the proposals of multimodal, adaptive and immersive interfaces to solve them.

2.1 Multi-robot Missions

In the last years, Multi-Robot Systems have become popular and have been implemented in several applications, since they can improve the capabilities of single robots without significantly increase their costs. Their main advantages are:

- Effectiveness: A MRS is more effective than a single robot, since it has more resources to perform the same tasks.
- Efficiency: A MRS is also more efficient than a single robot, since it can assign each task to the most appropriate robot.
- Flexibility: A MRS is more flexible than a single robot, because it is able to adapt to more scenarios.
- Fault tolerance: A MRS is more tolerant against failures than a single robot, because it can recover from a failure in one of the robots.

A diverse set of multi-robot missions from recent literature is collected in Table 1. These missions cover multiple tasks (e.g. surveillance, monitoring and tracking) in diverse scenarios (e.g. open fields, urban areas and disaster zones), integrating ground, aerial, surface and underwater robots, and considering from 0 to 6 operators (some works do not specify this information).

Most of the references collected by this table do not address the monitoring and commanding of the missions, they mainly use conventional interfaces, try to reach high levels of autonomy and/or employ a considerable number of operators. Some of these options are suitable for simulation and laboratory tests, but they are not able to fill the gap between these environments and the real world scenarios. In contrast to these approaches, our work addresses multi-robot missions from the perspective

Table 1 Some multi-robot missions reported by literature (picked up from [7] and completed with new information)

Reference	Mission	Robots	Operators	Scenario
[8]	Surveillance	2 UUVs, 2 USVs and 2 UAVs	No operator	Sea
[9]	Area coverage	3 UGVs	No operator	Open field
[10]	Area coverage	2 UGVs	No operator	Indoor
[11]	Surveillance	1 USV and 1 UAV	N operators	Sea
[12]	Monitoring	N UAVs	1 operator	Open field
[13]	People tracking	3 UGVs	No operator	Indoor
[14]	Surveillance	1 UGV and 1 UAV	1 operator	Urban area
[15]	Search and rescue	1 UGV, 1 UAV and 1 USV	N operators	Land and sea
[3]	Exploration	1 UGV and 1 UAV	No operator	Urban area
[16]	Search and rescue	N UGVs and N UAVs	N operators	Disaster area
[17]	Mapping	2 UGVs	1 operator	Disaster area
[1]	Area coverage	3 UGVs	1 operator	Open field
[18]	Mapping	2 UUVs	N operators	Ocean
[19]	Securing	1 UAV and 1 UUV	6 operators	Lake
[2]	Monitoring	1 UGV and 1 UAV	No operator	Greenhouse
[20]	Patrolling	20 UGVs	1 operator	Open field
[21]	Patrolling	3 UGVs	No operator	Indoor
[22]	Find and explore	6 UAVs	1 operator	Indoor
[23]	Exploration	6 UAVs	1 operator	Indoor
[7]	Fire surveillance	2 UAVs	1 operator	Indoor

of monitoring and commanding, searching to develop a new generation of interfaces that allow the operators to manage the MRS in real world scenarios, taking advantage of the recent developments in immersive and adaptive technologies.

Additionally, the references of Table 1 contain multiple MRS with different features. As reported in [24], there are several classifications for these fleets, which take into account different factors, such as the number of agents, the morphology and capabilities of robots and the type of coordination.

According to its size, a fleet may consist of a single unit, two units, multiple units (limited number of complex robots) or a swarm (large number of simple robots). Furthermore, the arrangement of a fleet can be static (the number of robots is fixed during the mission), coordinated (the number of robots can be modified manually to perform new tasks), and dynamic (the configuration of fleet can change automatically to adapt to the tasks).

According to the morphology and capabilities of robots, a fleet can be identical (any task can be performed by any robot), homogeneous (some tasks must be performed by a subgroup of robots), and heterogeneous (each task must be performed by a specific robot).

Table 2 A summary of human factors in multi-robot missions (originally from [25])

Issue	Problem	Detection	Solution
Situational Awareness	Lack: Inefficiency and errors	Actions and performance Test (SAGAT)	Immersive interface Filter information
Workload	Excessive: Inefficiency and errors	Physiological signals Test (NASA-TLX)	Adjust autonomy Transfer functions
Stress	Anxiety / Boredom: Human errors	Physiological signals Test (NASA-TLX)	Adjust autonomy Filter information
Trust	Mistrust: Human errors Overtrust: Machine errors	Reactions Survey	Adjust autonomy Train operators

Finally, according to its coordination, a fleet can be aware or unaware, if any robot know each other or not; strongly, weakly or not coordinated, if the coordination is established by a set of rules, emerges from reactive behaviors or directly does not exist; and centralized, distributed or hybrid, if the decisions are made by a single leader, all the agents or a set of leaders.

2.2   Human Factors

Multi-robot missions reveal that human operators face several challenges in the control and monitoring of autonomous systems. These challenges can sometimes be successfully addressed, but other times they can deteriorate performance or even safety. Table 2 collects the main problems due to human factors in multi-robot missions, as well as the most common methods to detect and prevent them. These issues are explained in the following paragraphs.

Situational awareness (SA) This term represents the perception of elements in an environment within a volume of time and space (level 1), the comprehension of their meaning (level 2), and the projection of their status in the near future (level 3) [26]. More generally, this concept involves five different types of awareness: human-human, human-robot, robot-human, robot-robot and human-mission [27].

The most common method to determine the SA of an operator is the Situational Awareness Global Assessment Technique (SAGAT) [28]. In this technique, operators are requested to supervise missions, at certain points of such missions the simulations are stopped, and the operators have to answer a questionnaire. Other methods estimate SA through operator performance in certain situations [29], introducing deviations in the missions and evaluating the operators' reactions.

Recent literature collects multiple strategies to improve situational awareness in multi-robot missions. Most of the proposals are related to the design of interfaces and focus on the management of information [30] and immersion of operator [31].

Workload This term is defined as the sum of the amount of work, the working time and the subjective experience of the operator [32]. This variable can be split into multiple attributes (input load, operator effort and work performance) [33] and dimensions (physical and mental demand) [34].

The most common method to determine the workload of a mission is the NASA Task Load Index (NASA-TLX) [34]. This is a questionnaire where the operators can perform a subjective evaluation of their workload by means of six variables: mental demand (lowhigh), physical demand (lowhigh), temporal demand (lowhigh), performance (goodpoor), effort (lowhigh) and frustration (lowhigh). Other methods estimate workload through actions of the operators (e.g. the distribution of attention among the robots or the time spent in monitoring and commanding tasks), as well as their physiological variables (e.g. cardiac response, eye movement, brain activity, blood pressure and galvanic skin response) [32].

Recent literature collects multiple strategies to reduce operator workload in multi-robot missions. Some works look for transferring functions from operators to robots to lighten their workload [35], whereas others propose to provide the operator with support in perception, comprehension, decision making and execution [36].

Stress Neither low nor high levels of stress are desirable: the first ones could indicate boredom, while the second ones could reveal anxiety [37]. Therefore, medium levels of stress are optimal in multi-robot operation, as they indicate that the operators are involved in the mission and are able to handle the objectives.

Although there are subjective techniques to evaluate stress, such as the questions from the NASA-TLX questionnaire related to effort and frustration, most of the studies of the literature propose objective measurements. Multiple physiological signals are used to estimate the stress, such as the heart rate variability [38], galvanic skin response [39] and brain readings [40].

Trust in automation Two extreme situations must be avoided: overtrust and mistrust [41]. The first case is more common when the level of automation is high, the error rate of automation is low and the operator does not have previous errors and experiences [21]. It can lead the operator to stop paying attention to the mission and, therefore, miss the detection of automation failures. The second case occurs when the level of automation is medium, the error rate of automation is significant and the operator can perceive the consequences of errors [21]. This can force the operator to reject the automation's support, increasing the workload and reducing the mission's performance.

2.3 Operator Interfaces

Operator performance and mission success strongly depend on the control interfaces. They influence not only the acquisition of relevant information regarding the mission, but also the efficiency in robot commanding. The interfaces for single robots are often developed for teleoperation, whereas the interfaces for MRS tend to lean towards supervisory control. Conventional interfaces usually integrate screens, keyboards, mouses, joysticks...

A review about the problems of conventional interfaces in the context of robot teleoperation shows that limited field of view, difficulty of perceiving the orientation, complexity of managing diverse cameras with different points of view, degradation in the perception of depth, degradation in the received video, time delays in perception and actuation, and reduction of accuracy with the motion are the main issues that might prevent operators form completing their tasks successfully [4].

Most of the proposals to address these challenges are based on multimodal interactions (to support the visual information), immersion (to improve the perception of the environment) and adaptive displays (to reduce and select the information displayed).

Multimodal interactions The effects of multimodal displays are diverse. The combination of visual and aural spatial information can lead to the enhancement of situational awareness [42]. Haptic displays can be applied to provide alerts regarding the location and speed of the robots [43]. The influence of visual, aural and haptic feedbacks on the spatial ability of operator is significant, but their effects in the performance of teleoperation are still not clear [44].

Furthermore, multimodal commands offer a greater deal of possibilities. The idea is to take advantage of voice, gestures and touch to reach natural, simple and fast interactions between operators and systems [45]. A review of the state of the art shows multiple approaches for commanding multi-robot missions by means of speech commands [46]. Additionally, commanding through gestures is common in this context, using not only hands [47], but also combinations of facial and manual gestures [48].

Adaptive interfaces These interfaces apply diverse types of techniques, such as data mining and machine learning, to adapt the information to the mission context. The objective is to provide agents with the adequate information for making decisions and executing actions, in order to improve their speed, accuracy, understanding, coordination and workload [6].

Adaptive interfaces require several types of models (e.g. knowledge, operators, robots, tasks and world) to properly manage the information of missions. Additionally, they require a certain adaption process: knowledge acquisition (what is happening?), attention (what adaptation is going to occur?), reasoning (why is it necessary?) and decision making (how is it happening?) [49].

Immersive technologies As explained in [25], immersive technologies search to introduce the operators in the robots' environment, in order to deal with problems such as the lack of situational awareness and the peaks of workload. These technologies integrate multiple devices (e.g. cameras and head displays) and disciplines (e.g. computer vision). Currently, there are three main immersive technologies: Virtual Reality (VR), Augmented Reality (AR) and Mixed Reality (MR); although some authors merge the latter two into a unique category.

Virtual Reality generates artificial environments based on real or imaginary scenarios and integrates the elements that take part in the mission: robots, targets, threats... This technology uses images, sounds and touch to simulate the presence of operators in scenarios and allow them to control and monitor robots. The first studies of VR interfaces pointed out they could improve the situational awareness of operators [5, 25].

Augmented Reality integrates relevant information about the mission in videos streamed by the robots or from the operator's point of view. This information usually includes maps, terrain elevation, obstacles, robot paths, target locations and other data [4]. Multiple experiments show the potential of AR video stream in comparison to conventional video stream in the context of multi-robot missions [50].

Mixed Reality is a combination of AR And VR that creates new scenarios and allows the interaction with both real and virtual objects. Although the application of this technology to immersive interfaces looks promising, there are only a few cases of use in the context of robotics [51].

3   Immersive Interfaces

The first question that should be asked before developing an immersive interface is "VR, AR or MR: when should we apply each technology?".

- Virtual Reality. This technology is the best option when the operator is away from the mission (e.g. spacial operations), the mission is dangerous (e.g. intervention in disaster areas), the scenario is large (e.g. open fields) or the operator has to manage many resources (e.g. multi-robot systems). In those contexts, the possibility of viewing the real scenarios and interacting with their elements does not add value to the operations; whereas the possibility of reproducing remote, dangerous or large scenarios and freely move there makes it easier to perform these tasks.
- Augmented Reality. This technology is the best option when the operator is close to the mission (e.g. indoor scenarios), the mission is complex (e.g. industrial applications), the scenario is small (e.g. rooms and corridors) or the operator has to interact with the robot (e.g. social robotics). In these contexts, in contrast to the previous ones, the view of the real scenario with virtual elements is more useful than the view of a virtual reproduction of the real scenario.
- Mixed Reality. This technology is useful in the same scenarios than AR and, specially, when human-robot collaboration is required. This is due to the fact that operators and robots can interact with the real and virtual objects of the scenario.

This chapter is focused on the development of VR interfaces, so the proposed architecture, resources and cases of use are only related to this technology. Once decided to work with VR, the second question is "Which resources of hardware and software can be used to develop the interface?"

The development of virtual reality systems has been fast in the last years, and new systems with improved features appear frequently. Table 3 collects the main VR headsets currently available in the market. Headsets can be classified into two

Table 3 Study of available VR headsets (originally from [25])

Name	Type	Hardware Required
Sony PlayStation VR	Tethered	PlayStation 4
HTC Vive	Tethered	PC
Oculus Rift	Tethered	PC
Google Daydream View	Mobile	Daydream compatible phone
Samsung Gear VR	Mobile	Latest samsung galaxy models
Homido VR	Mobile	Android and iOS phones
FreeFly VR	Mobile	Android and iOS phones
Google Cardboard	Mobile	Android and iOS phones

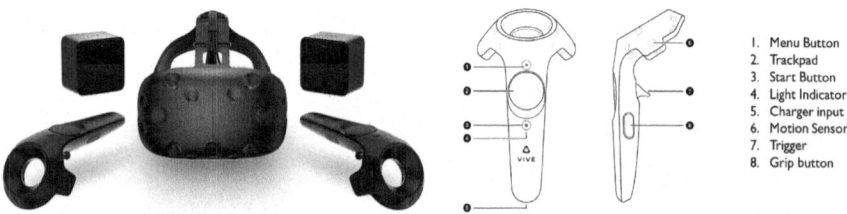

1. Menu Button
2. Trackpad
3. Start Button
4. Light Indicator
5. Charger input
6. Motion Sensor
7. Trigger
8. Grip button

Fig. 2 HTC Vive HMD, controllers and base stations (originally from [25])

categories: mobile headsets, which depend on the use of a phone for visualization, and tethered, which require a connection to a computer to work. The first ones can be more flexible and affordable, but only the second ones meet the requirements for being applied in the context of robotics, such as high frequency, resolution and field of view. Currently, the *HTC Vive*[8] and *Oculus Rift*[9] headsets are the most suitable options for the development of applications, since they are compatible with *Unity* and *Unreal Engine 4* game engines.

The interfaces explained in this chapter have been developed for the *HTC Vive* virtual reality headset. This system was originally conceived for gaming, but it is currently used for multiple purposes, including industry, services and research. As shown in Fig. 2, this system consists of a head-mounted display (HMD), two controllers and a tracking system. The HMD integrates two screens, each one of them with resolution of 1080 x 1200 pixels and frequency of 90 Hz, and an audio jack, which allows the integration of sounds to improve the immersion. The controllers allow interaction with elements of the virtual scenario, providing multiple types of buttons and various degrees of vibration. Finally, the tracking system consists of two base stations that are placed in two opposite vertices of the work area and provide the localization of the HMD and controllers with sub-millimeter accuracy and low latency [52].

---

[8]https://www.vive.com/.

[9]https://www.oculus.com/.

The most common method of programming applications for *HTC Vive* is based on the use of *Unity* and *SteamVR*. *Unity* is a game engine that provides a complete set of resources, such as realistic scenarios, kinematic and dynamic models, characters with artificial intelligence, etc. It allows the developers to write code in C#, which was used for all the developments of this chapter, and JavaScript. *SteamVR* is a plugin of *Unity* that provides resources related to virtual reality and allows to compile and deploy the developed applications. A current limitation of this plugin is that it only works under Windows-based operating systems, so an additional computer or a virtual machine is necessary in order to integrate the application with ROS-based programs.

Once the hardware and software resources have been selected, it is time to ask the third question: "How will the interface be?".

Obviously, the design of an interface will highly depend on the fleet, operators, mission and scenario. An interface for a single robot should be centered in this robot and its payload; for instance, adopting a first person view and giving priority to the images of its cameras. On the other hand, an interface for multiple robots should combine the visualization of the whole scenario with a closer view of each robot involved in the mission. In a similiar way, if there is only one operator to control the mission, the interface should provide them all the information required to perform it, whereas if there are multiple operators with different roles, the interface should integrate these roles and provide the operators with the adequate information according to them: this means that robot operators need to know the close environment of the robots, payload operators require to know the state and readings of the sensors, and mission commanders need to acquire a general view of the mission. Furthermore, the type of mission has a considerable impact on the information displayed by the interface. For instance, an interface for surveillance must show clearly the found targets, whereas another one for manipulation must give accurate information about the effector and objects. Finally, the scenario also impacts on the interface: small scenarios can be reproduced in 1:1 scale and traversed by the operator, whereas large ones need to be reduced in the virtual reality and may require the use of resources like teleporting.

Apart from these issues, a set of considerations for the development of immersive interfaces are summarized in Table 4. The virtual scenarios can be designed manually by using Unity and its assets, or reconstructed by means of 3D cameras or LIDAR sensors. In the latter case, this reconstruction can be performed offline before the mission, online or a combination of both (e.g. acquiring the scenario before the mission and detecting and mapping obstacles as it is performed). However, real-time reconstruction and immersion is more a goal than a reality, since the computational cost of the process of reconstructing the environment, improving the 3D map, creating a mesh and importing it in virtual reality is very high [53]. Furthermore, the interfaces must show diverse types of information: from the representation of the robots and their movements in the scenario, to the integration of their state, the readings of their sensors and the images of their cameras. For this purpose, an element can be added to each robot to represent the information that is relevant at any moment. When the fleets are large and the amount of information is vast, these elements can be shown or hidden at operator request. Additionally, the interfaces can integrate

Table 4 Main considerations for developing immersive interfaces

Issue	Objectives	Addressed
Scenarios	Manual design and load	Yes
	Offline reconstruction	No
	Obstacle detection and mapping	No
	Online reconstruction	No
Monitoring	Representation and movement	Yes
	Information of state	Yes
	Integration of sensors	Yes
	Integration of cameras	Yes
Commanding	Teleoperation	No
	WP commands	Yes
	Task commands	No
Robots	Mobile	Yes
	Manipulators	Yes
	Multi-robot systems	Yes

different methods for commanding the robots: from teleoperation, which is useful with a single robot, to task commands, which reduce the workload in single-operator multi-robot scenarios through waypoint (WP) commands, which are intuitive and require an intermediate effort. Finally, the integration of different types of robots and fleet will require developing different resources for their supervision and control.

And after the previous questions have been answered, only two questions remain: "How can we develop it?" and "Which resources can we use?".

As mentioned above, *Unity* provides a complete set of assets developed by its community, which can be bought and used in the projects. These assets facilitate the work for building the scenarios, managing the information and commands and developing human-robot interaction. Among the *Unity* assets, the *SteamVR* plugin is specially useful, since it contains the basic interactions required by virtual reality, such as visualizing, picking, placing and teleporting, among others. The main resources used in the development of the interfaces reported in this chapter are listed below:

- Teleport. The operator can be teleported in the scenario by pointing with the controller and pressing the trigger. This is actually useful when the scenario is large or complex and the task require accuracy.
- Selection of robot/goal. The operator can use the controller to shoot to a robot, selecting it to see information and enter commands, or a position in the scenario, to send it as a goal for the robot.
- Robot panel. A panel can be shown over a robot and oriented to the operator to show the relevant information of this robot: e.g. the current task, level of battery, images of camera...

- Robot mark. Some elements can be added to the interface in order to guide the attention of the operator to some robots: e.g. a spotlight that points the robot that performs the most relevant task and a smoke cloud which indicates that a robot is experiencing problems.
- Sound. The sounds of robots and other relevant elements (e.g. water and fire) help the operators to get proper orientations and find them in virtual scenario.
- Vibration. Controller vibration can be used to draw the attention of operators to the alerts, as well as to guide them in the robot commanding task.

4   Architecture and ROS Packages

This section presents the architecture developed in this work, which allows to connect multiple real and simulated robots to multiple conventional and immersive interfaces and reduces the amount of data that is exchanged between those components. Additionally, this section describes the architecture's components, which are ROS packages and Unity assets that must be properly configured and launched to work with MRS and VR.

4.1   Overview

An overview of the proposed architecture can be seen in Fig. 3. As shown in the figure, the architecture has been designed to be as general as possible to cover a wide set of MRS deployments. In fact, it has only one requirement: the existence of a common network for all robots and interfaces. Thus, the robots act as servers that receive commands and send telemetries, whereas the interfaces act as clients that receive telemetries and send commands. The fundamental elements of this architecture are *Unity* to run the immersive interface, *RosBridgeLib* and *RosBridge* to communicate this environment and ROS, and both the real and simulated robots. Additionally, the architecture can include other conventional interfaces powered by *Rviz*[10] or *Gazebo*. These elements of the architecture are described with further detail in the following subsections.

4.2   Robots

As mentioned in the previous section, the proposed architecture is able to manage both real and simulated robots. This feature can be useful in multiple scenarios: e.g. the operators can control simulated robots to perform the tasks and, if the operations

---

[10]http://wiki.ros.org/rviz.

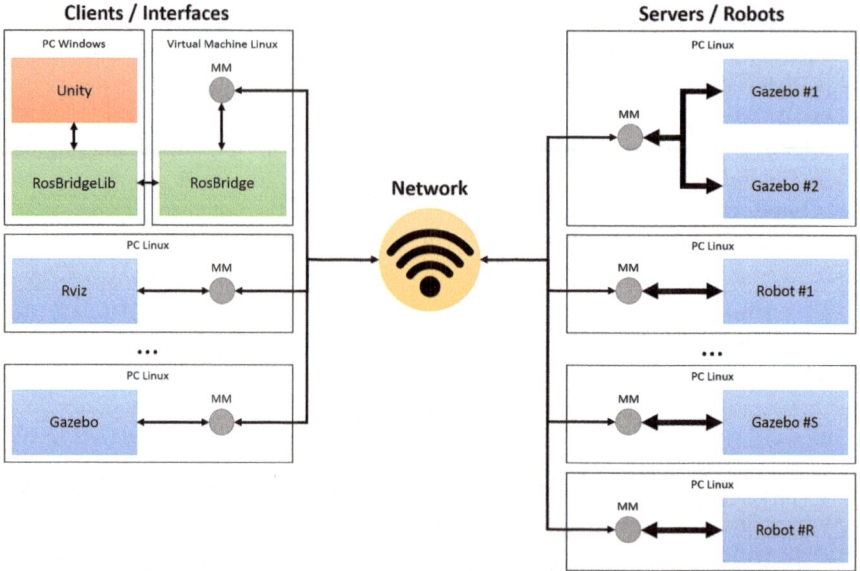

Fig. 3 The architecture that connects simulated and real robots to conventional and virtual reality interfaces. All the elements are connected to the same network, the robots act as servers, whereas the interfaces act as clients. The masters of the ROS multimaster architecture are marked with MM

are successful, the real robots can repeat the actions of the simulated ones. Although the tests presented in this chapter are performed in *Gazebo*, other simulators can be used as long as they can be integrated with ROS.

4.3 Multimaster_fkie

The use of multiple ROS masters provides some advantages, despite it can make the architecture harder to understand and configure. First, it allows to independently launch the ROS packages of robots even when the global network is not working. Second, it reduces the number of ROS topics that can be accessed from the global network, which simplifies the development of interfaces and avoids the errors in subscriptions or publications. Third, it provides the system with organization and modularity, since all the robots can send the same information to the interfaces, which simplifies its reception, processing and visualization.

The main packages for the management of MRS under ROS were analyzed in a previous work [24] and *multimaster_fkie*[11] was found the most promising due to its functionality, simplicity and resources (community, documentation and tutorials).

---

[11] http://wiki.ros.org/multimaster_fkie.

multimaster_fkie offers a series of resources to create and manage a multi-master ROS environment. This package allows to build a global ROS network that consists of multiple local ROS networks with their own masters, nodes, topics and services; as well as to control the exchange of information between these ROS networks by defining synchronized topics and services. Its main advantage is that it does not require a complex configuration and it facilitates the detection of mistakes and application of changes. A more detailed description of *multimaster_fkie* can be found in reference [54].

As shown in Fig. 3, this package must be installed in all the computers of the architecture (both servers - simulated and real robots - and clients - conventional and VR interfaces). For this purpose, assuming you are using the Ubuntu distribution of Linux, you just have to open a terminal and enter:

```
$ sudo apt-get install ros-%distribution%-multimaster-fkie
```

multimaster_fkie consists of two main nodes: *master_discovery*, which detects and notifies its existence to other ROS masters, and *master_sync*, which manages the exchange of topics and services between the masters. To configure this package, you just have to follow the instructions below.

Two requirements to use this package with MRS must be taken into account. First, all the computers used in robots and interfaces must work in the same network. And second, all these robots and interfaces must use the same coordinate system for their positions and orientations.

The first step is to configure the network where the robots and interfaces will work. For this purpose, you should assign static IP addresses, masks and gateways to every computer in the network. In addition to this, the \etc\hosts file of every computer must include the IP addresses of the rest.

Then, the masters of the local ROS networks must be configured to take part in the global ROS network. At this point there are two possibilities: if the computer has direct access to the global ROS network, it will run its own master, but if the computer accesses the global ROS network through other computer, it will export its master to this computer.

Therefore, in computers without direct access to global ROS network, you must open a terminal and enter:

```
$ export ROS_MASTER_URI=http://%hostname/IPaddress%:11311
```

to export the ROS master, as well as:

```
$ cat /proc/sys/net/ipv4/ip_forward
```

to set the default gateway.

On the other hand, in computers that access the global ROS network directly, you must check if the IP forwarding feature is enabled:

If this command returns 1, everything is correct. Otherwise, you have to enable this feature:

```
$ sudo sh -c "echo 1 >/proc/sys/net/ipv4/ip_forward"
```

Then, for all the computers of the architecture, you must check if the multicast feature is enabled:

```
$ cat /proc/sys/net/ipv4/icmp_echo_ignore_broadcasts
```

If this command returns 0, everything is correct. Otherwise, you have to enable this feature:

```
$ sudo sh -c "echo 0 >/proc/sys/net/ipv4/icmp_echo_ignore_broadcasts"
```

The second step is to configure the *multimaster_fkie*. For this purpose, you must define the topics and services that will be exchanged by the masters. This issue can be addressed by creating a *sync_mrs.sync* file. For instance, the following file was used in our tests reported in Sect. 5:

```
do_not_sync:
ignore_hosts:
sync_hosts:
ignore_topics:
ignore_publishers:
ignore_subscribers:
sync_topics:
 - /odometry
 - /commands
 - /robot
 - /image
ignore_services:
 - /*get_loggers
 - /*set_logger_level
sync_services:
```

This code configures the masters to exchange only four topics: one for robot odometries, other for operator commands, other for selecting robots and a last one for camera images.

Additionally, create *master_sync_mrs.launch* to launch the *master_sync* node of the *master_sync_fkie* package with the above described configuration. As an example, this file was used for our tests:

```
<launch>
  <node name="master_sync" pkg="master_sync_fkie" type="master_sync">
    <param name="interface_url" value="$(find %package%)/launch/sync_mrs.sync" />
  </node>
</launch>
```

Now you can launch the package by entering the following commands in the terminal:

```
# Window 1:
$ roscore # if the computer has master
# Window 2:
$ roslaunch master_discovery_fkie master_discovery.launch
# Window 3:
$ roslaunch master_discovery_fkie master_sync_mrs.launch
```

4.4 Rosbridge_suite

rosbridge_suite[12] is a ROS package that allows to communicate ROS networks with external non-ROS programs. For this purpose, this package converts ROS messages into JSON strings and provides a protocol to establish communications, subscribe and publish topics. Further information about *rosbridge_suite* can be found in the reference [55].

RosBridgeLib[13] is a Unity library for the communication with ROS through the *rosbridge_suite* package. This library implements a set of ROS messages in *Unity* and provides methods to establish the communication and generate subscribers and publishers. Further information about this library can be accessed in the reference [56].

Both *rosbridge_suite* and *RosBridgeLib* must be installed once per VR interface. As stated above, *Unity* and *SteamVR* must work under *Windows*. Nevertheless, *rosbridge_suite* must work under *Linux*. Our solution to this problem was to use a Windows computer for *Unity* with a Linux virtual machine for *rosbridge_suite*. We did not detect problems in the performance of the virtual reality interface and the communications between the interface and the robots.

The *RosBridgeLib* asset can be downloaded from its repository and directly copied in the *Unity* project, whereas *rosbridge_suite* can be installed in ROS by opening a terminal and entering:

```
$ sudo apt-get install ros-%distribution%-rosbridge-suite
```

---

[12]http://wiki.ros.org/rosbridge_suite.

[13]https://github.com/MathiasCiarlo/ROSBridgeLib.

Once both programs have been installed, they must be properly configured. Specifically, a script must be created in *Unity* and added to the scene to establish the communication, create the subscribers and publishers, and publish the messages. An example is shown below:

```
// Headers:

(...)

// ROSBridgeLib:
using ROSBridgeLib;
using ROSBridgeLib.geometry_msgs;
using ROSBridgeLib.nav_msgs;
using ROSBridgeLib.std_msgs;

(...)

// Variable for connection:
private ROSBridgeWebSocketConnection ros = null;

(...)

// Function for initialization:
void Start() {

    (...)

    // Configure IP and port:
    ros = new ROSBridgeWebSocketConnection("ws://%IPaddress%",
    9090);

    // Add subscribers:
    ros.AddSubscriber(typeof(OdometrySub)); // Robot odometries.
    ros.AddSubscriber(typeof(CameraSub)); // Camera images.

    // Add publishers:
    ros.AddPublisher(typeof(OdometryPub)); // Robot commands.
    ros.AddPublisher(typeof(SelectedRobotPub)); // Selection.

    // Connect to server:
    ros.Connect();

    (...)
}

// Function for loop:
void Update() {

    (...)

    // Generate messages:
```

```
47    OdometryMsg cmd = new OdometryMsg (header, frame_id, pose,
      twist);
      HeaderMsg msg = new HeaderMsg (index, time, robot_id);
49
      (...)
51
      // Publish messages:
53    ros.Publish (OdometryPub.GetMessageTopic (), cmd);
      ros.Publish (SelectedRobotPub.GetMessageTopic (), msg);
55
      (...)
57
      // Render:
59    ros.Render ();
61    (...)
}
```

Additionally, each subscriber requires a script with a callback function to receive and decode the messages, and each publisher needs a script with functions to encode the messages. The subscriber and publisher for odometry messages are shown below:

```
2   // Headers:

4   (...)

6   // RosBridgeLib:
    using ROSBridgeLib;
8   using ROSBridgeLib.geometry_msgs;
    using ROSBridgeLib.nav_msgs;
10  using ROSBridgeLib.std_msgs;

12  (...)

14  public class OdometrySub : ROSBridgeSubscriber {

16      // Function to get the topic name:
        public new static string GetMessageTopic() {
18          return "/odometry";
        }
20
        // Function to get the message type:
22      public new static string GetMessageType() {
            return "nav_msgs/Odometry";
24      }

26      // Function to parse received messages:
        public new static ROSBridgeMsg ParseMessage(JSONNode msg) {
28          return new OdometryMsg (msg);
        }
30
        // Function to process the parsed messages:
```

```
32    public new static void CallBack(ROSBridgeMsg msg) {
          OdometryMsg odom = (OdometryMsg) msg;
34
          (...)
36    }
    }
```

```
1
  // Headers:
3
  (...)
5
  // RosBridgeLib:
7 using ROSBridgeLib;
  using ROSBridgeLib.geometry_msgs;
9 using ROSBridgeLib.nav_msgs;
  using ROSBridgeLib.std_msgs;
11
  (...)
13
  public class OdometryPub : RosBridgePublisher {
15
      // Function to get the topic name:
17    public new static string GetMessageTopic() {
          return "/commands";
19    }

21    // Function to get the message type:
      public new static string GetMessageType() {
23        return "nav_msgs/Odometry";
      }
25
      // Function to convert to string:
27    public static string ToString(OdometryMsg msg) {
          return msg.ToString ();
29    }

31    // Function to convert to YAML string:
      public static string ToYAMLString(OdometryMsg msg) {
33        return msg.ToYAMLString ();
      }
35
      (...)
37    }
  }
```

Finally, it is time to execute both programs. To run the *Unity* part, you just have to press the play button, whereas to launch the ROS part, you should enter in the terminal:

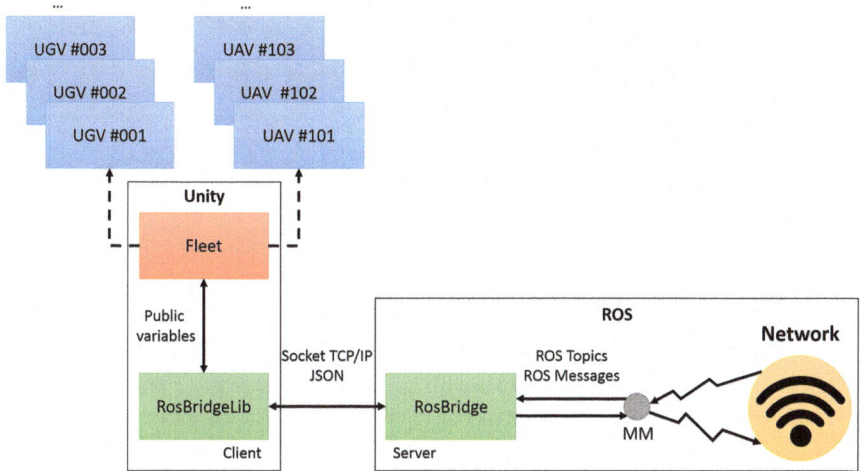

Fig. 4 The section of the architecture dedicated to VR interfaces in further detail. Again, the masters of the ROS multimaster architecture are marked with MM

```
Window #1:
$ roscore %(if the computer has master)%
Window #2:
$ roslaunch rosbridge_server rosbridge_websocket.launch
```

4.5 Interfaces

As depicted in Fig. 3, the architecture considers multiple types of interfaces, including conventional and VR ones. Among the conventional interfaces, we have worked with *Rviz*, which is a widely used visualization tool under ROS, and Gazebo, which can be used not only as simulator but also just as visualizer. However, the architecture can include any other interface that works under ROS or can be integrated with ROS through *rosbridge_suite*. A more detailed diagram of the section of the architecture dedicated to VR interfaces can be seen in Fig. 4.

5 Applications

This section summarizes our experiences with MRS, VR and ROS. These projects were developed in order to test the virtual reality interfaces in different applications and validate them under different conditions. The three following cases are explained in the next three subsections:

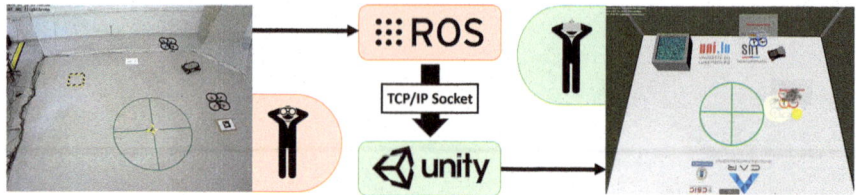

Fig. 5 Architecture to connect robot and VR worlds

- Case 1: Multi-robot monitoring interface (Sect. 5.1). It involves the development of an interface to supervise a mission of fire surveillance and extinguishing with a fleet of drones, which has to deal with an increased amount of information and a rapidly changing environment.
- Case 2: Manipulator commanding interface (Sect. 5.2). It involves the development of an interface to command a robot manipulator for pick and place tasks. Such interface must facilitate the control of a complex robot in terms of motion planning and joint control.
- Case 3: Multi-robot complete interface (Sect. 5.3). It involves the development of an interface to supervise and command a fleet of heterogeneous robots. It applies the architecture described in previous chapters to reach the requirements of flexibility and modularity of the mission.

5.1 Case 1: Multi-robot Monitoring Interface

For the monitorization of multi-robot missions, four interfaces were developed: a conventional, an adaptive conventional, a virtual reality and an adaptive virtual reality interface. The goal of this work was to study the effects of immersion and adaption on operators, specifically on their situational awareness and workload. Figure 5 shows the architecture developed to connect the real and virtual worlds. Further details of this work can be found in reference [25].

Robot world This work used a set of multi-robot missions of fire surveillance and extinguishing performed in previous works [7]. In these missions, two *Parrot AR.Drone 2.0*[14] aerial robots had to search and extinguish fires, as well as find and follow a suspicious *KUKA Youbot*[15] ground robot. The scenario had a robot base, a water well and a fire whose location varied from mission to mission.

All the telemetry generated by the robots, as well as the position and orientation obtained by an *Optitrack* motion capture system, were published in ROS during the real missions, and recorded in Bag files. This fact allowed to reproduce the missions multiple times during the development of the interfaces and the experiments with operators.

---

[14]https://www.parrot.com.

[15]https://www.kuka.com.

VR world The virtual reality interfaces were developed by using the *Unity* environment and *SteamVR* asset for the *HTC Vive* headset. These interfaces recreated the scenario with all the cited elements: some of them were fixed during the missions (base and water well), whereas the rest appeared when they were discovered by the drones (fire and ground robot).

Figure 6 shows captures of the non-adaptive and adaptive virtual reality interfaces. Both interfaces integrate panels with the drones' relevant information for the mission (e.g. the task that are performing and the level of battery). Additionally, the operator could use the VR controllers to teleport to any point of the scenario or an elevated observation platform, in order to reach the best point of view for monitoring the mission. Finally, the sounds of the drones and the vibration of controllers were used to reinforce the information displayed from the mission.

As shown in Fig. 6, the adaptive virtual reality interface was based on the non-adaptive one, with the incorporation of two additional elements: a spotlight that focused on the most relevant UAV at any time of the mission, and a smoke cloud that appeared when a UAV was at risk. The values of these variables (relevance and risk) were obtained through corresponding neural networks previously trained with data drawn from the analysis of operators [25].

Communications A custom TCP/IP socket was developed to connect robot and VR worlds. This communication is unidirectional: a client implemented in a ROS node sends the full information of the mission to a server integrated in the interface. This information includes the location, state and detections made by the robots during the missions.

Conclusions The virtual reality interfaces were compared against the conventional ones through an experiment involving twenty-four operators supervising the multi-robot missions. The results of the NASA-TLX workload and SAGAT situational awareness tests showed that virtual reality improved the situational awareness at least by 12% and reduced operator workload by 22%, whereas the effects of predictive components were not significant and depended on the implementation in the interface (as reported in [25]).

A summary of this application is shown in Table 5.

5.2 Case 2: Manipulator Commanding Interface

This work developed a virtual reality interface for monitoring and commanding a manipulator robot. This type of robot was selected for its complexity in terms of Degrees of Freedom (DoF) compared to mobile robots. Figure 7 shows the architecture developed to connect the real and virtual worlds. Further details of this work can be found in reference [57].

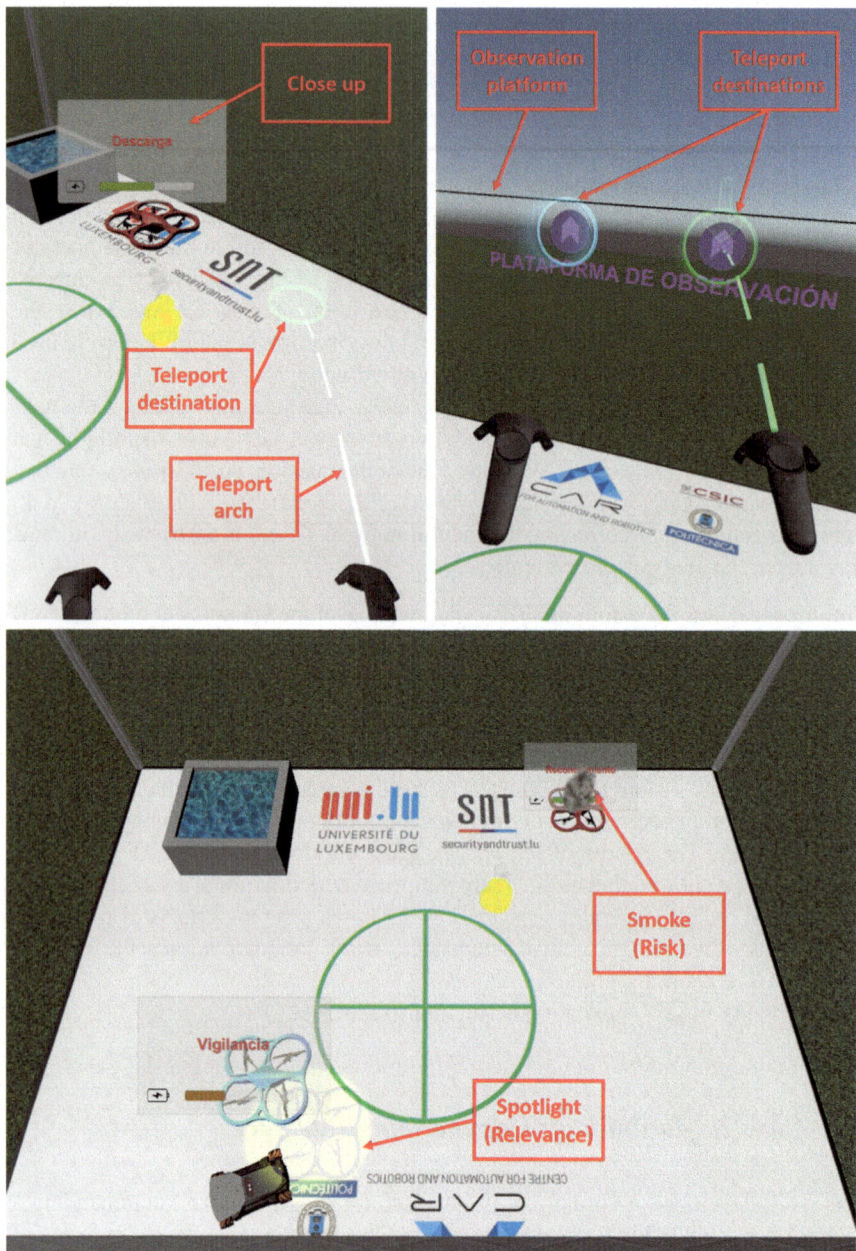

Fig. 6 Virtual reality interfaces: non-adaptive (top) and adaptive (bottom) (originally from [25])

Table 5 Summary of the application of VR to the control and monitoring of multiple robots

Mission	Fire surveillance and extinguishing
Robots	2 *Parrot AR.Drone* UAVs
Operators	1 Operator
Scenario	Recreation of mission field
Objective	Study the impact of operator immersion on situational awareness.
Resources	*HTC Vive, SteamVR, Unity, ROS, Rosbag*
Reference	Roldin et al. [25]

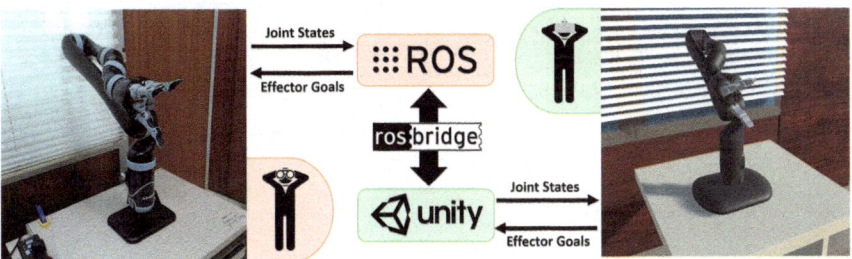

Fig. 7 Architecture to connect robot and VR worlds

Robot world The robot used for this application was a *Kinova Jaco*[216] with six degrees of freedom (shoulder, arm, forearm, wrist 1, wrist 2 and hand) and an effector with three fingers. This robot is able to manage a payload of 1.6 kilograms and reach distances of 90 mm.

The control was distributed between the robot, which performed low-level functions such as joint control, and a computer, which performed high-level tasks such as motion planning and obstacle avoidance, both connected via USB or Ethernet. The computer received the effector goals from the interface, computed the trajectory and sent joint values to the robot.

The motion planning was performed using *MoveIt!* [58], which is a software for mobile manipulation that provides kinematic functionalities such as collision checking. This tool integrates the *Open Motion Planning Library (OMPL)* [59], which implements multiple sampling-based algorithms: e.g. RRT (used in this work) [60], PRM and EST.

VR world The virtual reality interface was developed using the *Unity* environment and *SteamVR* asset for *HTC Vive* headset. The model of the robot was introduced in *Unity* split into eight parts: base, shoulder, arm, forearm, wrist 1, wrist 2, hand and fingers. Each part was implemented through two objects: one for assembling it with the previous one (considering the proper position and orientation), and another one to allow the rotation of its joint. The result was a virtual robot whose movements are identical to those of the real robot.

---

[16]http://www.kinovarobotics.com/.

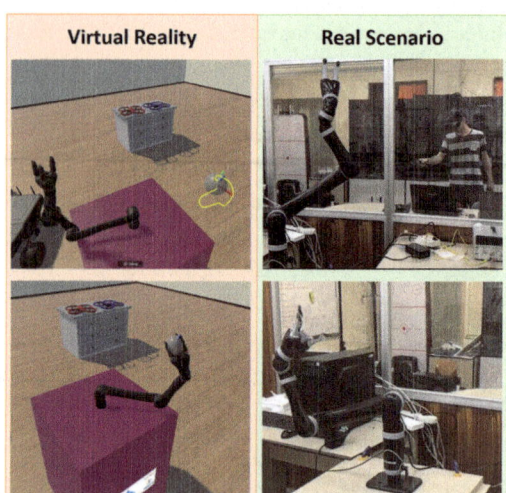

	The operator sends a goal by moving a sphere in the virtual reality interface.
1	

	The manipulator computes the path and moves to reach the goal.
2	

Fig. 8 Commanding of the manipulator

The robot commanding was performed by means of a virtual sphere with three axes. The sphere represented the position that the effector must reach, whereas the three axes defined the orientation that it must acquire. The process is shown in Fig. 8: when the sphere is in the effector, the robot remains at rest. The operator can pick the sphere, move it within the scene and place it in another location. When the operator leaves the sphere, the goal is sent to the robot. Additionally, the operator can teleport within the scene to reach the best point of view and select the goals more comfortably.

Communications In this case, the communication between robot and VR worlds was performed by means of the *Unity* asset *RosBridgeLib* and the ROS package *RosBridge*. Two ROS messages were used: *sensor_msgs/JointState* for sending joint angles from the robot to the interface, and *geometry_msgs/Pose* for sending goals from the interface to the robot.

Conclusions The results of the tests with operators reported in [57] showed that the interface enhances their situational awareness and provides an intuitive commanding method. The operators emphasized the advantages of being able to choose the point of view to monitor the robot, as well as commanding the robot by moving the goal for its effector. Although some of the operators were not familiar with the control of this kind of robots, they were able to move it to the desired locations efficiently and safely. In future works, more experiments with robot manipulator will be performed to objectively compare operator performance with immersive and conventional interfaces.

A summary of this application is shown in Table 6.

Table 6 Summary of the application of VR to the control of a manipulator

Mission	Pick and place of objects
Robots	1 *Kinova Jaco*^{2} manipulator
Operators	1 Operator
Scenario	Laboratory
Objective	Study the performance of commanding with a complex robot
Resources	*HTC Vive, SteamVR, Unity, RosBridge, ROS, MoveIt!*
Reference	Peña-Tapia et al. [57]

5.3 Case 3: Multi-robot Complete Interface

This use case involved the development of an interface for monitoring and controlling multiple robots based on virtual reality. The goal in this case was to implement and validate the previously explained architecture in complex environments. For this reason, a set of simulations were performed, involving aerial and ground robots in a critical infrastructure. The code of this work can be accessed in https://bitbucket.org/robcib/rosbookworld.

Robot world As mentioned before, this application integrated both ground and aerial robots to increase the difficulty of monitoring and commanding. The ground robots were *Robotnik Summit XL*,[17] simulated in *Gazebo* by using their corresponding package.[18] Further information about this platform and its start up under ROS can be found in references [61, 62]. The aerial robots were *Parrot Bebop 2*, also simulated in *Gazebo* by using the $rotor\_simulator$ package.[19]

The simulations had two particularities: on the one hand, the robots were simulated in multiple *Gazebo* simulators distributed in multiple computers, in order to reproduce the integration of real robots in the architecture; and on the other hand, the robots appeared and disappeared during the mission, in order to test the robustness of the architecture and the performance of the interface.

VR world As in the two previous applications, the VR interface was based on the reproduction of the scenario. Figure 9 tries to illustrate the immersion of the operator in this scenario, as well as the view of the robots and their environment.

This interface includes a module that continuously checks the robots that are exchanging messages. If a new robot sends a message to the interface, the robot is identified and spawned in the virtual scene. The identification is performed through an ID parameter: from 0 to 100 for ground robots and from 101–200 for aerial robots.

In addition to this, the interface provides the operator with a set of resources to ease robot control, as shown in Fig. 10. As in the other interfaces, the operator can

---

[17]http://wiki.ros.org/Robots/SummitXL.

[18]http://wiki.ros.org/summit_xl_sim.

[19]http://wiki.ros.org/rotors_simulator.

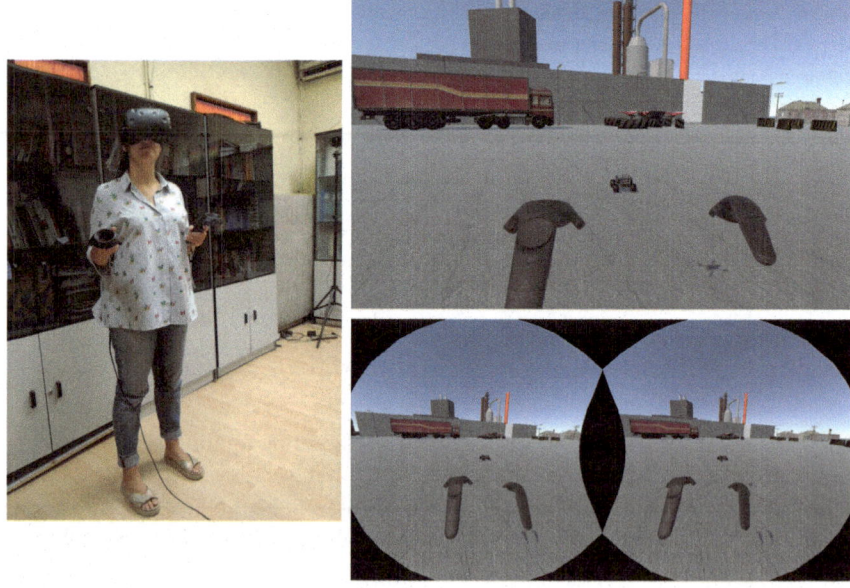

Fig. 9 Multi-robot complete interface: operator enjoying the immersion in the mission (left), 2D view (top-right) and headset view (bottom-right)

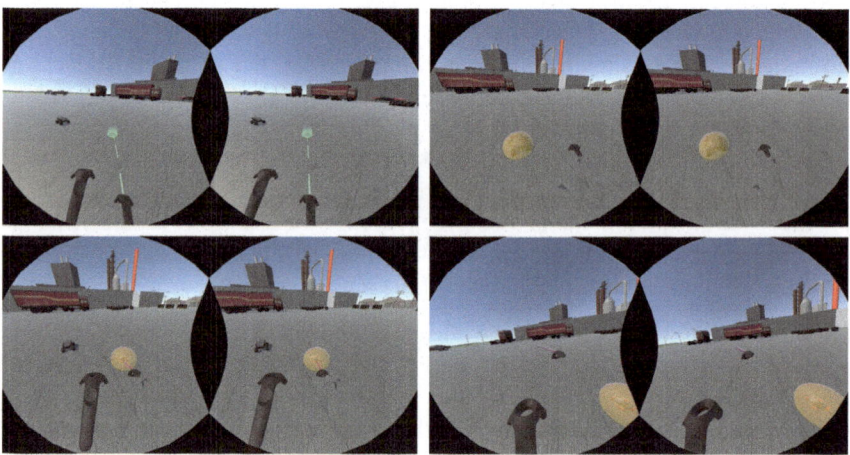

Fig. 10 Multi-robot complete interface: teleporting close to robots (top-left), selecting ground robot (top-right), selecting aerial robot (bottom-left) and generating a waypoint for it (bottom-right)

Table 7 Summary of the application of VR to the control and monitoring of multiple robots

Mission	Surveillance of critical infrastructure
Robots	Multiple *Robotnik Summit XL* UGVs and *Parrot Bebop 2* UAVs
Operators	1 Operator
Scenario	Simulation of critical infrastructure
Objective	Test the architecture and the interface.
Resources	*HTC Vive, SteamVR, Unity, multimaster_fkie, RosBridge, ROS, Gazebo*
Reference	This chapter

teleport throughout the scenario by pressing the controller's trigger and pointing to the desired destination. In this case, the operator can use the trackpad of the controller to select a robot and, then, send this robot a goal. The selected robot is surrounded by a translucent sphere, to point out its selection and avoid commanding mistakes. Finally, a pop-up panel is added to the selected robot to show the images of its camera.

Communications The communications between robot and VR worlds were addressed by following the architecture shown in Figs. 3 and 4. Nevertheless, there were two differences: first, there was only a VR interface, and second, all the robots were simulated by *Gazebo*. The installation and configuration of all the elements of the architecture were performed as described in Sect. 4.

The $nav\_msgs/Odometry$ messages were used for the odometries, which were generated by the robots and sent to the interface, and also the commands, which were generated by the interface and sent to the correspondent robots. The reason behind using this message is its flexibility: as shown in Fig. 11, it can integrate both waypoint and speed commands, as well as collect all the information of the robot state. Specifically, $frame\_id$ was used to name the robot that has to send or receive the message, the $pose$ message to exchange the position of the robot or the goal command, and the $twist$ message served to exchange the speed of the robot or the speed command. Additionally, the $std\_msgs/Header$ message was used to send the selected robot and the $sensor\_msgs/CompressedImage$ one was used to receive the images of its camera. A good idea for future works is to define custom ROS messages that can integrate more information, such as the level of battery and payload information.

Conclusions The results of the first tests with operators showed that the interface improves knowledge about the missions and allows a close control of the robots. The operators remarked the teleporting resource and intuitive commanding as advantages, but the size of scenario and the remoteness of robots were proven drawbacks. In future works, the possibility of changing the scale of player to alternate first person view and bird's eye view will be considered.

A summary of this application is shown in Table 7.

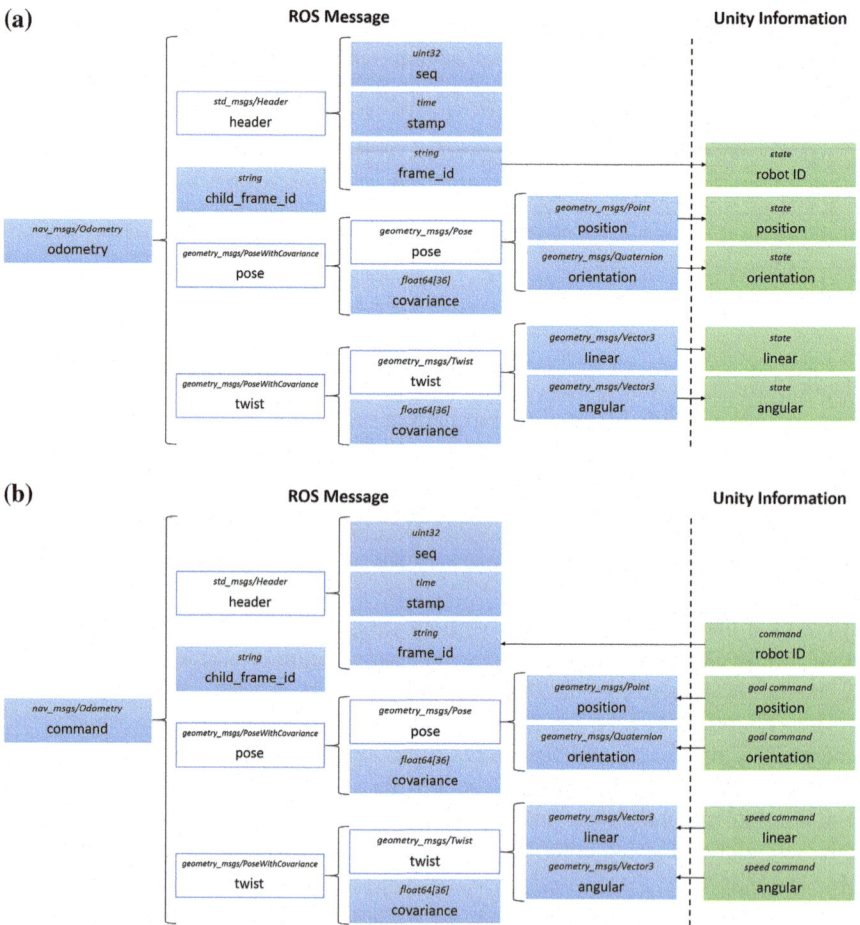

Fig. 11 Main messages of tests: **a** Odometry, from robots to interface, and **b** Command, from interface to robots

6   Results and Conclusions

This chapter presents some guidelines for the integration of Multi-Robot Systems (MRS), Virtual Reality (VR) and ROS. This integration is required for the development of future immersive interfaces to improve the situational awareness of operators.

The main contributions of this chapter are listed below:

- An architecture that allows to work with MRS and VR is proposed, developed, implemented and validated.
- Some guidelines for the development of VR interfaces are provided and some resources that can be used for it are analyzed.
- Three cases of use of VR interfaces with different missions, robots, operators and scenarios are described.

Some lessons learned from these applications are compiled below:

- VR is a powerful tool that improves the situational awareness of operators, but its effects on workload and performance may depend on the implementation for the interfaces, as well as on operator training.
- VR interfaces have advantages over AR and MR in large, remote or dangerous scenarios, as well as in MRS.
- The design of interfaces should vary according to the missions, robots, operators and scenarios. For instance, the best point of view depends on the size of scenario and the relevant information depends on the tasks that are performed.
- A set of resources can help in the development of interfaces, such as the teleport, robot panels and marks, spatial sounds and alert vibrations.
- VR interfaces are easier to manage for new operators than conventional ones, since they reproduce scenarios and robots and provide natural and intuitive methods for commanding.
- The integration of ROS-based robots and Unity-based VR requires the use of Linux and Windows operating systems, but there are some resources that manage this communication as $rosbridge\_suite$ and $RosBridgeLib$.
- The use of $multimaster\_fkie$ with the proposed communications allows to integrate more simulated and real robots without increasing the volume of exchanged data and the complexity of the interface.
- The modularity of the resources used in this work (ROS, Unity...) allows to easily adapt the interfaces to new robots, information and commands.

Finally, some future works are identified and proposed, such as the reconstruction of scenarios, use of high-level commands and integration of more types of sensors.

The code from the third experience can be accessed in our repository: https://bitbucket.org/robcib/rosbookworld

Acknowledgements This work is framed on SAVIER (Situational Awareness Virtual EnviRonment) Project, which is both supported and funded by Airbus Defence and Space. The research leading to these results has received funding from the RoboCity2030-III-CM project (Robótica aplicada a la mejora de la calidad de vida de los ciudadanos. Fase III; S2013/MIT-2748), funded by Programas de Actividades I+D en la Comunidad de Madrid and cofunded by Structural Funds of the EU, and from the DPI2014-56985-R project (Protección robotizada de infraestructuras críticas) funded by the Ministerio de Economía y Competitividad of Gobierno de España.

References

1. M. Garzón, J. Valente, J.J. Roldán, L. Cancar, A. Barrientos, J. Del Cerro, A multirobot system for distributed area coverage and signal searching in large outdoor scenarios. J. Field Robot. **33**(8), 1087–1106 (2016)
2. J.J. Roldán, P. Garcia-Aunon, M. Garzón, J. de León, J. del Cerro, A. Barrientos, Heterogeneous multi-robot system for mapping environmental variables of greenhouses. Sensors **16**(7), 1018 (2016)

3. M. Garzón, J. Valente, D. Zapata, A. Barrientos, An aerial-ground robotic system for navigation and obstacle mapping in large outdoor areas. Sensors **13**(1), 1247–1267 (2013)
4. J.Y. Chen, E.C. Haas, M.J. Barnes, Human performance issues and user interface design for teleoperated robots. IEEE Trans. Syst. Man Cybern. Part C (Applications and Reviews) **37**(6), 1231–1245 (2007)
5. J. Ruiz, A. Viguria, J. Martinez de Dios, A. Ollero, Immersive displays for building spatial knowledge in multi-uav operations, in *2015 International Conference on Unmanned Aircraft Systems (ICUAS)* (IEEE, 2015), pp. 1043–1048
6. J.T. Hansberger, Development of the next generation of adaptive interfaces. Technical report, DTIC Document (2015)
7. J.J. Roldán, M.A. Olivares-Méndez, J. del Cerro, A. Barrientos, Analyzing and improving multi-robot missions by means of process mining. Auton. Robot. **1**(1), 1–21 (2017)
8. P. Ulam, Y. Endo, A. Wagner, R. Arkin, Integrated mission specication and task allocation for robot teams-part 2: Testing and evaluation. Technical report, GEORGIA INST OF TECH ATLANTA COLL OF COMPUTING (2006)
9. S. Tully, G. Kantor, H. Choset, Leap-frog path design for multi-robot cooperative localization, in *Field and Service Robotics* (Springer, Berlin, 2010), pp. 307–317
10. A. Janchiv, D. Batsaikhan, G. hwan Kim, S.G. Lee, Complete coverage path planning for multi-robots based on, in *2011 11th International Conference on Control, Automation and Systems (ICCAS)* (IEEE, 2011), pp. 824–827
11. M. Lindemuth, R. Murphy, E. Steimle, W. Armitage, K. Dreger, T. Elliot, M. Hall, D. Kalyadin, J. Kramer, J. Palankar et al., Sea robot-assisted inspection. IEEE Robot. Autom. Mag. **18**(2), 96–107 (2011)
12. J. Valente, D. Sanz, A. Barrientos, Jd Cerro, Á. Ribeiro, C. Rossi, An air-ground wireless sensor network for crop monitoring. Sensors **11**(6), 6088–6108 (2011)
13. N.A. Tsokas, K.J. Kyriakopoulos, Multi-robot multiple hypothesis tracking for pedestrian tracking. Auton. Robot. **32**(1), 63–79 (2012)
14. L. Cantelli, M. Mangiameli, C.D. Melita, G. Muscato, Uav/ugv cooperation for surveying operations in humanitarian demining, in *2013 IEEE International symposium on Safety, Security, and Rescue Robotics (SSRR)* (IEEE, 2013), pp. 1–6
15. G. De Cubber, D. Doroftei, D. Serrano, K. Chintamani, R. Sabino, S. Ourevitch, The eu-icarus project: developing assistive robotic tools for search and rescue operations, in *2013 IEEE international symposium on Safety, Security, and Rescue Robotics (SSRR)* (IEEE, 2013), pp. 1–4
16. I. Kruijff-Korbayová, F. Colas, M. Gianni, F. Pirri, J. Greeff, K. Hindriks, M. Neerincx, P. Ögren, T. Svoboda, R. Worst, Tradr project: Long-term human-robot teaming for robot assisted disaster response. KI-Künstliche Intelligenz **29**(2), 193–201 (2015)
17. J. Gregory, J. Fink, E. Stump, J. Twigg, J. Rogers, D. Baran, N. Fung, S. Young, Application of multi-robot systems to disaster-relief scenarios with limited communication, in *Field and Service Robotics* (Springer, Berlin, 2016), pp. 639–653
18. A.C. Kapoutsis, S.A. Chatzichristofis, L. Doitsidis, J.B. de Sousa, J. Pinto, J. Braga, E.B. Kosmatopoulos, Real-time adaptive multi-robot exploration with application to underwater map construction. Auton. Robot. **40**(6), 987–1015 (2016)
19. C. Lesire, G. Infantes, T. Gateau, M. Barbier, A distributed architecture for supervision of autonomous multi-robot missions. Auton. Robot. **40**(7), 1343–1362 (2016)
20. N. Agmon, O. Maximov, A. Rosenfeld, S. Shlomai, S. Kraus, Multiple robots for multiple missions: architecture for complex collaboration
21. X.J. Yang, C.D. Wickens, K. Hölttä-Otto, How users adjust trust in automation: Contrast effect and hindsight bias, in *Proceedings of the Human Factors and Ergonomics Society Annual Meeting*, vol. 60 (SAGE Publications Sage CA: Los Angeles, CA, 2016), pp. 196–200
22. C. Sampedro, H. Bavle, J.L. Sanchez-Lopez, R.A.S. Fernández, A. Rodríguez-Ramos, M. Molina, P. Campoy, A flexible and dynamic mission planning architecture for uav swarm coordination, in *2016 International Conference on Unmanned Aircraft Systems (ICUAS)* (IEEE, 2016), pp. 355–363

23. T. Nestmeyer, P.R. Giordano, H.H. Bülthoff, A. Franchi, Decentralized simultaneous multi-target exploration using a connected network of multiple robots. Auton. Robot. **41**(4), 989–1011 (2017)
24. M. Garzón, J. Valente, J.J. Roldán, D. Garzón-Ramos, J. de León, A. Barrientos, J. del Cerro, Using ros in multi-robot systems: Experiences and lessons learned from real-world field tests, in *Robot Operating System (ROS)* (Springer, Berlin, 2017), pp. 449–483
25. J.J. Roldán, E. Peña-Tapia, A. Martín-Barrio, M.A. Olivares-Méndez, J. Del Cerro, A. Barrientos, Multi-robot interfaces and operator situational awareness: study of the impact of immersion and prediction. Sensors **17**(8), 1720 (2017)
26. M.R. Endsley, Design and evaluation for situation awareness enhancement, in *Proceedings of the human factors and ergonomics society annual meeting*, vol. 32 (SAGE Publications, 1988), pp. 97–101
27. J.L. Drury, J. Scholtz, H.A. Yanco, Awareness in human-robot interactions, in *IEEE International Conference on Systems, Man and Cybernetics*, vol. 1 (IEEE, 2003), pp. 912–918
28. M.R. Endsley, Situation awareness global assessment technique (sagat), in *Proceedings of the IEEE National Aerospace and Electronics Conference. NAECON* (IEEE, 1988), pp. 789–795
29. P. Salmon, N. Stanton, G. Walker, D. Green, Situation awareness measurement: a review of applicability for c4i environments. Appl. Ergon. **37**(2), 225–238 (2006)
30. J. Scholtz, J. Young, J.L. Drury, H.A. Yanco, Evaluation of human-robot interaction awareness in search and rescue, in *Proceedings of the ICRA'04 IEEE International Conference on Robotics and Automation*, vol. 3 (IEEE, 2004), pp. 2327–2332
31. N. Li, S. Cartwright, A. Shekhar Nittala, E. Sharlin, M. Costa Sousa, Flying frustum: a spatial interface for enhancing human-uav awareness, in *Proceedings of the 3rd International Conference on Human-Agent Interaction* (ACM, 2015), pp. 27–31
32. R.J. Lysaght, S.G. Hill, A. Dick, B.D. Plamondon, P.M. Linton, Operator workload: Comprehensive review and evaluation of operator workload methodologies. Technical report, DTIC Document (1989)
33. N. Moray, *Mental Workload: Its Theory and Measurement*, vol. 8 (Springer Science & Business Media, 2013)
34. S.G. Hart, L.E. Staveland, Development of nasa-tlx (task load index): Results of empirical and theoretical research. Adv. Psychol. **52**, 139–183 (1988)
35. S.R. Dixon, C.D. Wickens, D. Chang, Mission control of multiple unmanned aerial vehicles: a workload analysis. Hum. Factors J. Hum. Factors Ergon. Soc. **47**(3), 479–487 (2005)
36. B. Jacobs, E. De Visser, A. Freedy, P. Scerri, *Application of Intelligent Aiding to Enable Single Operator Multiple uav Supervisory Control*, Association for the advancement of artificial intelligence (Palo Alto, CA, 2010)
37. M.L. Cummings, C. Mastracchio, K.M. Thornburg, A. Mkrtchyan, Boredom and distraction in multiple unmanned vehicle supervisory control. Interact. Comput. **25**(1), 34–47 (2013)
38. D. McDuff, S. Gontarek, R. Picard, Remote measurement of cognitive stress via heart rate variability, in *36th Annual International Conference of the IEEE Engineering in Medicine and Biology Society (EMBC)* (IEEE, 2014), pp. 2957–2960
39. H. Kurniawan, A.V. Maslov, M. Pechenizkiy, Stress detection from speech and galvanic skin response signals, in *IEEE 26th International Symposium on Computer-Based Medical Systems (CBMS)* (IEEE, 2013), pp. 209–214
40. E.A. Kirchner, S.K. Kim, M. Tabie, H. Wöhrle, F. Maurus, F. Kirchner, An intelligent man-machine interfacemulti-robot control adapted for task engagement based on single-trial detectability of p300. Front. Hum. Neurosci. **10**(2016)
41. R. Parasuraman, T.B. Sheridan, C.D. Wickens, A model for types and levels of human interaction with automation. IEEE Trans. Syst. Man Cybern. Part A Syst. Hum. **30**(3), 286–297 (2000)
42. B.D. Simpson, R.S. Bolia, M.H. Draper, Spatial audio display concepts supporting situation awareness for operators of unmanned aerial vehicles, *Human Performance, Situation Awareness, and Automation: Current Research and Trends HPSAA II, Volumes I and II*, vol. 2 (2013), p. 61

43. S. Scheggi, M. Aggravi, F. Morbidi, D. Prattichizzo, Cooperative human-robot haptic navigation, in *IEEE International Conference on Robotics and Automation (ICRA)* (IEEE, 2014), pp. 2693–2698
44. C.E. Lathan, M. Tracey, The effects of operator spatial perception and sensory feedback on human-robot teleoperation performance. Presence Teleoper. Virtual Environ. **11**(4), 368–377 (2002)
45. V.M. Monajjemi, S. Pourmehr, S.A. Sadat, F. Zhan, J. Wawerla, G. Mori, R. Vaughan, Integrating multi-modal interfaces to command uavs, in *Proceedings of the 2014 ACM/IEEE International Conference on Human-Robot Interaction* (ACM, 2014), pp. 106–106
46. S. Kavitha, S. Veena, R. Kumaraswamy, Development of automatic speech recognition system for voice activated ground control system, in *International Conference on Trends in Automation, Communications and Computing Technology (I-TACT-15)*, vol. 1 (IEEE, 2015), pp. 1–5
47. T. Mantecón, C.R. del Blanco, F. Jaureguizar, N. García, New generation of human machine interfaces for controlling uav through depth-based gesture recognition, in *SPIE Defense+ Security, International Society for Optics and Photonics* (2014), pp. 90840C–90840C
48. J. Nagi, A. Giusti, G.A. Di Caro, L.M. Gambardella, Human control of uavs using face pose estimates and hand gestures, in *Proceedings of the 2014 ACM/IEEE International Conference on Human-Robot Interaction*, (ACM, 2014), pp. 252–253
49. M. Hou, H. Zhu, M. Zhou, G.R. Arrabito, Optimizing operator–agent interaction in intelligent adaptive interface design: a conceptual framework. IEEE Trans. Syst. Man Cybern. Part C (Applications and Reviews) **41**(2), 161–178 (2011)
50. J.L. Drury, J. Richer, N. Rackliffe, M.A. Goodrich, Comparing situation awareness for two unmanned aerial vehicle human interface approaches. Technical report, Mitre Corp Bedford MA (2006)
51. K. Foit, Mixed reality as a tool supporting programming of the robot, in *Advanced Materials Research*, vol. 1036 (Trans Tech Publ, 2014), pp. 737–742
52. D.C. Niehorster, L. Li, M. Lappe, The accuracy and precision of position and orientation tracking in the htc vive virtual reality system for scientific research. i-Perception **8**(3), 2041669517708205 (2017)
53. F. Navarro, J. Fdez, M. Garzon, J.J. Roldán, A. Barrientos, Integrating 3d reconstruction and virtual reality: a new approach for immersive teleoperation, in *Robot 2017: Third Iberian Robotics Conference* (Springer, 2018), pp. X–Y
54. S.H. Juan, F.H. Cotarelo, *Multi-Master Ros Systems*, Institut de robotics and industrial informatics (2015)
55. C. Crick, G. Jay, S. Osentoski, B. Pitzer, O.C. Jenkins, Rosbridge: Ros for non-ros users, in *Robotics Research* (Springer, Berlin, 2017), pp. 493–504
56. P. Codd-Downey, A.S.H.W. Mojiri Forooshani, M. Jenkin, From ros to unity: leveraging robot and virtual environment middleware for immersive teleoperation (2014)
57. E. Peña-Tapia, J.J. Roldán Gómez, M. Garzón, A. Martín-Barrio, A. Barrientos-Cruz, Interfaz de control para un robot manipulador mediante realidad virtual (2017)
58. I.A. Sucan, S. Chitta, Moveit!. http://moveit.ros.org (2013)
59. I.A. Şucan, M. Moll, L.E. Kavraki, The open motion planning library. IEEE Robot. Autom. Mag. **19**(4), 72–82 (2012)
60. J.J. Kuffner, S.M. LaValle, Rrt-connect: an efficient approach to single-query path planning, in *Proceedings of the ICRA'00 IEEE International Conference on Robotics and Automation*, vol. 2 (IEEE, 2000), pp. 995–1001
61. R. Guzman, R. Navarro, M. Beneto, D. Carbonell, Robotnikprofessional service robotics applications with ros, in *Robot Operating System (ROS)* (Springer, Berlin, 2016), pp. 253–288
62. R. Guzmán, R. Navarro, M. Cantero, J. Ariño, Robotnikprofessional service robotics applications with ros (2), in *Robot Operating System (ROS)* (Springer, Berlin, 2017), pp. 419–447

Juan Jesús Roldán (1988, Almería-Spain) studied BSc+MSc on Industrial Engineering (2006-2012) and MSc on Automation and Robotics (2013-2014) in Technical University of Madrid (UPM). Currently, he is a Ph.D. candidate of Centre for Automation and Robotics (CAR, UPM-CSIC) and Airbus Defence & Space, whose research is focused on the multi-UAV coordination and control interfaces. He was visiting student in the Interdisciplinary Centre for Security, Reliability and Trust (SnT) of the University of Luxembourg. He has researched about emergency detection and management in multirotors, surveillance of large fields with multiple robots and ground and aerial robots applied to environmental monitoring in greenhouses. Currently, he is interested in techniques for swarming control, treatment of information and interface.

Elena Peña-Tapia is a graduate of Industrial Technology Engineering at the Technical University of Madrid (UPM), with a specialty in Automation and Electronics. Fascinated by science from an early age, she has been a visiting student at UPenn and Stanford, where she grew an interest in the latest technologies. She is currently coordinating her studies (Masters Degree in Industrial Engineering) with research at the CAR (Centre for Automation and Robotics). Her fields of interest include VR, user interfaces and AI.

David Garzón-Ramos was born in Pasto, Colombia in 1990. He received his Electronic Engineer degree from the Universidad Nacional de Colombia (UN) in 2014. Then, he carried out his master's studies on Automation and Robotics in the Universidad Politcnica de Madrid (UPM), and Engineering - Industrial Automation in the Universidad Nacional de Colombia (UN) until 2016. In 2017 he started his PhD studies in the Université Libre de Bruxelles (ULB). His research activities started in 2011 as undergraduate research assistant in the Optical Properties of Materials (UN) research group. From 2014 to 2017 he did an internship in the Aerospace & Control Systems Section from SENER Ingeniería y Sistemas, S.A. In 2015 he joined to the Robotics and Cybernetics (UPM) group as postgraduate research assistant. Since 2017 he is a researcher in the Institut de Recherches Interdisciplinaires et de Dveloppements en Intelligence Artificielle (IRIDIA - ULB). His research interests (non-exhaustive) are Unmanned Ground Vehicles (UGVs), Multi-Robot systems (MRS), Guidance, Navigation, and Control (GNC) and Swarm Robotics.

Jorge de León was born in Barcelona in 1988. He received the Bachelor Degree in Industrial Electronics from University of La Laguna in 2013 and the Msc in Automation and Robotics from the Technical University of Madrid (UPM) in 2015. His main research interests are focused on designing new mobile robots and algorithms for searching and surveillance missions. He is currently a researcher at the Robotics and Cybernetics research group of the Centre for Automation and Robotics (CAR UPM-CSIC) where he is developing his PhD thesis.

Mario Garzón was born in Pasto, Colombia, in 1983. He is a researcher at the Robotics and Cybernetics research group at Centro de Automática y Robótica (CAR UPM-CSIC). He received the Electronics Engineer degree from Escuela Colombiana de Ingeniería in 2005 and both Msc and PhD degrees in Automation and Robotics from Universidad Politécnica de Madrid (UPM) in 2011 and 2016 respectively. He was also visiting researcher at INRIA (e-Motion team project) in Grenoble, France in the years 2013 and 2014. His main research interest is field robotics, specifically on navigation and cooperation between heterogeneous mobile robots as well as the detection and prediction of pedestrian trajectories. He has also worked on Bio-Inspired robotics, computer vision, machine learning and data mining.

Jaime del Cerro received his Ph.D. degree in Robotics and Computer vision at Polytechnic University of Madrid at 2007. He currently teaches Robotics, Guidance, Navigation and Control of autonomous robots and system programming in this University and also collaborates at UNIR (Universidad International de la Rioja) in the master of Management of Technological projects. He has participated in several European Framework projects and projects funded by ESA (European Space Agency) and EDA (European Defence Agency) as well as commercial agreements with relevant national companies.

Antonio Barrientos received the MSc Engineer degree in Automation and Electronics from the Polytechnic University of Madrid in 1982, and the PhD in Robotics by the same University in 1986. In 2002 he obtained de MSc Degree in Biomedical Engineering by Universidad Nacional de Educación a Distancia. Since 1988 he is Professor on robotics, computers and control engineering at the Polytechnic University of Madrid. He has worked for more than 30 years in robotics, developing industrial and service robots for different areas. His main interests are in air and ground field robotics. He is author of several textbooks in Robotics and Manufacturing automation, and also is co-author of more than 100 scientific papers in journals and conferences. Ha has been director o co-director of more than 20 PhD thesis in the area of robotics. Currently he is the head of the Robotics and Cybernetics research group of the Centre for Automatic and Robotics in the Technical University of Madrid Spanish National Research Council.

Part II
Unmanned Aerial Systems

Autonomous Exploration and Inspection Path Planning for Aerial Robots Using the Robot Operating System

Christos Papachristos, Mina Kamel, Marija Popović,
Shehryar Khattak, Andreas Bircher, Helen Oleynikova,
Tung Dang, Frank Mascarich, Kostas Alexis and Roland Siegwart

Abstract This use case chapter presents a set of algorithms for the problems of autonomous exploration, terrain monitoring and optimized inspection path planning using aerial robots. The autonomous exploration algorithms described employ a receding horizon structure to iteratively derive the action that the robot should take to optimally explore its environment when no prior map is available, with the extension to localization uncertainty–aware planning. Terrain monitoring is tackled

C. Papachristos · S. Khattak · T. Dang · F. Mascarich · K. Alexis (✉)
Autonomous Robots Lab, University of Nevada, 1664 N. Virginia St.,
Reno, NV 89557, USA
e-mail: kalexis@unr.edu

C. Papachristos
e-mail: cpapachristos@unr.edu

S. Khattak
e-mail: shehryar.khattak@nevada.unr.edu

T. Dang
e-mail: tung.dang@nevada.unr.edu

F. Mascarich
e-mail: fmascarich@nevada.unr.edu

M. Kamel · M. Popović · A. Bircher · H. Oleynikova · R. Siegwart
Autonomous System Lab, ETH Zurich, Leonhardstrasse 21,
8092 Zurich, Switzerland
e-mail: fmina@ethz.ch

M. Popović
e-mail: mpopovic@ethz.ch

A. Bircher
e-mail: abircher@ethz.ch; andreas.bircher@mavt.ethz.ch

H. Oleynikova
e-mail: holeynikova@ethz.ch; elena.oleynikova@mavt.ethz.ch

R. Siegwart
e-mail: rsiegwart@ethz.ch

© Springer International Publishing AG, part of Springer Nature 2019
A. Koubaa (ed.), *Robot Operating System (ROS)*, Studies in Computational
Intelligence 778, https://doi.org/10.1007/978-3-319-91590-6_3

by a finite–horizon informative planning algorithm that further respects time budget limitations. For the problem of optimized inspection with a model of the environment known a priori, an offline path planning algorithm is proposed. All methods proposed are characterized by computational efficiency and have been tested thoroughly via multiple experiments. The Robot Operating System corresponds to the common middleware for the outlined family of methods. By the end of this chapter, the reader should be able to use the open–source contributions of the algorithms presented, implement them from scratch, or modify them to further fit the needs of a particular autonomous exploration, terrain monitoring, or structural inspection mission using aerial robots. Four different open–source ROS packages (compatible with ROS Indigo, Jade and Kinetic) are released, while the repository https://github.com/ unr-arl/informative-planning stands as a single point of reference for all of them.

1 Introduction

Autonomous inspection, exploration, and mapping correspond to main and critical use cases for aerial robots. In a multitude of application domains, including infrastructure and industrial inspection [1], precision agriculture [2, 3], search and rescue [4], and security surveillance [5], such capabilities are enabling wide utilization of autonomous aerial systems. Their use leads to enhanced safety for human personnel, major cost savings, improved operation of the inspected facilities, enhanced security, and more. Recent advances in onboard sensing and processing, alongside a set of algorithmic contributions relating to the problems of Simultaneous Localization And Mapping (SLAM) and path planning, pave the way for aerial robots to conduct the aforementioned tasks autonomously, reliably, and efficiently.

In this chapter, we present a family of algorithmic contributions that address the problems of:

- autonomous area exploration and mapping when no prior knowledge of the environment is available,
- informative path planning for terrain monitoring,
- full coverage inspection planning provided a prior model of the environment.

For the problem of autonomous unknown area exploration, iterative online receding horizon algorithms are discussed that further account for the localization uncertainty of the robot. The terrain monitoring problem is addressed through a finite–horizon planner that optimizes for information gain subject to endurance constraints. For the problem of optimized coverage given a model of the environment, efficient derivation of an inspection path offline is presented. For all three cases, the methods account for the vehicle constraints and motion model.

Each method is presented in regards to the algorithmic concepts, key implementation details, user interface and developer information for the associated open–source contributions by our team, alongside relevant experimental data. By the end of the chapter, the reader should be able to implement and test these algorithms onboard real aerial robots or in simulation. The autonomous exploration, terrain monitoring, and inspection algorithms are open–sourced as Robot Operating System (ROS) packages and documented to enable and accelerate further development. For all algorithms, various implementation hints and practical suggestions are provided in this chapter. The methods have been experimentally verified and field demonstrated. The presented methods have enabled: (a) the autonomous exploration of railway and city tunnels in visually–degraded conditions (darkness, haze), (b) the mapping of a national park, (c) the exploration of several office room-like environments, (d) the monitoring of a mock–up relevant to precision agriculture tasks, and more. It is critical to highlight that ROS is not only the environment in which the planners are implemented but also the enabling factor for the presented experiments. The core loops of localization and mapping, as well as position control in all the described experiments are also ROS-based.

Section 2 presents the exact problem statements for each planning scenario above. Sections 3, 4 and 5 detail the algorithms for autonomous exploration, terrain monitoring, and inspection respectively and present indicative evaluation results. Finally, Section 6 details the open–source code contributions, followed by conclusions in Sect. 7.

2 Problem Definition

This section outlines the exact definitions for the three informative planning scenarios discussed in this chapter: unknown area exploration, terrain monitoring, and inspection planning. In total four algorithms are presented, namely the "Receding Horizon Next-Best-View Planner" (nbvplanner), "Localization Uncertainty–aware Receding Horizon Exploration and Mapping Planner" (rhemplanner), "Terrain Monitoring Planner" (tmplanner) and "Structural Inspection Planner" (siplanner). The associated problem statements are provided below.

2.1 Exploration Path Planning

The autonomous unknown area exploration problem considered in this chapter is that of exploring a bounded 3D volume $V^E \subset \mathbb{R}^3$ while possibly aiming to minimize the localization and mapping uncertainty as evaluated through a metric over the probabilistic belief of robot's pose and landmark positions. The exploration problem may be casted globally as that of starting from an initial collision free configuration and deriving viewpoints that cover the a priori unknown volume by determining which parts of the initially unexplored space $V^E_{une} \stackrel{init.}{=} V^E$ are free $V^E_{free} \subseteq V^E$ or occupied $V^E_{occ} \subseteq V^E$. For this process, the volume is discretized in an occupancy map $\mathcal{M}$ consisting of cubical voxels $m \in \mathcal{M}$ with edge length r. The operation is subject to vehicle dynamic constraints and limitations of the sensing system. As the perceptive capabilities of most sensors stop at surfaces, hollow spaces or narrow pockets can sometimes remain unexplored, leading to a residual volume:

Definition 1 (*Residual Volume*) Let Ξ be the simply connected set of collision free configurations and $\bar{\mathcal{V}}_m^E \subseteq \Xi$ the set of all configurations from which the voxel m can be perceived. Then the residual volume is given by $V_{res}^E = \bigcup_{m \in \mathcal{M}} (m \mid \bar{\mathcal{V}}_m^E = \emptyset)$.

The exploration problem is globally defined as:

Problem 1 (*Volumetric Exploration Problem*) Given a bounded volume V^E, find a collision free path σ starting at an initial configuration $\xi_{init} \in \Xi$ that leads to identifying the free and occupied parts V_{free}^E and V_{occ}^E when being executed, such that there does not exist any collision free configuration from which any piece of $V^E \setminus \{V_{free}^E, V_{occ}^E\}$ could be perceived. Thus, $V_{free}^E \cup V_{occ}^E = V^E \setminus V_{res}^E$.

While the exploration problem is globally defined, the optional sub–problem of localization and mapping uncertainty–aware planning in a priori unknown environments can only be casted locally as new features are tracked from the perception system. At every planning step, let $V^M \subset V^E$ be a local volume enclosing the current pose configuration and the next exploration viewpoint. The problem is that of minimizing the covariance of the propagated robot pose and tracked landmarks belief as the vehicle moves from its current configuration to the next exploration viewpoint. By minimizing the robot belief uncertainty, a good prior for high–quality mapping is achieved. More formally:

Problem 2 (*Belief Uncertainty–aware Planning*) Given a $V^M \subset V^E$, find a collision free path σ^M starting at an initial configuration $\xi_0 \in \Xi$ and ending in a configuration ξ_{final} that aims to maximize the robot's localization and mapping confidence by following paths of its optimized pose and covariance of tracked landmarks.

2.2 Path Planning for Terrain Monitoring

This chapter considers the problem of informative path planning for monitoring a flat terrain using an aerial robot equipped with an altitude-dependent sensor. The environment where measurements are to be taken is represented as a 2D plane $\mathcal{E} \subset \mathbb{R}^2$. The objective is to maximize the informative data collected within $\mathcal{E}$ while respecting a time budget. Thus, we seek a continuous trajectory ψ in the 3D space of all trajectories Ψ above the terrain for maximum gain in some information-theoretic measure:

$$\psi^* = \underset{\psi \in \Psi}{\operatorname{argmax}} \frac{I[\text{MEASURE}(\psi)]}{\text{TIME}(\psi)}, \tag{1}$$

$$\text{s.t. TIME}(\psi) \leq B,$$

where B denotes a time budget and $I[\cdot]$ defines the utility function quantifying the informative objective. The function MEASURE$(\cdot)$ obtains a finite set of measurements along the trajectory ψ and TIME$(\cdot)$ provides the corresponding travel time.

In Eq. 1, $I[\cdot]$ is formulated generically and can be defined to capture situation-specific interests with respect to the underlying terrain map representation (Sect. 4.3.3), e.g. entropy minimization. In this chapter, we study two different mapping scenarios involving (i) discrete and (ii) continuous target variables, as relevant for practical applications.

2.3 Inspection Path Planning

The problem of structural inspection path planning is considered in this chapter as that of finding a high quality path that guarantees the complete coverage of a given 3D structure S subject to dynamic constraints of the vehicle and limitations of its on–board sensor system. The 3D structure to be inspected may be represented in a computationally efficient way, such as a triangular mesh or voxel-based octomap, and is embedded in a bounded environment that could contain obstacle regions. The problem setup is to be such that that for each triangle in the mesh, there exists an *admissible* viewpoint configuration – that is, a viewpoint from which the triangle is visible for a specific sensor model. Then, for the given environment and with respect to the operational constraints, the aim is to find a path connecting all viewpoints which guarantees complete inspection of the 3D structure. Quality measures for paths are situation-specific, depending on the system and mission objectives, e.g. short time or distance.

3 Autonomous Unknown Area Exploration

This section overviews and details a set of sampling–based receding horizon algorithms enabling the autonomous exploration of unknown environments. Both methodologies, the "Receding Horizon Next-Best-View planner" (nbvplanner) and "Localization Uncertainty–aware Receding Horizon Exploration and Mapping planner" (rhemplanner), exploit a mechanism that identifies the next–best–view in terms of exploration gain along a finite–depth trajectory. The first viewpoint of this trajectory is then visited, while the whole process is repeated iteratively. The distinctive feature of the second algorithm is that it identifies the best path to go to this new viewpoint, such that the on–board localization uncertainty of the robot is minimized.

3.1 Receding Horizon Next–Best–View Planner

Considering the exploration of unknown space, the presented nbvplanner employs a sampling–based receding horizon path planning paradigm to generate paths that cover the whole, initially unknown but bounded, space V^E. A sensing system that perceives a subset of the environment and delivers depth/volumetric data (e.g. stereo camera, a LiDAR or a time-of-flight 3D camera) is employed to provide feedback on the explored space. All acquired information is combined in a map $\mathcal{M}$ representing an estimate of the current world state. This map is used for both collision–free navigation and determination of the exploration progress.

The representation of the environment is an occupancy map [6] dividing space V^E in cubical volumes $m \in \mathcal{M}$, that can either be marked as free, occupied or unmapped. For every voxel m in this set lies in the unexplored area V_{unm}^E, the direct Line of Sight (LoS) does not cross occupied voxels and complies with the sensor model. Within this work, the camera sensor is mounted with a fixed pitch, has a Field of View (FoV) described by vertical and horizontal opening angles $[a_v, a_h]$, and a maximum effective sensing distance. While the sensor limitation is d_{max}^{sensor}, the planner uses a value $d_{max}^{planner} < d_{max}^{sensor}$. The resulting array of voxels is saved in an octomap, enabling computationally efficient access for operations like integrating new sensor data or checking for occupancy. Paths are only planned through known free space V_{free}, thus inherently providing collision–free navigation. While a collision–free vehicle configuration is denoted by $\xi \in \varXi$, a path is given by $\sigma^E : \mathbb{R} \to \xi$, specifically from ξ_{k-1} to ξ_k by $\sigma_{k-1}^k(s)$, $s \in [0, 1]$ with $\sigma_{k-1}^k(0) = \xi_{k-1}$ and $\sigma_{k-1}^k(1) = \xi_k$. The path has to be such that it is collision–free and can be tracked by the vehicle, considering its dynamic and kinematic constraints. The corresponding path cost (distance) is $c(\sigma_{k-1}^k)$. For a given occupancy map representing the world $\mathcal{M}$, the set of visible and unmapped voxels from configuration ξ is denoted as **Visible**$(\mathcal{M}, \xi)$. Every voxel $m \in$ **Visible**$(\mathcal{M}, \xi)$ lies in the unmapped exploration area V_{unm}, the direct LoS does not cross occupied voxels and complies with the sensor model (in terms of FoV, maximum effective perception, and depth sensing range). Starting from the current vehicle configuration ξ_0, a geometric random tree (RRT [7]) $\mathbb{T}^E$ is incrementally built in the configuration space. The resulting tree contains $N_{\mathbb{T}}^E$ nodes n^E and its edges are given by collision–free paths σ^E, connecting the configurations of the two nodes. The quality - or collected information gain - of a node **Gain**(n) is the summation of the unmapped volume that can be explored at the nodes along the branch, e.g. for node k, according to

$$\textbf{ExplorationGain}(n_k^E) = \textbf{ExplorationGain}(n_{k-1}^E) + \textbf{VisibleVolume}(\mathcal{M}, \xi_k) \exp(-\lambda c(\sigma_{k-1,k}^E))$$

with the tuning factor λ penalizing high path costs and therefore indirectly giving preference to shorter paths as long as sufficient information–gathering is achieved [8].

After every replanning iteration, the robot executes the first segment of the branch to the best node as evaluated by **ExtractBestPathSegment**(n_{best}), where n_{best} is the node with the highest gain. Generally, tree construction is stopped when $N_{\mathbb{T}^E} = N_{max}^E$, but if the best gain g_{best} of n_{best} remains zero, this process continues until $g_{best} > 0$.

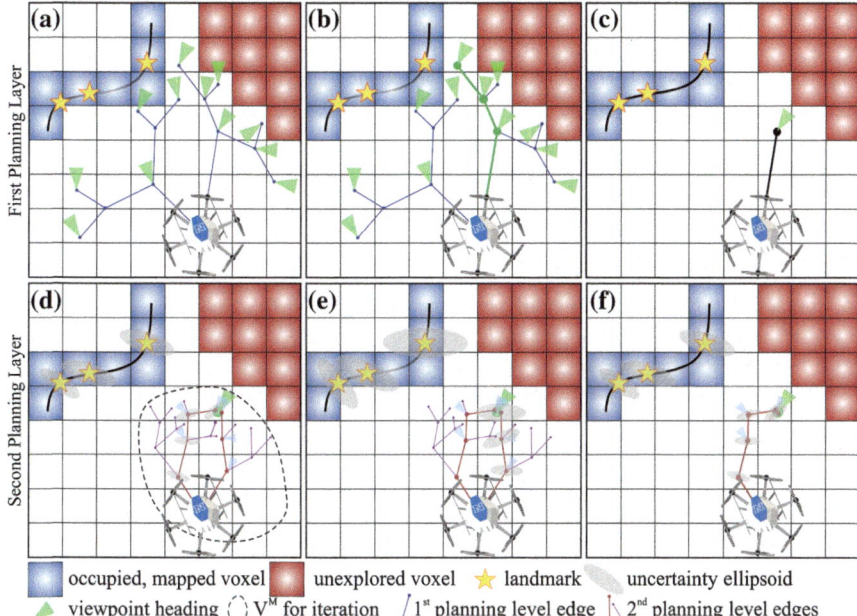

First Planning Layer

Second Planning Layer

| ▦ occupied, mapped voxel | ▦ unexplored voxel | ⭐ landmark | uncertainty ellipsoid |

▲ viewpoint heading ⟨ ⟩ V^M for iteration ╱ 1ˢᵗ planning level edge ╲ 2ⁿᵈ planning level edges

Fig. 1 2D representation of the two–step uncertainty–aware exploration and mapping planner. The first planning layer samples the path with the maximum information gain in terms of space to be explored. This also corresponds to a single iteration of the `nbvplanner`. The viewpoint configuration of the first vertex of this path becomes the reference to the second planning layer. This step samples admissible paths that arrive to this configuration and selects the one that provides minimum uncertainty of the robot belief about its pose and the tracked landmarks. To achieve this, the belief is propagated along the candidate edges. It is noted that the second step is only executed by the uncertainty–aware planner

For practical purposes, a tolerance value N_{TOL} is chosen significantly higher than N_{max} and the exploration step is considered solved when $N_{\mathbb{T}}^E = N_{TOL}^E$ is reached while g_{best} remains zero. Algorithm 1 and Fig. 1 summarize and illustrate the planning procedure.

3.1.1　Indicative Evaluation Study

A simulation–based evaluation study is presented to demonstrate intermediate steps of the exploration process commanded by the `nbvplanner`. Since the planner works in a closed loop with the robot's perception, estimation, and control loops, a detailed and realistic simulation is required. The Gazebo–based simulation environment RotorS has been used along with the provided model of the AscTec Firefly hexacopter MAV. It employs a stereo camera that gives real time depth images of the simulated environment. Its specifications are a FoV of $[60, 90]°$ in vertical and horizontal direction, respectively, while it is mounted with a downward pitch angle of

15°. For all simulations, the size of the box model for collision detection is assumed as $0.5 \times 0.5 \times 0.3$ m.

Algorithm 1 nbvplanner - Iterative Step

1: $\xi_0 \leftarrow$ current vehicle configuration
2: Initialize $\mathbb{T}^E$ with ξ_0 and, unless first planner call, also previous best branch
3: $g_{best} \leftarrow 0$ ▷ Set best gain to zero.
4: $n_{best} \leftarrow n_0^E(\xi_0)$ ▷ Set best node to root.
5: $N_{\mathbb{T}}^E \leftarrow$ Number of initial nodes in $\mathbb{T}^E$
6: **while** $N_{\mathbb{T}}^E < N_{\max}^E$ **or** $g_{best} = 0$ **do**
7:     Incrementally build $\mathbb{T}^E$ by adding $n_{new}^E(\xi_{new})$
8:     $N_{\mathbb{T}}^E \leftarrow N_{\mathbb{T}}^E + 1$
9:     **if** **ExplorationGain**$(n_{new}^E) > g_{best}$ **then**
10:         $n_{best} \leftarrow n_{new}^E$
11:         $g_{best} \leftarrow$ **ExplorationGain**(n_{new}^E)
12:     **end if**
13:     **if** $N_{\mathbb{T}}^E > N_{TOL}^E$ **then**
14:         Terminate exploration
15:     **end if**
16: **end while**
17: $\sigma^E \leftarrow$ **ExtractBestPathSegment**(n_{best})
18: Delete $\mathbb{T}^E$
19: **return** σ^E

The simulation scenario considers an indoor room–like environment involving significant geometric complexity and, specifically, a room with a very narrow corridor entrance. Figure 2 visualizes the environment and instances of the exploration mapping based on the reconstructed occupancy map.

The open–sourced ROS package of nbvplanner further supports multi–robot systems. In Fig. 3, an additional result is presented to demonstrate the significant improvement in terms of exploration time when appropriately pre–distributed multiple agents are used. Note that the code release currently assumes that each vehicle explores independently with only local knowledge about its environment, and can be improved by integrating more complex collaborative strategies.

A set of experimental studies are presented in [9–11] and relate to experimental evaluation in indoor–like environments of varying geometric complexity. The video in https://youtu.be/D6uVejyMea4 is indicative of the planner experimental operation.

3.2 Localization Uncertainty–Aware Receding Horizon Exploration and Mapping Planner

The abovementioned planning paradigm ensures efficient exploration of unknown volume but neglects the fact that robots often operate subject to uncertain localization and mapping. In fact, especially when GPS–denied operation is considered, the view-

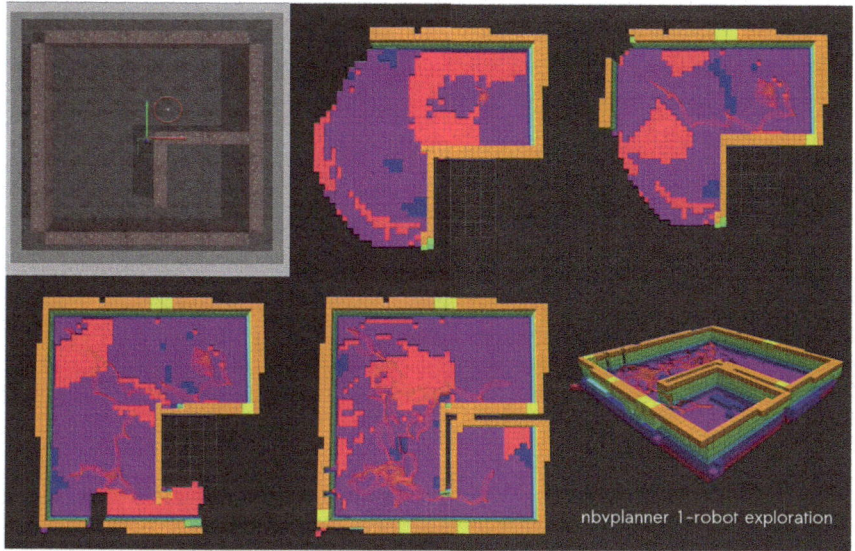

Fig. 2 Single robot exploration of a room–like environment with challenging geometry (narrow corridor to secondary room) using the `nbvplanner`. Incremental steps of exploration are presented while the identical simulation scenario (in terms of environment and robot configuration - paths will be different due to the randomized nature of the solution) can be executed using the open–source ROS package for this planner

point and trajectory selection mechanisms not only influence the exploration process but also the accuracy of the robot pose estimation and map quality. To address this problem, the receding horizon exploration paradigm was extended into a second algorithm accounting for the propagated robot belief regarding its localization uncertainty. This new algorithm, called `rhemplanner`, employs a two–step sampling–based receding horizon planning paradigm to iteratively generate paths that can cover the whole space V^E, while locally aiming to minimize the localization uncertainty of the robot in terms of the covariance of its pose and tracked landmarks. The first planning step replicates the strategy of the `nbvplanner` and identifies a finite–steps path that maximally explores new, previously unmapped space. Subsequently, the first new viewpoint configuration of this path becomes the goal point of a second, nested, planning step that computes a new path to go to this viewpoint while ensuring that low localization uncertainty belief is maintained. The robot follows this path and the entire process is iterated. A visualization of this procedure is shown in Fig. 1.

A sensor, e.g. a camera or a stereo visual–inertial unit, as in the presented experimental studies, is used to enable robot localization and map building. More specifically, estimates for robot pose and tracked landmarks as well as the associated covariance matrix are provided, along with a volumetric map of the environment derived using the stereo depth sensing and the estimated robot poses. Any other sensing system that provides landmarks–based localization and a depth estimate can

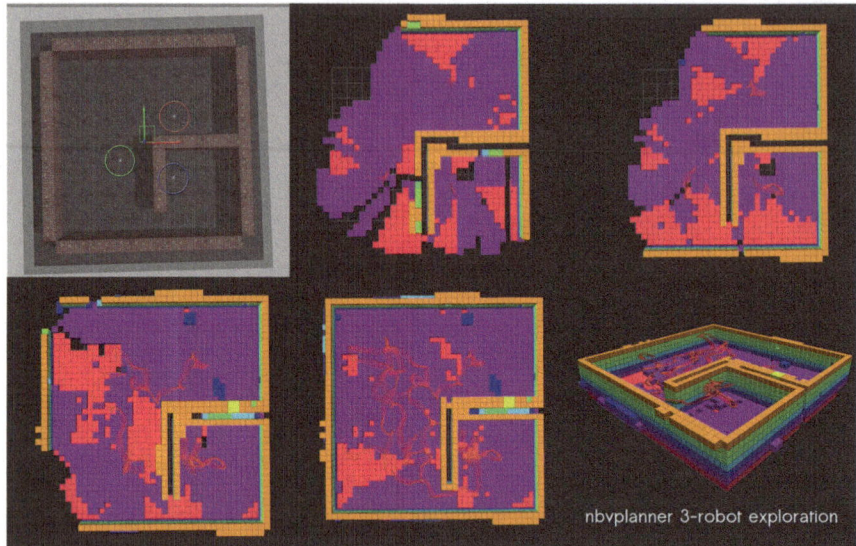

Fig. 3 Multi robot exploration of a room–like environment with challenging geometry (narrow corridor to secondary room) using the nbvplanner. Incremental steps of exploration are presented while the identical simulation scenario (in terms of environment and robot configuration - paths will be different due to the randomized nature of the solution) can be executed using the open–source ROS package for this planner. Although the support for for multi–agent systems in the code release aims to support the community of multi–robot collaboration, it also demonstrates the significant savings in time if multiple and appropriately pre–distributed robotic systems are employed for the same environment

also be used without any change in the planner functionality. The robot belief and the volumetric map are used for collision–free navigation and exploration path planning.

The same volumetric representation of an occupancy map $\mathcal{M}$ [6] (implemented in our system with the support of ROS) dividing the space V^E in cubical volumes $m \in \mathcal{M}$ is employed. Each voxel is marked as free or occupied with a probability $\mathcal{P}(m)$ or unmapped. Similarly, for planning purposes, a robot configuration is defined by the state $\xi = [x, y, z, \psi]^T$ with roll (ϕ) and pitch (θ) considered to be close to zero. The following details the role of each planning step.

3.3 Exploration Planning

At each iteration, for the map representing the world $\mathcal{M}$, the set of visible but unmapped voxels given a robot configuration ξ is denoted as **VisibleVolume**$(\mathcal{M}, \xi)$. Beyond the volumetric exploration gain as in the case of the nbvplanner, this extended planner also considers the probability of a voxel being occupied $\mathcal{P}(m) < p_{thres}$ and calculates the **ReobservationGain**$(\mathcal{M}, \mathcal{P}, \xi)$ which refers to the voxels' volume weighted by $(1 - \mathcal{P}(m))$. Starting from the current configuration ξ_0, a RRT $\mathbb{T}^E$ with maximum edge length ℓ_E is incrementally built. The resulting tree contains $N_{\mathbb{T}}^E$ nodes n^E and its edges are given by collision–free paths σ^E. A minimum $N_{\max}^E$ and maximum N_{TOT}^E threshold on the amount of nodes sampled are set. The collected information gain of a node **ExplorationGain**(n^E) is the cumulative unmapped volume that can be explored from the nodes along the branch, added to the **ReobservationGain**$(\mathcal{M}, \mathcal{P}, \xi)$. For node k:

$$\textbf{ExplorationGain}(n_k^E) = \textbf{ExplorationGain}(n_{k-1}^E) + \tag{2}$$
$$\textbf{VisibleVolume}(\mathcal{M}, \xi_k) \exp(-\lambda c(\sigma_{k-1,k}^E)) +$$
$$\textbf{ReobservationGain}(\mathcal{M}, \mathcal{P}, \xi_k) \exp(-\lambda c(\sigma_{k-1,k}^E))$$

At the end of this planning level iteration, the first segment σ_{RH}^E of the branch to the best node **ExtractBestPathSegment**(n_{best}^E), the vertex at the end of this segment n_{RH}^E, and the associated pose configuration ξ_{RH} are extracted. If no positive gain can be found, the exploration process is terminated.

3.4 Localization Uncertainty–Aware Planning

The second, nested, planning step serves to minimize the localization uncertainty of the robot and tracked landmarks uncertainty given that the vehicle is equipped with a relevant sensor system and algorithms. At every call of this nested planning level, a new RRT $\mathbb{T}^M$ with maximum edge length ℓ_M is spanned within a continuous local volume $V^M \subset V^E$ that includes the current robot pose and ξ_{RH}. A total set of $N_{\mathbb{T}}^M$ vertices are sampled with $N_{\mathbb{T}}^M < N_{\max}^M$. This allows the derivation of admissible paths (with σ_{RH}^E treated as one of them) that (a) start from the current robot configuration and arrive in a local set $\mathbb{S}_{\xi_{RH}}$, around ξ_{RH} provided from the first exploration planning layer, and (b) have an overall length $c \leq \delta c(\sigma_{RH}^E)$ where $\delta \geq 1$ a tuning factor. Since an admissible path is one that arrives into the local set $\mathbb{S}_{\xi_{RH}}$ $(\mathbb{V}(\mathbb{T}^M) \cap \mathbb{S}_{\xi_{RH}} \neq \emptyset)$ and not necessarily on ξ_{RH} an additional connection takes place to guarantee that the robot arrives exactly on the configuration sampled by the previous planning layer. For each of the $N_{\mathbb{T}}^M - 1$ edges of the tree (including the additional connections to n_{RH}), belief propagation takes place to estimate the expected robot belief about its state and the tracked landmarks along the sampled paths. Given the updated estimates of the state and landmarks covariance, the admissible tree branches are evaluated regarding their "belief gain" measured using the D-optimality [12] operation over the robot pose and tracked landmarks covariance matrix at the end–vertex of each admissible path. The admissible branch with the best belief gain is then selected to be executed by the robot. Algorithm 2 details the steps of the complete nbvplanner process, while Sects. 3.4.1–3.4.3 detail the localization framework employed from a generalized perspective, the belief propagation, and the belief gain calculations, respectively.

3.4.1 Visual–Inertial Localization

For the purposes of GPS–denied robot navigation, a visual–inertial odometry framework is employed due to the high robustness and accuracy provided by these methods. In particular, the open–source and ROS-based Robust Visual Inertial Odometry (ROVIO) is utilized [13]. Below, a necessarily very brief summary will be provided as its formulation is also used for belief propagation. It is important that the par-

---

Algorithm 2 `rhemplanner` - Iterative Step

1: $\xi_0 \leftarrow$ current vehicle configuration
2: Initialize $\mathbb{T}^E$ with ξ_0
3: $g_{best}^E \leftarrow 0$ ▷ Set best exploration gain to zero.
4: $n_{best}^E \leftarrow n_0(\xi_0)$ ▷ Set best exploration node to root.
5: $N_{\mathbb{T}}^E \leftarrow$ Number of initial nodes in $\mathbb{T}^E$
6: **while** $N_{\mathbb{T}}^E < N_{\max}^E$ **or** $g_{best}^E = 0$ **do**
7:      Incrementally build $\mathbb{T}^E$ by adding $n_{new}^E(\xi_{new})$
8:      $N_{\mathbb{T}}^E \leftarrow N_{\mathbb{T}}^E + 1$
9:      **if** **ExplorationGain**$(n_{new}^E) > g_{best}^E$ **then**
10:         $n_{best}^E \leftarrow n_{new}^E$
11:         $g_{best}^E \leftarrow$ **ExplorationGain**(n_{new}^E)
12:     **end if**
13:     **if** $N_{\mathbb{T}}^E > N_{TOL}^E$ **then**
14:         Terminate planning
15:     **end if**
16: **end while**
17: $\sigma_{RH}^E, n_{RH}^E, \xi_{RH} \leftarrow$ **ExtractBestPathSegment**(n_{best}^E)
18: $\mathbb{S}_{\xi_{RH}} \leftarrow$ **LocalSet**(ξ_{RH})
19: Propagate robot belief along σ_{RH}^E
20: $\alpha \leftarrow 1$ ▷ number of admissible paths
21: $g_\alpha^M \leftarrow$ **BeliefGain**(σ_{RH}^E)
22: $g_{best}^M \leftarrow g_\alpha^M$ ▷ straight path belief gain
23: $\sigma_{best}^M \leftarrow \sigma_{RH}^M$ ▷ Set best belief path.
24: **while** $N_{\mathbb{T}}^M < N_{\max}^M$ **or** $\mathbb{V}(\mathbb{T}^M) \cap \mathbb{S}_{\xi_{RH}} = \emptyset$ **do**
25:     Incrementally build $\mathbb{T}^M$ by adding $n_{new}^M(\xi_{new})$
26:     Propagate robot belief from current to planned vertex
27:     **if** $\xi_{new} \in \mathbb{S}_{\xi_{RH}}$ **then**
28:         Add new vertex n_{new}^M at ξ_{RH} and connect
29:         $\alpha \leftarrow \alpha + 1$
30:         $\sigma_\alpha^M \leftarrow$ **ExtractBranch**(n_{new}^M)
31:         $g_\alpha^M \leftarrow$ **BeliefGain**(σ_α^M)
32:         **if** $g_\alpha^M < g_{best}^M$ **then**
33:             $\sigma^M \leftarrow \sigma_\alpha^M$
34:             $g_{best}^M \leftarrow g_\alpha^M$
35:         **end if**
36:     **end if**
37: **end while**
38: **return** σ^M

---

ticular visual–inertial system does not have to be utilized for the actual localization of the robot: as long as a landmarks–based localization system is considered and the expected robot pose and the covariance of its pose and tracked landmarks are calculated, then those can be provided to the algorithm without any modification requirement. ROVIO tracks referringmultilevel image patches. The estimated landmarks are decomposed into a bearing vector, as well as a depth parametrization. The Inertial Measurement Unit (IMU) fixed coordinate frame ($\mathcal{B}$), the camera fixed frame ($\mathcal{V}$) and the inertial frame ($\mathcal{W}$) are considered and the employed state vector with dimension l and associated covariance matrix Σ_l is:

$$\mathbf{x} = [\underbrace{\overbrace{\mathbf{r}\,\mathbf{q}}^{\text{pose,}} l_p \boldsymbol{v}\,\mathbf{b}_f\,\mathbf{b}_\omega\,\mathbf{c}\,\mathbf{z}}_{\text{robot states,}l_s}\,\Big|\,\underbrace{\boldsymbol{\mu}_0,\ \cdots\ \boldsymbol{\mu}_J\,\rho_0\,\cdots\,\rho_J}_{\text{features states,}l_f}]^T \tag{3}$$

where l_p, l_s, l_f are dimensions, $\mathbf{r}$ is the robocentric position of the IMU expressed in $\mathcal{B}$, $\boldsymbol{v}$ represents the robocentric velocity of the IMU expressed in $\mathcal{B}$, $\mathbf{q}$ is the IMU attitude represented as a map from $\mathcal{B} \to \mathcal{W}$, $\mathbf{b}_f$ represents the additive accelerometer bias expressed in $\mathcal{B}$, $\mathbf{b}_\omega$ stands for the additive gyroscope bias expressed in $\mathcal{B}$, $\mathbf{c}$ is the translational part of the IMU–cameras extrinsics expressed in $\mathcal{B}$, $\mathbf{z}$ represents the rotational part of the IMU–cameras extrinsics and is a map from $\mathcal{B} \to \mathcal{V}$, while $\boldsymbol{\mu}_j$ is the bearing vector to feature j expressed in $\mathcal{V}$ and ρ_j is the depth parameter of the jth feature such that the feature distance d_j is $d(\rho_j) = 1/\rho_j$. The relevant state propagation and update steps are briefly summarized in Table 1, and are detailed in [13].

3.4.2 Belief Propagation

To enable localization uncertainty–aware planning at the second layer of the presented `rhemplanner`, belief propagation takes place such that the sampled paths contain the expected values and covariance estimates of both the robot state and the landmarks corresponding to the latest tracked features. This is achieved through the following: First, for every path segment (an edge $\sigma_{k-1,k}$ of $\mathbb{T}^M$), the expected IMU trajectories are derived by simulating the closed–loop robot dynamics $\dot{\chi} = f(\chi, \chi^r)$ (χ being the robot pose $[\mathbf{r}\,\mathbf{q}]^T$ expressed in $\mathcal{W}$ and χ^r its reference) sampled every T_s. The closed–loop dynamics can be derived through first–principles methods or (as in this work) grey–box system identification. A second–order system was considered for the purposes of grey–box system identification. Using these IMU trajectories, prediction of the robot belief takes place by running the state propagation step in the EKF–fashion of ROVIO shown in Table 1, Eq. 3. In order to then conduct the filter update step, the landmarks corresponding to the features tracked at the initialization of the planner are projected in world coordinates $\mathcal{V} \to \mathcal{W}$ using the estimated feature distance, as well as the extrinsic and intrinsic camera parameters. For feature $\boldsymbol{\mu}_j$ this takes the form:

Table 1 ROVIO state propagation and filter update steps

State propagation step - Eq. 3

$$\dot{\mathbf{r}} = -\hat{\boldsymbol{\omega}}^{\times}\mathbf{r} + \boldsymbol{v} + \mathbf{w}_r$$

$$\dot{\boldsymbol{v}} = -\hat{\boldsymbol{\omega}}^{\times}\boldsymbol{v} + \hat{\mathbf{f}} + \mathbf{q}^{-1}(\mathbf{g})$$

$$\dot{\mathbf{q}} = -\mathbf{q}(\hat{\boldsymbol{\omega}})$$

$$\dot{\mathbf{b}}_f = \mathbf{w}_{bf}$$

$$\dot{\mathbf{b}}_{\omega} = \mathbf{w}_{bw}$$

$$\dot{\mathbf{c}} = \mathbf{w}_c$$

$$\dot{\mathbf{z}} = \mathbf{w}_z$$

$$\dot{\boldsymbol{\mu}}_j = \mathbf{N}^T(\boldsymbol{\mu}_j)\hat{\boldsymbol{\omega}}_V - \begin{bmatrix} 0 & 1 \\ -1 & 0 \end{bmatrix}\mathbf{N}^T(\boldsymbol{\mu}_j)\frac{\hat{\boldsymbol{v}}_V}{d(\rho_j)} + \mathbf{w}_{\mu,j}$$

$$\dot{\rho}_j = -\boldsymbol{\mu}_j^T \hat{\boldsymbol{v}}_V / d'(\rho_j) + w_{\rho,j}$$

$$\hat{\mathbf{f}} = \tilde{\mathbf{f}} - \mathbf{b}_f - \mathbf{w}_f$$

$$\hat{\boldsymbol{\omega}} = \tilde{\boldsymbol{\omega}} - \mathbf{b}_{\omega} - \mathbf{w}_{\omega}$$

$$\hat{\boldsymbol{v}}_V = \mathbf{z}(\boldsymbol{v} + \hat{\boldsymbol{\omega}}^{\times}\mathbf{c})$$

$$\hat{\boldsymbol{\omega}}_V = \mathbf{z}(\hat{\boldsymbol{\omega}})$$

Filter update step - Eq. 4

$$\mathbf{y}_j = \mathbf{b}_j(\boldsymbol{\pi}(\hat{\boldsymbol{\mu}}_j)) + \mathbf{n}_j$$

$$\mathbf{H}_j = \mathbf{A}_j(\boldsymbol{\pi}(\hat{\boldsymbol{\mu}}_j))\frac{d\boldsymbol{\pi}}{d\boldsymbol{\mu}}(\hat{\boldsymbol{\mu}}_j)$$

By stacking the above terms for all visible features, standard EKF update step is directly conducted to derive the new estimate of the robot belief for its state and tracked features.

Notation

$^{\times} \rightarrow$ skew symmetric matrix of a vector, $\tilde{\mathbf{f}} \rightarrow$ proper acceleration measurement, $\tilde{\boldsymbol{\omega}} \rightarrow$ rotational rate measurement, $\hat{\mathbf{f}} \rightarrow$ biased corrected acceleration, $\hat{\boldsymbol{\omega}} \rightarrow$ bias corrected rotational rate, $\mathbf{N}^T(\boldsymbol{\mu}) \rightarrow$ projection of a 3D vector onto the 2D tangent space around the bearing vector, $\mathbf{g} \rightarrow$ gravity vector, $\mathbf{w}_{\star} \rightarrow$ white Gaussian noise processes, $\boldsymbol{\pi}(\boldsymbol{\mu}) \rightarrow$ pixel coordinates of a feature, $\mathbf{b}_i(\boldsymbol{\pi}(\hat{\boldsymbol{\mu}}_j)) \rightarrow$ a 2D linear constraint for the jth feature which is predicted to be visible in the current frame with bearing vector $\hat{\boldsymbol{\mu}}_j$

$$\begin{bmatrix} X_j \\ Y_j \\ Z_j \end{bmatrix} = \mathbf{R}_C^W \left(\mathbf{K}_I^{-1}\frac{1}{\rho_j}\begin{bmatrix} u \\ v \\ 1 \end{bmatrix} \right) + \mathbf{T}_C^W \tag{4}$$

where X_j, Y_j, Z_j are their calculated world coordinates, $\mathbf{R}_C^W$ and $\mathbf{T}_C^W$ represent extrinsic camera pose in world coordinates, $\mathbf{K}_I$ represents the intrinsic camera matrix, and $\boldsymbol{\pi}(\boldsymbol{\mu}_j) = [u, v]$ are the feature pixel coordinates. At each step of the belief propagation, only visible landmarks are considered. To evaluate the visibility of a certain landmark, an appropriate check given the robot pose, the map $\mathcal{M}$ and the camera frustrum model is performed (using ray–casting and checking collisions with the map). As bearing vectors for all the landmarks are propagated in the state propagation step, the algorithm subsequently utilizes those predicted to be visible in the new planned configuration in order to update the covariance estimates through the filter update step. The calculated Jacobians are used to update the robot covariance estimate in standard EKF fashion, that is $\Sigma_{t,l} = (\mathbf{I} - \mathbf{K}_t\mathbf{H}_t)\bar{\Sigma}_{t,l}$ with $\mathbf{K}_t$ being the current gain, $\mathbf{H}_t$ the Jacobians, $\mathbf{I}$ the identity matrix, $\Sigma_{t,l}$ the updated covariance, and $\bar{\Sigma}_t$ denoting the covariance estimate of the state propagation step. This is analogous to ROVIO's covariance update operation following the filter update of Table 1, Eq. 4. By interacting this process for all simulated trajectories and the corresponding sampled locations along the tree $\mathbb{T}^M$ branches, the belief is propagated to obtain the estimates for the covariance matrix Σ_l and its pose–features subset $\Sigma_{p,f}$. These estimates are then used in the belief–based gain calculation as shown below. Finally, from an implementation standpoint it is important to highlight that the belief propagation steps rely on a modification of the ROVIO framework. A specific ROS package called rovio_bsp has been developed and is organized as a submodule on the RHEM planner open–source repository. This is another example of the instrumental role of ROS in the implementation and deployment of the presented framework for autonomously exploring robots.

3.4.3 Belief Uncertainty–Based Gain Calculation

Among the key requirements for a robust visual–inertial odometry system are that it (a) must reobserve landmarks with good confidence, while (b) preferably following trajectories that excite the inertial sensors and are therefore informative to the estimation process. This serves to improve both the feature location and the robot pose estimates due to the statistical correlations linking the vehicle to the features. Especially when the robot explores an unknown environment, new features are initialized into the map. This imposes the need to reobserve previous features in order to reduce the growth in localization error. To encode this behavior into the planning process, each admissible branch of $\mathbb{T}^M$ is evaluated regarding the uncertainty of the robot belief about its pose and the tracked landmarks along its edges as explained in Sect. 3.4.2. Then, for the derived pose and tracked landmarks covariance matrix $\Sigma_{p,f}$ (subset of Σ_l), **BeliefGain** computes its *D-optimality* (*D-opt*) [12] metric which measures the area of the uncertainty ellipsoids. Therefore, for two robot path policies σ_1^M and σ_2^M, D-opt is used to evaluate which of the two $(l_p + l_f) \times (l_p + l_f)$ covariance matrices $\Sigma_{p,f}(\sigma_1^M)$ and $\Sigma_{p,f}(\sigma_2^M)$ corresponds to a more confident belief state once the robot arrives at the *end–vertex* of each path. Following the unifying uncertainty criteria definitions of Kiefer [14], the formulation of [12] is employed:

$$D_{opt}(\sigma^M) = \exp(\log([\det(\Sigma_{p,f}(\sigma^M)]^{1/(l_p+l_f)}))) \tag{5}$$

Accordingly, **BeliefGain** computes the D-opt at the end–vertex ξ_{RH} for each admissible branch σ_α^M(where α indexes the admissible branches extracted using the **ExtractBranch** function):

$$\mathbf{BeliefGain}(\sigma_\alpha^M) = D_{opt}(\sigma_\alpha^M) \tag{6}$$

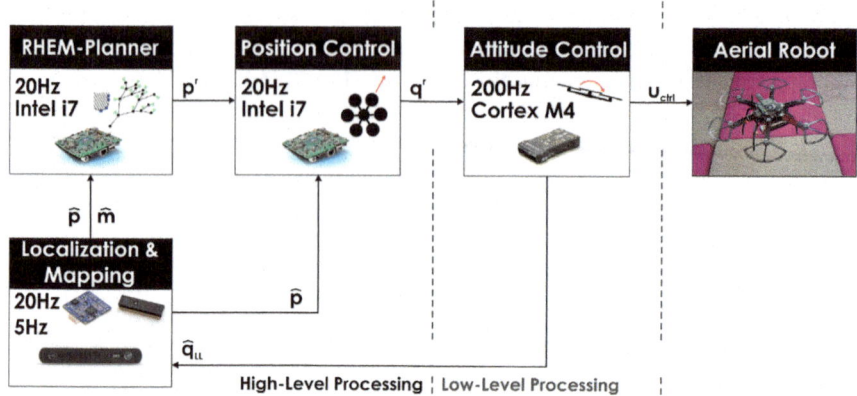

Fig. 4 Diagram of the main functional loops of the robot. The planning, localization and mapping, as well as position control loops (high–level processing) are implemented and interconnected using ROS. Finally, the position controller provides reference attitude and thrust commands to a PX4 autopilot through the MAVROS package

This procedure has the effect of selecting the path which arrives to the viewpoint provided by the exploration planning level with minimal expected localization uncertainty.

3.4.4   Indicative Evaluation Study

A first experimental evaluation of the described planner took place using a custom–built hexarotor platform which has a weight of 2.6 kg. The system is relying on a Pixhawk–autopilot for its attitude control, while further integrating an Intel NUC5i7RYH, and a stereo visual–inertial sensor. The planning, localization and mapping, and position control loops are running on the NUC5i7RYH. The system performs visual–inertial odometry using ROVIO and dense mapping through the depth point cloud from the stereo camera and the robot pose estimates. For the position control task, the linear model predictive controller described in [15] is utilized. The complete set of high–level tasks run with the support of the ROS. In particular, in the utilized set-up, the localization pipeline based on ROVIO, the position model predictive control, as well as the exploration planner are implemented through ROS which facilitates their easier integration. Topics for the map and the pose are seamlessly shared among these core functional modules of the robot and enable its operational autonomy. Furthermore, the ROS-based implementation of ROVIO allowed the convenient modification of its propagation and correction pipeline in order to develop the uncertainty–aware capabilities of the RHEM planner. Our team has used similar systems before for relevant operations [16–19]. Figure 4 presents the main functional loops of the robot and highlights the importance of ROS in its realization.

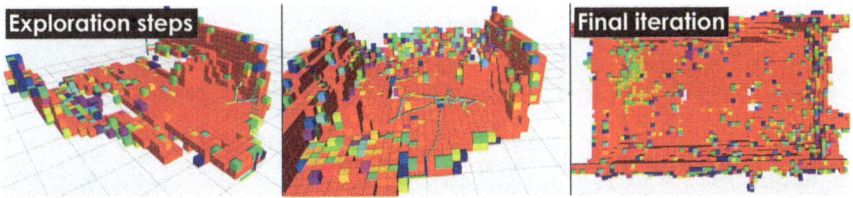

Fig. 5 Exploration steps of the uncertainty–aware receding horizon exploration and mapping planner alongside with mapping

The experimental scenario refers to the mapping of an indoors environment with dimensions $12 \times 6.5 \times 2$ m. Using 300 boxes with size $0.4 \times 0.3 \times 0.3$ m, vertical and T–shaped walls, as well as other structural elements are created to complexify the robot exploration and mapping mission. During the experiment, the robot is constrained to fly with a maximum velocity of $v_{max} = 0.75$ m/s and maximum yaw rate $\dot{\psi}_{max} = \pi/4$ rad/s. Figure 5 presents instances of the localization uncertainty–aware mission, while the colormap of the occupancy map relates to the probability of a voxel being occupied.

A second experimental study was conducted using the aerial robot now equipped with the visual–inertial localization module enhanced with flashing LEDs synchronized with the camera shutter, therefore enabling localization and mapping in dark, visually-degraded environments. Furthermore, a depth time-of-flight camera sensor is installed on the system to deliver reliable point cloud information in conditions of darkness. The experimental scenario refers to a subset of the same environment and the derived results are depicted in Fig. 6. In such a scenario, the ability of the proposed planner to account for the localization uncertainty is particularly important due to the increased challenges the perception system faces in such conditions.

The abovementioned results are indicative. An extensive set of studies using the `rhemplanner` are presented in [20, 21] and include indoor office–like environments in well–lit conditions, (b) dark rooms, as well as (c) dark and broadly visually–degraded (e.g. haze) city tunnel environments that particularly demonstrate the importance of accounting for the localization uncertainty of the robot. The video in https://youtu.be/iveNtQyUut4 presents an indicative experimental demonstration of the operation of the presented planner. Furthermore, the video in https://youtu.be/1-nPFBhyTBM presents the planner operation for a robot exploring its environment in conditions of darkness using a Near IR stereo camera.

4 Terrain Monitoring

This section overviews the Terrain Monitoring Planner (`tmplanner`): a generic finite-horizon informative planning method for monitoring flat terrain using aerial robots [22, 23]. This strategy aims for the timely and cost-effective delivery of high-

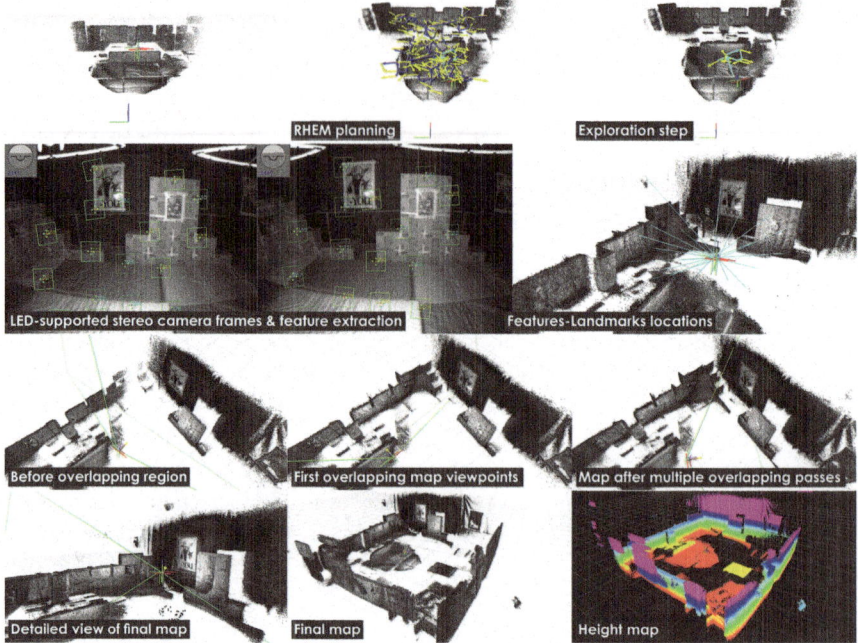

Fig. 6 Experimental demonstration of the proposed planning in darkness supported by a specialized stereo visual–inertial sensor synchronized with flashing LEDs. First row: RHEM exploration steps enabled the volumetric mapping derived from the combination of visual–inertial odometry estimates and the dense point cloud from the ToF depth sensor. Second row: robust stereo localization based on the LED–assisted visual–inertial odometry, while operating in darkness. Third row: Mapping consistency demonstrated after multiple passes from overlapping viewpoints without loop closure - as shown 3D reconstruction consistency is achieved. Fourth row: Final map result and detailed views

quality data as an alternative to using portable sensors in arduous and potentially dangerous campaigns. The key challenge we address is trading off image resolution and FoV to find most useful measurement sites, while accounting for limited endurance and computational resources. The presented framework is capable of probabilistically mapping either discrete or continuous target variables using an altitude-dependent sensor. During a mission, the terrain maps maintained are used to plan information-rich trajectories in continuous 3D space through a combination of grid search and evolutionary optimization.

As a motivating application throughout this section, we consider the use-case of agricultural monitoring employing aerial robots, where the objective may be to detect the presence of an infectious growth (binary) or to monitor the distributions of green biomass level (continuous) on a field. This workflow enables quickly finding precision treatment targets, procuring detailed data for management decisions to reduce chemical usage and optimize yield.

Sections 4.1 and 4.2 present methods of map representation for mapping discrete and continuous targets, respectively, as the basis of our framework. The informative planning algorithm is then outlined in Sect. 4.3.

4.1 Mapping Discrete Variables

The task of monitoring a discrete variable is considered as an active classification problem. For simplicity in the following description, we consider binary variable monitoring. The mapped environment $\mathcal{E}$ (Sect. 2.2) is discretized at a certain resolution and represented using a 2D occupancy map $\mathcal{M}$ [24], where each cell is associated with a Bernoulli random variable indicating the probability of target occupancy (e.g., weed occupancy in agricultural monitoring). Measurements are taken with a downwards-looking camera supplying inputs to a classifier unit. The classifier provides occupancy updates for cells within FoV from an aerial configuration $\mathbf{x}$ above the terrain. At time t, for each observed cell $\mathbf{m}_i \in \mathcal{M}$, a log-likelihood update is performed given an observation z:

$$L(\mathbf{m}_i | z_{1:t}, \mathbf{x}_{1:t}) = L(\mathbf{m}_i | z_{1:t-1}, \mathbf{x}_{1:t-1}) + L(\mathbf{m}_i | z_t, \mathbf{x}_t), \tag{7}$$

where $L(\mathbf{m}_i | z_t, \mathbf{x}_t)$ denotes the altitude-dependent inverse sensor model capturing the classifier output.

Figure 7 shows an example sensor model for a hypothetical binary classifier labeling observed cells as "1" (occupied by target) or "0" (target-free). For each of the two classes, curves are defined to account for poorer classification with lower-resolution images at higher altitudes. In practice, these curves can be obtained through a Monte Carlo–type analysis of raw classifier data by averaging the number of true- and false-positives (blue and orange curves, respectively) recorded at different altitudes.

The described approach can be easily extended to mapping multiple class labels by maintaining layers of occupancy maps.

4.2 Mapping Continuous Variables

To monitor a continuous variable, our framework leverages a more sophisticated mapping method using Gaussian processes (GPs) as a natural way of encoding spatial correlations common in environmental distributions [25]. In brief, a GP is used to initialize a recursive procedure of Bayesian data fusion with probabilistic sensors at potentially different resolutions. A key benefit of this approach is that it replaces the computational burden of applying GPs directly with constant processing time in the number of measurements. This sub-section describes our method of creating prior maps before detailing the multiresolution data fusion technique.

Fig. 7 Probabilistic inverse sensor model for a typical snapshot camera-based binary classifier. The blue and orange curves depict the probability of label "1" given that label "1" or "0" was observed, respectively, i.e., $p(\mathbf{m}_i|z, \mathbf{x})$. As altitude increases, the curves approach unknown classification probability (0.5)

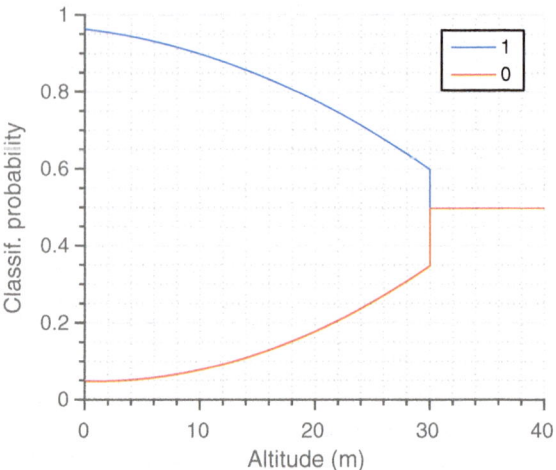

4.2.1 Gaussian Processes

A GP is used to model spatial correlations on the terrain in a probabilistic and non-parametric manner [25]. The target variable for mapping is assumed to be a continuous function in 2D space: $\zeta : \mathcal{E} \to \mathbb{R}$. Using the GP, a Gaussian correlated prior is placed over the function space, which is fully characterized by the mean $\mu = E[\zeta]$ and covariance $\mathbf{P} = E[(\zeta - \mu)(\zeta^\top - \mu^\top)]$ as $\zeta \sim \mathcal{GP}(\mu, \mathbf{P})$.

Given a pre-trained kernel $K(\mathcal{X}, \mathcal{X})$ for a fixed-size terrain discretized at a certain resolution with a set of n locations $\mathcal{X} \subset \mathcal{E}$, we first specify a finite set of new prediction points $\mathcal{X}^* \subset \mathcal{E}$ at which the prior map is to be inferred. For unknown environments, the values at $\mathbf{m}_i \in \mathcal{X}$ are initialized uniformly with a constant prior mean. For known environments, the GP can be trained from available data and inferred at the same or different resolutions. The covariance is calculated using the classic GP regression equation [26]:

$$\mathbf{P} = K(\mathcal{X}^*, \mathcal{X}^*) - K(\mathcal{X}^*, \mathcal{X})[K(\mathcal{X}, \mathcal{X}) + \sigma_n^2\mathbf{I}]^{-1} \times K(\mathcal{X}^*, \mathcal{X})^\top, \qquad (8)$$

where $\mathbf{P}$ is the posterior covariance, σ_n^2 is a hyperparameter representing signal noise variance, and $K(\mathcal{X}^*, \mathcal{X})$ denotes cross-correlation terms between the predicted and initial locations.

The kernel function is chosen to describe the properties of the target distribution. For instance, a commonly used kernel in geostatistical analysis (e.g., for vegetation mapping) is the isotropic Matérn 3/2 function, defined as [25]:

$$k_{Mat3}(\mathbf{m}, \mathbf{m}^*) = \sigma_f^2 \left(1 + \frac{\sqrt{3}d}{l}\right) \exp\left(-\frac{\sqrt{3}d}{l}\right), \qquad (9)$$

where d is the Euclidean distance between inputs $\mathbf{m}$ and $\mathbf{m}^*$, and l and σ_f^2 are hyperparameters representing the lengthscale and signal variance, respectively.

The resulting set of fixed hyperparameters $\theta = \{\sigma_n^2, \sigma_f^2, l\}$ controls relations within the GP. These values can be optimized using various methods [25] to match the properties of ζ by training on multiple maps previously obtained at the required resolution.

Once the correlated prior map $p(\zeta | \mathcal{X})$ is determined, independent noisy measurements at possibly different resolutions are fused as described in the following.

4.2.2   Sequential Data Fusion

Measurement updates are performed based on a recursive filtering procedure [27]. Given a uniform mean and the spatial correlations captured with Eq. 8, the map $p(\zeta | \mathcal{X}) \sim \mathcal{GP}(\mu^-, \mathbf{P}^-)$ is used as a prior for fusing new uncertain sensor measurements.

Let $\mathbf{z} = [z_1, \ldots, z_m]^\top$ denote a set of m new independent measurements received at points $[\mathbf{m}_1, \ldots, \mathbf{m}_m]^\top \subset \mathcal{X}$ modelled assuming a Gaussian sensor as $p(z_i | \zeta_i, \mathbf{m}_i) = \mathcal{N}(\mu_{s,i}, \sigma_{s,i})$. To fuse the measurements $\mathbf{z}$ with the prior map $p(\zeta | \mathcal{X})$, the maximum a posteriori estimator is used:

$$\underset{\zeta}{\mathrm{argmax}}\ p(\zeta | \mathbf{z}, \mathcal{X}) \tag{10}$$

The Kalman Filter (KF) update equations are applied directly to compute the posterior density $p(\zeta | \mathbf{z}, \mathcal{X}) \propto p(\mathbf{z} | \zeta, \mathcal{X}) \times p(\zeta | \mathcal{X}) \sim \mathcal{GP}(\mu^+, \mathbf{P}^+)$ [26]:

$$\mu^+ = \mu^- + \mathbf{K}\mathbf{v} \tag{11}$$

$$\mathbf{P}^+ = \mathbf{P}^- - \mathbf{K}\mathbf{H}\mathbf{P}^-, \tag{12}$$

where $\mathbf{K} = \mathbf{P}^- \mathbf{H}^\top \mathbf{S}^{-1}$ is the Kalman gain, and $\mathbf{v} = \mathbf{z} - \mathbf{H}\mu^-$ and $\mathbf{S} = \mathbf{H}\mathbf{P}^- \mathbf{H}^\top + \mathbf{R}$ are the measurement and covariance innovations. $\mathbf{R}$ is a diagonal $m \times m$ matrix of altitude-dependent variances $\sigma_{s,i}^2$ associated with each measurement z_i, and $\mathbf{H}$ is a $m \times n$ matrix denoting a linear sensor model that intrinsically selects part of the state $\{\zeta_1, \ldots, \zeta_m\}$ observed through $\mathbf{z}$. The information to account for variable-resolution measurements is incorporated straightforwardly through the measurement model $\mathbf{H}$ as detailed in the following sub-section.

The constant-time updates in Eqs. 11 and 12 are repeated every time new data is registered. Note that, as all models are linear in this case, the KF update produces the optimal solution. This approach is also agnostic to the type of sensor as it permits fusing heterogeneous data into a single map.

4.2.3 Altitude-Dependent Sensor Model

This sub-section details an altitude-dependent sensor model for the downward-facing camera used to take measurements of the terrain. As opposed to the classification case in Scct. 4.1, our model needs to express uncertainty with respect to a continuous target distribution. To do this, the information collected is considered as degrading with altitude in two ways: (i) noise and (ii) resolution. The proposed model accounts for these issues in a probabilistic manner as described below.

We assume an altitude-dependent Gaussian sensor noise model. For each observed point $\mathbf{m}_i \in \mathcal{X}$, the camera provides a measurement z_i capturing the target field ζ_i as $\mathcal{N}(\mu_{s,i}, \sigma_{s,i})$, where $\sigma_{s,i}^2$ is the noise variance expressing uncertainty in z_i. To account for lower-quality images taken with larger camera footprints, $\sigma_{s,i}^2$ is modelled as increasing with vehicle altitude h using:

$$\sigma_{s,i}^2 = a(1 - e^{-bh}), \tag{13}$$

where a and b are positive constants.

Figure 8 shows an example sensor noise model for a hypothetical snapshot camera. The measurements z_i denote the levels of the continuous target variable of interest, derived from the images. In the agricultural monitoring case, for example, green biomass level can be computed from calibrated spectral indices in the images to map areas on a farmland with excessive growth [3]. As for the binary classifier, this model can be obtained from a raw analysis of sensor data.

To cater for poorer-quality sensors, we define altitude envelopes corresponding to different image resolution scales with respect to the initial points $\mathcal{X}$. This is motivated from the fact that an area per pixel ratio on the altitude of the sensor and its implemented resolution. At higher altitudes and lower resolutions, adjacent grid cells $\mathbf{m}_i$ are indexed by a single sensor measurement z_i through the sensor model $\mathbf{H}$. At the maximum mapping resolution, $\mathbf{H}$ is simply used to select the part of the state observed with a scale of 1, i.e., one-to-one mapping. However, to handle lower-

Fig. 8 Sensor noise model for a snapshot camera providing measurements as $\mathcal{N}(\mu_{s,i}, \sigma_{s,i})$ with $a = 0.2$, $b = 0.05$ in Eq. 13. The uncertainty $\sigma_{s,i}^2$ increases with h to represent degrading image quality. The dotted line at $h = 10$ m indicates the altitude above which (to the right) image resolution scales down by a factor of 2

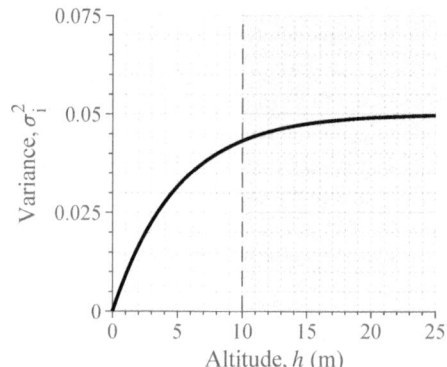

resolution data, the elements of $\mathbf{H}$ are used to map multiple ζ_i to a single z_i scaled by the square inverse of the resolution scaling factor s_f. Note that the fusion described in the previous section is always performed at the maximum mapping resolution, so that the proposed model $\mathbf{H}$ considers low-resolution measurements as a scaled average of the high-resolution map.

4.3 Planning Approach

This sub-section details the proposed planning scheme for terrain monitoring, which generates fixed-horizon plans for maximizing an informative objective. To do this efficiently, an evolutionary technique is applied to optimize trajectories initialized by a 3D grid search in the configuration space. We begin with an approach to parametrizing trajectories before detailing the algorithm itself.

4.3.1 Trajectory Parametrization

A polynomial trajectory ψ is represented with a sequence of N control waypoints to visit $\mathcal{C} = [\mathbf{c}_1, \ldots, \mathbf{c}_N]$ connected using $N - 1$ k-order spline segments for minimum-snap dynamics [28]. The first waypoint $\mathbf{c}_1$ is clamped as the initial vehicle position. As discussed in Sect. 2.2, the function MEASURE($\cdot$) in Eq. 1 is defined by computing the spacing of measurement sites along ψ given a constant sensor frequency.

4.3.2 Algorithm

A fixed-horizon approach is used to plan adaptively as data are collected. During a mission, we alternate between replanning and execution until the elapsed time t exceeds the budget B.

The replanning approach consists of two stages. First, an initial trajectory, defined by a fixed N control points $\mathcal{C}$ is derived through a coarse grid search (Lines 3-6) in the 3D configuration space. This step proceeds in a greedy manner, allowing a rough solution to quickly be obtained. Then, the trajectory is refined to maximize the informative objective using an evolutionary optimization routine (Line 8).

This procedure is described generally in Algorithm 3, where $\mathcal{M}$ symbolizes a discretized environmental model capturing a discrete or continuous target variable. The following sub-sections discuss possible objectives for informative planning, and outline the key steps of the algorithm.

Algorithm 3 REPLAN_PATH procedure

Input: Current model of the environment $\mathcal{M}$, number of control waypoints N, lattice points $\mathcal{L}$
Output: Waypoints defining the next polynomial plan $\mathcal{C}$
1: $\mathcal{M}' \leftarrow \mathcal{M}$ ▷ Create local copy of the map.
2: $\mathcal{C} \leftarrow \emptyset$ ▷ Initialize control points.
3: **while** $N \geq |\mathcal{C}|$ **do**
4: $\quad$ $\mathbf{c}^* \leftarrow$ Select viewpoint in $\mathcal{L}$ using Eq. 1
5: $\quad$ $\mathcal{X}' \leftarrow$ UPDATE_MAP$(\mathcal{M}', \mathbf{c}^*)$ ▷ Predict a measurement from this point.
6: $\quad$ $\mathcal{C} \leftarrow \mathcal{C} \cup \mathbf{c}^*$
7: **end while**
8: $\mathcal{C} \leftarrow$ CMAES$(\mathcal{C}, \mathcal{M})$ ▷ Optimize polynomial trajectory.

4.3.3 Utility Defintion

The utility (information gain) function I in Eq. 1 is generic and can be set to express application-specific interests with respect to the map representation. For mapping a binary variable, we consider two definitions of I for evaluating a measurement from a configuration $\mathbf{x}$ (Line 4). To encourage exploration, the reduction of Shannon's entropy H in the occupancy map $\mathcal{M}$ is maximized:

$$I[\mathbf{x}] = \mathrm{H}(\mathcal{M}^-) - \mathrm{H}(\mathcal{M}^+), \qquad (14)$$

where the superscripts denote the prior and posterior maps given a measurement from $\mathbf{x}$.

To encourage classification, $\mathcal{M}$ is divided into "occupied" and "free" cells using thresholds δ_o and δ_f, leaving an unclassified subset $U = \{\mathbf{m}_i \in \mathcal{M} \mid \delta_f < p(\mathbf{m}_i) < \delta_o\}$. The reduction of $\mathcal{U}$ is then maximized:

$$I[\mathbf{x}] = |\mathcal{U}^-| - |\mathcal{U}^+|. \qquad (15)$$

For mapping a continuous variable using a GP-based map, we consider maximizing uncertainty reduction, as measured by the covariance $\mathbf{P}$:

$$I[\mathbf{m}] = \mathrm{Tr}(\mathbf{P}^-) - \mathrm{Tr}(\mathbf{P}^+), \qquad (16)$$

where $\mathrm{Tr}(\cdot)$ is the trace of a matrix.

In an adaptive-planning setting, a value-dependent objective can be defined to focus on higher- or lower-values regions of interest. To this end, Eq. 16 is modified so that the elements of $\mathrm{Tr}(\mathbf{P})$ mapping to the mean of each cell μ_i via location $\mathbf{m}_i \in \mathcal{X}$ are excluded from the computation, given that they exceed specified thresholds [23].

Note that, while Eqs. 14–16 define I for a single measurement site $\mathbf{x}$, the same principles can be applied to determine the utility of a trajectory ψ by fusing a sequence of measurements and computing the overall information gain.

Fig. 9 A visualization of an 3D lattice grid $\mathcal{L}$ example (30 points) for obtaining an initial trajectory solution. The length scales are defined for efficiency given the environment size and computational resources available

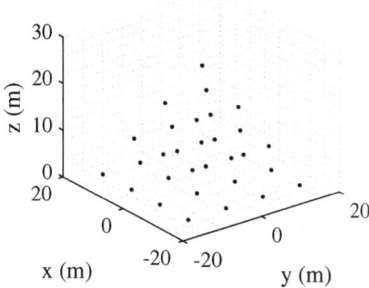

4.3.4   3D Grid Search

The first replanning step (Lines 3–6) supplies an initial solution for optimization in Sect. 4.3.5. To achieve this, the planner performs a 3D grid search based on a coarse multiresolution lattice $\mathcal{L}$ in the vehicle configuration space (Fig. 9). A low-accuracy solution neglecting sensor dynamics is obtained efficiently by using the points in $\mathcal{L}$ to represent candidate measurement sites and assuming constant velocity travel. Unlike in frontier-based exploration commonly used in cluttered environments [29], selecting measurement sites using map boundaries is not applicable in a terrain monitoring set-up. Instead, we conduct a sequential greedy search for N waypoints (Line 3), where the next-best point $\mathbf{c}^*$ (Line 4) is found by evaluating Eq. 1 with the chosen utility I in over $\mathcal{L}$. For each $\mathbf{c}^*$, a fused measurement is simulated (Line 5) via Eqs. 7 or 12 for a discrete or continuous mapping scenario, respectively. This point is then added to the initial trajectory solution (Line 6).

4.3.5   Optimization

The second replanning step (Line 7) optimizes the grid search solution by computing I (Sect. 4.3.3) for a sequence of measurements taken along the trajectory. For global optimization, we propose the Covariance Matrix Adaptation Evolution Strategy (CMA-ES) [30]. This choice is motivated by the nonlinearity of the objective space (Eq. 1) as well as previous results in similar applications [22, 23, 31].

4.3.6   Indicative Evaluation Studies

Extensive simulation studies were conducted for mapping both discrete and continuous scenarios in realistic environments. The results obtained show that the proposed approach performs better than state-of-the-art strategies ("lawnmower" coverage planning, sampling-based planning) in terms of informative metrics. Applications of various utility functions and optimization methods in our framework were also compared. Finally, proof of concept experiments were demonstrated using rotorcraft aerial robots to map different artificial agricultural farmland set-ups, including (a) AprilTag distributions and (b) color variations, corresponding to discrete and continuous targets, respectively. These experiments were performed in real-time with a preliminary version of tmplanner, the released open–source ROS release of our terrain monitoring framework. The employed robots rely on the ROS-based localization and position control methods utilized also in the framework of the previously mentioned experiments. In addition, we are working on adapting our framework and sensing interfaces for deployment in field trials on the farm.

Further details concerning the experimental results above can be found in our relevant publications [22, 23]. The video in https://youtu.be/1cKg1fjT54c is indicative of the planner experimental operation.

5   Optimized Inspection Path Planning

Unlike the exploration planners, the method described in this section aims to derive an optimized full coverage inspection path when a model of the structure of interest is known a priori. A triangular mesh–based representation of the structure is considered. Within the literature, a set of algorithms have been proposed to address the problem of 3D structural inspection.

In particular, approaches employing a two–step optimization scheme are renowned for their increased performance and versatility against increasing structural complexity. Such methods initially compute the minimal set of viewpoints that cover the whole structure (first step), which corresponds to solving an Art Gallery Problem (AGP). Subsequently (second step), the shortest connecting tour over all these viewpoints has to be computed, which corresponds to the Traveling Salesman Problem (TSP). Fast algorithms to approximately solve both the AGP and the TSP are known despite their inherent NP–hard properties. Indicative examples include the works in [32, 33] for the AGP, and [34, 35] for the TSP. Following a different approach, the works in [17, 36] concentrate on a truly optimal solution at the cost of expensive computational steps to a level that challenges their scalability. The fast inspection path planner described in this work retains a two–step optimization structure but, instead of attempting to find a minimal set of guards in the AGP, tries to sample them such that the connecting path is short, while ensuring full coverage. This alternative strategy is motivated from the fact that, given a continuously sampling sensor such as a camera or a LiDAR on a robot, it is not the number of viewpoints that makes the inspection path more efficient but rather their configuration and how this influences the travel cost of the overall inspection trajectory.

Given this fact, the presented Structural Inspection Planner (siplanner) selects one admissible viewpoint for every facet of the mesh of the structure to be inspected. In order to compute viewpoints allowing low–cost connections, an iterative resampling scheme is employed. Between each resampling, the best path connecting the current viewpoints is recomputed. The quality of the viewpoints is assessed by the cost to connect to their respective neighbours on the latest tour. This cost is minimized in the subsequent resampling, resulting in locally optimized paths. Initialization of the viewpoints is arbitrarily done such that full coverage is provided with, at this stage, non–optimized viewpoints. A fast implementation of the Lin–Kernighan–Helsgaun Heuristic (LKH) TSP solver [37] is employed to compute the best tour, while the cost of the interconnecting pieces of path is calculated by means of a Boundary Value Solver (BVS). An overview of the proposed inspection path planning procedure is presented in Algorithm 4.

The following sub-sections outline the path generation approach and the optimization problem formulation to sample the viewpoints for a rotorcraft aerial robot. Then, we detail how the method deals with cluttered environments.

Algorithm 4 `siplanner`

1: $k \leftarrow 0$
2: Sample initial viewpoint configurations
3: Compute cost matrix for the TSP solver (Sect. 5.1)
4: Solve the TSP problem to obtain initial tour
5: **while** running **do**
6:    Resample viewpoint configurations (Sect. 5.2)
7:    Recompute the cost matrix (Sect. 5.1)
8:    Recompute best tour T_{best} using the LKH and update best tour cost c_{best} if applicable
9:    $k \leftarrow k + 1$
10: **end while**
11: **return** T_{best}, c_{best}

5.1 Path Computation and Cost Estimation

In order to find the best tour among the viewpoints, the TSP solver requires a cost matrix containing the connection cost of all pairs of viewpoints. A two–state BVS is employed to generate paths between different viewpoints and estimate the respective costs. The BVS is either employed directly to connect the two viewpoints or as a component in a local planner. The latter applies in case the direct connection is not collision–free. In that case, our implementation uses the RRT*-planner [38] to find a collision–free connection. For rotorcraft aerial robots, the flat state consisting of position as well as yaw, $\xi = \{x, y, z, \psi\}$ is considered. Roll and pitch angles are assumed to be near zero as slow maneuvering is desired to achieve increased inspection accuracy. The path from configuration ξ_0 to ξ_1 is given by $\xi(s) = s\xi_1 + (1 - s)\xi_0$, where $s \in [0, 1]$. The single limitation considered is the speed limit, both in terms of maximum translational velocity v_{max}, and maximum yaw rate $\dot{\psi}_{max}$. To allow sufficiently accurate tracking of paths with corners, both values are small and should be adjusted for the scenario and the system being used. The resulting execution time is $t_{ex} = \max(d/v_{max}, \|\psi_1 - \psi_0\| / \dot{\psi}_{max})$, with d the Euclidean distance. The considered cost metric of a path segment corresponds to the execution time t_{ex}. The fact that time of travel, and not the Euclidean distance, is used as the cost function is particularly important given that the planner also supports fixed–wing Unmanned Aerial Vehicles. The open–source repository details how the method can be used for fixed–wing systems and account for the bank angle and velocity limits of the vehicle.

5.2 Viewpoint Sampling

Considering a mesh–based representation of the environment, a viewpoint has to be sampled for each of its facets (triangular meshes are assumed). According to the method of this planner, the position and heading of each viewpoint is determined sequentially, while retaining visibility of the corresponding triangle. First, the position is optimized for distance w.r.t. the neighbouring viewpoints using a convex problem formulation and, only then, the heading is optimized. To guarantee a good result of this multi–step optimization process, the position solution must be constrained such as to allow finding an orientation for which the triangle is visible.

More specifically, the constraints on the position states $g = [x, y, z]$ consist of the inspection sensor limitations of minimum incidence angle, and minimum and maximum effective range (d_{min}, d_{max}) (depicted in Fig. 10). They are formulated as a set of planar constraints:

$$\begin{bmatrix} (g - x_i)^T n_i \\ (g - x_1)^T a_N \\ -(g - x_1)^T a_N \end{bmatrix} \succeq \begin{bmatrix} 0 \\ d_{min} \\ -d_{max} \end{bmatrix}, i = \{1, 2, 3\} \tag{17}$$

where x_i are the corners of the mesh triangle, a_N is the normalized triangle normal and n_i are the normals of the separating hyperplanes for the incidence angle constraints as shown in Fig. 10.

Furthermore, as in the other planners presented in this chapter, the sensor is considered to have a limited FoV with certain vertical and horizontal opening angles and is mounted to the vehicle with a fixed pitch angle and relative heading. The imposed constraint on the sampling space resulting from the vertical camera opening is not convex (a revoluted 2D-cone, the height of which is depending on the relevant corners of the triangle over the revolution). Nonconvexity at a step that is iteratively computed multiple times can lead to significant computational cost for the planner. To approximate and convexify the problem, the space is divided in N_C equal convex pieces according to Fig. 10. The optimum is computed for every slice in order to find the globally best solution. Multiple sensors with different vertical FoVs are handled equally, resulting in a multitude of N_C convex pieces that possibly overlap. The constraints for piece j are derived as follows: Left and right boundaries of the sampling space are the borders of the revolution segment and the cone top and bottom are represented by a single plane tangential to the centre of the slice. Angular camera constraints in horizontal direction are not encoded and instead d_{min} is chosen high enough to allow full visibility of the triangle. This leaves some space

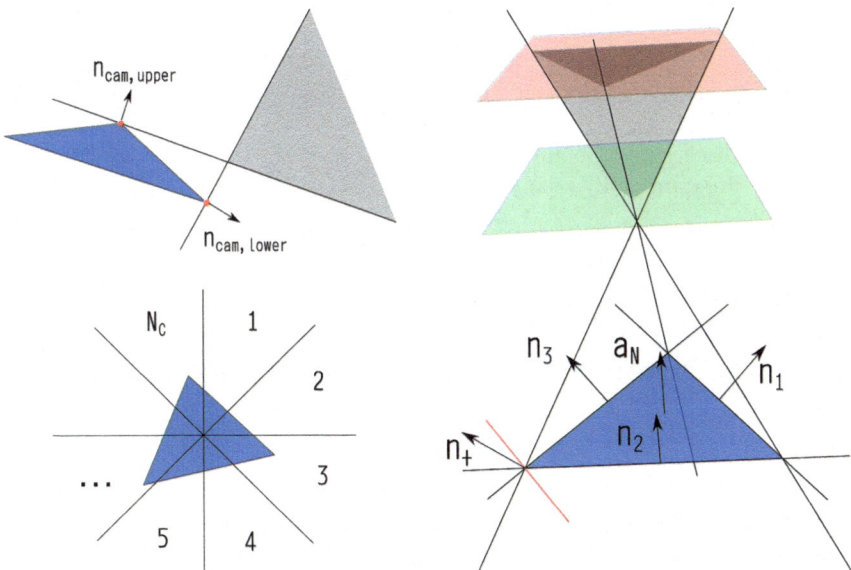

Fig. 10 Left: The vertical camera angle constraints with the relevant corners of the triangle in red are depicted in the upper part, while the partition of the space for convexification is depicted beneath. Right: The figure depicts the three main planar angle of incidence constraints on all sides of the triangle. For a finite number of such constraints, the incidence angle is only enforced approximately. The red line (and n_+) remarks a sample orientation for a possible additional planar constraint at a corner. Minimum (green plane) and maximum (red plane) distance constraints are similar planar constraints on the sampling area. These constraints bound the sampling space, where g can be chosen, on all sides (gray area)

for variation in the sampling of the heading, where the horizontal constraints are enforced. Specifically, the abovementioned constraints take the following form:

$$\begin{bmatrix} (g - x_{lower}^{rel})^T n_{lower}^{cam} \\ (g - x_{upper}^{rel})^T n_{upper}^{cam} \\ (g - m)^T n_{right} \\ (g - m)^T n_{left} \end{bmatrix} \succeq \begin{bmatrix} 0 \\ 0 \\ 0 \\ 0 \end{bmatrix}, \tag{18}$$

where $x_{lower}^{rel}, x_{upper}^{rel}$ are the respective relevant corners of the mesh triangle, m the middle of the triangle and $n_{lower}^{cam}, n_{upper}^{cam}, n_{right}$ and n_{left} denote the normal of the respective separating hyperplanes.

The optimization objective for the viewpoint sampling in iteration k, in the case of a rotorcraft aerial robot, is to minimize the sum of squared distances to the preceding viewpoint g_p^{k-1}, the subsequent viewpoint g_s^{k-1} and the current viewpoint g^{k-1} in the old tour. The former two parts potentially shorten the tour by moving the viewpoints closer together, while the latter limits the size of the improvement step, as g_p^{k-1} and g_s^{k-1} potentially move closer as well. The weighting matrix B for the neighbour distance is given by $\mathbf{diag}(b_{const}, b_{const}, a_{const} + b_{const})$, where b_{const} is the general weight for distance to neighbours, while a_{const} additionally punishes changes in height. The distance to the current viewpoint in the old tour is weighted by the matrix $D = \mathbf{diag}(d_{const}, d_{const}, d_{const})$.

The resulting convex optimization problem is given below. Its structure as a Quadratic Program (QP) with linear constraints allows the use of an efficient solver [39].

$$\min_{g^k}\ (g^k - g_p^{k-1})^T B(g^k - g_p^{k-1}) + (g^k - g_s^{k-1})^T B(g^k - g_s^{k-1}) + \tag{19}$$

$$(g^k - g^{k-1})^T D(g^k - g^{k-1})$$

$$\text{s.t.}\quad
\begin{bmatrix}
n_1^T \\
n_2^T \\
n_3^T \\
a_N^T \\
-a_N^T \\
n_{lower}^{cam\ T} \\
n_{upper}^{cam\ T} \\
n_{right}^T \\
n_{left}^T
\end{bmatrix}
g^k \succeq
\begin{bmatrix}
n_1^T x_1 \\
n_2^T x_2 \\
n_3^T x_3 \\
a_N^T x_1 + d_{min} \\
-a_N^T x_1 - d_{max} \\
n_{lower}^{cam\ T} x_{lower}^{rel} \\
n_{upper}^{cam\ T} x_{upper}^{rel} \\
n_{right}^T m \\
n_{left}^T m
\end{bmatrix}
\tag{20}$$

For the computed optimal position, the heading is determined according to the criterion $\min_{\psi^k} = \left(\psi_p^{k-1} - \psi^k\right)^2 / d_p + \left(\psi_s^{k-1} - \psi^k\right)^2 / d_s$, subject to $\mathbf{Visible}(g^k, \psi^k)$, where $\mathbf{Visible}(g^k, \psi^k)$ means that from the given configuration, g^k and ψ^k, the whole triangle is visible for at least one of the employed sensors. d_p and d_s are the Euclidean distances from g^k to g_p^{k-1} and g_s^{k-1} respectively. For simple sensor setups establishing the boundaries on ψ^k for $\mathbf{Visible}(g^k, \psi^k) = \text{TRUE}$ makes the solution explicit. Otherwise a grid search can be employed. It is highlighted that one of the important abilities of the `siplanner` from an applications standpoint is its ability to incorporate and account for multiple sensor models (e.g. multiple cameras).

5.3 Obstacle Avoidance

The `siplanner` performs efficiently in cluttered environments. Viewpoints are checked for occlusion by parts of the mesh that is to be inspected or meshes that represent obstacles of the environment for which sensor coverage is not desired. If interference is detected, the QP is reformulated to include an additional planar constraint defined by a corner x^I of the interfering facet triangle, the centre of which is closest to the facet, as well as its normal a_N^I:

$$a_N^{I\ T} g^k \geq a_N^{I\ T} x^I \tag{21}$$

For meshes that do not selfintersect, this is a conservative exclusion of cluttered subareas. Furthermore, paths are checked for interference with the mesh to ensure collision–free connections between the viewpoints. This is easily achieved by integrating the collision check in the BVS. In order to shape a scenario according to special requirements, cuboidal obstacle regions can be defined being either transparent or non transparent. Both are respected in the planning of the connecting paths, while the latter are also included in the visibility check. When, such an obstacle is encountered during viewpoint sampling, the optimization problem is executed individually over eight subregions. Each is bounded away from the cuboid on one of its sides, similar to Eq. 21.

5.4 Indicative Evaluation Study

A simulation evaluation study aiming to identify the optimized inspection trajectory for the case of a solar park covering an area of more than $4100\,m^2$ is presented in Figs. 11 and 12. The total cost of the mission corresponds to 1259.36 s given a robot with FoV angles $[a_v, a_h] = [90, 120]°$ and a mounting pitch of 20° downwards. The presented study is indicative, while more a set of experimental tests are presented in [16, 18, 40–42]. These include (a) the inspection of structures of the ETH Zurich Polyterrasse, (b) the inspection of a Wattmeter, (c) the mapping of a National Park in Switzerland using the fixed–wing AtlantikSolar solar–powered UAV, and more. It is noteworhy that beyond the planner, also the localization, and position control functionalities of the employed multirotor are also based on ROS and specifically on ROVIO [13] and Linear MPC [15] respectively. The videos in https://youtu.be/5kI5ppGTcIQ and https://youtu.be/qbwoKFJUm4k are indicative of the operation of the inspection planner.

6 Open–Source Code and Implementation Details

Towards enabling the research and development community, as well as the end–users to utilize and further develop the described autonomous exploration and inspection path planning algorithms, a set of open–source contributions have taken place. The "Receding Horizon Next-Best-View Planner" (nbvplanner), the "Localization Uncertainty–aware Receding Horizon Exploration and Mapping Planner" (rhemplanner), the "Terrain Monitoring Planner" (tmplanner) as well as the "Structural Inspection Planner" (siplanner) are already released as open–sourced Robot Operating System packages. Together with this chapter, we also organize and release a comprehensive single repository that links all the planners, further documents them and relates them in terms of their functionality and the relevant problems they address. This can be found at https://github.com/unr-arl/informative-planning.

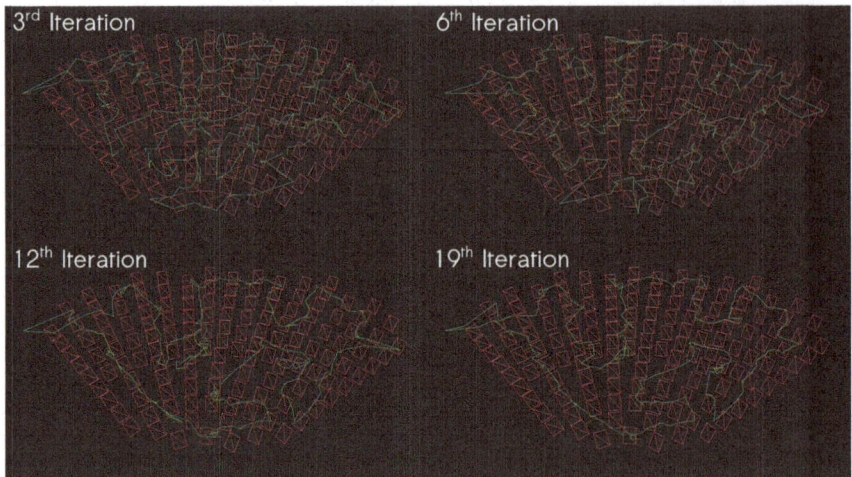

Fig. 11 Iterations of the `siplanner` in order to identify the most optimized trajectory for the inspection of a solar power park covering an area of more than $4100 \, m^2$

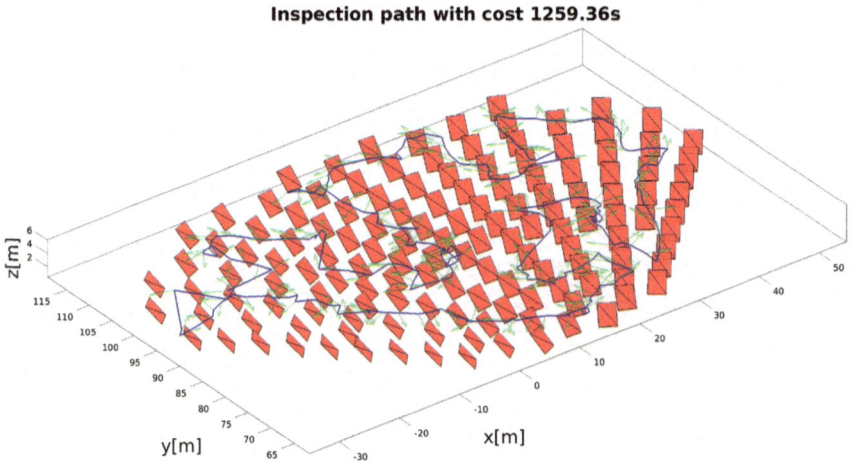

Fig. 12 Final optimized inspection path for covering a solar power park covering an area of more than $4100 \, m^2$. The total travel time is $1259.36 \, s$ given a robot with a sensor characterized by FoV angles $[a_v, a_h] = [90, 120]°$ and a mounting pitch of $20°$ downwards and subject to dynamic constraints of maximum forward velocity of $0.5 \, m/s$ and maximum yaw rate of $0.5 \, rad/s$

Additionally, a set of associated datasets from real flight tests and demo scenarios are released.

6.1 ROS Environment Configuration

In terms of ROS support, the released packages are tested to work with three of the latest ROS versions, namely Indigo, Jade, and Kinetic, while continuous maintenance takes place. As thoroughly explained in the Wiki and Readme sections of the respective open–source repositories the catkin build for the nbvplanner, rhemplanner, and tmplanner. Several package dependencies are required for the compilation of each package which are documented in detail within the respective Wkis for nbvplanner, rhemplanner, and siplanner, while a rosinstall file for the tmplanner.

6.2 Exploration Planners

For nbvplanner and rhemplanner, the command to the robot is provided iteratively in the form of a new reference waypoint to be tracked by the onboard position controller. For the implementation of both planners, the environment is represented using the octomap framework [6] and planning takes place within the free space mapped at every planning iteration. Both planners initialize each iteration from the currently estimated robot pose, while the rhemplanner further considers the latest set of tracked landmarks by ROVIO (or an alternative pipeline that tracks landmarks to estimate the robot pose) and the associated robot pose and landmarks covariance matrix.

The planners were developed having aerial robots in mind but can be extended to any robotic configuration that affords a BVS such that sampling can take place in the configuration space. The flat state $\xi = [x, y, z, \psi]^T$ considered in the work is an implementation detail that can be modified. Slow maneuvering aerial robots of the rotorcraft class are considered and the connection cost of a path σ_{k-1}^k is considered to be the Euclidean distance $c(\sigma_{k-1}^k) = \sqrt{(x_k - x_{k-1})^2 + (y_k - y_{k-1})^2 + (z_k - z_{k-1})^2}$. The maximum speed and the yaw rate limit are enforced when sampling the path to generate a reference trajectory.

Both planners consider fixed–mounted cameras, limited by a certain horizontal and vertical FoV $[a_v, a_h]$, as well as a maximum distance of effective perception. For a sensor with a practical limitation of maximum sensing distance $d_{\max}^{\text{sensor}}$, the algorithm uses a value $d_{\max}^{\text{planner}} \leq d_{\max}^{\text{sensor}}$ to determine the volumetric gain of a configuration. Using a lower $d_{\max}^{\text{planner}}$ enhances the robustness against suboptimal sensing conditions.

6.2.1 nbvplanner Open–Source Release

The nbvplanner is implemented and released as an open–source ROS package at https://github.com/ethz-asl/nbvplanner and further documented through https://github.com/unr-arl/informative-planning. The planner is verified to work on ROS

Indigo, Jade and Kinetic. Experimental datasets relevant with the planner can also be acquired through its repository under "Wiki-Example Results". The simplest way to test and evaluate the planner is through its integration with the RotorS Simulator [43]. Provided that the installation instructions have been thoroughly followed, execution of the simplest scenario refers to the execution of the following command

```
$ roslaunch interface_nbvp_rotors flat_exploration.launch
```

During its operation, the planner is called via a rosservice. To embed the open-sourced `nbvplanner` in your own C++ project, add the following snippet:

```cpp
#include <ros/ros.h>
#include <nbvplanner/nbvp_srv.h>

...

int iteration = 0;
while (ros::ok()) {
    nbvplanner::nbvp_srv planSrv;
    planSrv.request.header.stamp = ros::Time::now();
    planSrv.request.header.seq = iteration;
    planSrv.request.header.frame_id = ros::this_node::
      getNamespace();
    if (ros::service::call("nbvplanner", planSrv)) {

        ...

        // Process the computed path in planSrv.response.
      path to make
        // the robot track it


        ...

        iteration++;
    }
    } else {
        ROS_WARN_THROTTLE(1, "Planner not reachable");
        ros::Duration(1.0).sleep();
    }
}
```

A set of parameters can be defined to account for limitations of the particular scenario and adjust the behavior of the planner. Those include system–related parameters and model constraints (e.g. maximum velocity and heading rate), octomap parameters (e.g. voxel edge length size, probability of a hit), exploration gain computation parameters (e.g. gain for visible free volumes, maximum distance of volumes to be considered for the gain computation), path planning parameters (e.g. maximum extension range for new tree branches when sampling for a holonomic system, time step for path sampling), logging parameters (e.g. time interval to log values), and scenario parameters (e.g. minimum coordinates values of the scenario bounding only the gain computation and not the path planning tree expansion). Further parameters are documented in the code repository and its associated wiki, alongside installation instructions and demo scenarios.

In addition, it is highlighted that the released `nbvplanner` implementation provides a demo example for a multi–robot system exploring a common space. This can be triggered by running the following command:

```
1  $ roslaunch interface_nbvp_rotors multiagent_area_exploration.
2  launch
```

6.2.2 `rhemplanner` Open–Source Release

The `rhemplanner` is implemented and released as an open–source ROS package at https://github.com/unr-arl/rhem_planner and further documented in https://github.com/unr-arl/informative-planning. Accompanying datasets are provided at the planner's repository under "Wiki-How to use Demo". The planner is verified to work on ROS Indigo, Jade and Kinetic. The planning pipeline is implemented in the `bsp_planner` which incorporates the two layered–planning procedure. As mentioned, the `rhemplanner` conducts propagation of the robot's localization belief which relies on the internally incorporated estimation and propagation pipeline. This is based on the ROVIO visual–inertial framework [13] and a respectively modified version is incorporated in the code release. It calculates and maintains the real–time localization and mapping state and associated statistics (estimation), and forward–simulates the process given an initial state and its associated statistics and a sample planned trajectory (propagation). These two key pipelines (planning, estimation and propagation) are then accordingly fused to realize the iteratively planning process of the `rhemplanner`. As in the case for the `nbvplanner`, the `rhemplanner` is called via a rosservice. The easiest way to test the planner is through the released dataset, as getting feedback from onboard sensors and the localization pipeline is critical for the operation of the planner. To do so, one has to execute:

```
1  $ cd rhem_planner
2  $ wget –P rhem_demo https://www.cse.unr.edu/%7Ekalexis/datasets/
3  icra2017–datasets/icra17_visensor_data.bag
```

Then the relevant launch file has to be fired:

```
1  $ roslaunch rhem_demo/rhem_demo_icra17.launch
```

As this refers to a recorded demo, executions of the planner should be triggered through the rosservice:

```
1  $ rosservice call /bsp_planner '{header: {stamp: now, frame_id:
2      world}}'
```

Extended details for the subscribed topics and the algorithm parameters are provided within the corresponding code repositories. The algorithm parameters refer to the system constraints and the sensor model, the resolution and fidelity of the estimation and propagation pipeline, octomap parameters, planning parameters for the first–layer of planning, and further parameters for the random tree expansion and associated gains of the uncertainty–aware second layer of planning. Finally, visualization and further system parameters can be set.

6.3 Structural Inspection Planner

For the exact implementation of the `siplanner`, a set of heuristic improvements were implemented to further optimize the behavior of the algorithm. Having rotorcraft–type aerial robots in mind, faster and more rigorous ordering of the viewpoints can be achieved. In particular, the initial iterations of the algorithm conduct viewpoint allocation by not only considering the nearest neighbor on the tour to minimize the distance to, but also the neighbors that are $N_{Neighbour}$ away on both sides. This allows for extending the locality of the viewpoint sampling considerations to a broader neighborhood and therefore better optimization within this locally generated graph. In every iteration, $N_{Neighbour}$ is then decremented to finally reach 1.

6.3.1 `siplanner` Open–Source Release

The `siplanner` open–source release can be found directly at https://github.com/ethz-asl/StructuralInspectionPlanner or linked and further documented through https://github.com/unr-arl/informative-planning. The first open–source release was in the form of a ROS Indigo package, while its use is verified for both Jade and Kinetic versions. The open–sourced algorithm is developed for both rotorcraft and fixed–wing systems and supports triangled–meshes for the representation of the inspection structure of interest. Accompanying experimental results are provided at the planner's repository under "Wiki-Example Results". The easiest way to test the planner is through one of the provided demos. For example, one may run two instances of a terminal and execute the following two commands:

```
$ roslaunch koptplanner hoaHakanaia.launch
```

```
$ rosrun request hoaHakanaia
```

In order to write a client based on `siplanner`, the following code may be used as an example. The key steps of the code example below refer to: (a) setting up a publisher for obstacles and mesh visualization. (b) delaying the publishing of display elements to avoid buffer overflow, (c) instantiating the datastructure to call the service with, (d) defining the bounding box of the operating space, (e) filling the vector of `requiredPoses` with all fixed poses that should be included in the path, (f) specifying the minimum incidence angle, the minimal and maximal inspection distances, as well as the number of iterations of the planner, (g) reading and visualizing the mesh, (h) definition of the obstacle regions, (i) calling the `siplanner` service, and (j) reading the mesh from a non–binary STL file.

```
1  #include <ros/ros.h>
2  #include "koptplanner/inspection.h"
3  #include "shape_msgs/SolidPrimitive.h"
4  #include <cstdlib>
5  #include <visualization_msgs/Marker.h>
6  #include <fstream>
7  #include <geometry_msgs/PolygonStamped.h>
8  #include <nav_msgs/Path.h>
9  #include <std_msgs/Int32.h>
10 #include <ros/package.h>
11 #include "tf/tf.h"
12
13 std::vector<nav_msgs::Path> * readSTLfile(std::string name);
14
15 int main(int argc, char **argv)
16 {
17   ros::init(argc, argv, "requester");
18   ROS_INFO("Requester is alive");
19   if (argc != 1)
20   {
21     ROS_INFO("usage: plan");
22     return 1;
23   }
24
25   ros::NodeHandle n;
26   ros::Publisher obstacle_pub = n.advertise<visualization_msgs::
27     Marker>
28     ("scenario", 1);
29   ros::Publisher stl_pub = n.advertise<nav_msgs::Path>("stl_mesh",
30     1);
30   ros::ServiceClient client = n.serviceClient<koptplanner::
     inspection>
31     ("inspectionPath");
32
33   ros::Rate r(50.0);
34   ros::Rate r2(1.0);
35   r2.sleep();
36
37   /* define the bounding box */
38   koptplanner::inspection srv;
39   srv.request.spaceSize.push_back(1375);
40   srv.request.spaceSize.push_back(2165);
41   srv.request.spaceSize.push_back(0.001);
42   srv.request.spaceCenter.push_back(1375.0/2.0);
43   srv.request.spaceCenter.push_back(2165.0/2.0);
44   srv.request.spaceCenter.push_back(200.0);
45
46   geometry_msgs::Pose reqPose;
47
48   /* starting pose (comment the push_back if no explicit starting
     pose
49      is desired) */
50   reqPose.position.x = 300.0;
```

```cpp
51    reqPose.position.y = 300.0;
52    reqPose.position.z = 200.0;
53    tf::Quaternion q = tf::createQuaternionFromRPY(0.0, 0.0, 0.0);
54    reqPose.orientation.x = q.x();
55    reqPose.orientation.y = q.y();
56    reqPose.orientation.z = q.z();
57    reqPose.orientation.w = q.w();
58    srv.request.requiredPoses.push_back(reqPose);
59
60    /* final pose (remove if no explicit final pose is desired) */
61    reqPose.position.x = 400.0;
62    reqPose.position.y = 300.0;
63    reqPose.position.z = 200.0;
64    q = tf::createQuaternionFromRPY(0.0, 0.0, 0.0);
65    reqPose.orientation.x = q.x();
66    reqPose.orientation.y = q.y();
67    reqPose.orientation.z = q.z();
68    reqPose.orientation.w = q.w();
69    srv.request.requiredPoses.push_back(reqPose);
70
71    /* parameters for the path calculation (such as may change
72    during mission) */
73    srv.request.incidenceAngle = M_PI/6.0;
74    srv.request.minDist = 40.0;
75    srv.request.maxDist = 300.0;
76    srv.request.numIterations = 20;
77
78    /* read STL file and publish to rviz */
79    std::vector<nav_msgs::Path> * mesh = readSTLfile(ros::package::
        getPath(
80        "request")+"/meshes/regularPlanes/rPlane.stl");
81    for(std::vector<nav_msgs::Path>::iterator it = mesh->begin();
82        it != mesh->end() && ros::ok(); it++)
83    {
84        stl_pub.publish(*it);
85        geometry_msgs::Polygon p;
86        geometry_msgs::Point32 p32;
87        p32.x = it->poses[0].pose.position.x;
88        p32.y = it->poses[0].pose.position.y;
89        p32.z = it->poses[0].pose.position.z;
90        p.points.push_back(p32);
91        p32.x = it->poses[1].pose.position.x;
92        p32.y = it->poses[1].pose.position.y;
93        p32.z = it->poses[1].pose.position.z;
94        p.points.push_back(p32);
95        p32.x = it->poses[2].pose.position.x;
96        p32.y = it->poses[2].pose.position.y;
97        p32.z = it->poses[2].pose.position.z;
98        p.points.push_back(p32);
99        srv.request.inspectionArea.push_back(p);
100       r.sleep();
101   }
102
103   /* define obstacle regions as cuboids that are coordinate
```

```
104    system aligned */
105    shape_msgs::SolidPrimitive body;
106    body.type = shape_msgs::SolidPrimitive::BOX;
107    body.dimensions.push_back(40.0);
108    body.dimensions.push_back(50.0);
109    body.dimensions.push_back(4.0);
110    srv.request.obstacles.push_back(body);
111    geometry_msgs::Pose pose;
112    pose.position.x = 600.0;
113    pose.position.y = 600.0;
114    pose.position.z = 200.0;
115    pose.orientation.x = 0.0;
116    pose.orientation.y = 0.0;
117    pose.orientation.z = 0.0;
118    pose.orientation.w = 1.0;
119    srv.request.obstaclesPoses.push_back(pose);
120    srv.request.obstacleIntransparancy.push_back(0);
119
120    // publish obstacles for rviz
121    visualization_msgs::Marker marker;
122    marker.header.frame_id = "/kopt_frame";
123    marker.header.stamp = ros::Time::now();
124    marker.ns = "obstacles";
125    marker.id = 0; // enumerate when adding more obstacles
126    marker.type = visualization_msgs::Marker::CUBE;
127    marker.action = visualization_msgs::Marker::ADD;
128
129    marker.pose.position.x = pose.position.x;
130    marker.pose.position.y = pose.position.y;
131    marker.pose.position.z = pose.position.z;
132    marker.pose.orientation.x = pose.orientation.x;
133    marker.pose.orientation.y = pose.orientation.y;
134    marker.pose.orientation.z = pose.orientation.z;
135    marker.pose.orientation.w = pose.orientation.w;
136
137    marker.scale.x = body.dimensions[0];
138    marker.scale.y = body.dimensions[1];
139    marker.scale.z = body.dimensions[2];
140
141    marker.color.r = 0.0f;
142    marker.color.g = 0.0f;
143    marker.color.b = 1.0f;
144    marker.color.a = 0.5;
145
146    marker.lifetime = ros::Duration();
147    obstacle_pub.publish(marker);
148    r.sleep();
149
150    if (client.call)
151    {
152      ROS_INFO("Successfully planned inspection path");srv))
153    }
154    else
155    {
```

```
156      ROS_ERROR("Failed to call service planner");
157      return 1;
158    }
159
160    return 0;
161  }
162
163  std::vector<nav_msgs::Path> * readSTLfile(std::string name)
164  {
165    std::vector<nav_msgs::Path> * mesh = new std::vector<nav_msgs::
         Path>;
166    std::fstream f;
167    f.open(name.c_str());
168    assert(f.is_open());
169    int MaxLine = 0;
170    char* line;
171    double maxX = -DBL_MAX;
172    double maxY = -DBL_MAX;
173    double maxZ = -DBL_MAX;
174    double minX = DBL_MAX;
175    double minY = DBL_MAX;
176    double minZ = DBL_MAX;
177    assert(line = (char *) malloc(MaxLine = 80));
178    f.getline(line, MaxLine);
179    if(0 != strcmp(strtok(line, " "), "solid"))
180    {
181      ROS_ERROR("Invalid mesh file! Give in ascii-format.");
182      ros::shutdown();
183    }
184    assert(line = (char *) realloc(line, MaxLine));
185    f.getline(line, MaxLine);
186    int k = 0;
187    while(0 != strcmp(strtok(line, " "), "endsolid") && !ros::
         isShuttingDown())
188    {
189      int q = 0;
190      nav_msgs::Path p;
191      geometry_msgs::PoseStamped v1;
192      for(int i = 0; i<7; i++)
193      {
194        while(line[q] == ' ')
195          q++;
196        if(line[q] == 'v')
197        {
198          // used to rotate the mesh before processing
199          const double yawTrafo = 0.0;
200          // used to scale the mesh before processing
201          const double scaleFactor = 1.0;
202          // used to offset the mesh before processing
203          const double offsetX = 0.0;
204          // used to offset the mesh before processing
205          const double offsetY = 0.0;
206          // used to offset the mesh before processing
207          const double offsetZ = 0.0;
```

```
208
209        geometry_msgs::PoseStamped vert;
210        char* v = strtok(line+q," ");
211        v = strtok(NULL," ");
212        double xtmp = atof(v)/scaleFactor;
213        v = strtok(NULL," ");
214        double ytmp = atof(v)/scaleFactor;
215        vert.pose.position.x = cos(yawTrafo)*xtmp-sin(yawTrafo)*
      ytmp;
216        vert.pose.position.y =  sin(yawTrafo)*xtmp+cos(yawTrafo)*
      ytmp;
217        v = strtok(NULL," ");
218        vert.pose.position.z =  atof(v)/scaleFactor;
219        vert.pose.position.x -= offsetX;
220        vert.pose.position.y -= offsetY;
221        vert.pose.position.z -= offsetZ;
222        if(maxX<vert.pose.position.x)
223          maxX=vert.pose.position.x;
224        if(maxY<vert.pose.position.y)
225          maxY=vert.pose.position.y;
226        if(maxZ<vert.pose.position.z)
227          maxZ=vert.pose.position.z;
228        if(minX>vert.pose.position.x)
229          minX=vert.pose.position.x;
230        if(minY>vert.pose.position.y)
231          minY=vert.pose.position.y;
232        if(minZ>vert.pose.position.z)
233          minZ=vert.pose.position.z;
234        vert.pose.orientation.x =  0.0;
235        vert.pose.orientation.y =  0.0;
236        vert.pose.orientation.z =  0.0;
237        vert.pose.orientation.w =  1.0;
238        p.poses.push_back(vert);
239        if(p.poses.size() == 1)
240          v1 = vert;
241      }
242      assert(line = (char *) realloc(line, MaxLine));
243      f.getline(line, MaxLine);
244    }
245    p.poses.push_back(v1);
246    p.header.frame_id = "/kopt_frame";
247    p.header.stamp = ros::Time::now();
248    p.header.seq = k;
249    mesh->push_back(p);
250    k++;
251  }
252  free(line);
253  f.close();
254  return mesh;
255 }
```

As with the other two planners, further details on the planning parameters are provided in the corresponding code repository. Among the critical parameters to be set is the flag for rotorcrafts or fixed–wing aircraft, the model constraints for each, as well as algorithm parameters that influence its behavior (e.g. collision–check interval, iterations of RRT* if no connection could be established, minimum distance for viewpoints to an obstacle and more).

6.4   Terrain Monitoring Planner

The `tmplanner` is implemented and operates in a finite–horizon alternating between replanning and plan execution, while taking new sensor data into account. The replanning stage consists of two steps, namely: (a) coarse 3D grid search in the UAV configuration space and (b) optimization of this trajectory for maximized information/exploratory gain using an evolutionary scheme.

6.4.1   `tmplanner` Open–Source Release

The `tmplanner` open–source release can be found directly at https://github.com/ethz-asl/tmplanner or through the additional and unifying repository https://github.com/unr-arl/informative-planning. The release has been tested to operate with ROS Kinetic, while back compatibility with ROS Indigo and Jade has been verified. The repository contains two stand-alone ROS packages corresponding to discrete (`tmplanner_discrete`) and continuous (`tmplanner_continuous`) variable monitoring, as described in Sects. 4.1 and 4.2, respectively.

The implementation of both planners includes the following key components:

- Node interface for ROS communication (`tmplanner_node` in the respective packages)
- Planning unit
- Mapping unit

A third package, `tmplanner_tools`, provides convenience functions used by both planners, in addition to the infrastructure needed to run examples set-up in the Gazebo-based RotorS environment [44]. As per the examples, the algorithms are intended for use on-board a rotorcraft-type aerial robot equipped with a downward-facing camera. All parameters relevant for mapping (e.g. environment dimensions, grid resolution), sensing (e.g. measurement frequency, FoV angles), and planning (e.g. time budget, optimization parameters) are exposed in seperate file. Note that the complexity of the programs can be easily adapted for online operation even with limited computational resources, e.g. by using a coarser grid for 3D search or limiting the number of optimization iterations in the first or second replanning stages, respectively.

The following steps outline the terminal commands to run an illustrative demo in RotorS for the application of mapping a continuous variable. The same procedure can be applied to map discrete variables, by replacing for the `tmplanner_discrete` package.

1. Start the simulation environment by running:

```
$ roslaunch tmplanner_continuous
  monitoring_example.launch
```

2. Start the planner node:

```
$ roslaunch tmplanner_continuous tmplanner.launch
```

Note that the appropriate parameters must be loaded in the launch file.

3. Initialize the finite-horizon planning routine via the rosservice:

```
$ rosservice call /start planning
```

Complete documentation, featuring extended details for the subscribed topics and algorithm parameters, are provided within the code repository, in addition to visualization tools.

7 Conclusion

This chapter presented a path planning ensemble addressing the problems of autonomous exploration, terrain monitoring and optimized coverage, alongside their utilization as ROS packages and experimental studies using aerial robots. All the algorithms presented are already open–sourced. In their combination, these tools represent a path planning ensemble capable of supporting important application domains for aerial robots, including those of infrastructure and industry inspection, precision agriculture, security monitoring, and more. The open–source contributions, along with the limited set of assumptions considered, permit their generic utilization for any robotic system for which a boundary value solver is available. Moreover, extensive experimental evaluation and field demonstration using aerial robots support the vision for a collective contribution in the field of path planning that can have a transformable impact, better equip end–users of robotic systems and further enable developers. Finally, experimental datasets are released to allow systematic scientific comparisons.

References

1. L. Yoder, S. Scherer, Autonomous exploration for infrastructure modeling with a micro aerial vehicle, *Field and Service Robotics* (Springer, Berlin, 2016), pp. 427–440
2. P. Lottes, R. Khanna, J. Pfeifer, R. Siegwart, C. Stachniss, Uav-based crop and weed classification for smart farming, in *2017 IEEE International Conference on Robotics and Automation (ICRA)* (IEEE, 2017), pp. 3024–3031

3. F. Liebisch, M. Popovic, J. Pfeifer, R. Khanna, P. Lottes, A. Pretto, I. Sa, J. Nieto, R. Siegwart, A. Walter, Automatic UAV-based field inspection campaigns for weeding in row crops, in *EARSeL SIG Imaging Spectroscopy Workshop, Zurich* (2017)
4. H. Balta, J. Bedkowski, S. Govindaraj, K. Majek, P. Musialik, D. Serrano, K. Alexis, R. Siegwart, G. Cubber, Integrated data management for a fleet of search-and-rescue robots. J. Field Robot. **34**(3), 539–582 (2017)
5. C. Bolkcom, Homeland security: Unmanned aerial vehicles and border surveillance. DTIC Document, 2004
6. A. Hornung, K.M. Wurm, M. Bennewitz, C. Stachniss, W. Burgard, Octomap: an efficient probabilistic 3d mapping framework based on octrees. Auton. Robot. **34**(3), 189–206 (2013)
7. S.M. LaValle, Rapidly-exploring random trees a new tool for path planning, 1998
8. H.H. González-Banos, J.-C. Latombe, Navigation strategies for exploring indoor environments. Int. J. Robot. Res. **21**(10–11), 829–848 (2002)
9. A. Bircher, M. Kamel, K. Alexis, H. Oleynikova, R. Siegwart, Receding horizon "next-best-view" planner for 3d exploration, in *IEEE International Conference on Robotics and Automation (ICRA)* (2016)
10. C. Papachristos, K. Alexis, Autonomous detection and classification of change using aerial robots, in *IEEE Aerospace Conference* (2017)
11. A. Bircher, M. Kamel, K. Alexis, H. Oleynikova, R. Siegwart, Receding horizon path planning for 3d exploration and surface inspection. Auton. Robot. 1–16 (2016)
12. H. Carrillo, I. Reid, J.A. Castellanos, On the comparison of uncertainty criteria for active slam, in *2012 IEEE International Conference on Robotics and Automation (ICRA)* (IEEE, 2012)
13. M. Bloesch, S. Omari, M. Hutter, R. Siegwart, Robust visual inertial odometry using a direct ekf-based approach, in *2015 IEEE/RSJ International Conference on Intelligent Robots and Systems (IROS)* (IEEE, 2015). (to appear)
14. J. Kiefer, General equivalence theory for optimum designs (approximate theory). Ann. Stat. **2**, 849–879 (1974)
15. M. Kamel, T. Stastny, K. Alexis, R. Siegwart, Model predictive control for trajectory tracking of unmanned aerial vehicles using ros, *Robot Operating System (ROS)* (Springer, Cham, 2017)
16. A. Bircher, M. Kamel, K. Alexis, M. Burri, P. Oettershagen, S. Omari, T. Mantel, R. Siegwart, Three-dimensional coverage path planning via viewpoint resampling and tour optimization for aerial robots. Auton. Robot. 1–25 (2015)
17. A. Bircher, K. Alexis, U. Schwesinger, S. Omari, M. Burri, R. Siegwart, An incremental sampling-based approach to inspection planning: the rapidly-exploring random tree of trees, 2016
18. A. Bircher, K. Alexis, M. Burri, P. Oettershagen, S. Omari, T. Mantel, R. Siegwart, Structural inspection path planning via iterative viewpoint resampling with application to aerial robotics, in *IEEE International Conference on Robotics and Automation (ICRA)* (2015), pp. 6423–6430, https://github.com/ethz-asl/StructuralInspectionPlanner
19. K. Alexis, C. Papachristos, R. Siegwart, A. Tzes, Uniform coverage structural inspection path-planning for micro aerial vehicles, 2015
20. C. Papachristos, S. Khattak, K. Alexis, Autonomous exploration of visually-degraded environments using aerial robots, in *2017 International Conference on Unmanned Aircraft Systems (ICUAS)* (IEEE, 2017)
21. C. Papachristos, S. Khattak, K. Alexis, Uncertainty–aware receding horizon exploration and mapping using aerial robots, in *IEEE International Conference on Robotics and Automation (ICRA)* (2017)
22. M. Popovic, G. Hitz, J. Nieto, I. Sa, R. Siegwart, E. Galceran, Online informative path planning for active classification using UAVs, in *IEEE International Conference on Robotics and Automation* (IEEE, Singapore, 2017)
23. M. Popovic, T. Vidal-Calleja, G. Hitz, I. Sa, R. Y. Siegwart, J. Nieto, Multiresolution mapping and informative path planning for UAV-based terrain monitoring, in *IEEE/RSJ International Conference on Intelligent Robots and Systems* (IEEE, Vancouver, 2017)

24. A. Elfes, Using occupancy grids for mobile robot perception and navigation. Computer **22**(6), 46–57 (1989)
25. C.E. Rasmussen, C.K.I. Williams, *Gaussian Processes for Machine Learning* (MIT Press, Cambridge, 2006)
26. S. Reece, S. Roberts, An introduction to gaussian processes for the Kalman filter expert, in *FUSION* (2013), pp. 1–9
27. T. Vidal-Calleja, D. Su, F. D. Bruijn, J.V. Miro, Learning spatial correlations for bayesian fusion in pipe thickness mapping, in *IEEE International Conference on Robotics and Automation* (IEEE, Hong Kong, 2014)
28. C. Richter, A. Bry, N. Roy, Polynomial trajectory planning for quadrotor flight, in *International Conference on Robotics and Automation* (Springer, Singapore, 2013)
29. B. Charrow, S. Liu, V. Kumar, N. Michael, Information-theoretic mapping using Cauchy–Schwarz quadratic mutual information, in *IEEE International Conference on Robotics and Automation* (IEEE, Seattle, 2015), pp. 4791–4798
30. N. Hansen, The CMA evolution strategy: a comparing review. Stud. Fuzziness Soft Comput. **192**(2006), 75–102 (2006)
31. G. Hitz, E. Galceran, M.-È. Garneau, F. Pomerleau, R. Siegwart, Adaptive continuous-space informative path planning for online environmental monitoring, 2016
32. J. O'rourke, *Art Gallery Theorems and Algorithms*, vol. 57 (Oxford University Press, Oxford, 1987)
33. H. González-Baños, A randomized art-gallery algorithm for sensor placement, in *Proceedings of the Seventeenth Annual Symposium on Computational Geometry* (ACM, 2001), pp. 232–240
34. G. Dantzig, R. Fulkerson, S. Johnson, Solution of a large-scale traveling-salesman problem. J. Oper. Res. Soc. Am. **2**(4), 393–410 (1954)
35. S. Lin, B.W. Kernighan, An effective heuristic algorithm for the traveling-salesman problem. Oper. Res. **21**(2), 498–516 (1973)
36. G. Papadopoulos, H. Kurniawati, N.M. Patrikalakis, Asymptotically optimal inspection planning using systems with differential constraints, in *2013 IEEE International Conference on Robotics and Automation (ICRA)* (IEEE, 2013), pp. 4126–4133
37. K. Helsgaun, An effective implementation of the lin-kernighan traveling salesman heuristic. Eur. J. Oper. Res. **126**(1), 106–130 (2000)
38. S. Karaman, E. Frazzoli, Sampling-based algorithms for optimal motion planning. Int. J. Robot. Res. **30**(7), 846–894 (2011)
39. H. Ferreau, C. Kirches, A. Potschka, H. Bock, M. Diehl, qpOASES: a parametric active-set algorithm for quadratic programming. Math. Program. Comput. **6**(4), 327–363 (2014)
40. C. Papachristos, K. Alexis, L.R.G. Carrillo, A. Tzes, Distributed infrastructure inspection path planning for aerial robotics subject to time constraints, in *2016 International Conference on Unmanned Aircraft Systems (ICUAS)* (IEEE, 2016), pp. 406–412
41. C. Papachristos, K. Alexis, Augmented reality-enhanced structural inspection using aerial robots, in *2016 IEEE International Symposium on Intelligent Control* (IEEE, 2016), pp. 1–6
42. P. Oettershagen, T. Stastny, T. Mantel, A. Melzer, K. Rudin, G. Agamennoni, K. Alexis, R. Siegwart, Long-endurance sensing and mapping using a hand-launchable solar-powered uav, 2015
43. F. Furrer, M. Burri, M. Achtelik, R. Siegwart, Rotorsa modular gazebo mav simulator framework, *Robot Operating System (ROS)* (Springer, Cham, 2016), pp. 595–625
44. RotorS: An MAV gazebo simulator, https://github.com/ethz-asl/rotors_simulator

A Generic ROS Based System for Rapid Development and Testing of Algorithms for Autonomous Ground and Aerial Vehicles

Pawel Ladosz, Matthew Coombes, Jean Smith
and Michael Hutchinson

Abstract This chapter presents a Robot Operating System (ROS) framework for development and testing of autonomous control functions. The developed system offers the user significantly reduced development times over prior methods. Previously, development of a new function from theory to flight test required a range of different test systems which offered minimal integration; this would have required great effort and expense. A generic system has been developed that can operate a large range of robotic systems. By design, a developed controller can be taken from numerical simulation, through Software/Hardware in the loop simulation to flight test, with no adjustment of code required. The flexibility and power of ROS was combined with the Robotic Systems toolbox from MATLAB/Simulink, Linux embedded systems and a commercially available autopilot. This affords the user a low cost, simple, highly flexible and reconfigurable system. Furthermore, by separating experimental controllers from the autopilot at the hardware level, flight safety is maintained as manual override is available at all times, regardless of faults in any experimental systems. This chapter details the system and demonstrates the functionality with two case studies.

P. Ladosz · M. Coombes (✉) · J. Smith · M. Hutchinson
Department of Aeronautical and Automotive Engineering, Loughborough
University, Loughborough, Leicestershire LE11 3TU, United Kingdom
e-mail: M.J.Coombes@lboro.ac.uk

P. Ladosz
e-mail: P.Ladosz@lboro.ac.uk

J. Smith
e-mail: J.Smith5@lboro.ac.uk

M. Hutchinson
e-mail: M.Hutchinson2@lboro.ac.uk

© Springer International Publishing AG, part of Springer Nature 2019
A. Koubaa (ed.), *Robot Operating System (ROS)*, Studies in Computational
Intelligence 778, https://doi.org/10.1007/978-3-319-91590-6_4

1 Introduction

The advantages of autonomous vehicles are well known. They offer comfort, safety and reliability for vehicles, replace humans in mundane and dangerous jobs, and provide access to a range of data at a significantly reduced cost. Due to the recent miniaturisation of electronics and increasing maturity of autonomous systems, there is an ever increasing demand from the civilian market for such systems. This has led to a huge demand for autonomous vehicle functions of ever increasing complexity. Taking these functions through the full development cycle, from numerical simulation to software and hardware in the loop testing and real world experiments is complex, expensive and time-consuming. This will usually require a number of variations of code and systems to test the same function across all stages. Typically, this complexity and the requirement to be familiar with multiple systems and programming languages can be an prohibitive barrier to conducting comprehensive testing.

Real world algorithm testing is often fraught with difficulties, from poor code to unmodelled external factors not included in simulation. If real world testing is conducted with both experimental and safety control systems running on the same computing platform, this introduces a single point of failure where the (not unlikely) failure of development software can jeopardise any safety overrides. This is especially dangerous for aerial systems where an out of control vehicle can cause significant damage or harm. To mitigate this, it is important for the vehicle operator to have complete and reliable manual override at all times.

To this end, we have developed a reliable system which enables the rapid prototyping of autonomous vehicle algorithms using a single accessible programming language, and provides the pilot with full manual override of the vehicle at the "flick of a switch". The system is capable of controlling several configurations of systems including wheeled ground vehicles and rotary or fixed wing aircraft. This is achieved through close integration between the Robot Operating System (ROS), MATLAB, and a Commercially Off The Shelf (COTS) autopilot.

Developing a full system from scratch that handles vehicle control, communications, on-board processing, and MATLAB integration would be expensive and very time-consuming without ROS. This is why the system we have developed use COTS hardware and software. Also, many people are already familiar with MATLAB and ROS; this lowers the barriers to entry while providing access to the huge number of functions and packages already developed for use with these programs.

2 Development Cycle

Before performing real world testing, it is important to verify code functionality, which is commonly done through Software In the Loop (SIL) and then Hardware In the Loop (HIL) testing. Putting algorithms through the full development cycle of SIL, HIL, and then flight testing is a good systematic debugging method, which

Fig. 1 The development
cycle for autonomous control
algorithms

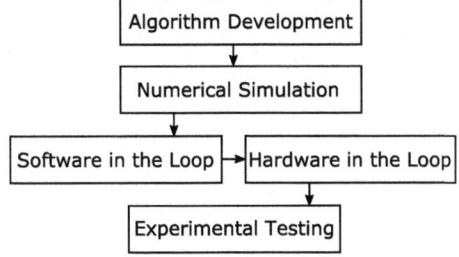

Fig. 1 The development cycle for autonomous control algorithms

significantly reduces risk at each stage. This process is shown in Fig. 1. Separation of the various stages allows for any issues to be pinpointed easily. For example, without this method, an algorithm performing poorly when taken straight to flight test could be the result of coding errors, communication problems, sensor quality issues or any number of other causes. By using the full development cycle, numerical simulation can be used to study issues in the control design itself, SIL would highlight compatibility issues, HIL might highlight issues such as processing power limitations. When it comes to finally flight testing, the number of potential causes has been substantially reduced. The virtues of going through this testing procedure are discussed in [1].

2.1 Background

A range of systems have been used before this framework was developed. By examining some of these, we can select some of their desirable functionality. Furthermore, some of the shortcomings can be identified and avoided.

2.1.1 Study of Prior Systems

A number of organisations have developed environments for testing a huge range of algorithms on a number of platforms for a range of applications. Generally, these systems are developed for a single set of tests, for example in [2] an autonomous hobbyist grade helicopter platform is developed to conduct testing of obstacle avoidance and mapping algorithms outdoors. The vehicle has a number of on board sensors with all control and high-level algorithms ran aboard two PC-104 embedded computers. The aircraft was designed with a safety pilot manual override, ensuring that in the event of failure of the tested algorithm the aircraft can be landed safely. However, as all these functions are on the PC104, any major single hardware faults would cause loss of the vehicle. While putting an x64 PC on board a vehicle allows great flexibility for algorithm development, it significantly increases costs and complexity of the system. This includes not only the significant computer cost but also the additional cost associated with the requirement of carrying the additional heavy payload.

In [3] a system is developed to control a number of remote controlled fixed wing aircraft from the ground to test cooperative control algorithms for Unmanned Aerial Vehicle (UAV) teams. They used a commercially available fixed wing only autopilot with HIL testing functions. The autopilot enabled them to use the pre-existing low-level attitude control and mid level path following functions of the autopilot and update its way points from a ground station using a long range serial modem. This method meant they could concentrate on the high-level aspects of the mission planning and vehicle co-ordination. However, this system has no ability to further expand on its sensors or capability for more advanced on-board computation.

Reference [4] concludes that a more generic test bed system is needed. They wanted a system that enabled testing of small scale heterogeneous vehicles indoors. The system would use the VICON motion capture system to replicate Global Positioning System (GPS) indoors, and feedback attitude data for control. It uses standard remote controlled vehicles which are controlled directly over standard remote control transmitters. While this system is very flexible in the different tests it can conduct, it has huge limitations. All algorithms needs to be run on the ground with no on board sensor feedback, while being limited to an indoor environment thereby also limiting the testing volume.

In [5] a test system is developed which utilises both quadrotors and differential drive ground robots to test mesh network communications. All vehicles run ROS for communication, command and control of small scale robots. The vehicles all communicate with each other using XBee serial modems which use the ZigBee protocol in mesh mode. Vehicles that run ROS can take considerable time to set up, or if purchased pre-configured incur additional cost. XBee modems make wireless serial communication easy and can be set to a few different network modes. However they have short range and low bandwidth, so are not flexible enough to use for a large range of testing applications, especially long range outdoor tests.

Each of these systems has its own strengths and weakness, and a few techniques are used to lower costs, or development times. However, our generic system will aim to take all the best aspects of these systems and combine them into one highly flexible system. This system will provide a framework and set of tools capable of performing all the tests discussed in the literature, using a single generic system. This system should lower costs and reduce development time, while also improving reliability and safety.

The most important feature identified is a reliable safety pilot manual override. Ideally the fail-safe flight software is running on separate hardware such as a Pixhawk or Navio, while high level tasks are running on a more powerful (but less reliable) embedded system. This introduces a separation layer to keep critical tasks running despite any failure in the more complex embedded system which is running the developed algorithm.

To satisfy the identified strengths and weaknesses, the following system requirements were defined:

- Guaranteed manual vehicle override.
- A standardised embedded computer structure on board each vehicle.

- Compatibility with multiple vehicle configurations and types.
- Simultaneous control of multiple heterogeneous vehicles.
- Indoor and outdoor testing.
- Low cost.
- Compatibility with common development environments i.e. MATLAB/Simulink.
- SIL, HIL testing functions.
- Allow functions to be run either off or on board vehicles.

The final point highlights that the main aim of this system is to provide the capability of testing high and low level algorithms, without the usual time investment in hardware specific code adaptation. The system was designed for use with MATLAB and Simulink as these are commonly used programming environments in the industry which have features that enable simple interfacing with embedded hardware.

2.2 COTS Autopilots

There are already several examples of autopilot systems developed by research organisations, usually designed to assist with their own research goals such as [1, 6–8]. Producing a bespoke autopilot offers the chance to tailor it to the research requirements exactly, adding only the functionality required. However, there are many drawbacks to this approach. Long development times and considerable investment are required.

In the last decade, the miniaturisation of sensor technology from mobile phones and great increase in community led open source software and hardware have resulted in remote controlled vehicle autopilots which are reliable, feature rich and cheap. This all means that in house development is unnecessary.

The system we developed was built to utilise the PX4 autopilot software stack. This is open source autopilot software that can be flashed on to a number of autopilot devices, which supports fixed wing, rotary wing, ground vehicles and aquatic vessels. It has been designed with development and research in mind, as it has a so called "offboard" mode, which allows commands to be sent to the vehicle from a companion computer (e.g. Raspberry Pi) with fool proof manual override.

The autopilot software is written in C++, and is completely open source with an active development community, ensuring that it will continue to be developed. It is due to the flexibility, reliability, and multi-platform capabilities that PX4 was chosen as the basis of the system. Although PX4 is usually referred to as an autopilot, the degree of authority it has over the vehicle can be varied at any point through an input command. There are various modes depending on the vehicle type. Typical modes include attitude control, height and speed control and way point following. Under manual override mode, any vehicle can be controlled directly by a human pilot.

Levels of increasing autonomy can then be applied; for example when PX4 is aboard an aircraft, Fly by Wire (FBW) mode enables a human pilot to give pitch and roll angle demands to the autopilot, rather than directly controlling servos. As fixed limits on the maximum roll and pitch angles can be set within PX4 inner loop control, this makes the aircraft much easier to fly, with the aircraft returning to straight and level flight should control be relinquished. Finally, in guided mode PX4 will handle all high level control itself, allowing a vehicle to to fly or drive a series of predefined waypoints autonomously.

The PX4 autopilot includes a companion computer interface, which can send sensor data and receive commands over a serial link connected to an embedded computer. This is discussed in further detail in Sect. 3. Each of the aforementioned modes are available to be controlled from the embedded system, this variable degree of control authority makes it an extremely useful tool upon which higher order control strategies can be built.

2.3   Embedded Systems

The world's first Computer on Module (COM) system was Gumstix's ARM based Linux Machine, released September 9, 2003. This was made possible by the new generation of small low cost processors made for the growing smartphone market. Ever since, a huge number of small, low power COM embedded systems have been developed and released. It is due to the small size, low weight and relatively powerful embedded computers that this generic system is possible. Previously, embedded systems would have been large, expensive, power hungry x86 computers which would mean that only a small number of large test vehicles would be capable of carrying them.

A few examples of common COM embedded systems:

- Gumstix
- Raspberry Pi
- Odroid
- Beaglebone
- Panda Board

Any one of these boards would be appropriate as our choice of companion computer, as the algorithms that run on board tend to be computationally cheap. However, if the required computational load were to increase, the embedded system could swapped to a more powerful system with relative ease. We have decided to use the Raspberry Pi 2 as our embedded system, based mainly on its' ease of use. Furthermore, it is one of the cheapest, most ubiquitous embedded systems currently available, with vast amounts of online support and tutorials. This made it the obvious choice for a simple, easy to integrate system.

2.3.1   Network Architecture

As we will be using a Linux based embedded computer with both wired and wireless networking, this gives massive flexibility in the type and size of data transmitted

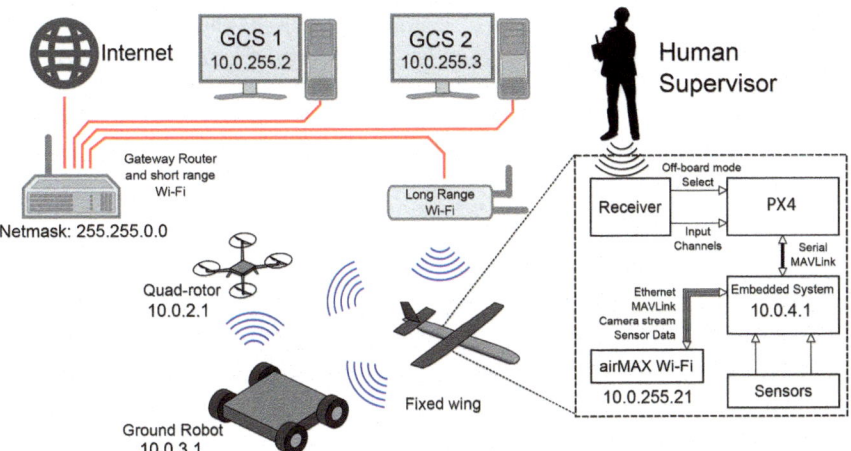

Fig. 2 Example network layout

due to the high bandwidth. In addition, a range of different network topologies and routing configurations can used. This has two advantages: the user can select the best configuration that gives the coverage needed to keep a reliable connection. This means certain topologies, or hardware can be tested as a part of the whole system with plug and play style simplicity.

An example network configuration is shown in Fig. 2, where a mesh network topology is employed in order to operate three different vehicles at distance from the Ground Control Station (GCS). The GCS consists of two computers connected directly to a gateway router, which in turn is directly connected to a long range Wi-Fi access point. The GCS does not have line of sight to the quadrotor or the ground robot so any data transferred between the GCS and these two vehicles is relayed through the fixed wing aircraft which, is within line of sight of all nodes. The mesh network enables packets to be routed efficiently so when the quadrotor or the ground robot wish to communicate with one another, they can do so directly.

2.4   Chapter Contents

By the end of this chapter, the reader will be armed with all the tools and knowledge to create their own ROS based test system. They should be able to implement and test various autonomous vehicle functions running both on and off the vehicle quickly and simply.

In Sect. 3 we give an overview of the system and describe the functions and setup with justification of our choices. We then proceed to detail two experimental case studies. Section 4 assesses advanced Disturbance Observer Based Control (DOBC) on a fixed wing UAV for rejection of internal (modelling) and external (e.g. wind) disturbances. Section 5 details an indoor experiment using ground vehicles to locate and reconstruct the source of a simulated hazardous atmospheric release.

3 System Overview

To fulfil the requirements outlined in Sect. 2 a combination of ROS and Simulink was chosen. Simulink provides an easy to use and widely adopted means of programming control algorithms, while also offering excellent support for ROS and the Raspberry Pi. A key part of the ROS/Simulink integration is the Robotics System toolbox of Simulink.

3.1 System Framework

In general, there are two types of robotics system used in our research lab: ground and air. Ground systems are robots such as Turtlebot, which are used as a platform to either test algorithms such as path planning or sensors such as a camera. Ground systems are mostly used in an indoor area, where a VICON motion capture system is installed to provide the robots position and attitude information, which is shared using the ROS VICON package. The position data is subscribed to in Simulink, where a controller can be designed to obtain the desired behaviour. For air systems, a variety of configurations are used which can be largely divided into fixed wing and rotary wing aircraft. The MAVROS package is used as an interface between ROS and MAVlink; MAVlink is a common protocol which is understood by many COTS autopilots. For all air platforms, the PX4 autopilot software is used alongside a Pixhawk autopilot. PX4 has support for many airframes as well as many research friendly features such as off-board mode. If the air vehicle is used indoors it can also benefit from integration with VICON, which can replace GPS for position information in PX4. A summary of this layout is given in Fig. 3. It is worth noting however, that the air systems framework could likewise be used for ground vehicles, provided that the relevant PX4 firmware is uploaded to the Pixhawk to control ground based systems.

With usage of ROS and Simulink, a work flow which is consistent across all systems could be established to enable rapid development and deployment. First, the user must ensure that any information needed by the control scheme to be designed is available as a ROS topic. An initial prototype in Simulink can then be developed, ready for simulation testing. Generally, numerical simulations are conducted first if a suitable model of the system is available. Subsequently, the algorithms can be tested using more advanced simulation tools such as: (i) X-Plane with a Pixhawk in the loop for aerial vehicles; or (ii) Gazebo running on a Ubuntu virtual machine that is connected to Simulink for ground based vehicles. Given satisfactory vehicle behaviour in the simulations, the algorithms will be ready for the physical systems.

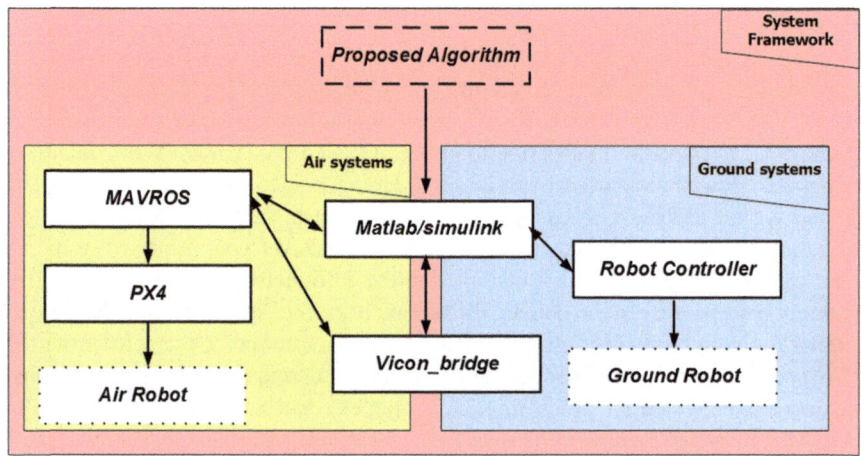

Fig. 3 Full framework overview. Dotted lines around squares indicate physical system, dashed represent input to system, and solid lines are software and ROS packages. Please note that robots could be virtual in either HIL or SIL simulation for example using gazeboo for ground robot

Deployment to physical systems happens in two steps. Firstly, the algorithm is ran in Simulink on a PC, using a Wi-Fi network to communicate with either the ground or flying vehicle. In this state, rapid changes to the algorithm are possible as needed, although some additional latency in transmission of data is incurred. Once the algorithm is performing well, the auto code generation feature of Simulink is used to compile the algorithm on the desired platform. This then removes the requirement of a stable Wi-Fi connection and improves latency.

The system also allows great flexibility in adding and removing components as needed. These components could be additional sensors or even whole vehicles. Provided that the sensors or equipment have ROS packages, they can be very easily and quickly integrated with the rest of the system. For new sensors, which may not have their own ROS packages, it is possible to integrate them into the ROS network. A system is needed which can read the sensor data and transfer it to a serial output. This can then be connected to the Raspberry Pi, which can publish the data to the ROS network. This is described further in Sect. 5.

3.2 Simulink

Simulink is a graphical programming language widely used in engineering practice. It is characterized by the availability of many libraries such as computer vision, machine learning, model predictive control and communication toolboxes. In this section particular focus is on robotics system toolbox and auto code generation

3.2.1 Robotics System Toolbox

For additional functionality different toolboxes can be added to Simulink. Simulink has good compatibility with ROS, and support through a dedicated robotic systems toolbox. It can publish or subscribe to both topics and parameters. While handling strings has been an issue in earlier releases, it has been fixed as of release 2016a. The robotic system toolbox consists of five blocks: publish, subscribe, blank message, set parameter and get parameter. Publish is a block used to publish to ros topic. A topic can be chosen from a list of topics in the ROS network from an interactive menu, or a new topic can be defined. Subscribe is used to subscribe to a ROS topic. Similar to the publish block, topics can be either user defined or selected from the ROS network. The blank message block is used to define a message type, so that the data can be added for publishing. Get parameter and Set parameter are used to access the ROS parameter server. With these five blocks, Simulink can be integrated into ROS with ease. Additionally, as they control only the inputs and outputs to the Simulink model, and everything else is mainly unchanged, designing new models or adapting an existing one is quick process.

3.2.2 Auto Code Generation

One of the key features of Simulink is auto code generation from models. Auto code generation means transferring a model into C/C++ code so that it can be deployed on devices which do not support Simulink models. Auto code generation is a powerful feature often utilised in our work due to its many attractive features. First of all, by putting code on board of vehicles, communication delays are greatly reduced. This is of particular importance to low level controllers, such as the work in Sect. 4; delay in receiving data and sending commands can cause instability. Another improvement comes from increased robustness of the system. While a model is run on the ground, it is extremely sensitive to even the smallest connection issues, sometimes resulting in unsafe behaviour. With the algorithm running on board the target system, connection issues are mitigated thereby increasing the robustness and safety of the system.

Auto code generation has a major limitation; not all blocks available in Simulink are compatible with auto code generation. A list of supported blocks can be found on the MathWorks website. During development, referring to this list when adding new blocks is a good habit as it avoids these compatibility issues. Also, a quick compilation test can be performed on a specific block, to confirm whether it will compile correctly. In the case where incompatible blocks have to be used, Simulink can still be used with ROS, as a standalone node, and communicate with the target systems over Wi-Fi. It is worth noting that the list of supported blocks is continually updated and it covers the majority of blocks needed for control systems. All blocks within the robotic systems toolbox are fully compatible with auto code generation.

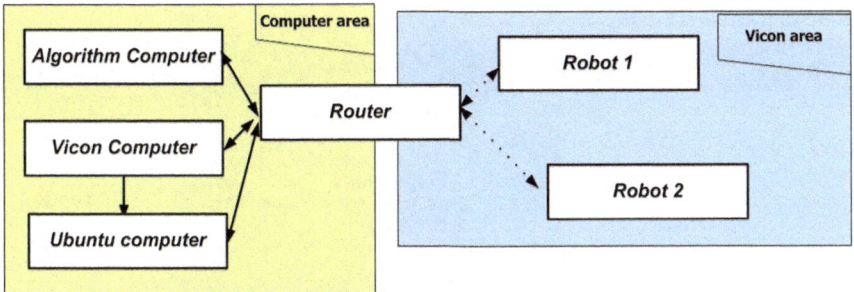

Fig. 4 A system overview of indoor area system, showing data flow. Solid lines are wired connections, with dashed lines being wireless

3.3 Ground

Ground vehicles are mostly used indoors as a sensor carrier or as an initial test platform for algorithms. To be able to utilise them effectively, a common indoor laboratory area is used (Fig. 4).

3.3.1 Indoor Laboratory Overview

The lab setup consists of three computers, a VICON motion capture system and a wireless network router. One of the computers is running Ubuntu 14.04 and is used for running ROS packages, testing and development. All three computers are connected to the router via Ethernet cable for speed and reliability. One of the computers is connected directly to the VICON system, and is used to run VICON tracker. The other is the main computer where Simulink models are run. The main reason for such a split is system safety and redundancy. Algorithms tested are highly experimental and their behaviour is largely unknown, so computer slow downs and freezes are not uncommon. Such a split guarantees that even in case of algorithm causing unexpected problems, the VICON position data can still be obtained reliably.

A VICON motion tracking system is used to obtain position indoors, where GPS data is unavailable or poor, while outdoors GPS is used for robot position. The system uses a set of infrared cameras to obtain 3D position and orientation of an object with special infrared markers. VICON tracker software is used to define each object as a ridged body to be tracked. This information can then be easily used in robotic applications.

For most robots, the Ubuntu computer acts as the ROS master. Making the master permanent and the same for each robot makes connecting Simulink to each of them very easy, as the master IP address remains unchanged. Additionally, this allows for simple addition of future systems, which would be available to all systems on the network.

Fig. 5 Turtlebot in standard
configuration with laptop
and 3D camera

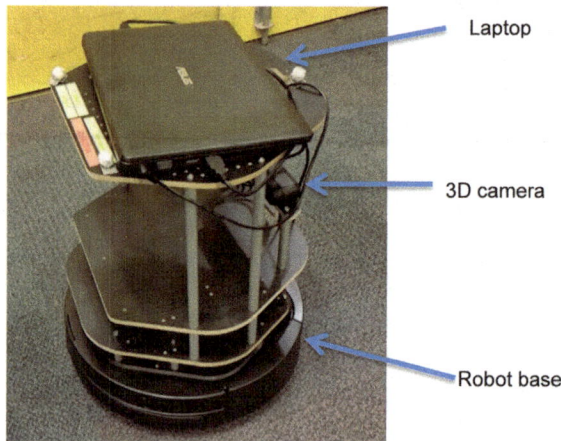

3.3.2 Turtlebot

The Turtlebot, shown in Fig. 5, is a very versatile robotic platform for autonomous
systems research, with its own ROS package [9]. As standard it is equipped with
a netbook and a 3D camera. The netbook is an x86 computer, guaranteeing high
compatibility with other devices, systems and ROS packages. The 3D camera can be
used for mapping and navigation. The Turtlebot system is very flexible and highly
compatible with other sensors.

3.3.3 VICON Package

To the best of the author's knowledge there exist two packages to transfer VICON
positional data into ROS: VICON_bridge [10] and vrpn_client_ros [11]. Although
both packages perform similar roles, there are some differences between them.
VICON_bridge is using an API provided by VICON to obtain data from the VICON
system. Conversely, vrpn_client_ros uses a commonly defined system called Virtual-
Reality Peripheral Network. This means that VICON_Bridge package will work with
only VICON positional system, while vrpn_client_ros will work with many other
positional systems, for example OptiTrack. Another advantage of vrpn_client_ros is
that it can be compiled for an ARM based system, meaning packages can be run on
embedded devices such as a Raspberry Pi. This is beneficial, as it reduces delay in
providing accurate position information to the systems. It is also worth considering
that more and more robots (i.e. turtlebot 3) are using ARM based devices due to
energy efficiency and cost. The VICON_bridge package has its own benefits. First
of all it offers additional options such as simulating bad data (to make VICON data
more realistic) and calibrating exact origin of objects automatically. The latter option
is particularly appealing as in VICON Tracker setting objects centre accurately can

Listing 1 vrpn launch file

```
1  <launch>
2
3    <arg name="server" default="0.0.0.0"/>
4
5    <node pkg="vrpn_client_ros" type="vrpn_client_node" name="vrpn_client_node"
          output="screen">
6      <rosparam subst_value="true">
7        server: $(arg server)
8        port: 3883
9
10       update_frequency: 100.0
11       frame_id: world
12
13       # Use the VRPN server's time, or the client's ROS time.
14       use_server_time: false
15       broadcast_tf: true
16
17       # Must either specify refresh frequency > 0.0, or a list of trackers to
             create
18       refresh_tracker_frequency: 1.0
19       #trackers:
20       #- FirstTracker
21       #- SecondTracker
22     </rosparam>
23     <remap from="/vrpn_client_node/miniQuad/pose" to="/mavros/mocap/pose"/>
24   </node>
25
26 </launch>
```

be a long and laborious manual process. In VICON_bridge it can be done by simply matching the middle of an object with the centre of the VICON area, making the process much easier.

Both packages follow a similar setup procedure. In the appropriate .launch (see included vrpn launch file in Listing 1) file, the IP address of a computer running VICON tracker needs to be specified. With this information both packages should work. Once an object is enabled in VICON tracker, the appropriate topic for it will start publishing. For example if a robot is named "robot_1" in VICON, the corresponding topic name should be "/VICON/robot_1/robot_1". The topic contains three translational (x, y and z) and three rotational (around x, y and z axis) elements. Rotations are in quaternion form (which is a standard and accepted way to present angles in ROS).

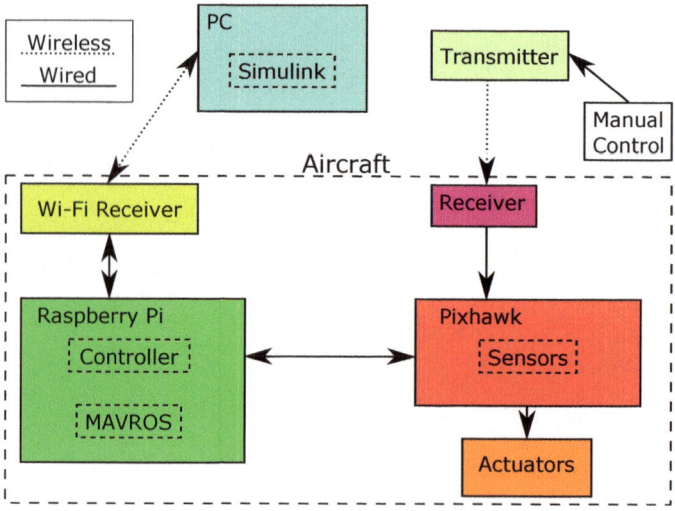

Fig. 6 System overview for fixed or rotary with aircraft

3.4 Air

Both fixed and rotary wing aircraft are used in the lab. The same system can be used with either configuration with only minor adjustments. The system overview which is common to both fixed and rotary wing aircraft is shown in Fig. 6.

What follows is a detailed breakdown of system components which complete the overall ROS system. All subsystems are common to both fixed and rotary wing platforms, although specific components such as actuators and Wi-Fi receivers differ between platforms.

Raspberry Pi - A popular micro computer running Ubuntu 14.04 and ROS Indigo. MAVROS facilitates communication with the Pixhawk and any other ROS enabled device, such as Simulink running on a PC. The Pi on board the aircraft serves two main purposes. Firstly, it hosts and runs a controller which has previously been designed in Simulink. Secondly, it acts as the interpreter and relay for messages between various sources which would not be able to communicate directly. For example, MAVlink messages sent out by the Pixhawk on the serial port have to be interpreted and made available to the ROS network so that it can be used by the controller and read by the PC. Similarly, messages from the controller or PC need to be converted to a suitable format to be sent to and understood by the Pixhawk. The Pi is powered via its' Micro USB port and communicates with the Pixhawk via a serial port, as shown in Fig. 7.

MAVROS- [12] This is a ROS package which acts as a bridge between ROS and the autopilot. It takes messages emitted by the autopilot in the form of MAVLink messages and transfers them to ROS topics. MAVROS topics include, but are not limited to: transmitter output, position in global and local frame, aircraft attitude and

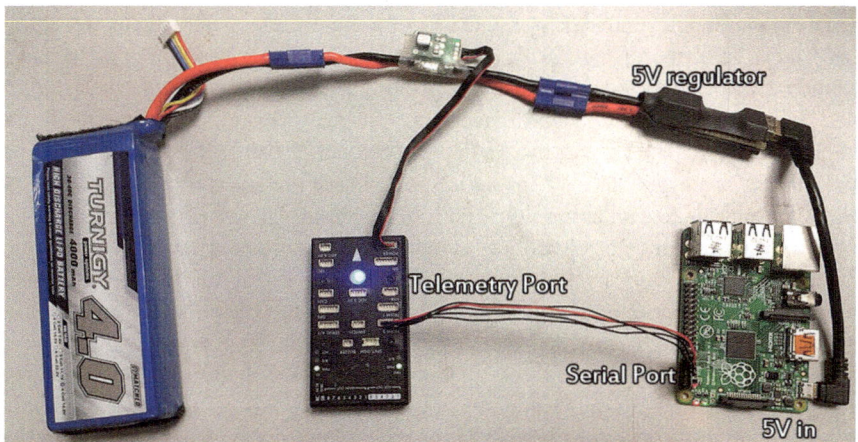

Fig. 7 Hardware overview of common ROS/Autopilot system components

speed. One of the key features of MAVROS is the ability to send commands to a vehicle through a facility called "offboard mode". There are specific topics which allow control of attitude, position and velocities namely

- */mavros/setpoint/attitude*
- */mavros/setpoint/position_local*
- */mavros/setpoint/velocity*

where velocities can be both linear and rotational. Offboard mode is fully compatible with the PX4 autopilot stack. It can be activated by a switch which can be setup on a transmitter using QGroundControl, which will enable and disable off board mode. With offboard mode disabled, the transmitter has full control over vehicle. While enabled, offboard mode allows the vehicle to be controlled through one of the topics mentioned above. As an added safety feature, if the offboard mode switch is enabled and no information is provided on any of the control topics, or if the message rate is too low, offboard mode will not engage. Such an approach means that the pilot can always take over control if a tested algorithm does not behave in desired or safe manner. Not being allowed to start off-board mode unless data to appropriate topics is published is another useful safety feature. Two files used to launch mavros can be seen in Listings 2 and 3.

Wi-Fi Receiver - A Wi-Fi interface is used to provide connectivity between the the aircraft and other systems. The specific adapter is chosen based on range requirements; the indoor system for example can use a small USB powered adapter, while the outdoor platform requires a more powerful solution for additional range. Depending on the test configuration, it may not be required for this connection to be active in flight. If, for example, the controller is compiled to run on-board the Pi and real time data monitoring is not needed, the Wi-Fi connection is not required.

Pixhawk - The Pixhawk is a COTS autopilot which is popular in the small UAV area. In terms of hardware, the Pixhawk provides a large range of sensors including a 6 Degree of Freedom (DOF) Inertial Measurement Unit (IMU), pitot static airspeed sensor, GPS and barometer. On the software side, the PX4 flight stack is used. With PX4 and the MAVlink protocol, most of the PX4 data inputs and outputs are available over the serial port. These are interpreted by MAVROS on the Pi and made available to the controller. One of the key advantages of using ROS to test experimental controllers is safety. Adding custom controllers into the Pixhawk firmware is possible through source code modification. However, as well as requiring good programming knowledge, this also introduces risk to the aircraft. A bug in custom code can lead to crashing of the Pixhawk software, which would likely lead to crashing of the aircraft. Placing custom code on the separate Raspberry Pi and not modifying Pixhawk source code means there is a permanent and reliable fail-safe if an issue occurs with the custom code. Control can be handed back to the Pixhawk with a single button toggle on the transmitter.

PC - This ground station computer is running Simulink with both the Raspberry Pi and ROS support packages installed. A Wi-Fi link is used to connect to the target system. Due to the flexibility of ROS, initial ground testing of the controller can be done by running the controller in Simulink while connected to the Pixhawk. The radio transmitter can still be used to control the aircraft, allowing for quick assessment and verification of the various controller settings specific to the aircraft/transmitter flight testing setup. Rapid changes can be made to the controller as needed. Although the aircraft could be controlled in this way, this requires the Wi-Fi link to be maintained at all times. To improve safety and reliability, the controller can be transferred to the Raspberry Pi. Once controller performance is verified, Simulink is used to compile and upload the controller to the Raspberry Pi.

Transmitter/Receiver - The transmitter and receiver provide a long range wireless control link between the aircraft and the pilot. The transmitter interprets manual control inputs and sends them using radio signals to the receiver, which passes these on to the Pixhawk. The main consideration for the transmitter and receiver is signal quality, which dictates maximum operation range. On-board the aircraft, multiple "satellite" (auxiliary) receivers are connected to the main receiver to improve signal quality by removing dead zones which can arise from electrical and signal interference. As so many electrical components are contained within a small area (especially the Ubiquiti which is a very high powered source of radio waves), interference is likely to occur. The use of multiple receivers in different locations and orientations mitigates this risk. In the case of loss of control link, the Pixhawk is configured to autonomously return to the launch location and circle until communication is restored.

Actuators - these are the components which affect the physical motion of the test platform. For a multirotor, the motors would be considered actuators whereas for a fixed wing aircraft the control surfaces would be considered as actuators.

Listing 2 mavros launch file

```xml
<launch>
        <!-- vim: set ft=xml noet : -->
        <!-- example launch script for PX4 based FCU's -->

        <arg name="fcu_url" default="/dev/ttyAMA0:921600" />
        <arg name="gcs_url" default="" />
        <arg name="tgt_system" default="1" />
        <arg name="tgt_component" default="1" />
        <arg name="log_output" default="screen" />

        <include file="$(find mavros)/launch/node.launch">
                <arg name="pluginlists_yaml" value="$(find mavros)/launch/
                        px4_pluginlists.yaml" />
                <arg name="config_yaml" value="$(find mavros)/launch/
                        px4_config.yaml" />
                <arg name="fcu_url" value="$(arg fcu_url)" />
                <arg name="gcs_url" value="$(arg gcs_url)" />
                <arg name="tgt_system" value="$(arg tgt_system)" />
                <arg name="tgt_component" value="$(arg tgt_component)" />
                <arg name="log_output" value="$(arg log_output)" />
        </include>
</launch>
```

4 Test Case 1 - Outdoor UAV Flight Tests

4.1 *Introduction*

The growth in popularity of small fixed wing Unmanned Aerial Vehicles (UAVs) has generally been accompanied by a reduction in size and weight of the aircraft and internal components. These UAVs are significantly smaller and lighter than their full sized counterparts; this makes them far more susceptible to external disturbances such as wind. UAVs are generally controlled by generic commercially available autopilots, such as an ArduPilot Mega (APM) or Pixhawk. The control algorithms aboard these autopilots must be generic enough to be applied to a range of aircraft, with tuning which can be conducted by the operator. This restricts the application of more advanced control methods, such as optimal control, which relies on a good model of the system to be controlled.

These problems have been addressed by applying Linear Quadratic Regulator (LQR) optimal control, coupled with DOBC, to flight testing of a fixed wing UAV. The practical flight testing component was made possible by the ROS framework outlined in the previous sections. The control methods were extensively tested in simulation, including SIL/HIL; this was easily possible with the ROS framework.

Application of an optimal control technique, such as a LQR relies on a model of the controlled system. While this is likely available for full scale aircraft as part

Listing 3 mavros node launch file

```xml
 1  <launch>
 2          <!— vim: set ft=xml noet : —>
 3          <!— base node launch file—>
 4
 5          <arg name="fcu_url" />
 6          <arg name="gcs_url" />
 7          <arg name="tgt_system" />
 8          <arg name="tgt_component" />
 9          <arg name="pluginlists_yaml" />
10          <arg name="config_yaml" />
11          <arg name="log_output" default="screen" />
12
13          <node pkg="mavros" type="mavros_node" name="mavros" required="true"
                clear_params="true" output="$(arg log_output)">
14                  <param name="fcu_url" value="/dev/ttyAMA0:921600" />
15                  <param name="gcs_url" value="udp://10.0.0.121:14551@10.0.0.130
                        :14551" />
16                  <param name="target_system_id" value="$(arg tgt_system)" />
17                  <param name="target_component_id" value="$(arg tgt_component)"
                        />
18                  <!—<param name="use_sim_time" value="true" />—>
19                  <!—remap from="/mavros/vision_pose/pose" to="/vicon/miniQuad/
                        miniQuad"/—>
20
21
22                  <!— load blacklist, config —>
23                  <rosparam command="load" file="$(arg pluginlists_yaml)" />
24                  <rosparam command="load" file="$(arg config_yaml)" />
25          </node>
26  </launch>
```

of the design process, it is not commonly available for small UAVs. Furthermore, a significant amount of variation exists between various UAVs, depending on how they are built or what equipment is on board. Choice of servos, for example, is down to the operator meaning actuator dynamics will vary between aircraft. Any model acquired by the operator is likely to contain some small errors as ideal system identification can be difficult with the low cost sensors used on small UAVs. Generally, an error feedback term is added to the control scheme to account for such errors e.g. proportional integral (PI) control or integral augmented LQR (LQI). While such solutions can account for these errors, as well as external disturbances, a more advanced solution is available, DOBC, which is described and assessed in this section.

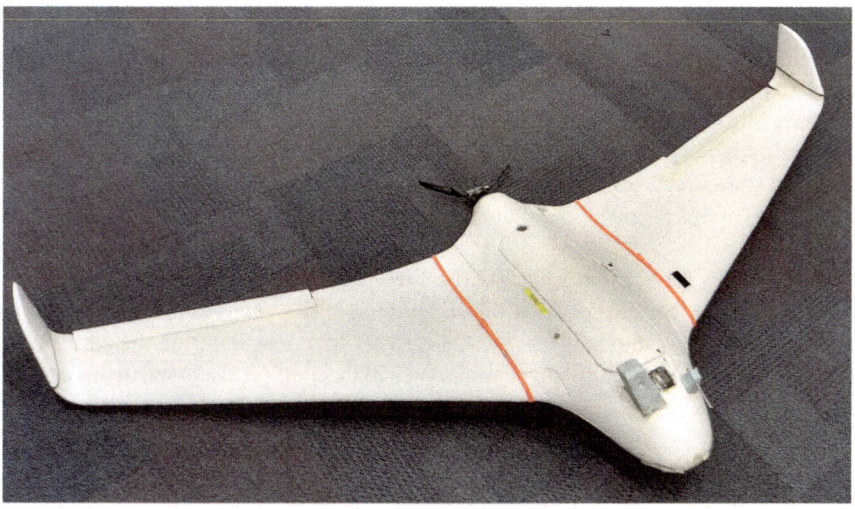

Fig. 8 The Skywalker X8 used in the outdoor flight testing

4.2 Preliminaries

4.2.1 Aircraft

The aircraft used in this work is a Skywalker X8, shown in Fig. 8. In flight configuration the aircraft weights 2.95 kg, with a wingspan of 2.12 m. The aircraft is equipped with a Pixhawk autopilot, Raspberry Pi computer and a Ubiquiti Rocket M5 Wi-Fi receiver.

4.2.2 System Description

While most of the components of the system described in Sect. 3.4 are used on the X8 platform, it is worth describing two platform dependant components, namely the actuators and Wi-Fi receiver:

Wi-Fi Receiver - For the X8 flight testing, a high powered and long range Wi-Fi solution is needed; a Ubiquiti Rocket M5 is ideal for this purpose. In the work detailed herein, this connection is not used for flight control, although it does offer the facility of live data reporting if this were required. It is connected to the Pi via Ethernet.

Actuators - On the X8, a total of 3 actuators are available: an identically operated pair of elevons for aerodynamic control and an electric motor for thrust. Control demands are supplied to the actuators from the Pixhawk. The elevons are actuated by servos which take angular position inputs. An internal control scheme is then responsible for achieving the demanded position. The servos have a maximum angular velocity limit which is significantly higher than the rates offered by most full sized aircraft, capable of around 450°/s. The motor is driven by a speed controller which is given a throttle percentage command. The speed controller is then responsible for converting battery power into the required 3 phase power for the motor. The speed controller also features internal limitations on the ramp up rate. Despite this, the bandwidth of this propulsion system is also significantly higher than that of most full sized aircraft.

One of they key features enabled by the employed control system is access to direct control of the actuator positions. By publishing to "*/mavros/actuator_control*", actuator position commands can be sent directly to any actuator on the aircraft. For the study of low level control algorithms, such as the method discussed in this chapter, this is essential.

4.3 System Dynamics

For this case study, the lateral channel of the X8 is considered. Given in (1) is the standard lateral state space model. Here, the state x_f is comprised of the sideslip velocity v, roll rate p, yaw rate r, roll angle ϕ and yaw angle ψ. $L_{\delta a}$ and N_{δ_a} are the roll and yaw moments resulting from aileron input, δ_a. This will be referred to as the *full* model, as it includes all the states conventionally used for lateral state space aircraft modelling. This model assumes that it is produced around a $\phi = 0$ trim condition. These are defined as

$$\dot{x}_f = \underbrace{\begin{bmatrix} Y_v & Y_p & Y_r & g\cos\theta^* & 0 \\ L_v & L_p & L_r & 0 & 0 \\ N_v & N_p & N_r & 0 & 0 \\ 0 & 1 & 0 & 0 & 0 \\ 0 & 0 & 1 & 0 & 0 \end{bmatrix}}_{A_f} \underbrace{\begin{bmatrix} v \\ p \\ r \\ \phi \\ \psi \end{bmatrix}}_{x_f} + \underbrace{\begin{bmatrix} 0 \\ L_{\delta a} \\ N_{\delta a} \\ 0 \\ 0 \end{bmatrix}}_{B_f} \underbrace{[\delta_a]}_{u_\delta},$$

$$y_f = \underbrace{[0\ 0\ 0\ 1\ 0]}_{C_f} x_f,$$

(1)

with the output y_f defined as ϕ, the roll angle. However, due to sensor availability, this model is not easily implementable on the aircraft. The standard instrumentation for the aircraft does not provide aerodynamic flow angles, so the sideslip velocity is not directly measurable. For this reason, flight testing shown in this work will use a reduced state space model, given in (2).

$$\dot{x} = \underbrace{\begin{bmatrix} L_p & 0 \\ 1 & 0 \end{bmatrix}}_{A} \underbrace{\begin{bmatrix} p \\ \phi \end{bmatrix}}_{x} + \underbrace{\begin{bmatrix} L_{\delta a} \\ 0 \end{bmatrix}}_{B} \underbrace{[\delta_a]}_{u_\delta},$$

$$y = \underbrace{\begin{bmatrix} 0 & 1 \end{bmatrix}}_{C} x.$$

(2)

This represents a substantially simplified model, although it is fully available for measurement. This model will make the control problem for DOBC much tougher as modelling inaccuracies are certain in this case. Again, the C matrix is defined such that the output is ϕ.

Model parameters for the X8 are still required. For this, an AeroProbe Micro Air Data probe was fitted to the aircraft. This 5 hole pitot probe provided angle of attack and sideslip data, which when fused with the Pixhawk IMU data provided all the information needed to identify the full model in (1), from which the reduced order model is easily extracted. This demonstrates that with this additional sensor, the full order model could be implemented for control on the aircraft, if the AeroProbe data could be added into the ROS network, where it could be combined with the Pixhawk data in real time. Indeed, through methods described in Sect. 5.2.2 this would be possible. While this may be implemented at a later time, the initial flight testing was done without the additional sensors while the new system was tested.

4.3.1 Actuator Modelling

For accurate simulation of the control scheme, actuator dynamics should be included. For this, VICON motion tracking was used to measure the true deflection of the actuators when subjected to reference commands. The X8 elevon shown in Fig. 9 shows the preparation for motion capture.

As ROS features straight forward VICON interfacing, it was trivial to obtain both the demanded and true actuator positions in a single Simulink model; this allowed for accurate assessment of not only the actuator dynamics but also any time delays which may be present in the system. Figure 10 shows the resulting actuator time response. This response could then be modelled easily with a first or second order system; when implemented into the simulation environment, the resulting response would be more representative of the response of the aircraft in flight.

4.4 Control and Performance Objectives

To assess the performance of the controller, and any improvement offered by DOBC augmentation, some control and performance objectives must be defined.

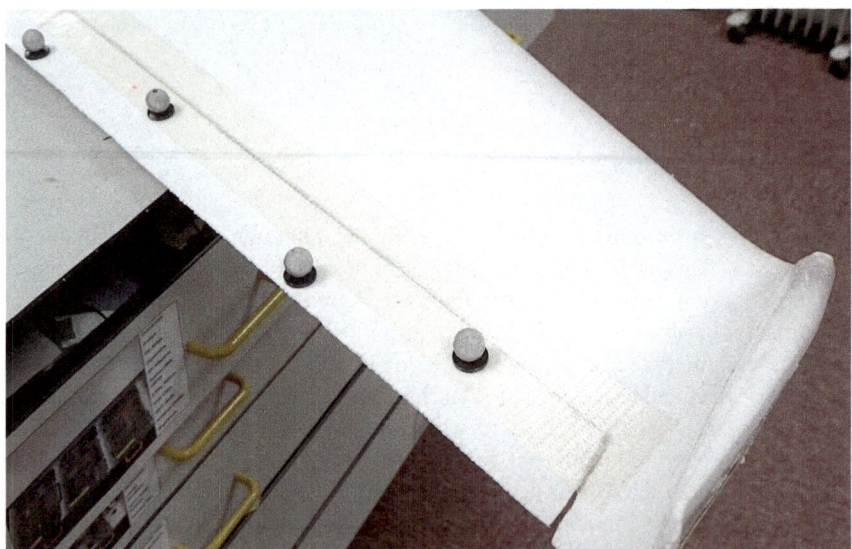

Fig. 9 An X8 elevon ready for VICON motion capture

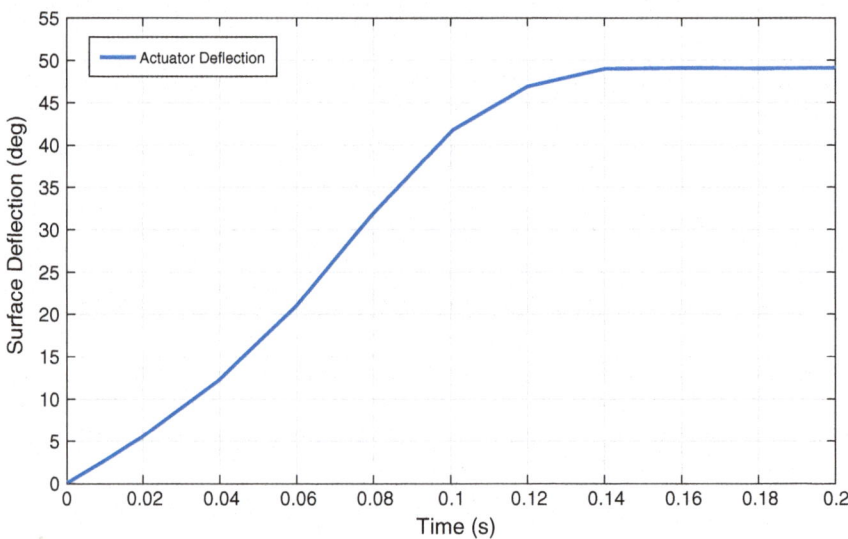

Fig. 10 The response of the elevon to a step command to maximum deflection

1. Track ϕ about a given reference command

The main objective of the control scheme is to allow for tracking of reference commands. This inner loop control forms the basis of any later developed outer loop control and as such accurate and timely reference command tracking is essential.

2. Maintain suitable performance in the presence of disturbances

It is required that the control scheme is able to complete the first objective even in the presence of unknown and unpredictable disturbances. This includes both internal disturbances, such as modelling inaccuracies and external disturbances such as wind.

4.5 Control Design

With the system detailed and performance objectives outlined, the design of the control schemes is now discussed. This begins with the baseline LQR controller, which will later be augmented with DOBC. Performance assessment will be carried out by comparing performance of the LQR with and without DOBC active. As the DOBC is an augmentation added to the LQR, the same controller is used throughout making direct comparison feasible.

4.5.1 Feedback Control Design

For the system described by (2), an LQR with reference command tracking takes the form shown in Fig. 11, where the physical aircraft is represented by the system G_{AC}.

This requires a control law in the form of

$$u = -k_x x + Nr, \qquad (3)$$

where k_x is the state regulation gain and N is the DC gain of the system, required to track the reference command r. To calculate k_x, we formulate the standard LQR cost function as

$$J = \frac{1}{2} \int_0^\infty (x^T Q x + u_\delta^T R u_\delta) \mathrm{dt}, \qquad (4)$$

Fig. 11 The layout of the LQR with reference tracking ability

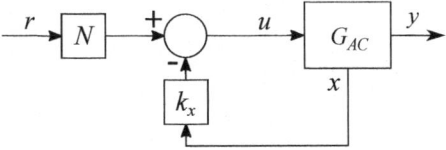

where Q and R are the state and control weighting matrices, respectively. The solution of this problem is not addressed here for brevity; further detail can be found in [13] and the references therein.

To obtain the DC gain of the system, first the transfer function from reference to output, G_{ry} is formed

$$G_{ry} = C(sI - (A - Bk_x))^{-1}BN.$$

This can be solved for N by setting G_{ry} to the appropriately sized identity matrix and letting $s \to 0$ for the steady state solution, leaving

$$N = \left[C(-(A - Bk_x))^{-1}B\right]^{-1}.$$

At this point, all terms required for the control law described by (3) are known. However, it should also be clear as to how modelling errors will affect this system. As all the gains are based on the system matrices, it can be said that the resulting performance relies on the accuracy of the model. An error in the model will result in steady state errors, particularly in reference tracking. This is one of the primary reasons why DOBC augmentation is a favourable addition to LQR control. This is discussed in more detail in the following section.

4.6 Disturbance Observer Augmentation

As shown, the LQR performance depends directly on any unknown modelling errors. In this case, there will be some unknown error due errors in the model parameters. Furthermore, there is some known, but unquantifiable, error in the model as it has been heavily simplified for implementation purposes. Fortunately such errors, as well as other disturbances, can be attenuated by DOBC augmentation [14]. First, (2) is rewritten to include disturbances.

$$\dot{x} = A_t x + B_t u_\delta + d_x, \tag{5}$$

here, A_t and B_t are the true matrices which would ideally describe the modelled system and d_x is the external disturbances defined as

$$d_x = \begin{bmatrix} p_d & \phi_d \end{bmatrix}^T.$$

To also account for the internal modelling errors, we define the lumped disturbances as

$$d_{lx} = (A_t - A)x + (B_t - B)u_\delta + d_x. \tag{6}$$

Now the difference between the true, unknown matrices (A_t, B_t) and those used for modelling (A, B) are *lumped* together with any external disturbances, d_x. However, d_{lx} is not directly measurable. As such, before the disturbance can be dealt with, it must first be obtained. This is done by following the disturbance observer methodology in [15], which results in the disturbance observer

$$\begin{cases} \dot{z} = -L(z + Lx) - L(Ax + Bu_\delta), \\ \hat{d}_{lx} = z + Lx. \end{cases}$$

This observer is vulnerable to windup due to actuator saturation. As the control input u does not take into account the maximum deflections of the physical actuators, it is possible for large disturbance estimates to build up internally from high control demands. To eliminate this issue, the anti-windup method developed in [13] is implemented, resulting in the final observer

$$\begin{cases} \dot{z} = -L(z + Lx) - L(Ax + B\bar{u}_\delta), \\ \hat{d}_{lx} = z + Lx, \end{cases} \tag{7}$$

where $\bar{u}_\delta$ represents the control demand u, saturated at its minimum and maximum values. Finally, to reject the estimated disturbance, the control law must be updated to include a disturbance rejection term

$$u_\delta = -k_x x + Nr + k_{dx}\hat{d}_{lx}, \tag{8}$$

where k_{dx} is the compensation gain which must be designed. This can be calculated as

$$k_{dx} = -\left[C(A - Bk_x)^{-1}B\right]^{-1} \times C(A - Bk_x)^{-1}. \tag{9}$$

The DOBC augmented control structure is given in Fig. 12.

4.7 Results

In this section, a comparison of the LQR and DOBC augmented control systems is conducted. First, state space simulations are used to verify performance of the controllers and study disturbance rejection characteristics. This is followed by ver-

Fig. 12 Layout of the anti-windup modified DOBC augmented LQR control scheme

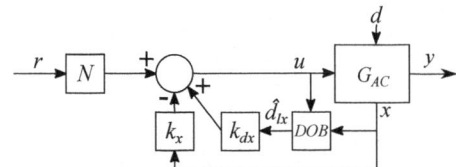

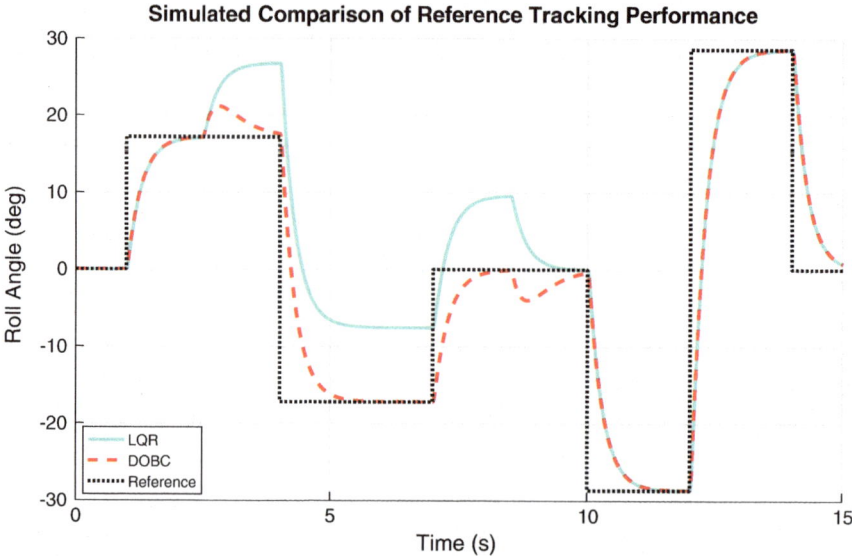

Fig. 13 Comparison of the reference tracking performance of the two controllers. A disturbance is active $2.5s \leq t \leq 8.5s$

ification of the controller within the ROS framework using the SIL environment. Finally, a flight test is also conducted to study performance in real world application.

4.7.1 State Space Simulation

Initial controller assessment is performed using the state space model developed in Sect. 4.3. This provides an ideal case for the controller where all parameters are fully known with no uncertainty. The result is shown in Fig. 13

This simulation demonstrates several key points of the technique. First, notice that when the disturbance is not active, the DOBC scheme has no effect on the system. This shows that in the absence of disturbance, the DOBC scheme reverts to the optimal LQR design. When the disturbance is added, the LQR system is not able to remove its' effect from the output, while the DOBC augmentation removes the effect of the disturbance entirely, restoring nominal performance to reference tracking. This is a very positive result which indicates that the designed system is working as intended.

4.7.2 Software in the Loop Simulation

With good controller performance demonstrated in the ideal state space environment, the next step was to verify performance of the control schemes using the ROS frame-

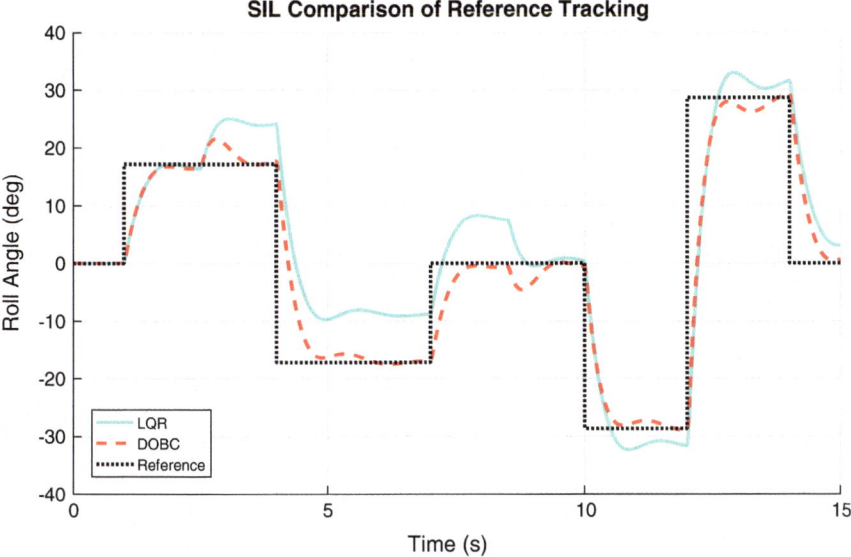

Fig. 14 Comparison of the reference tracking performance of the two controllers using the SIL environment. A disturbance is active $2.5s \le t \le 8.5s$

work. For this, the SIL environment was employed. For this result, some disturbances are expected due to modelling errors. Furthermore, the same disturbance used in the state space modelling is added to this simulation to study performance. The main aim of this simulation is to verify proper configuration of the controller within the ROS network. The result is shown in Fig. 14.

Here it can be seen that, overall, the controller performance is the same as within the state space simulation, which is encouraging. It is also clear that some modelling errors are present, as both controllers show some error when the known disturbance is not active. However, the DOBC augmentation has again significantly improved performance to all the disturbances. Furthermore, it has been demonstrated that the designed controllers work as expected within the ROS system.

4.7.3 Flight Test Result

Finally, to demonstrate the performance of the system on the aircraft, flight tests were conducted; a video comparison of one test flight is available online.[1] Figure 15 shows the resulting performance of the baseline LQR as well as the DOBC augmented LQR.

Clearly, there are significantly more disturbances present in this result than the simulations, as would be expected. The data is noisier on the whole; this is expected due to ever present wind disturbances. A clear and constant bias is present in the LQR,

---

[1] https://youtu.be/Czz8hqTjgAE.

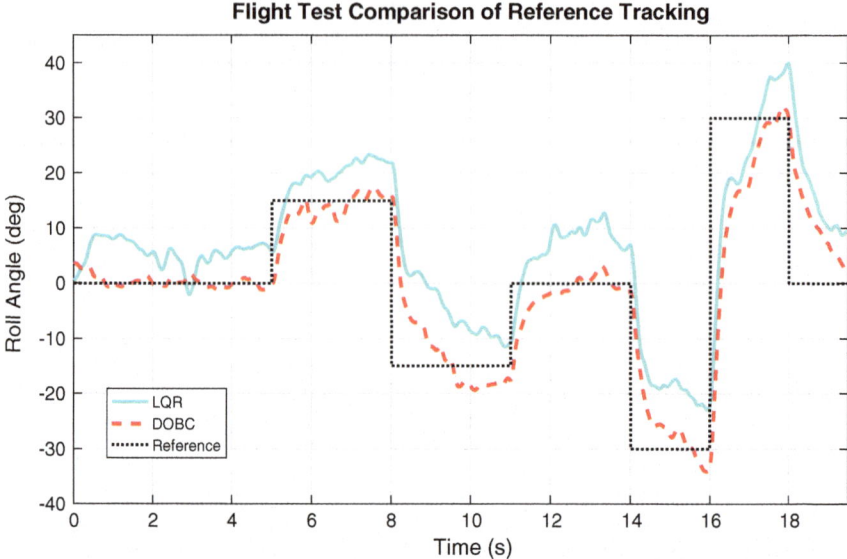

Fig. 15 Flight test comparison of reference tracking performance of the two controllers

which is likely due to a physical disturbance on the aircraft. There are a range of possible sources such as unbalanced weight or small damage to a lifting surface. What is obvious is that the DOBC augmentation accounts for this disturbance extremely well. The scheme has settled to 0° reference command within 1 s and offers vastly improved reference tracking throughout the flight. Furthermore, although difficult to quantify, the DOBC result appears smoother throughout; this could be attributed to continuous rejection of small wind perturbations. On the whole, this result demonstrates very well why DOBC is a powerful technique in the application of control techniques to physical systems. It offers excellent and fast disturbance rejection and restoration of nominal performance in physical application, where disturbances are always present.

4.8   Conclusions

This case study has demonstrated clearly the benefits of applying DOBC control in both simulation and application to a physical plant. Importantly, this work has been made possible by the extensive utilisation of the ROS framework. From initial system identification for modelling and control design, through simulation demonstrations and finally flight testing, the work has relied on the performance and usability of ROS. Control schemes designed and tested in simulation environments can be transferred easily and automatically to run on board the physical plant, with the same performance as demonstrated in the simulation environment.

5   Test Case 2 - Indoor Source Localisation Experiment Using an Autonomous Ground Robot

5.1   Introduction

The ROS software framework has been used to test a gas source localisation algorithm using a robot equipped with a low cost metal oxide (MOX) gas sensor. The goal of source localisation is to find the source of some potentially hazardous material that is dispersing into the atmosphere. There are a diverse range of applications, for example: (i) Finding the source of accidental or deliberate releases of chemical, biological, radiological or nuclear (CBRN) material; (ii) Searching for emissions of greenhouse gases; or iii) Locating sources of odour. Where possible, source localisation is performed using visual methods. For substances difficult to detect visibly, concentration sensors are used to observe point-wise observations of the material of interest. The source can be localised using model based estimation methods or by directly tracking the concentration of the dispersing material towards its source [16]. Several sensors can be used to form a static network, or they can be placed upon mobile platforms.

Tracking the concentration towards its source can be problematic, especially where turbulence and sensing issues can cause intermittent detections. Some positive experimental results have been achieved in the literature, however, there is still research required to increase the reliability of the methods and to prepare them for real applications [17].

Model based source localisation methods are used in conjunction with static as well as mobile sensors as the source location can be inferred from a stand-off position. These approaches use a model to predict the dispersion from a source, this can be used to compare the concentration observed by a sensor with that expected from the model. Algorithms have been developed to optimise a point estimate or to approximate a probability density function of the source location [16].

Most of the model based source localisation work has been assessed using simulated data or with experimental datasets. Collecting the datasets is often an expensive and time consuming process. For example, a dispersion study conducted in 2003, known as the Joint Urban experiment, was a huge process involving over 150 people from more than 20 organisations and hundreds of sensors [18].

In this work, we assessed a source localisation algorithm with a simple experiment, where smoke produced by a burning incense stick was used to simulate some hazardous dispersing material. An autonomous robot is used to execute a systematic sweep search pattern, stopping at regular intervals to take measurements of the concentration. Point-wise measurements of the concentration were made by a low cost size weight and power (SWaP) MOX gas sensor. Probability density functions of the two dimensional coordinates of the source location were approximated using Bayes' theorem, implemented using a particle filter based algorithm. The estimates were updated recursively in response to the new sensory information. The experiments were set up quickly and easily by virtue of the ROS software framework and the openly available packages. Some of the useful packages that were used include the Turtlebot [9], Rosserial_arduino [19], and Vicon_bridge [10] ROS packages. ROS enabled rapid development, on-line data visualisation, and strong situational awareness and control throughout the experiments.

5.2 Methodology

An autonomous ground robot equipped with a MOX gas sensor executes a systematic sweep search of a predefined area. Bayes' theorem is used to estimate the location of the source in response to noisy sensor readings, uncertain dispersion parameters and imperfect models. Implemented as a particle filter [20], posterior probability distributions of the source parameters are updated in a recursive fashion, as new sensory information is made available. The ROS framework is used for communication between the autonomous vehicle, the indoor positioning system and a ground control station (GCS). Various ROS packages are employed to support quick and reliable experimentation and on-line visualisation of the data.

5.2.1 Source Estimation Algorithm

Bayes' theorem is used to estimate the location of the source with state vector $\Theta_k = \begin{bmatrix} p_s^\mathsf{T} & a_s & u_s & \phi_s \end{bmatrix}^\mathsf{T}$. Where $p_s^\mathsf{T} = (x_s, y_s)$ refers to the location of the source in Cartesian coordinates, a_s is the scaled release rate, and (u_s, ϕ_s) are the mean wind speed and direction. The latter three parameters are included to account for uncertainty in the source strength and meteorological conditions, which are important inputs to model the dispersion. Posterior distributions of the source parameters Θ_k are updated recursively in response to new sensor observations D_k at each discrete time step k as:

$$Posterior \propto \frac{Prior \times Likelihood}{Evidence} \rightarrow p(\Theta_k|D_{1:k}) = \frac{p(\Theta_k|D_{1:k-1})p(D_k|\Theta_k)}{p(D_k|D_{1:k-1})}$$
(10)

where

$$P(D_k|D_{1:k-1}) = \int P(\Theta_k|D_{1:k-1})P(D_k|\Theta_k)\,d\Theta_k.$$
(11)

At each time step, the prior distribution is replaced by using the posterior distribution from the previous iteration. At the beginning of the experiments, the initial prior distributions must be set for each of the parameters in the state vector. For the location of the source, the prior distributions are set to uniform within the bounds of the experimental area. Likewise, prior distributions for the remaining parameters are wide uniform distributions to account for large amounts of uncertainty.

The likelihood function in (10) provides a numeric measure of the probability that a hypothesised source vector Θ_k^i is true, by comparing the measured data D_k with that expected by running the source in a model $V(p_k|\Theta_k^i)$. In this work a Gaussian relationship has been used, leading to the following likelihood function:

$$P(D_k|\Theta_k^i) = \frac{1}{\sigma_k(D_k)\sqrt{2\pi}} \exp\left[-\frac{(D_k - V(p_k|\Theta_k^i))^2}{2(\sigma_k(D_k))^2}\right], \qquad (12)$$

where $\sigma_k(D_k) = 0.03 + (0.1 \times D_k)$ is the standard deviation of the error as a function of the current observation. The values used in the function were chosen during experimentation.

The output of the MOX gas sensor is a voltage reading v_k^{tot} which will vary due to a change in resistance of the sensor, which is caused by contact with atmospheric contaminants [21]. The sensor can be calibrated to relate the reading to a value with a physical meaning to the sensed material, such as parts per million (ppm). In this work, the sensor is not calibrated, instead, by assuming the concentration is directly proportional to the sensor reading we can estimate a scaled release a_s mass in the dispersion model. The sensor is not specific to a material and there exists a positive reading in clean air which is modelled as the background reading v_k^b. The observational data used in the likelihood function is then given as:

$$D_k = v_k^{tot} - v_k^b - v_k^e \qquad (13)$$

where v_k^e represents other errors in the reading that can arise from several sources such as internal noise or atmospheric changes.

The standard Pasquill Gaussian model is used to obtain the expected observations $V(p_k|\Theta_k^i)$, assuming a constant source release strength and meteorological conditions [22]. When the y-axis coincides with the direction of the wind, the expected observation of the sensor at position $p_k = (x_k, y_k)$ from a source with parameters Θ_k^i is given as:

$$V(p_k|\Theta_k^i) = \frac{a_s}{4\pi D||p_k - p_s||} \exp\left[\frac{-||p_k - p_s||}{\lambda}\right] \exp\left[\frac{-(y_k - y_s)U_0}{2D}\right]. \qquad (14)$$

where D is the diffusivity of the material in air, τ is its lifetime in seconds and:

$$\lambda = \sqrt{\frac{D\tau}{1 + \frac{u_s^2\tau}{4D}}}. \qquad (15)$$

An example plot of the dispersion from a source described by (14) is shown in Fig. 16.

Under the framework of the particle filter [20], the posterior distribution is approximated by N weighted random samples $\{\Theta_k^{(i)}, w_k^{(i)}\}_{i=1}^N$, where $\Theta_k^{(i)} = \left[x_{s,k}^{(i)} \ y_{s,k}^{(i)} \ a_{0,k}^{(i)} \ u_{s,k}^{(i)} \ \phi_{s,k}^{(i)}\right]^T$ is a point estimate of the source parameters and $w_k^{(i)}$ is the corresponding normalised

Fig. 16 An illustrative run of the dispersion model from (14) with parameters: $x_s = -0.5, y_s = 1.8, a_s = 0.1, u_s = 1, \phi_s = 160°, d = 0.1$ and $\tau = 2$. The shading represents the sensor reading expected from the sensor at the corresponding location

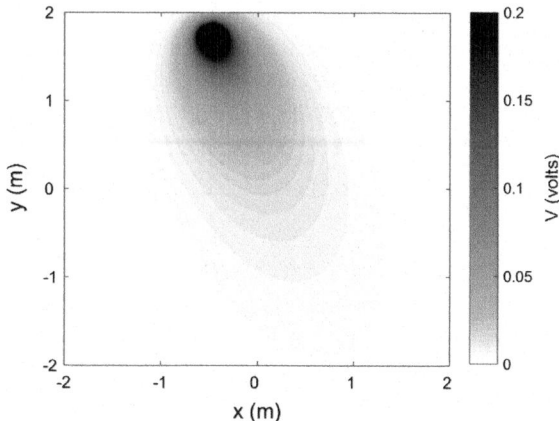

weighting such that $\sum_{i=1}^{N} w_k^{(i)} = 1$. Given the weighted samples, the posterior distribution is approximated as:

$$p(\Theta_k | \boldsymbol{D}_{1:k}) \approx \sum_{i=1}^{N} w_k^{(i)} \delta(\Theta_k - \Theta_k^{(i)}). \tag{16}$$

Assuming a static state vector and ignoring process noise, the weights are updated according to [20]:

$$\bar{w}_{k+1}^{(i)} = w_k^{(i)} \cdot p(D_{k+1} | \Theta_{k+1}^{(i)}). \tag{17}$$

At each time step the number of effective random samples are estimated as $N_{eff} = \sum_{i=1}^{N} 1/(w_k^{(i)})^2$. To avoid sample degeneracy, the samples are re-sampled when this value falls below a threshold η. To increase diversity among the samples, the new samples are regularised using a Gaussian kernel followed by a Markov chain Monte Carlo (MCMC) acceptance step [23].

At the end of each iteration summary statistics can be produced such as the mean and modal estimates of the source location and measures of variance. Marginal posterior densities can also be plotted for each of the individual parameters which are shown in the results in Sect. 5.3.2.

5.2.2 Implementation

A Turtlebot, as described in Sect. 3.3.2, was adapted for the gas sensing experiments as is shown in Fig. 17a. An MQ135 MOX gas sensor was used to sense the smoke, which is shown in Fig. 17b. This sensor was chosen for the experiments due to its reported sensitivity towards smoke, however, early experimentation found it not to be very sensitive. In order to improve the response of the sensor during the experiments,

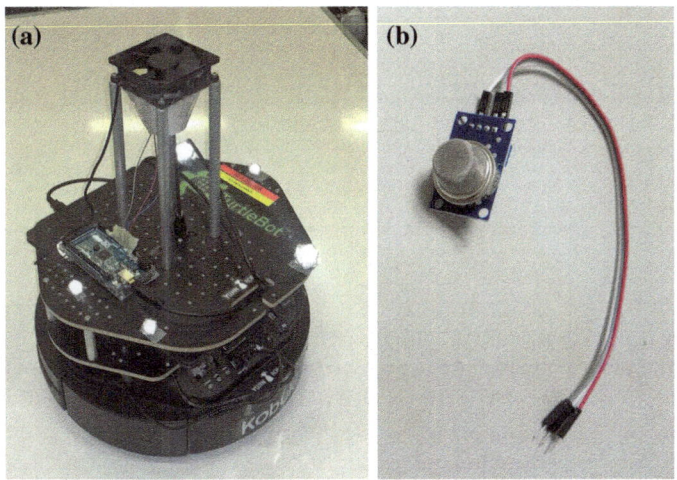

Fig. 17 **a** Modified Turtlebot robot for gas sensing experiments. **b** MQ135 metal oxide gas sensor

Fig. 18 Flow diagram for the source localisation experiment. Green arrows indicate ROS messages, red arrows are raw data and blue are commands

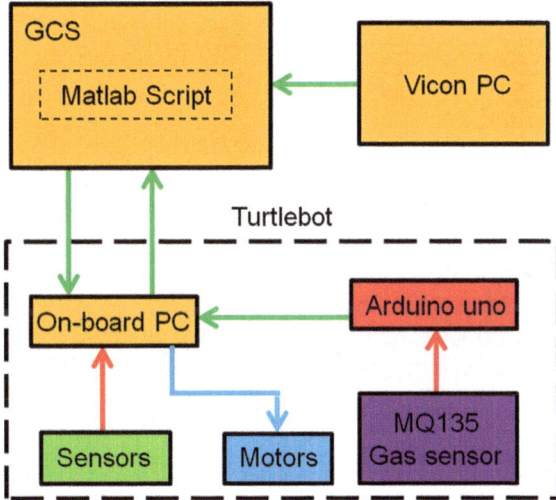

a cone and a CPU cooling fan were added to suck air into the sensing instrument as illustrated on-board the Turtlebot in Fig. 17a.

A flow diagram of the system is shown in Fig. 18. The MOX gas sensor is wired to an Arduino uno micro controller which supplies power and reads the output voltage from the sensor at an analogue input pin. A ROS message is created using the rosserial_arduino package and the ROS libraries for arduino. A simple float message is created to store the sensor data, which is sent over a serial connection to the PC on-board the Turtlebot. The Turtlebot reads the pre-formatted message and publishes it to the ROS network using the rosserial_arduino package. The launch file

Listing 4 rosserial_arduino launch file

```
1  <launch>
2    <node pkg="rosserial_python" type="serial_node.py" name="serial_node">
3      <param name="port" value="/dev/ttyACM0"/>
4      <param name="baud" value="57600"/>
5    </node>
6  </launch>
```

for the package is quite simple, and is shown in Listing 4 for clarity. The position of the Turtlebot was obtained from a Vicon motion capture system and published to the ROS network using the Vicon_bridge package as described in Sect. 3.3.3. The sensory and positional data are subscribed to by the ground control station (GCS), where a Matlab script runs the source estimation algorithm. After the robot has published the new sensor message at its current sampling position, a separate Matlab script is used to navigate the Turtlebot to the next location in the uniform sweep. A pure pursuit controller from Matlab's robotics toolbox is used to carry out the navigation to the new position. The outputs of the controller are velocity commands that are published by the GCS. The Turtlebot ROS package is used to subscribe to the navigation commands and implements them accordingly.

5.3 Experimental Study

The source estimation algorithm was tested in experimental conditions within the Loughborough University centre for autonomous systems (LUCAS) lab. Smoke produced from burning an incense stick was used to simulate a dispersing hazardous release. Ventilation in the room produced a slight draft which caused the smoke to follow a plume shape somewhat similar to that described by (14). Three runs of the algorithm are performed, the results of the source parameter estimates at the end of each run are summarised and discussed.

5.3.1 Set-Up

The experimental runs were performed in an indoor area where a VICON motion capture system was used to provide position coordinates of the Turtlebot. The set-up of the experiments is shown in Fig. 19a. The recognisable windows logo shown in the figure was produced by a down-facing projector which was used for data visualisation during the experiments. The smoke source was a burning incense stick as shown in Fig. 19b. The starting position of the Turtlebot, and the location of the burning incense sticks during all the experimental runs are indicated by the green and red circles in Fig. 19a.

(a)   (b)

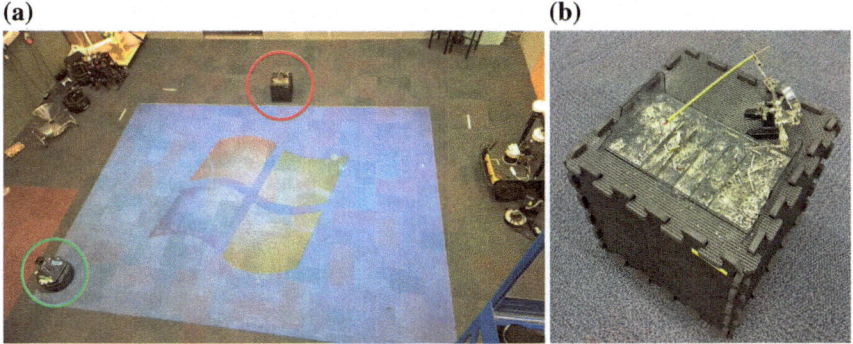

Fig. 19 **a** Experimental set-up during the 3 test runs. The starting position of the robot and the location of the smoke source are indicated by the green and red circles. **b** A burning incense stick used to simulate a hazardous source

At the end of each run, a suitable amount of time was waited for the smoke to clear before starting the next. The indoor area was fairly large and the experiments would take roughly five minutes. This limited the effect of factors such as wall reflectance on the dispersion of the smoke, which is not handled in the model used for estimation.

5.3.2 Experimental Runs

The experiment was performed three times using the described set-up. During each run the robot would navigate to the new position, stop for 5 s to take an averaged reading from the MOX gas sensor, and then publish the reading to the ROS network using the rosserial_arduino package. This data was fed to a ground station which ran the Bayesian estimation of the source parameters and published velocity commands to the Turtlebot. The robot stopped to take measurements every 50 cm during the first run and every 40 cm during the second and third.

The first run is summarised at various time steps in Fig. 20. Figure 20a–d show the random sample approximation of the posterior estimates of the source parameters at the current time steps, represented by the cyan dots. The true position of the source from the set-up in Fig. 19a is indicated by a red cross. The robot starts the search at $(-2, -1.3)$ and follows the path represented by the blue line, where the dots indicate locations where the robot stopped to take a measurement. The current position of the Turtlebot is given by the large blue dot. In Fig. 20e the sensor measurements are shown throughout the search which were used as the inputs to the source estimation algorithm. In Fig. 20a the random samples are yet to be re-sampled, so they still represent the uniform prior distribution which was given for the location estimate of the source. By Fig. 20b, several zero or very small measurements had been recorded, resulting in lower weights around the visited locations. When the algorithm re-sampled, the new samples were moved towards the higher weighted particles. By the fourth sweep, larger measurements of smoke were recorded, leading to the random

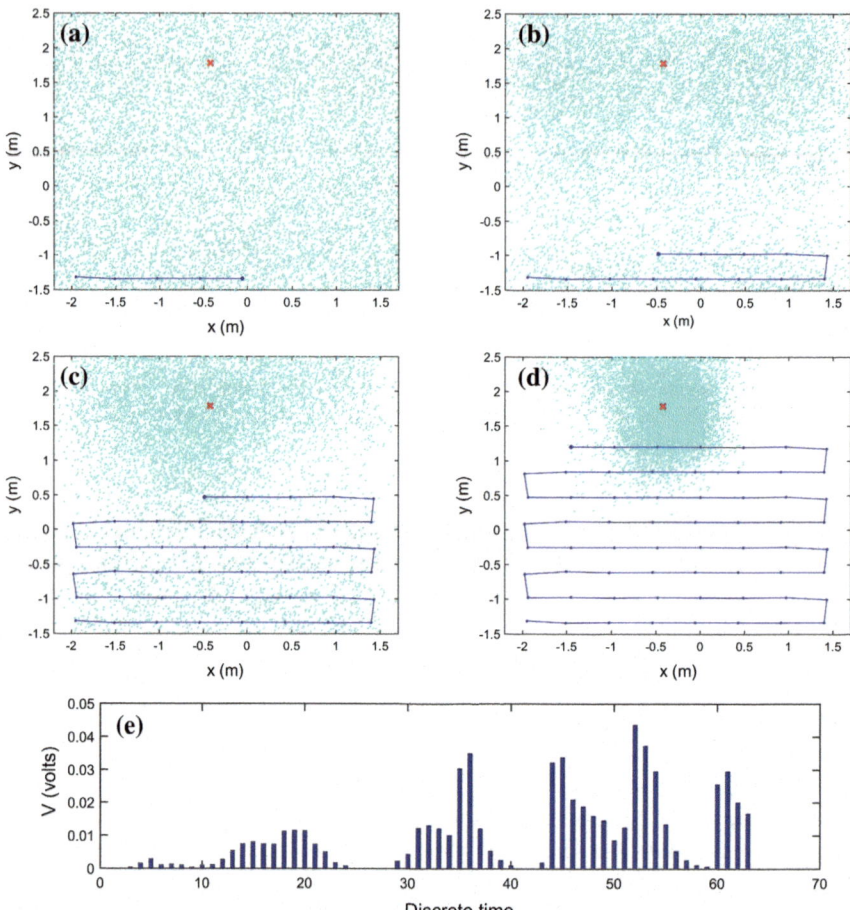

Fig. 20 Experimental run 1 at discrete time-steps: **a** $k = 1$, **b** $k = 12$, **c** $k = 45$ and **d** $k = 63$. A single burning incense stick releasing smoke, located at $(-0.5, 1.8)$, indicated by the red cross. The robot starts at $(-2, -1.3)$ and performs a sweeping search path represented by the blue line. Blue dots represent positions where the robot stops to take a measurement, the values of which are shown in **e**. The cyan dots represent the random samples used in the particle filter to estimate the source parameters

samples to begin to converge onto the source position in Fig. 20c. At the end of the search, an accurate estimate of the source position was produced. Histograms of the final estimate of the source position are shown in Fig. 21, where the true value is given by the red line. A photo of the robot at the end of the search is shown in Fig. 22, where the path followed and posterior estimates of the source are projected onto the ground at real scale. This provides some scale to the accuracy of the source position estimate.

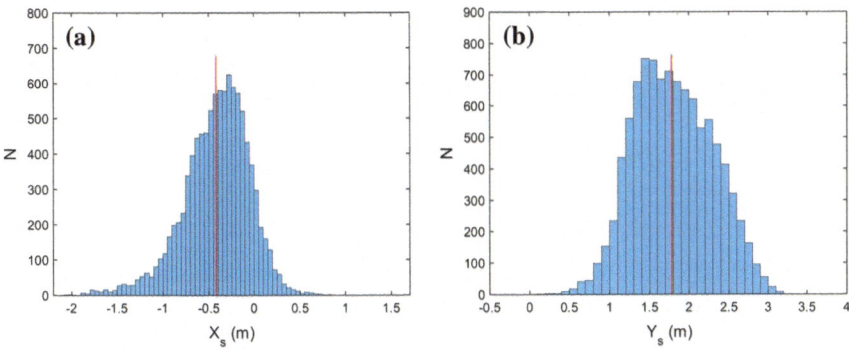

Fig. 21 Posterior histograms for the first run at the end of the sweep search for the location estimates of the source in the **a** x and **b** y coordinates

Fig. 22 Photo at the end of the first experimental run

The second and third runs are summarised in Figs. 23 and 24. The blue dots in these figures are sized proportionally to the value of the concentration measured by the MOX gas sensor. Similarly to the first run, the histograms show the final estimates of the algorithm, where it can be seen the true values are not far from the mode or mean. Run 3 had a larger spread in the y direction than the others, this is explained by the comparatively large observations from the sensor during the first and third sweeps performed by the Turtlebot, as illustrated by the size of the blue dots in Fig. 24.

The mean estimates and euclidean position errors at the end of each of the experimental runs are given in Table 1. All the estimates were accurate to within half a meter. The x_s estimate was typically more accurate than y_s. The x_s coordinate axis is more crosswind than the y_s coordinate axis, crosswind estimates of the source position are typically more accurate in source estimation work as there is less correlation among uncertain variables. For example, the upwind position estimate of the source and the wind speed are highly related, where a source with a large release rate in a strong wind can look similar to a small release rate in weak wind conditions. Overall,

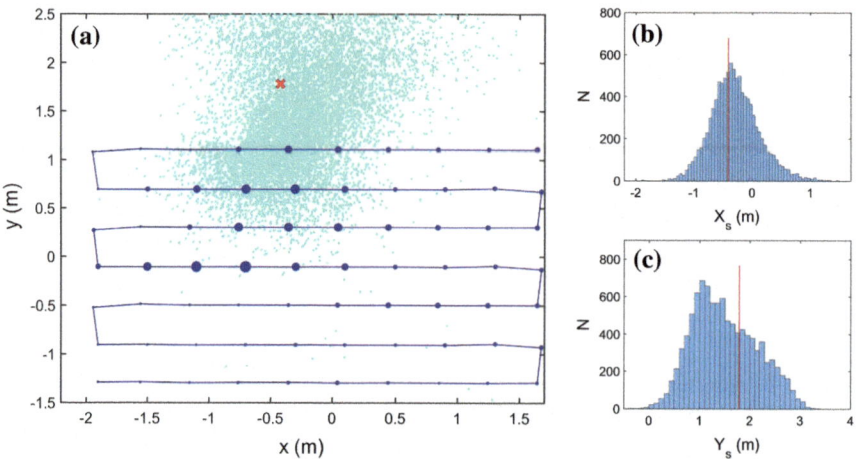

Fig. 23 **a** Illustration of the second run. The smoke source is indicated by the red cross and blue dots represent sensor readings where the size of the dot is indicative of the value. The cyan dots represent the random samples used in the particle filter. **b–c** Posterior histograms for the location estimates of the source

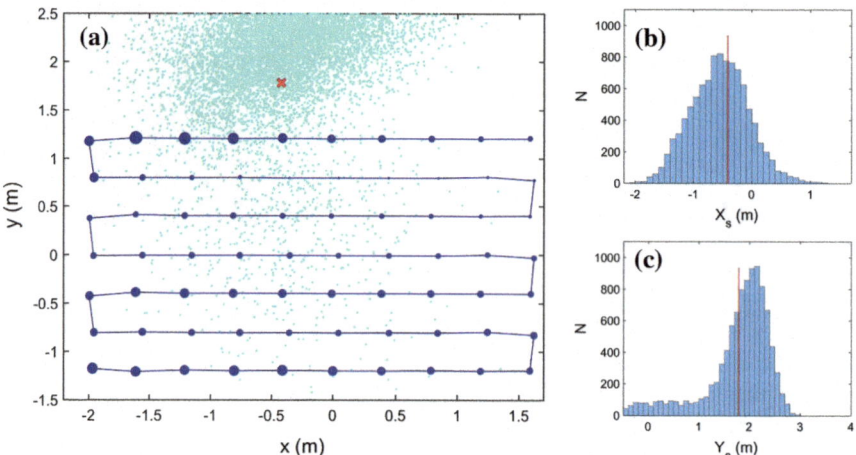

Fig. 24 **a** Illustration of the third run. The smoke source is indicated by the red cross. Blue dots represent sensor measurements where the size of the dot is indicative of the value. The cyan dots represent the random samples used in the particle filter. **b–c** Posterior histograms for the location estimates of the source

Table 1 Results of the mean source parameter estimates after the three experimental runs of the sweep search algorithm

	Truth	Run 1	Run 2	Run 3
x_s (m)	−0.5	−0.37	−0.34	−0.52
y_s (m)	1.8	1.73	1.36	1.61
Euclidean position error (m)		0.08	0.43	0.19

the position estimates of the source were satisfactory for all the runs and the true parameters were generally within a single standard deviation of the final posterior distributions.

5.4 Summary

A source estimation algorithm has been tested in experimental conditions. Smoke released from a burning incense stick was used to simulate a hazardous release. A low cost metal oxide gas sensor mounted on a Turtlebot robot was used to take measurements of the smoke concentration. Smoke measurements, position coordinates and velocity commands were shared between the robot and a ground control station using a ROS network. This enabled a fast set-up, and on-line data visualisation throughout the experimental runs. The Turtlebot was set as the 'Master' during the experiments making it easy to command from different ground stations or instances of Matlab by adding a new ROS node. The results of the experiments demonstrate successful use of the system, and how the location of a dispersing source can be estimated using a low cost gas sensor.

6 Conclusions

A ROS based system for autonomous vehicle algorithm development was detailed in this chapter. It was shown that by combining MATLAB/Simulink with ROS, the simplicity of programming in the familiar MATLAB/Simulink environment is easily combined with the flexibility and adaptability of ROS.

The system is highly flexible, being able to run on both ground and air vehicles, as well as x86 or ARM based systems with similar configuration and interfaces. The system flexibility is demonstrated by the ease of adding extra sensors and components, such as VICON data.

Two test cases serve as great examples of how complex algorithms can be quickly and simply implemented on vehicles for testing. The fixed wing DOBC disturbance rejection case study demonstrated that the system can be used to develop low level control algorithms through the entire development cycle rapidly and safely. Furthermore, the auto code generation feature of Simulink was demonstrated by compiling a ROS node to run on the embedded system itself. The ground robot source detection case study is a good demonstration of indoor testing, and the ease of integrating the gas sensor and ease of providing positional information from VICON to the system by using freely available ROS packages and libraries. The overall system has greatly enhanced the lab capabilities and efficiency, affording us huge flexibility in what and how we test.

Acknowledgements The authors would like to thank Prof Wen-Hua Chen and Dr Cunjia Liu of Loughborough University for their valuable discussions during this work.

References

1. G. Cai, B.M. Chen, T.H. Lee, M. Dong, Design and implementation of a hardware-in-the-loop simulation system for small-scale UAV helicopters. Mechatronics **19**(7), 1057–1066 (2009)
2. D. Shim, H. Chung, H.J. Kim, S. Sastry, Autonomous exploration in unknown urban environments for unmanned aerial vehicles, in *Proceedings of the AIAA GN&C Conference*, 2005
3. J. How, E. King, Y. Kuwata, Flight demonstrations of cooperative control for UAV teams, in *AIAA 3rd Unmanned Unlimited Technical Conference, Workshop and Exhibit*, 2004
4. J.P. How, B. Bethke, A. Frank, D. Dale, J. Vian, Real-time indoor autonomous vehicle test environment. IEEE Control Syst. **28**(2), 51–64 (2008). April
5. N. Bezzo, B. Griffin, P. Cruz, J. Donahue, R. Fierro, J. Wood, A cooperative heterogeneous mobile wireless mechatronic system. IEEE/ASME Trans. Mechatron. **19**(1), 20–31 (2014). Feb
6. D. Kingston, R. Beard, A. Beard, T. McLain, M. Larsen, W. Ren, Autonomous vehicle technologies for small fixed wing UAV, in *AIAA Journal of Aerospace Computing, Information, and Communication* (2003), pp. 2003–6559
7. Y.C. Paw, G.J. Balas, Development and application of an integrated framework for small UAV flight control development. Mechatronics **21**, 789 – 802 (2011)
8. A.M. Mehta, K.S.J. Pister, Warpwing: a complete open source control platform for miniature robots, in *IEEE/RSJ International Conference on Intelligent Robots and Systems (IROS), 2010* (2010), pp. 5169 –5174
9. turtlebot. http://wiki.ros.org/turtlebot. Accessed 14 Dec 2017
10. Vicon_bridge. http://wiki.ros.org/vicon_bridge. Accessed 14 Dec 2017
11. vrpn_client_ros. http://wiki.ros.org/vrpn_client_ros. Accessed 14 Dec 2017
12. Mavros. http://wiki.ros.org/mavros. Accessed 14 Dec 2017
13. J. Smith, C. Liu, W.-H. Chen, Disturbance observer based control for gust alleviation of a small fixed-wing UAS, in *2016 International Conference on Unmanned Aircraft Systems (ICUAS)*. IEEE (2016), pp. 97–106
14. W.-H. Chen, D.J. Ballance, P.J. Gawthrop, J. O'Reilly, A nonlinear disturbance observer for robotic manipulators. IEEE Trans. Ind. Electron. **47**(4), 932–938 (2000)
15. J. Yang, A. Zolotas, W.-H. Chen, K. Michail, S. Li, Robust control of nonlinear MAGLEV suspension system with mismatched uncertainties via DOBC approach. ISA Trans. **50**(3), 389–396 (2011)
16. M. Hutchinson, O. Hyondong, W.-H. Chen, A review of source term estimation methods for atmospheric dispersion events using static or mobile sensors. Inf. Fus. **36**, 130–148 (2017)
17. H. Ishida, Y. Wada, H. Matsukura, Chemical sensing in robotic applications: a review. IEEE Sens. J. **12**(11), 3163–3173 (2012)

18. K.J. Allwine, M.J. Leach, L.W. Stockham, J.S. Shinn, R.P. Hosker, J.F. Bowers, J.C. Pace, J7.
    1 overview of joint urban 2003–an atmospheric dispersion study in oklahoma city (2004)
19. rosserial. http://wiki.ros.org/rosserial. Accessed 14 Dec 2017
20. M.S. Arulampalam, S. Maskell, N. Gordon, T. Clapp, A tutorial on particle filters for online
    nonlinear/non-gaussian bayesian tracking. IEEE Trans. Signal Process. **50**(2), 174–188 (2002)
21. C. Wang, L. Yin, L. Zhang, D. Xiang, R. Gao, Metal oxide gas sensors: sensitivity and influ-
    encing factors. Sensors **10**(3), 2088–2106 (2010)
22. M. Vergassola, E. Villermaux, B.I. Shraiman, 'Infotaxis' as a strategy for searching without
    gradients. Nature **445**(7126), 406 (2007)
23. B. Ristic, S. Arulampalam, N.J. Gordon, *Beyond the Kalman Filter: Particle Filters for Tracking
    Applications* (Artech House, Boston, 2004)

ROS-Based Approach for Unmanned Vehicles in Civil Applications

Abdulla Al-Kaff, Francisco Miguel Moreno and Ahmed Hussein

Abstract Unmanned vehicle is the term that describes any platform without a human operator on-board. These vehicles can be either tele-operated remotely through a control station, or autonomously driven using on-board sensors and controllers. With the advances in micro and nano electronics, the increase in computing efficiency, and the ability to work in dull, dirty and dangerous environments, modern unmanned vehicles aim at higher levels of autonomy. This is through development of accurate control systems and a high-level environment understanding, in order to perform complex tasks. The main part of autonomous vehicles is the navigation system, along with the supporting subsystems. The navigation system utilizes information from various sensors, in order to estimate the position and orientation of the vehicle, sense the surrounding environment and perform the correct maneuver to achieve its assigned task. Accordingly, this chapter presents a ROS-based architecture for two different unmanned vehicles to be used in civil applications, which are constrained by Size, Weight and Power (SWap). This architecture includes the algorithms for control, localization, perception, planning, communication and cooperation tasks. In addition, in order to validate the robustness of the presented vehicles, different experiments have been carried out in real world applications with two different types of Unmanned Aerial Vehicle (UAV). The experiments cover applications in various fields; for instance, search and rescue missions, environment exploration, transportation and inspection. The obtained results demonstrates the effectiveness of the proposed architecture and validates its functionality on actual platforms.

A. Al-Kaff (✉) · F. M. Moreno · A. Hussein
Intelligent Systems Lab (LSI) Research Group, Universidad Carlos III
de Madrid (UC3M), Av. de la Universidad 30, 28911 Madrid, Spain
e-mail: akaff@ing.uc3m.es

F. M. Moreno
e-mail: franmore@ing.uc3m.es

A. Hussein
e-mail: ahussein@ing.uc3m.es

© Springer International Publishing AG, part of Springer Nature 2019
A. Koubaa (ed.), *Robot Operating System (ROS)*, Studies in Computational
Intelligence 778, https://doi.org/10.1007/978-3-319-91590-6_5

1   Introduction

The field of unmanned vehicles was limited to military industries, this is due to the complexity and the costs of designing and constructing these vehicles. However, during the last ten years, the world of civil unmanned vehicles gained a great momentum driven by technological development, which is pushed by private institutes, universities and other entities with various goals and aims [1]. Through these technological advances, the aerial unmanned vehicles are used in several academic and civil applications; which are but not limited to, hurricane and storm data collection [2], agricultural inspection [3], wild life monitoring and analysis [4], delivery [5], search and rescue [6], 3D mapping [7] and infrastructure inspection [8].

Unmanned vehicles are successfully being used both in research and for commercial applications in recent years, moreover, a significant progress in the design of robust control software and hardware has been noticed. The current approaches for autonomous unmanned vehicles can be divided into two main groups; behavior-based and deliberative-based [9]. On the one hand, behavior-based approaches use a network of interacting high-level behaviors to perform a task; on the other hand, deliberative-based focus on developing specific actions for each task. A comprehensive review regarding the vehicles behavior is presented in [10].

There are architectures which use ground-station as one the system nodes. In [11], DragonFly UAV software architecture is presented. This architecture computes both high and low level planning strategies on the same hardware. Moreover, its inter-communication is managed through a ground-station computer. In [12], MITs UAV testbed is presented, where they use Cloudcap Piccolo technology for low-level control and sensing. And all high level planning is carried out on ground-station computers, along with the inter-communication.

Other architectures apply all computations on-board. In [13], Berkeley UAV software architecture is presented. It uses separate off-the-shelf hardware to compute low-level control commands via Cloudcap Piccolo. Moreover, it uses on-board computer for the high level planning and decisions. In [14], CAPECON UAV software architecture is presented. Similar to the Berkeley UAV platform in their use of Cloudcaps Piccolo technology and also the addition of having an on-board computer. Moreover, several researchers use the multi-agent systems paradigm in their behavior-based architecture, such as in [9, 15, 16].

However, since the capabilities of these unmanned vehicles are increasing, it is essential to unify software architecture to allow for rapid development of modular software. Moreover, in order to investigate the behavior of a heterogeneous multiple of unmanned vehicles, a unified software architecture is required [13].

Accordingly, the main objective of this chapter is to present a Robot Operating System (ROS) based software architecture, which is implemented on different unmanned vehicles to prove the adaptability of the architecture. This architecture must be generic to consider the different tasks and vehicle capabilities. This means, it must be able to handle different sensors and controller processes. Last but not least, it must be able to consider the concept of multiple vehicles for coordination and collaboration. Several experiments were carried out in different scenarios of civil applications, and the obtained results demonstrates the effectiveness of the proposed architecture.

The remainder of the chapter is organized as follows: Sect. 2 highlights the proposed approach with detailed architecture explanation and highlights for the organized ROS packages. The platform used in the experiments, and the obtained results are discussed in Sect. 4. Finally, the conclusion and future recommendations are summarized in Sect. 5.

2 Proposed Approach

A well designed, high quality architecture is key for robust, modular and reusable applications, hence the need to design a general software architecture for unmanned vehicles focused on the performance of civil applications. In this section, a generic software architecture is presented, and its different modules are described.

When using ROS for this kind of purpose, the design process becomes easier due to the intrinsic modularity inside the ROS ecosystem. The general architecture designed, which schema is presented in Fig. 1, is composed of nine different groups. Each group is formed by one or more modules, which are responsible for the performance of several specific tasks and which are connected to other modules from either the same group, or a different one. Although this structure is designed for a full multi-autonomous vehicle application, simpler applications may not require any module in several groups, for instance, an application with only one vehicle may not use any cooperation module, or a purely reactive application will not need planning algorithms.

One of the main purposes of this design is to strictly define the tasks to be performed in each group and the connections between groups. This way, the different modules implemented shall perform small, specific tasks, thus increasing the flexibility and modularity of the system. Furthermore, this approach makes the integration of this architecture in a ROS environment become a simple task, since the groups boundaries and the connection types are well defined beforehand.

The following subsections describe the tasks performed by each group, together with the inputs and outputs that enable the communication between groups.

2.1 Acquisition

The data acquisition is a step performed in every system equipped with on-board sensors. In this very first stage, the raw data is obtained from the sensors and processed to be used in other system modules. This preprocessing may be composed, for example, of data filtering algorithms to remove noise from the sensors, or a format

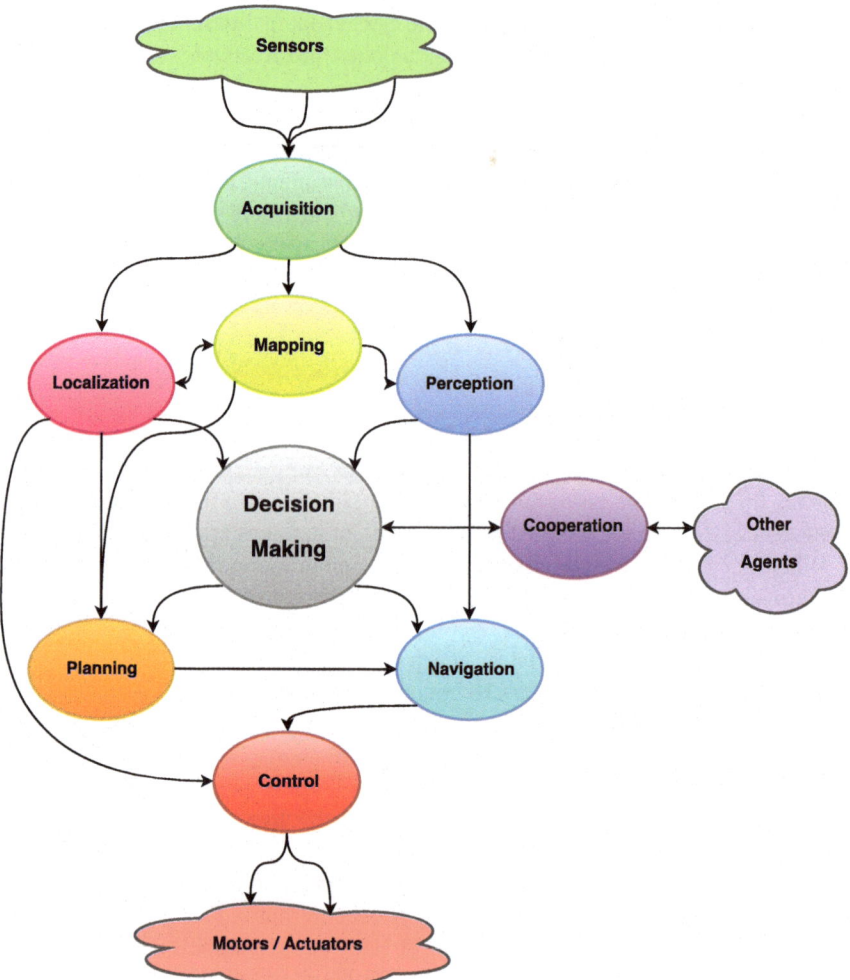

Fig. 1 General architecture for an autonomous vehicle

conversion to adapt and transform the information into a different and preferably standard type, thus it can be used by other modules.

This group is where the sensor drivers are included. The amount of processing in the acquisition modules will mostly depend on each sensor. For instance, some sensors such as monocular cameras will have simple drivers to only perform the data acquisition. However, other sensors such as ToF cameras (Kinect and similar) and LIDAR sensors, require the driver to also compute and convert the depth information into a point-cloud format such as *sensor_msgs/PointCloud2*.

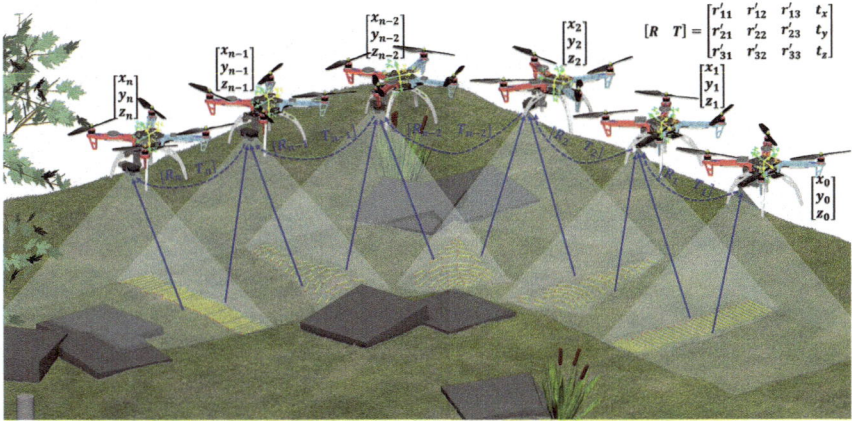

Fig. 2 6D Pose estimation strategy for aerial vehicles: the motion from point to point is based on the rotation and translation matrices, represented by R and T respectively

2.2 Localization

Localization is the process of estimating the local or the global position and the orientation of the UAV during the motion, based on the information generated by one or more sensors; such as IMU, GPS, vision, laser, ultrasonic, among others. The information can be generated by each sensor separately or by fusing the data from different sensors. Localization is considered as a fundamental phase for any navigation or mapping processes. The global localization methods require a map of the environment, in order to estimate the position of the vehicle within this map. Whilst, local localization methods, also known as odometries, compute the relative location to the vehicle initial position.

In this approach, as it is shown in Fig. 2, a monocular vision-based algorithm is presented; in order to estimate the pose of the UAV (position and orientation). By solving the different homographies *world-to-frame* and *frame-to-frame*, from the tracked feature points that are extracted by a downward looking camera [17].

2.3 Mapping

When performing autonomous navigation tasks, the vehicle needs to know which area around it is free and which area is occupied. This information is stored as a map, which can be local, including only information of the area around the vehicle, or global, including information of the whole operation area. The most common way to represent this information in ROS is with occupancy grid maps for two dimensional maps [18], and octree-based maps [19] for three dimensional maps.

The global maps can be obtained in two different ways: on the one hand, they can be computed online while the vehicle is moving. This ensures that the information from the environment is always up to date, however, this method needs more computational resources, which are not always available. On the other hand, in case that the area can be explored beforehand, these maps can be computed and stored just once to be used in later operations. The main drawback of this method is that, the map may need to be computed again if the environment changes.

Depending on the mapping algorithm used, the vehicle localization may be required as an input, or it can even be computed while mapping the environment (SLAM). The map of the vehicle surroundings is useful later on for planning and navigation tasks as well as for perception analysis in order to find near obstacles and other threats.

2.4 Perception

Perception is the process which is responsible of detecting and classifying the environment surrounding the vehicle. In the proposed approach, different submodules are presented; in order to perform several tasks to detect obstacles and patterns.

The obstacle detection algorithm [20] is based on estimating the size changes of the detected feature points, combined with the expansion ratios of the convex hull constructed around the detected feature points from the sequence of monocular frames; as it is shown in Fig. 3.

In addition, another perception approach is implemented to detect and track the helipad to perform autonomous landing, as it is shown in Fig. 4. In which the vehicle is able to detect the pre-designed pattern on the helipad, in terms of position, altitude and orientation. These data are transmitted to the planning node in order to estimate the necessary waypoints to land over the helipad.

2.5 Decision Making

The core modules inside an autonomous vehicle are those performing the decision making. Without them, only some kind of automation level can be achieved. These modules usually have a high number of inputs and outputs, since they need to monitor the status of the vehicle, location, target and threat detections, among others. Moreover, these modules must give the proper commands to planning, navigation and control modules in order to execute the current task. The tasks to perform may be stored and loaded from inside the vehicle, or entered as an external command from an external operator or even another vehicle or agent through the cooperation modules, as will be introduced in Sect. 2.9.

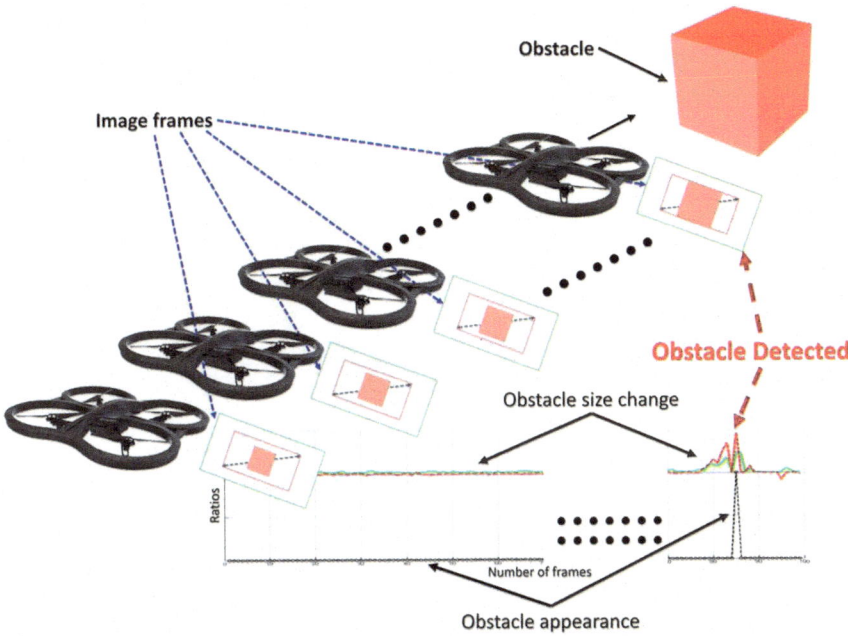

Fig. 3 The concept of approaching obstacle detection

Fig. 4 Helipad detection and tracking

2.6 Planning

The path planning algorithm enables the vehicle to find a collision free path between two given locations, taking into consideration the map of the operation environment. The proposed approach exploits many solutions to reach the best path in terms of the minimum distance, minimum time and lowest computational cost. Therefore, the planning solution is implemented as metaheuristic optimization-based method to solve the multiple objective function as a minimization problem. The used algorithm is well explained in [21].

For any given occupancy grid map, a generic optimization approach is applied to solve the path planning problem to generate a path for the vehicle to follow. This path is divided into several waypoints, which are separated with a dynamic distance based on the vehicle dimensions and look-ahead distance [22].

The planning of the vehicle can be executed in either local maps or the global maps. In order to do that, the inputs of planning modules are the maps, the current vehicle position and the destination point. The output generated will be the path that the vehicle should follow. Accordingly, this planning algorithm can be used both online and offline, in other words, online to always update the path of the vehicle based on the surroundings obstacles and obtain the required maneuvers, whilst the offline is for the pre-saved routes between points of interests in the environment.

2.7 Navigation

The main difference between planning and navigation modules is that the navigation algorithms work in a reactive way. Each navigation module implements a reactive behavior. For example, a *path_follower* module will take the path received from the planning algorithm and output the next waypoint to achieve. This output may be modified if an obstacle is in the path, so the navigation behavior overrides the planning output if needed. Another different module may produce points to approach an objective or to maintain a certain distance to it. All these different navigation behaviors are selected or mixed by the decision making modules, which will choose the best navigation behavior in order to perform the current task.

2.8 Control

In the previous subsections, different algorithms have been presented; in order to achieve several tasks that are required for autonomous vehicles. These algorithms covered the areas of the localization, mapping, sensing the surrounding, as well as navigation, planning and making decisions. This subsection aims to include the data obtained from the previous algorithms as an input to the high level control loop of the vehicle, in order to perform the correct maneuvers for the autonomous navigation purposes. Based on the processed information and the type of the task, a semi-automatic control is designed.

The flight controller ($Pixhawk$) is responsible for the low level control and has two control modes (angle and position), whose selection depends on the flight mode chosen, which sweep the controllability level from manual to full autonomous mode. The set of different flight modes are shown in Fig. 5. This section presents the semi-automatic object approaching module, which is implemented with the *Position Control* flight mode, that receives velocity feed-forwards as input.

The proposed semi-automatic control is based on the force feedback, which is obtained by directly modify the reference velocity in higher levels [23]. Based on the tunable parameters of the problem, a repulsive velocity field is created, and divided

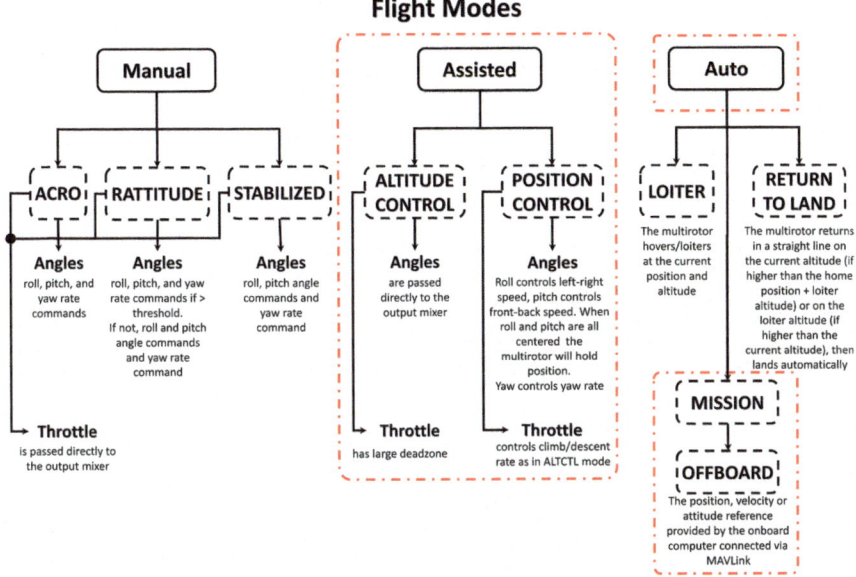

Fig. 5 Pixhawk autopilot flight modes

into distinct zones, that express different levels of risk in terms of collision, and depend on the relative position and velocity with respect to the object.

In order to obtain the distance to the detected objects, the 3D information received from the depth sensor ($Kinect V2$), taking the advantages of its ability to obtain data in outdoor environment. Therefore, the actual depth of each pixel in the image is known. So the relative position of the object from the UAV can be computed. This color-depth correlation is performed using the camera intrinsic and extrinsic parameters, as in [8]. Let h is the distance to the object, then four danger zones are defined as it is shown in Fig. 6:

- *Safe zone*: $[h > h_{max}]$ where no risk of impact thus no opposing velocity perceived.
- *Warning zone*: $[h_{max} > h > h_{mid}]$, low levels of opposing velocity.
- *Transition zone*: $[h_{mid} > h > h_{min}]$, which makes the pilot aware of imminent impact through a sudden increase in repulsion.
- *Collision zone*: $[h < h_{min}]$, within which the impact is unavoidable.

To achieve this, the shape of the repulsive curve is customized to achieve optimal deceleration and parametrized, so that its behavior can be easily tuned to the needs of the pilot. There are four parameters that determine the curve shape; the distance at which the UAV should stop $h_{desired}$, the distance at which the maximal velocity feedback is verified h_{min}, the maximal velocity feedback v_{max}, and the velocity feedback $v_{desired}$ at a distance $h_{desired}$. In addition, h_{max}, which defines the starting of the repulsive behavior, and the minimal velocity feedback v_{min}.

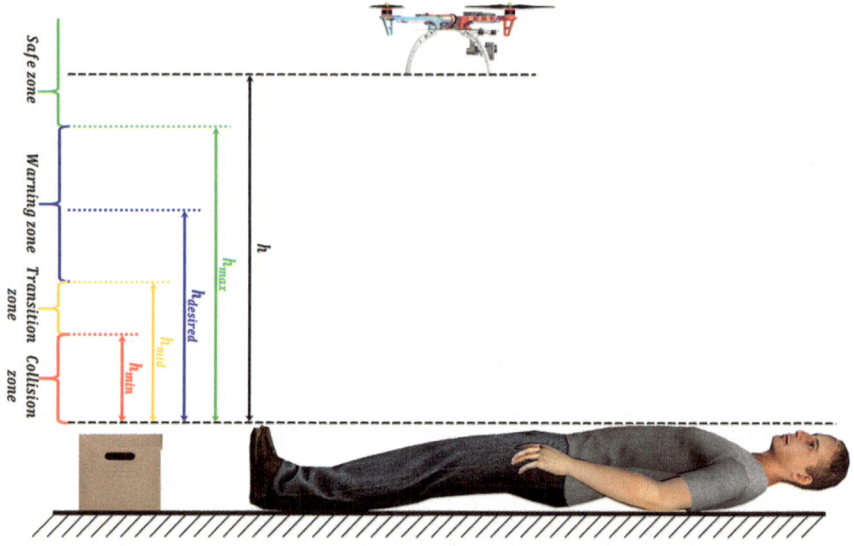

Fig. 6 Dynamic parametric field algorithm danger zones

The change of velocity is provided when the UAV approaches to an obstacle is defined for the curve shape as follows:

$$v_{Repulsion} = \frac{a}{h} + \frac{b}{h^2} \tag{1}$$

where, $v_{Repulsion}$ is the output velocity, the parameters a and b allow to weight the importance of the two components, and are set so that the curve passes through the points $(h_{desired}, v_i)$ and (h_{min}, v_{max}), where:

$$a = v_i \cdot h_{desired} - \frac{b}{h_{desired}}$$

$$b = \frac{v_{max} - v_i \cdot \frac{h_{desired}}{h_{min}}}{\frac{1}{h_{min}^2} - \frac{1}{h_{min} h_{desired}}} \tag{2}$$

The choice for the curve to pass by the $(h_{desired}, v_i)$ point allows the algorithm to make the UAV stop at $h_{desired}$. This strategy actuates a simple braking of the UAV until reaching the desired distance as it is shown in Fig. 7.

As a result, the effect on the RC command is the least invasive possible. A repulsion that overcomes the RC command is provided only when this threshold is passed. This is in order to correct small overshoots due to the time taken by the position controller to converge to zero velocity reference. Scaling the curve plot for all the possible approach velocities it is obtained the surface plot in Fig. 8.

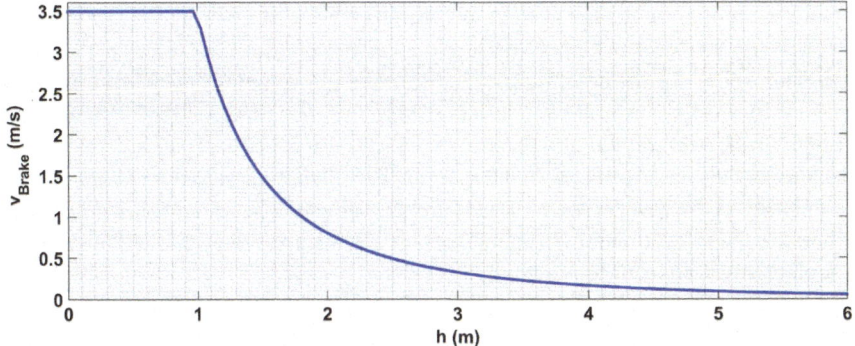

Fig. 7 Repulsion curve for $v_i = 1.5\,\text{m/s}$ and $h_{desired} = 1.5\,\text{m}$

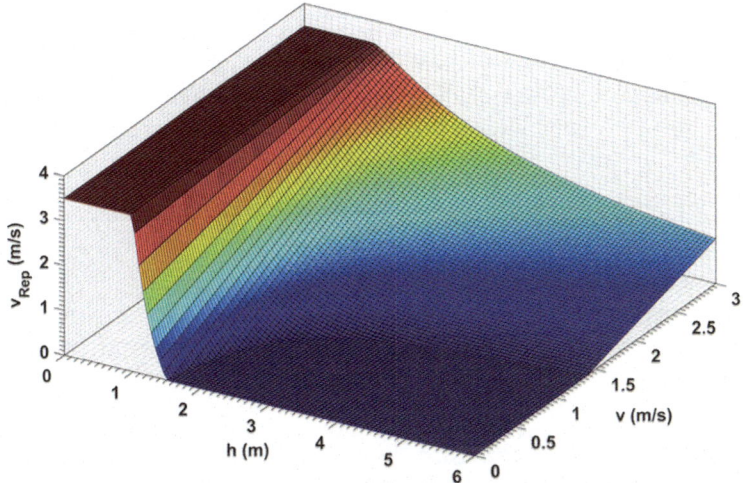

Fig. 8 Velocity braking output as a function of velocity and distance

2.9 Communication and Cooperation

Communication and cooperation among a team of unmanned vehicles is essential and crucial for multiple reasons. In case of long duration tasks, or vehicle failure, it is unlikely that a single vehicle is enough to complete the task. Moreover, many civil applications tasks require multiple vehicles, in order to achieve the task in a efficient, rapid and stable manner [24].

For the communication all vehicles must share a common network. Furthermore, a decentralized multi-master approach is preferred, so the group of vehicles can continue operating if any member fails.

For the cooperation, the system is formulated as Multi-Robot Task Allocation (MRTA) problem, which is represented in the architecture through the *mrta* node. This node take different inputs that describe the vehicle battery level through *mavros/battery* topic, overall vehicle status through *status* topic, and the vehicle pose through *localization/global_pose* topic. The allocation process depends on the vehicle status and all other vehicles in the system. The allocation output *mrta/allocation* is published to the *task_executor* node of each vehicle. The idea of this node is to decompose the allocation into executable tasks according to the vehicles capabilities through *task_executor/next_task* topic.

3   Architecture Integration

In this section, the selected platforms are described with their specifications. Then, a description of how the proposed architecture has been implemented and integrated into these platforms follows, including details and instructions in order to replicate a similar setup in any other vehicle (Fig. 9).

3.1   *Platforms*

Hexacopter Carbon fiber hexacopter of total weight 3.5 Kg, based on *Pixhawk 2.1* autopilot is used. This UAV is equipped with GNSS, magnetometer, IMU (accelerom-

Fig. 9 SkyOnyx, the hexacopter platform used for experiments

eters, gyroscopes) and barometer. In addition to a downward looking camera *Kinect V2*, which provides 1920 × 1080 RGB images, and 512 × 424 infrared pattern is used as the main perception sensor for obstacle detection and more. Furthermore, a front-looking SJCAM SJ4000 camera that provides 640 × 480 RGB images is mounted.

All the processing is performed on-board by an Intel NUC embeded computer, which has an Intel i7-7567U CPU at 3.5 GHz CPU and 8 GB RAM. The software has been developed and integrated with ROS Kinetic, under Ubuntu 16.04 LTS operating system.

Quadcopter 3D printed quadcopter (Fig. 10), which constructed with PLA plastic of a total weight 1.5Kg. The autopilot used with this quadcopter is the *Pixhawk*, with with GPS, magnetometer, IMU and barometer. For the perception purposes, SJCAM SJ4000 camera is used and mounted on Walkera G-2D gimbal, which provides 640 × 480 RGB images.

All the processing is performed on-board by ODROID-XU4 embedded computer, which has Samsung Exynos 5422 CortexTM-A15 at 2.0 GHz CPU, 2 GB LPDDR3 RAM, and eMMC 5.0 HS400 Flash Storage.

Fig. 10 3D printed quadcopter used for experiments

3.2   ROS Implementation of Proposed Architecture

In order to integrate the proposed software architecture into the ROS ecosystem, several new packages have been created, while also using other popular ROS packages. Thanks to the use of ROS standard messages and services, the packages implemented can be easily replaced, as long as their inputs and outputs have the same type. For the sake of an example, Fig. 11 shows a rearranged output of *rqt_graph*, including all nodes and topics presented in the Quadcopter platform. The figure follows the same color schema as Fig. 1, in order to help differentiate the different architecture groups. Below, the details of each part of this system is explained.

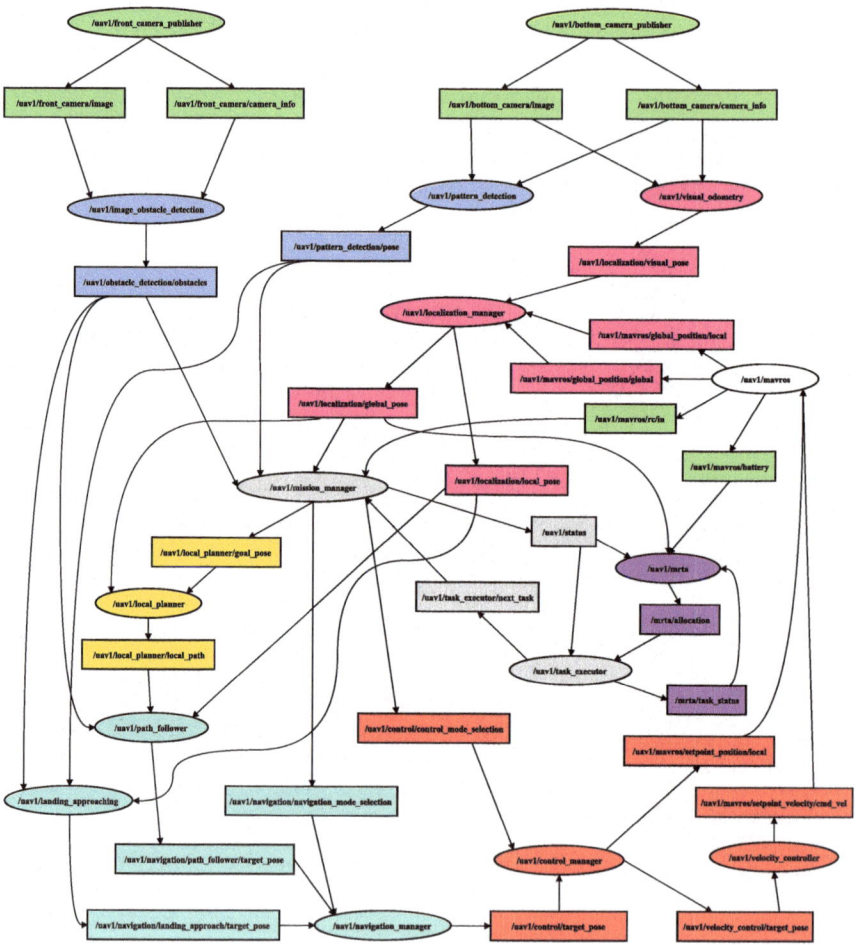

Fig. 11 Proposed ROS-based architecture graph

One of the main characteristics of the presented platforms is that both use the *Pixhawk* autopilot, which has different roles in the proposed architecture. Firstly, it acts as a data acquisition module, in order to get the information from IMU, GPS, barometer and magnetometer sensors. Secondly, the PX4 firmware includes different localization algorithms for data filtering and fusion. Finally, the flight controller also has different control modes, which were previously defined in Sect. 2.8. In order to communicate all this information with the outside, the PX4 firmware uses the *MAVLINK* protocol. For this reason, the *mavros* package is used as a bridge between the on-board computer and the flight controller. Thus, the *mavros* node will act as a node for data acquisition, localization and control, although those different methods are actually inside the *Pixhawk* flight controller.

Acquisition: For the acquisition group, the Quadcopter platform has three different nodes. On the one hand, the *mavros* node is used to get the information about the battery levels. On the other hand, the UAV has two monocular cameras, one pointing downwards and another looking forward. The driver nodes used to capture the images and publish them are custom *webcam_publisher* nodes, which use OpenCV libraries to capture the images and *image_transport* to publish them in ROS. These nodes can be replaced by any other USB camera driver already published in ROS, such as *usb_cam*, *cv_camera* or *libuvc_camera*. If the camera used is not USB, or uses a different protocol, the chosen driver must output the images and camera parameters following the conventions of *image_transport* package. Figure 12 is a portion of the full graph from Fig. 11, which shows the ROS nodes in the acquisition group.

Localization: The localization group is composed of three different nodes, as shown in Fig. 13. The first one is the *visual_odometry* node, which performs the visual odometry algorithm presented in Sect. 2.2. The second one is the *mavros* node, which outputs the filtered and fused localization coming from the *Pixhawk*, including data from the GPS, IMU, barometer and magnetometer sensors. These modules may generate different localization values. Therefore, an extra node, the *localization_manager* has been implemented to deal with multiple localization sources. This node implements a fusion algorithm to combine the different inputs. Also, the *localization_manager* node can act as a switcher, thus enabling the selection between one of the localization

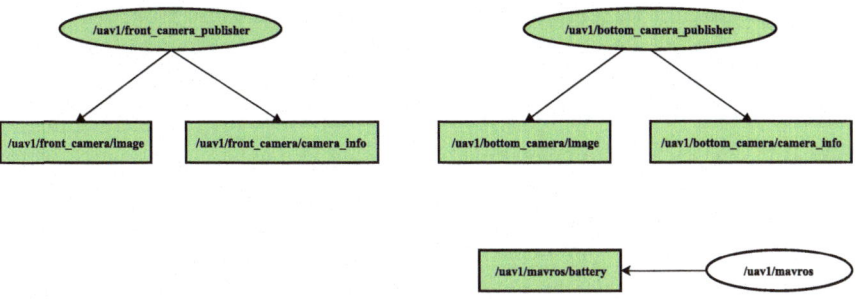

Fig. 12 Acquisition part from the ROS-based architecture

Fig. 13 Localization part from the ROS-based architecture

sources or the fusion output. This node is needed when multiple localization sources coexist. In cases where there is only one input, such as systems relying only in the localization provided by the flight controller, just the *mavros* node would suffice.

One important detail regarding the localization packages is that the output format must be always timestamped. This condition is needed later on for message synchronization in other nodes. Additionaly, if any fusion algorithm wants to be used, the corresponding covariances of the localization data must be included in the messages. For this reason, the *localization_manager* node only accepts inputs of type *nav_msgs/Odometry*. However, in order to integrate the localization coming from the *Pixhawk*, the type *geometry_msgs/PoseWithCovarianceStamped* needed to be accepted, since is the type used by *mavros* to publish the localization data. Regarding the *TF* library [25], it is recommended that the node in charge of publishing the final localization (in this case, the *localization_manager*) also broadcasts the *TF* transforms with the pose of the vehicle in the global coordinates frame. However, for the input sources this is not needed, since their output values will only be used by the manager node (Fig. 14).

Perception: Since the Quadcopter platform is equipped only with monocular cameras, all the perception nodes in this group implement vision-based algorithms. In the graph presented in Fig. 11 there are two nodes responsible of perception tasks.

The first one, *obstacle_detection*, performs one of the most important tasks in an autonomous vehicle: detecting the obstacle threats in the vehicle surroundings. In this case, this node uses only visual information; however, a different node using

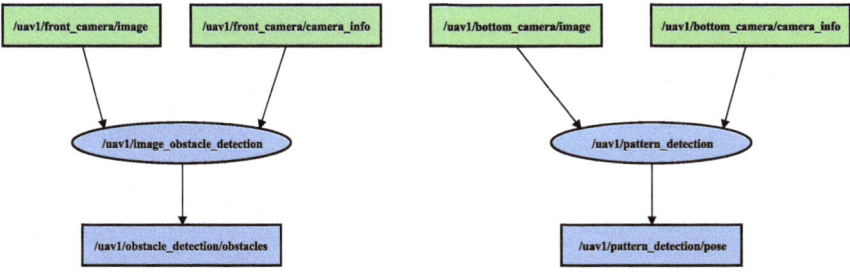

Fig. 14 Perception part from the ROS-based architecture

data from a LIDAR sensor may also be used to provide information about the near obstacles. Note that this node may not output a map of the environment (used for planning), but a list of the obstacles that may result in a threat to the vehicle (used for navigation and decision making). Nowadays, a standard way in ROS for representing this kind of information does not exist. Hence, these objects are modeled by a custom message which contains the obstacle's pose, velocity, size and its class, in case it was classified.

The *pattern_detection* is the second implemented node for perception purposes. This node is responsible for detecting an helipad for both, ground vehicle localization and auto-landing operations. This node simply implements a pattern matching algorithm for detection of the helipad, which is followed by a median-flow tracking algorithm to increase the robustness of the method. Note that this node is task specific, and is not a part of the base architecture, but a node implemented to perform a subtask required in the civil application to carry out. Other nodes of this type (which are not included in this example), can be responsible of detecting other specific objects, such as humans, for rescue operations or boxes, for transportation purposes. The output information of these nodes is sent to the decision making modules, in order to perform the current mission (Fig. 15).

Decision Making: The nodes responsible for the decision making in the vehicle are completely specific to each application. This is because each mission requires a different behavior and actions from the vehicle. In the example system, the decision making group is composed of two different nodes: *mission_manager* and *task_executor*.

The *task_executor* node is the one responsible for selecting the next task to perform (move to a location, find an object, land, etc). The list of tasks to perform can be inputted to this node in different ways, depending on the application configuration. For instance, the Quadcopter platform is designed to cooperate with other vehicles, and a task allocation algorithm is running to send the next task to each vehicle. If only one vehicle is operating, the tasks to perform in the mission can be specified in a text file, or the user can indicate the next task in a guided manner.

The *mission_manager* node is the core node of the vehicle. It has access to all information relevant for the mission, such as localization, obstacle threats, target detections, etc. This node receives the next task to be executed from the *task_executor*

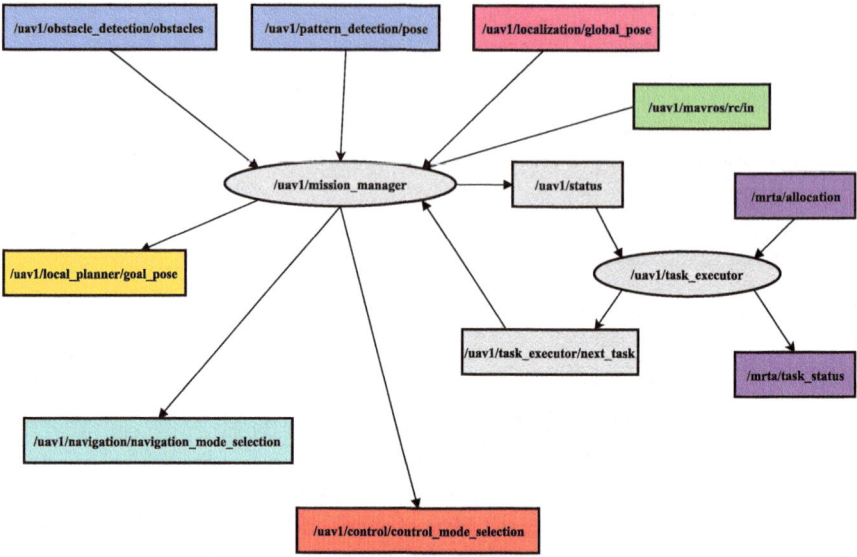

Fig. 15 Decision making part from the ROS-based architecture

and gives the proper commands to the planning, navigation and control modules to perform it. Moreover, the feedback of the current task is always sent back to the *task_executor* node, which will update the task status to the user or the task allocation node.

One very important factor about the *mission_manager* node, is that it receives a direct input from the user in order to enable and disable the autonomous mode. In the example case, this input comes from the RC controller connected to the *Pixhawk*, which is available in the *mavros/rc/in* topic provided by mavros. In the Quadcopter and Skyonyx platforms, the RC controller has a switch configured to enable and disable the *mission_manager*. This configuration is critical and must be checked before each operation, to ensure that manual control can be taken back at any moment.

In order to link *mission_manager* and *task_executor* nodes, *action_lib* is used. This way, each task is implemented as an *action*, and the whole set of tasks that *mission_manager* can perform is known by *task_executor* beforehand. This mechanism allows to easily sent the task status feedback to the *task_executor* node (Fig. 16).

Planning: The path planning group in the proposed example is simple, because in the Quadcopter platform, there is no mapping at all. This is due to sensor limitation and the simplicity of the tasks to perform with this platform. In this case, the only planning node present in this platform is the *local_planner* node, which provides paths for search and exploration without taking any map information into account. Note that with this architecture, a local planner with such simplicity can exist, because the obstacle threats are still considered in the navigation layer before sending the final

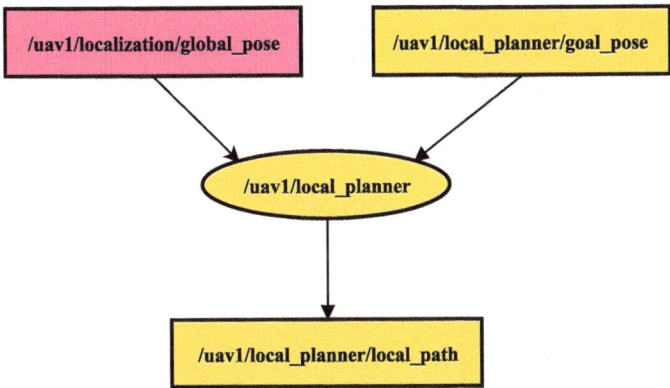

Fig. 16 Planning part from the ROS-based architecture

target pose to the control. This does not justify the absence of mapping, but still makes the operation possible without it.

Navigation: In the navigation group, several nodes coexist to perform different navigation behaviors. For example, in the Quadcopter platform, the *path_follower* node takes the path received from the planning algorithm and outputs the next waypoint to achieve. This output may be modified if an obstacle is in the path, so the navigation behavior overrides the planning output if needed. Another navigation node is the *landing_approaching* node, which performs a reactive approach to the landing platform. All these outputs are directed to the *navigation_manager* node, which works like a multiplexer, as Fig. 17 represents.

The *navigation_manager* node has access to all available navigation modules and switches from one to another in order to send the corresponding target pose to the control modules. This selection is given by the *mission_manager*, which decides which navigation behavior should be used at each moment for the given task (Fig. 18).

Control: The low-level control of the vehicle is performed by the *Pixhawk* flight controller. However, *mavros* can be used to send *setpoint* commands to the *Pixhawk*. These *setpoint* commands can be used to set position, velocity, and attitude targets, which will be achieved by the flight controller. Given this, the control implemented in the architecture is focused on high-level control, and sends the corresponding *setpoint* commands to *mavros*. In the Quadcoper platform there are two control nodes. The first one is the *velocity_control* node, which receives a target position and computes the corresponding velocities for a safe approaching. This node implements the algorithm presented in Sect. 2.8. The second one is the *control_manager* node, which acts as a switcher, similarly to the *navigation_manager* node, as Fig. 19 represents. In this case, the *mission_manager* node selects the most suitable control mode for the current task among the available modes. For example, a simple go-to task would use the position control mode, and an object approaching or an auto landing maneuver will use the velocity controller (Fig. 20).

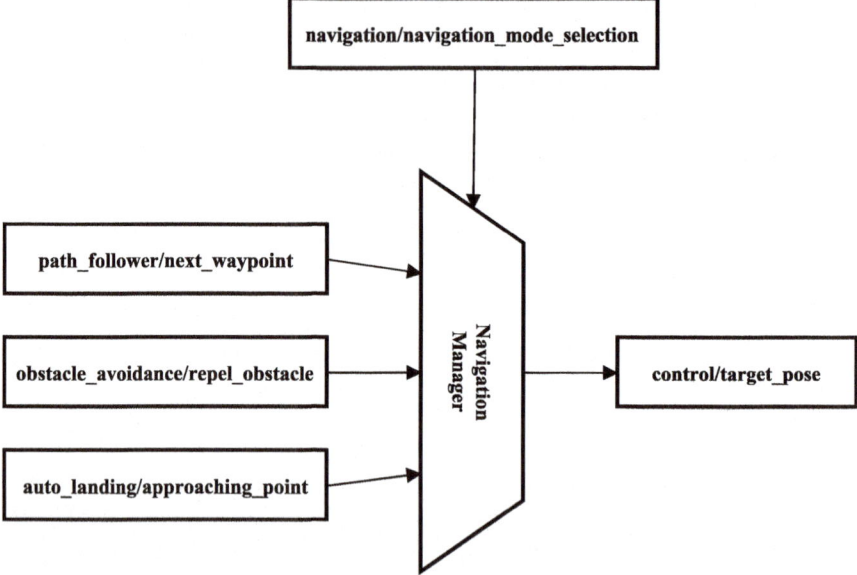

Fig. 17 Navigation manager

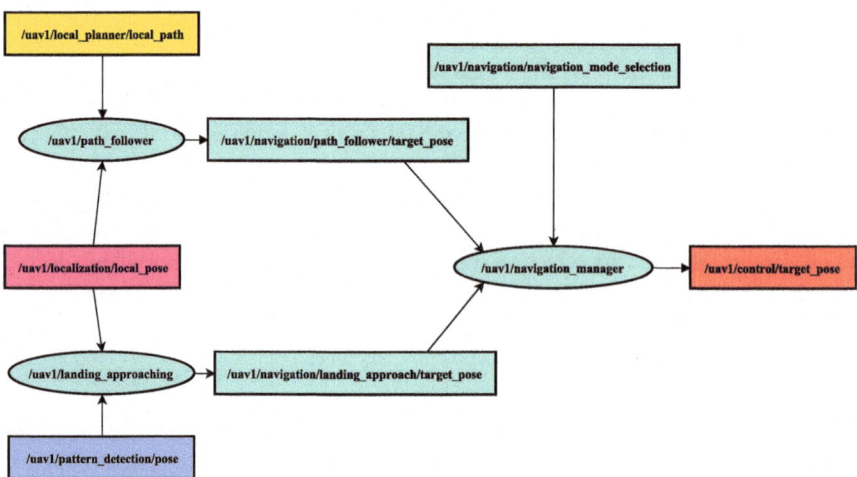

Fig. 18 Navigation part from the ROS-based architecture

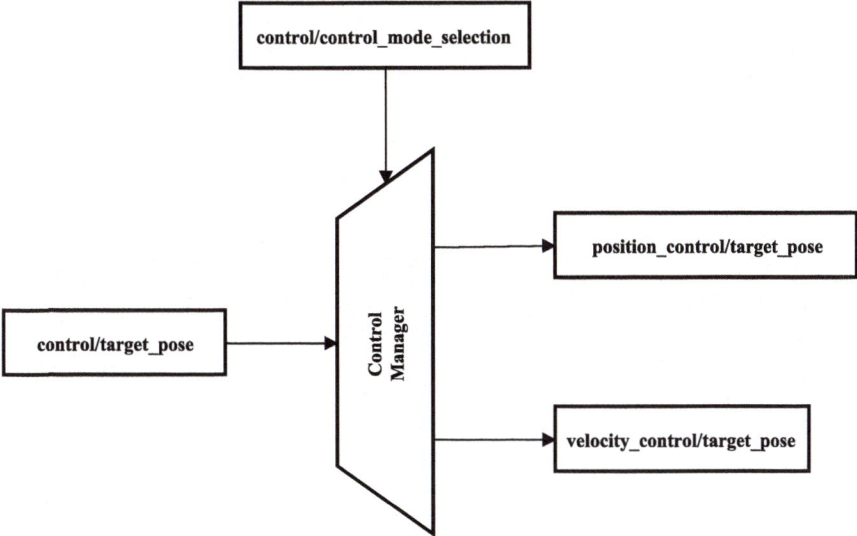

Fig. 19 Control manager

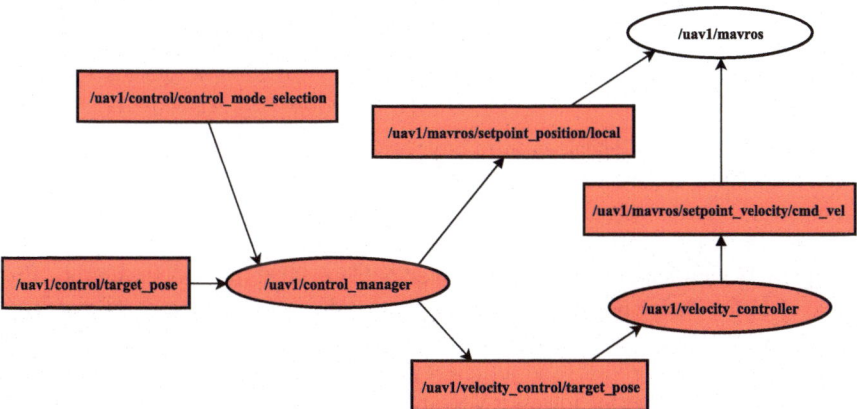

Fig. 20 Control part from the ROS-based architecture

Communication and Cooperation: For the communication, the *multimaster_fkie* package is used [26]. It combines the nodes required to establish and manage a multi-master network over ROS-based architectures. In order to use the aforementioned package, vehicles must operate under a common network. Thus, for secure and stable paradigm, a Virtual Private Network (VPN) is used to overlay the commonly available one, the detailed description for using this communication scheme is explained in [27].

For the cooperation, the system is formulated as Multi-Robot Task Alloca-tion (MRTA) problem, which is represented in the architecture through the *mrta* node. This node take different inputs that describe the vehicle battery level from *mavros*, overall vehicle status from *mission_manager*, the vehicle pose from *localization_manager* and the task status from the *task_executor* node. The allocation process depends on the vehicle status and all other vehicles in the system. The allocation output *mrta/allocation* is published to the *task_executor* node of each vehicle. The idea of this node is to decompose the allocation into executable tasks according to the vehicles capabilities, which are defined as actions in the *mission_manager* node, as described before.

4 Experimental Work

In this section, the results from the carried-out experiments in localization, perception and control are discussed.

4.1 Results

The results sub-section is divided into three categories, the localization results using the quadcopter, the perception results using the hexacopter, and finally the control results using the simulator for the proposed architecture.

Localization Results The localization algorithm was verified with data gathered from real flights in outdoor arenas, with at total distance of 3742.1 m, and various paths with different visual conditions such as ground texture, illumination, flight altitudes (between 1–4 m).

Figure 21 shows an example of A-shaped outdoor flight trajectory with total distance of 14.5 m. It can be seen from the figure that the visual odometry estimates the trajectory and obtains good data compared to the ground truth; this is because of the good illumination conditions and sufficient texture. Whilst, the trajectory obtained by the fusion approach has the lowest RMS error, due to the correction of the orientation by the IMU.

From the outdoor experiments, the maximum error generated by the vision system is 50 cm in (x,y)-axis and 30 cm in z-axis. Table 1 shows a comparison in terms of Mean drift, Maximum error, deviation and accuracy of the different flights. On the other hand, the estimated processing time is 64:2 ms. This is due to work with SIFT-FREAK combination.

Perception Results The use of UAVs for transportation applications is a topic that is gaining popularity recently. In this subsection, a perception approach is presented to detect boxes from the vehicle using the *Kinect V2* depth sensor and estimate their maximum weight in order to determine if they can be transported. The validation of the presented algorithm has been performed in a urban-like scenario. In which, 100 different tests have been carried out with a total number of 531.

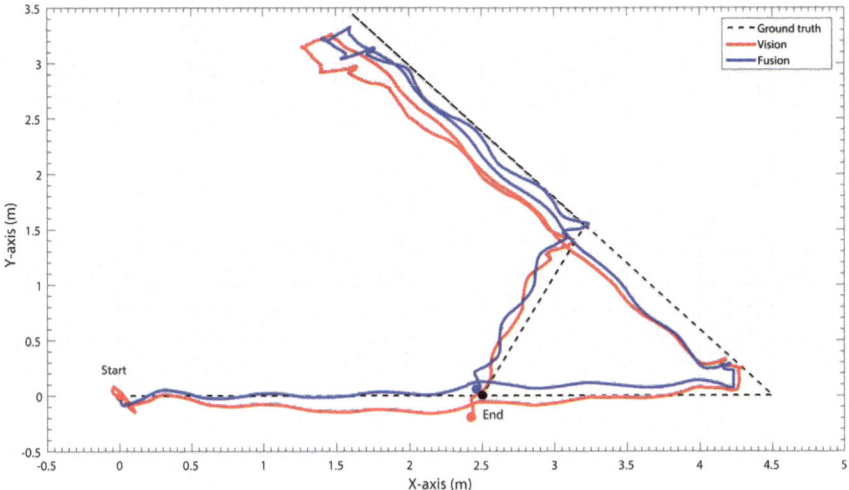

Fig. 21 A-shaped outdoor flight with a total distance of 14.5 m

Table 1 Accuracy of localization algorithm

Number of flights	30
Total distance (m)	3752.1
Mean (x,y) drift(m)	±0.38
Mean (z) drift(m)	±0.30
Mean position error (%)	0.53
Max position error (%)	1.38
Standard deviation	±0.19
Accuracy (%)	99.4

The proposed approach is divided in three steps: First, the pointcloud is used to detect the floor plane and extract the outlier points. Then, a clustering algorithm is performed over the outlier points to compute their sizes. Finally, those sizes are matched with standard delivery boxes sizes to filter wrong detection and to estimate the maximum weight of the package.

During the flight, the 3D information is processed to segment the floor plane and extract the outliers on top of it. This method relies on the whole floor area, so it is more robust against outliers and works with angled planes. For this segmentation method, RANSAC algorithm is applied to compute the coefficients of the mathematical model of the plane. The fitted model is defined as a plane perpendicular to the camera. This method allows to apply an empirical angle threshold $\epsilon = \pm 15°$ and a distance threshold $d = \pm 10$ cm to estimate the inliers, which increases the robustness of the floor detection, even with inclined planes. Figure 22 shows both, the color image of the floor with three boxes and the results from the RANSAC segmentation algorithm to segment the floor plane, which is colored in red.

(a) Color image   (b) Plane segmentation

Fig. 22 Box perception results

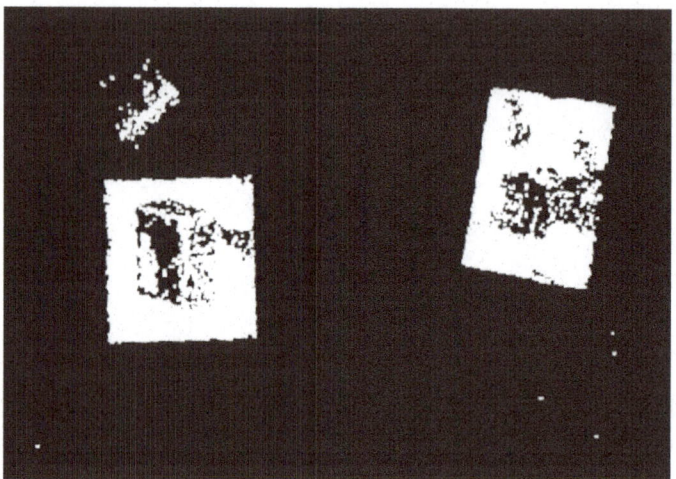

Fig. 23 3D object clustering

Once the floor points have been removed from the point-cloud, a clustering algorithm is performed over the rest of the outliers to detect the different objects in
the scene. The clustering method implemented is based on the euclidean distance
between points. Then, for each cluster, its 3D bounding box is computed, obtaining
the length, width and depth of the possible box. The output of this step is presented
in Fig. 23, where each detected object is colored in a different color.

The last step involved in this perception approach is to recognize the type of
box and estimate its maximum weight. For this, the different types of boxes must be
known beforehand. For example, *Amazon* has different types of boxes with fixed sizes
and labels [28]. With this information it is possible to filter wrong objects from the
clustering process. For each cluster, only those objects that match one of the possible

standard sizes are taken into account. Moreover, common carrier companies, such as *UPS* and *FedEx*, currently use the same maximum weight calculation, which is the cubic size of a package divided by 166 for domestic shipments [29]. Given this, the maximum weight of the detected package can be estimated with the Eq. 3, where *Weight* is the package weight in pounds, and $\{L, W, H\}$ are the length, width and height of the package in inches.

$$Weight \leq \frac{L \cdot W \cdot H}{166} \tag{3}$$

The weight estimation is useful to determine if the object can actually be carried out by the vehicle or if more vehicles would be needed to perform the transportation task.

The accuracy obtained in the box detection is shown in Table 2.

Control Results In this simulation experiment, as shown in Fig. 24, the proposed control algorithms has been compared with two known strategies; Dynamic Parametric Field (DPF) and Time to Impact (TTI).

Table 2 Box detection algorithm performance

Number of flights	Total boxes	Detected boxes	Accuracy (%)
100	531	515	96.98

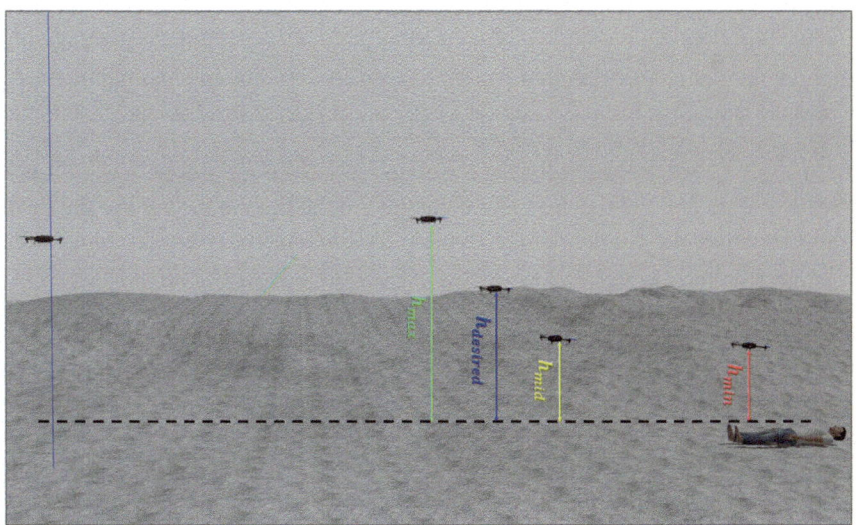

Fig. 24 Reference velocity with respect to the distance

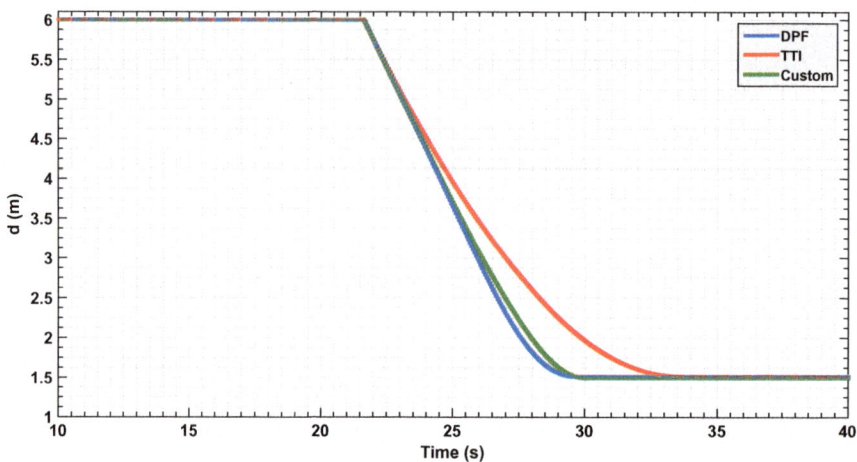

Fig. 25 Distance estimation

In the experiment, the input is given at a starting distance to the object greater than d_{max}, which allows the UAV to attain the requested velocity before starting the avoidance procedure.

The parameters of the three algorithms have been calibrated. So that, in the three flights in the simulation, the UAV would stop at the same distance to the obstacle chosen to be $d_{desired} = 1.5$ m. Figures 25, 26 and 27 illustrate the evaluation of the three algorithms during the first experiment.

From Figs. 25 and 26, it is illustrated that, for the chosen settings, DPF is the first to reach the desired distance, followed by the proposed solution, and finally by TTI that is the slowest to reach $d_{desired}$. The main reason lies in the distance of activation of the obstacle avoidance module.

In Fig. 27, it is shown that DPF starts the repulsion far closer to the obstacle than the other two approaches. Consequently, the deceleration is more uniformly distributed over the avoidance range for TTI followed by the proposed solution that, after an initial, progressively increasing deceleration, reaches a constant steepness zone in the $v(t)$ curve, that means constant deceleration until $d_{desired}$ is reached. The convexity of the curves in Fig. 26 depicts a negative jerk. As a result, the curve from the proposed algorithm has the least violent change in velocity at the beginning of the maneuver, oppositely to DPF that, due to the late activation, takes action with a strong jerk at its activation. In the three figures, the proposed parametrized curve shows a behavior that falls between the extremities defined by the other two approaches, thus representing a good compromise between them.

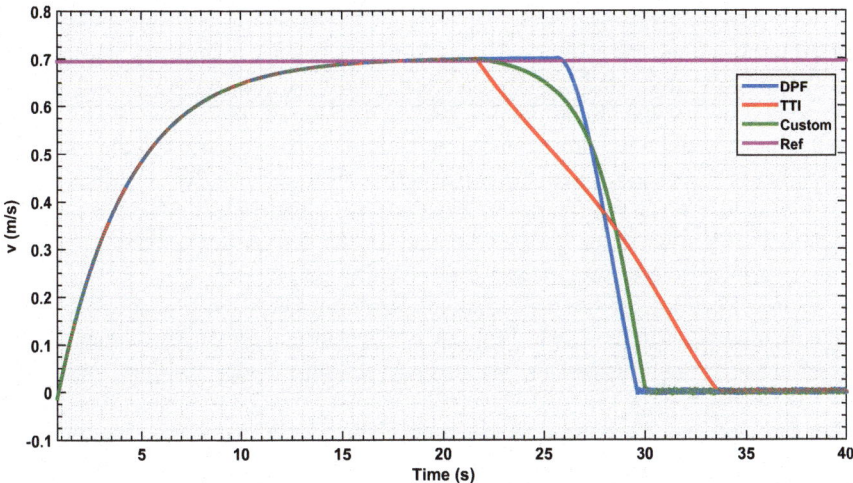

Fig. 26 Velocity estimation

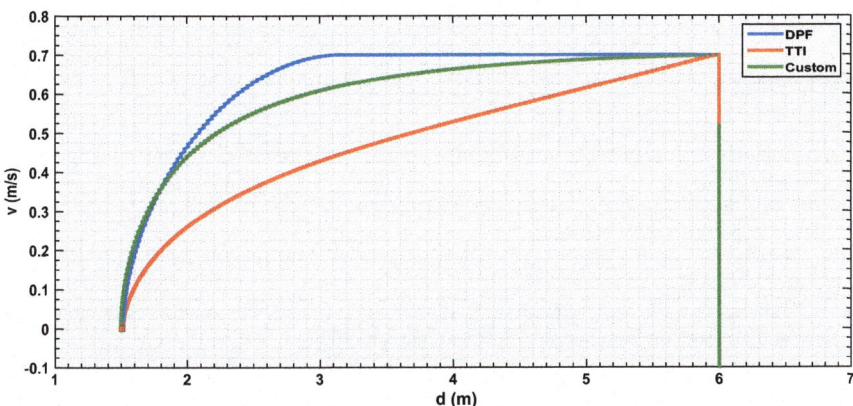

Fig. 27 Reference velocity with respect to the distance

5 Conclusion and Future Work

The research about unmanned vehicles is increasing on a daily basis and demonstrates the importance of using them in civil applications along side military ones. However, the teleoperation of such vehicles taking into consideration the human error is not recommended, thus the introduction of autonomous vehicles. The autonomy of the vehicle includes but not limited to sensor acquisition, self-localization, environment mapping and perception, high-level decision making, planning and navigation and finally control of actuators. In order to manage all of the aforementioned systems, a solid and stable architecture is required.

Accordingly, this chapter presented a complete ROS-based software architecture for different unmanned vehicles. The architecture is generic in a way that considers the different tasks and various vehicle capabilities. The proposed architecture approach is able to handle different sensors and controller processes, in addition to the flexibility in utilizing information from multiple heterogeneous vehicles. In order to evaluate the proposed system, several experiments were carried out in different scenarios of civil applications, using the two platforms (quadcopter and hexacopter)and implemented simulator for UAVs. Obtained results demonstrates the effectiveness of the proposed architecture under various conditions.

Future directions for this work include several aspects. One, all proposed and implemented algorithms should be further tested in real-world experiments using different scenarios to ensure the stability and viability under multiple conditions. Two, the concept of heterogeneous vehicles can be extended to include the cooperation with unmanned ground vehicles, at which the UGVs and UAVs coordinate the tasks together for a better overall performance. Three and lastly, development and testing of advanced control systems, with high capabilities, which include the use of fuzzy logic controller and other advanced AI approaches.

References

1. A. Toma, Use of Unmanned Aerial Systems in Civil Applications. Ph.D. thesis, Politecnico di Torino (2015)
2. NASA: Global hawk high-altitude, long endurance science aircraft. National Aeronautics and Space Administration (2014)
3. I. Simelli, A. Tsagaris, The use of unmanned aerial systems (uas) in agriculture. HAICTA, 730–736 (2015)
4. J.C. Hodgson, S.M. Baylis, R. Mott, A. Herrod, R.H. Clarke, Precision wildlife monitoring using unmanned aerial vehicles. Scientific reports 6 (2016)
5. S. Muramatsu, K. Takahashi, T. Tomizawa, S. Kudoh, T. Suehiro, Development of the robust delivery robot system with the unknown object in indoor environment, in *Industrial Electronics Society, IECON 2013–39th Annual Conference of the IEEE* (IEEE, 2013), pp. 8271–8276
6. T. Tomic, K. Schmid, P. Lutz, A. Domel, M. Kassecker, E. Mair, I.L. Grixa, F. Ruess, M. Suppa, D. Burschka, Toward a fully autonomous uav: research platform for indoor and outdoor urban search and rescue. IEEE Robot. Autom. Mag. **19**(3), 46–56 (2012)
7. G. Zhou, A. Liu, K. Yang, T. Wang, Z. Li, An embedded solution to visual mapping for consumer drones, in*Proceedings of the IEEE Conference on Computer Vision and Pattern Recognition Workshops* (2014), pp. 656–661
8. A. Al-Kaff, F.M. Moreno, L.J. San José, F. García, D. Martín, A. de la Escalera, A. Nieva, J.L.M. Garcéa, Vbii-uav: vision-based infrastructure inspection-uav, in *World Conference on Information Systems and Technologies* (Springer, Berlin, 2017), pp. 221–231
9. M. Mataric, Issues and approaches in the design of collective autonomous agents. Robot. Auton. Syst. **16**, 321–331 (1995)
10. J. Meyer, A. Sendobry, S. Kohlbrecher, U. Klingauf, O. von Stryk, Comprehensive simulation of quadrotor uavs using ros and gazebo, in *Simulation, Modeling, and Programming for Autonomous Robots* (Springer, Berlin, 2012), pp. 401–411
11. J.S. Jang, C.J. Tomlin, Design and implementation of a low cost, hierarchical and modular avionics architecture for the dragonfly uavs, in *AIAA GNC Conference* (Monterey, 2002)

12. J. How, E. King, Y. Kuwata, Flight demonstrations of cooperative control for uav teams, in *AIAA Unmanned Unlimited Technical Conference, Workshop and Exhibit* (Chicago, 2004)
13. J. Tisdale, A. Ryan, M. Zennaro, X. Xiao, D. Caveney, S. Rathinam, J.K. Hedrick, R. Sengupta, The software architecture of the berkeley uav platform. Int. Conf. Control Appl. 1420–1425 (2006)
14. S. Tsach, A. Peled, D. Penn, D. Touitou, The capecon program: Civil applications and economical effectivity of potential uav configurations, in *Proceedings of the 3rd AIAA Unmanned Unlimited Technical Conference, Workshop, and Exhibit* (2004)
15. T. Balch, R.C. Arkin, Behavior-based formation control for multi-robot teams. IEEE Trans. Robot. Autom. (1999)
16. P. Pirjanian, M. Mataric, Multi-robot target acquisition using multiple objective behavior coordination. Int. Conf. Robot. Autom. (ICRA) (2000)
17. A. Al-Kaff, A. de la Escalera, J.M. Armingol, Indoor and outdoor navigational system for uavs based on monocular onboard camera, in *Advanced Concepts for Intelligent Vision Systems* (Springer, Berlin, 2017)
18. P. Marin-Plaza, J. Beltran, A. Hussein, B. Musleh, D. Martin, A. de la Escalera, J.M. Armingol, Stereo vision-based local occupancy grid map for autonomous navigation in ros, in *Joint Conference on Computer Vision, Imaging and Computer Graphics Theory and Applications (VISIGRAPP)*, vol. 3 (2016), pp. 703–708
19. A. Hornung, K.M. Wurm, M. Bennewitz, C. Stachniss, W. Burgard, OctoMap: an efficient probabilistic 3D mapping framework based on octrees. Auton. Robot. (2013). http://octomap.github.com
20. A. Al-Kaff, F. García, D. Martín, A. De La Escalera, J.M. Armingol, Obstacle detection and avoidance system based on monocular camera and size expansion algorithm for uavs. Sensors **17**(5), 1061 (2017)
21. A. Hussein, H. Mostafa, M. Badreldin, O. Sultan, A. Khamis, Metaheuristic optimization approach to mobile robot path planning, in *International Conference on Engineering and Technology (ICET)* (2012), pp. 1–6
22. A. Hussein, A. Al-Kaff, A. de la Escalera, J.M. Armingol, Autonomous indoor navigation of low-cost quadcopters, in *IEEE International Conference on Service Operations And Logistics, And Informatics (SOLI)* (2015), pp. 133–138
23. A. Al-Kaff, F.M. Moreno, A. de la Escalera, J.M. Armingol, Intelligent vehicle for search, rescue and transportation purposes, in *2017 IEEE International Symposium on Safety, Security and Rescue Robotics (SSRR)* (IEEE, 2017), pp. 110–115
24. A. Khamis, A. Hussein, A. Elmogy, Multi-robot task allocation: a review of the state-of-the-art, in *Cooperative Robots and Sensor Networks* (Springer International Publishing, 2015), pp. 31–51
25. T. Foote, tf: The transform library, in *2013 IEEE International Conference on Technologies for Practical Robot Applications (TePRA)* (Open-Source Software workshop, 2013), pp. 1–6
26. S.H. Juan, F.H. Cotarelo, Multi-master ros systems. Institut de Robotics and Industrial Informatics (2015)
27. A. Kokuti, A. Hussein, P. Marín-Plaza, A. de la Escalera, F. García, V2x communications architecture for off-road autonomous vehicles, in *IEEE International Conference on Vehicular Electronics and Safety (ICVES)* (2017), pp. 69–74
28. Startsolid: A listing of amazon box sizes, https://www.startsolid.com/pages/a-catalog-of-amazon-box-sizes. Accessed 13 Sep 2017
29. S. Bulger, How do i calculate dimensional weight? (2012), https://efulfillmentservice.com/2012/11/how-to-calculate-dimensional-weight. Accessed 13 Sep 2017

A Quadcopter and Mobile Robot Cooperative Task Using Visual Tags Based on Augmented Reality ROS Package

Alvaro Rogério Cantieri, Ronnier F. Rohrich, André Schneider de Oliveira, Marco Aurélio Wehrmeister, João Alberto Fabro, Marlon de Oliveira Vaz, Magnus Eduardo Goulart and Guilherme Hideki

Abstract The objective of this chapter is to provide a simple tutorial on how to use a virtual reality tag (VR-TAG) tool and a Robot Operating System–compatible simulated multirotor vehicle to achieve the position of a small mobile ground robot, making possible the creation of a cooperative schema among them. The great novelty of the proposed architecture is that the ground robots do not have any onboard odometry, and all the position information is provided by the multirotor using a camera and the VR-TAGs to evaluate it. This kind of architecture poses value for real-world cooperative multiple robot research, in which the cost of constructing a large number of small robots makes practical applications inviable. In such cases, simple robots with minimal control hardware and sensors are a good alternative, and offboard positioning and control of these robots can be effective.

A. R. Cantieri (✉) · R. F. Rohrich · M. de Oliveira Vaz · M. E. Goulart
Federal Institute of Parana, Rua Joao Negrao, 1285, Curitiba, Brazil
e-mail: alvaro.cantieri@ifpr.edu.br
URL: http://www.ifpr.edu.br/

R. F. Rohrich
e-mail: rohrich@utfpr.edu.br

M. de Oliveira Vaz
e-mail: marlon.vaz@ifpr.edu.br

M. E. Goulart
e-mail: magnus.goulart@ifpr.edu.br

A. S. de Oliveira · M. A. Wehrmeister · J. A. Fabro · G. Hideki
Federal University of Technology—Parana, Av. Sete de Setembro, 3165, Curitiba, Brazil
e-mail: andre.oliveira@dainf.com
URL: http://www.utfpr.edu.br

M. A. Wehrmeister
e-mail: wehrmeister@utfpr.edu.br

J. A. Fabro
e-mail: fabro@dainf.ct.utfpr.edu.br

G. Hideki
e-mail: guilhermekaik@hotmail.com

© Springer International Publishing AG, part of Springer Nature 2019
A. Koubaa (ed.), *Robot Operating System (ROS)*, Studies in Computational
Intelligence 778, https://doi.org/10.1007/978-3-319-91590-6_6

185

Keywords Multirotor simulation · Multirotor ROS · Virtual reality tag · AR-TAG

1 Introduction

Unmanned aerial vehicles (UAVs) are currently one of most interesting areas of robotics, with many applications in several areas, such as agriculture, security, logistics, structure inspection, and the military. The decrease in equipment and hardware costs are making possible the development of intelligent aircraft with several kinds of sensors that make tasks simple and secure in situations where previously human presence would be required.

Some applications demand cooperation of a group of robots to reach some specific objectives. These are called cooperative multirobot applications, and sometimes a mix of aircraft and mobile ground robots are used to perform some function. A big challenge confronting the development of cooperative systems is achieving the correct position among all the robots when they move within the environment. Without precise position data, the intelligent algorithms used to control the cooperative robot group will not work properly.

A traditional example of cooperative robotic systems is the Robot Cup Soccer League (SSL) competition [1]. In this kind of competition, a central ground pointed camera achieves images from a group of small robots in an arena, and image-processing software estimates the position of all them and the ball, controlling the movements during the game. The task mixes many different challenges, enabling the correct evaluation position to implement strategies for making a goal.

The correct placement and movement control of the mobile robots within the environment are still a challenge owing to the great variability of obstacles and the difficulty in getting precise location during displacements. Embedded sensors, GPS, and cameras are often used to get position data in many ways. Data fusion and pose estimation are not simple tasks, and many errors can occur, making proper operation of the robot systems difficult. Faults in the fixed absolute referential system for the environment create some difficulty in this kind of application. In external environments, GPS can be used. Unfortunately, this kind of solution is not appropriate for indoor applications.

Over the past years, the use of image systems has become easier, cheaper, and largely available. The decrease in the cost of image-processing technology allows new image-based algorithms for mobile robotic systems, including flying ones. A visual tag is a tool that allows an object position and angle estimation by using a special figure tag, a camera, and image-processing hardware. The tag is created to present an easily recognizable form, decreasing the processing time and performing a good position estimation.

In this chapter, we show how to assemble a cooperative system that joins a quadcopter and a small ground robot to perform a collaborative task. The robot is the "dummy," which means that it does not have any embedded sensors. The position of the robot on the ground is provided by the quadcopter that performs a position

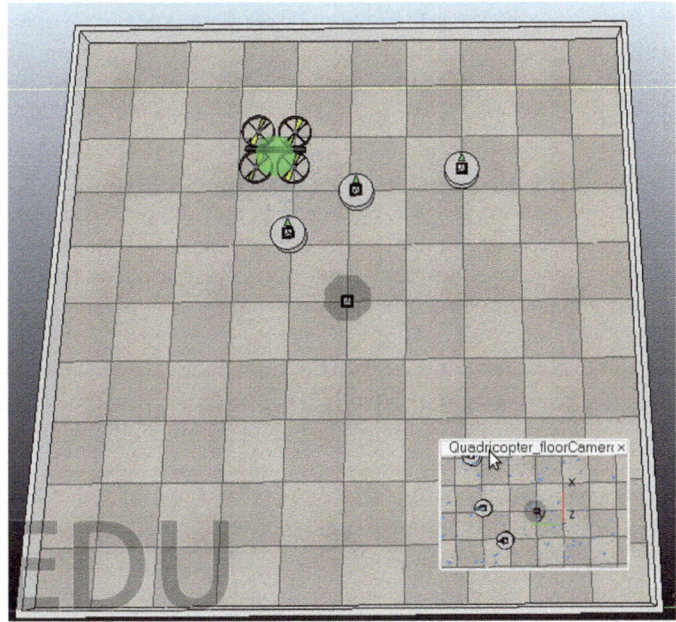

Fig. 1 V-REP environment

estimation based on a virtual reality tag (VR-TAG) and a camera system. The robot is controlled by a C++ script that subscribes a Robot Operating System (ROS) topic, where the position data are published by the quadcopter camera ROS node. All the positions are relative to the coordinate center of the quadcopter, posing an additional challenge in calculating position information in the environment. Figure 1 shows an image of the simulation environment.

The advantage of using augmented reality tags (AR-TAGs) to provide position information for a group of robots is that this kind of tool offers an easy way to locate the robot on the ground and provide position information to perform its control. The number of available AR-TAG tools available from the Internet is quite extensive, making it easy to choose the correct one to apply to each desired problem. Several applications in which control robots use a camera and AR-TAGs in different ways were found in literature, proving that the schema is technically viable. The major contribution of this work is to provide a starting point for researchers to create a cooperative system between a quadcopter using AR-TAGs.

The application is simulated using the Virtual Robot Experimentation Platform (V-REP) simulation software and the ROS. The ARTrack Alvar ROS package provides position information to the control application, performing image processing on all environment visible tags and publishing them on an ROS node. The images are captured by a ground-pointing camera fixed to the underside of the quadricopter.

The objective of this chapter is to provide a tutorial application example of how to use the specific group of tools to build a practical simulated application. The chosen tools are not the only possible way to create this kind of application, and similar ones can be used to build this kind of architecture as well. This is an initial application work that provides an initial starting point for technical or scientific research based on this kind of schema. There are numerous real-world limitations that must be investigated before bringing this application to the real world. In a simulation environment, image noise problems, quadcopter orientation errors, hardware-processing performance, and other practical problems are minimized; a practical evaluation of these tools must take into account these limitations.

This chapter is divided in six sections. Section 1 presented provides a general vision of the application area, the proposal description, a glimpse of the tools used, and the benefits and limitations of using this kind of schema. Section 2 reviews some similar works to provide a view of some AR-TAGs and robotics integration systems and also briefly describes some of the several available AR-TAGs tools cited in the literature. Section 3 describes the V-REP software interface and the creation of the simulated components and environment used to simulate the application. Section 4 explains the control scripts created to perform the robot actions and how to run the ROS packages used on the proposed architecture. Section 5 presents some initial experimental results, created to show the viability of using this kind of architecture for practical applications. On Sect. 6 final remarks are commented.

2 Application of Augmented Reality Tools to Mobile Robot Control

Robot position estimation using visual data is not a novelty, but the high cost of the hardware necessary made real-world application difficult until recent years. Some systems use environmental visual information to provide position data to the robots, but this is a difficult task owing to the great degree of object variation and the high processing information level necessary. The use of special generated images called "tags" is a good way to provide visual information to image systems without significant hardware processing. Such tags are generated to be easily identified in the environment, standing out from the other objects captured in the images. These tags also offer the possibility of including additional data on their draw, such as a QR code for example.

The tools that provide estimation position of an object based on visual processing using tags are generally call AR-TAGs. Several similar AR-TAG architectures have been proposed in research works, with some small differences among them. To provide an adequate background on the subject, some commonly used virtual reality tags tools are cited and briefly explained in Table 1.

In this work, the ALVAR Tag development tool was used to provide position data for robot control. The Alvar were created by the VTT Technical Research Centre of

Table 1 Some of AR-Tags Tools

Tool	OS compatible	ROS package
ARToolkit [2, 3]	iOS, Android, Linux, Windows and Mac OS X	artoolkit [4]
ALVAR [5]	Android, iOS, Linux, Windows, Mac	ar_track_alvar [6]
ArUco [7]	Android, iOS, Linux, Windows, Mac	ar_sys [8]
AprilTag [9]	Windows, Linux, and Mac	apriltags_ros [10]
ARTag [11]	Windows, Linux, and Mac	No ROS Package

Finland [5]. This tool provides position and orientation estimation for a special tag image, and it allows the use of a group of tags simultaneously within the environment. A package developed for ROS is available [6], facilitating applications in robotics. The ROS package is very easy to use, reducing the time necessary for application development.

Several other AR-TAG tools are available on the Internet, with some having interesting characteristics. It is not possible to describe all of them in this brief survey, so the authors suggest some additional research to verify the best AR-TAG tool to use for each desired application.

Possible applications of AR-TAGs in robotics are extensive and flexible. The tags can be used to improve position data, to create checkpoints within the environment, to send additional information to the robots' map localization, or to work like an identifier placed on the mobile robots to facilitate visual cooperative work among them. Some interesting examples found in the literature are cited here to illustrate the kind of possible applications of this schema.

A working group at the Okayama University of Science in Japan [12] presented a visual simultaneous localization and mapping (vSLAM®) algorithm assisted by artificial landmarks. The landmarks are used to improve the positioning of the robot during navigation. Some AR tracking markers are fixed onto specific locations within the environment to provide additional information on positioning and orientation, improving the calculation of the robot's position during simultaneous localization and mapping (SLAM).

A group from Aldebaran-Robotics Enterprise (Paris, France) [13] used a set of bar codes to provide additional position and orientation information to an NAO humanoid robot. The bar codes are fixed to the environment walls in visual range of NAO cameras. Navigation is performed by traditional means by the robot, using information of a group of detected bars at each time to improve the position data on a global map.

Wang and Hu from the College of Mechatronics and Automation, National University of Defense Technology, in China [14] implemented a visual tag navigation system based on the AprilTag algorithm to provide position information to an AR Drone 2.0. The drone's camera captures frames from the environment in which the tags are fixed and off-board software performs computations to identify and establish the drone relative position, providing good position feedback. This work is interesting because the use of a drone's camera to achieve and process this position feedback presents numerous demands, such as noise treatment, drone stabilization, and adequate image capture. It is also stimulating because the implementation uses a multirotor and guides it, which is not a simple task. The image captured by the aircraft is subjected to vibration noises and to variations of the horizontal and vertical positioning of the quadcopter among other practical problems, but the schema could work well and demonstrate that the use of AR-TAGs captured by an aircraft camera to provide position data is possible and technically viable.

The robot navigation system presented by Limosani et al. [15] employs an interesting use of AR tracking markers tags. The robot navigates within the environment and, when a tag is found and recognized, its ID is used to search and download specific map information of the actual robot space. In this application, the tag does not provide direct positioning or orientation information, but it is an auxiliary system to provide information about the robot's surroundings.

All these examples show some possible applications of AR-TAGs to mobile robotics, especially to multirotor aircraft. The inspiration for the work described on this chapter came from all the previous examples and from others from the literature.

3 Creation of Environment and Simulation Objects on V-REP

The application example proposed in this chapter uses the AR-TAG tools and a simulated quadcopter with an integrated camera to estimate the position and orientation of a small mobile robot on the ground, providing information for controlling it. A small and simple mobile robot called BOB was assembled for this application. It consisted of a disk with two motorized independent wheels and one other spherical nonmotorized one. The AR-TAG is fixed to the top of the robot, providing individual identification and position data for each one. The robot tag is processed by the AR-TAG package and the position data are published on an ROS node.

The simulation software chosen for this application was V-REP [16], developed by Copellia Robotics, an enterprise from Zürich, Switzerland. V-REP is a flexible robot software simulation that allows the development of several kinds of robotic mechanisms, sensors, control systems, and script codes in an easy way. This is a useful tool for reducing development time for complex robotic systems. The user interface is easy to learn and intuitive, and the drawing capabilities are adequate for a small number of robot model creations. Some integrated components such as sensors, actuators, and motors offer a good set of basic tools for the creation of complex robotic systems for both research or for industrial development. The educational version of V-REP is available free on the Copellia Robotics website for Windows, Linux, and Mac operational systems. For details of how to install and configure the software, we recommend following the steps described in the software manual. It is usually a very simple process.

The main component of the V-REP simulation is a scene. The scene contains all the components, environment, scripts, and other elements necessary to run an application. In this section, we show all the necessary steps to create and integrate all the needed components for the example application scene.

Once started, the V-REP software will show a workspace like the one shown in Fig. 2. This workspace shows a central simulation area with a "ground" where the components will stay during the simulation. The left size presents the components menu, where the component models available are displayed. On the top of the window are the simulation tools, such as simulation time step configuration; play, stop, and pause buttons; and the component manipulations tools, such as displacement buttons, rotation buttons, scale, etc. The complete description of all the V-REP components and simulation parameter are beyond the scope of this chapter. The authors strongly recommend studying the V-REP tutorials before starting its implementation for an adequate understanding of the software tools and its uses. For these applications, to facilitate quick implementation at the beginning of testing, the scene and all its components and scripts are available from the on-line chapter repository [17].

To run the desired simulation the following components must be added to the scene:

- a quadcopter capable of controlled movements within the environment and of staying in static flight at one desired point,
- a small ground mobile robot controlled by an ROS script,
- a global camera fixed to the bottom of the quadricopter pointing to the ground, publishing the captured images on an ROS node, and
- AR-TAG figures fixed to the top of each ground robot.

First, let us add the quadcopter to the scene. For this application, the V-REP quadcopter model available from the mobile robots file will be used. The model was provided by Eric Rohmer, and the propellers are courtesy of Lyall Randell. To add the quadcopter to the scene, just navigate to the *Mobile Robots* file on the *Model Browser* window, shown on the left side of the workspace, and drag and drop the model to the center of the environment.

This quadcopter has horizontal and vertical stabilization performed by a Lua script and a target that can be moved within the space. When the target moves to another position, the quadcopter position controller performs a set of commands that leads into a new position and stabilizes it at this static point.

Now it is necessary to add a camera to the base of the quadcopter. A vision sensor is used to capture images from the simulated environment, emulating the work of a small camera coupled to the aircraft. To add this vision sensor to the workspace, click on

Add –> Vision Sensor –> Perspective type

Move the camera to the center of the quadcopter base, pointing it to the ground. To associate the camera to the quadcopter, select it and the quadcopter base model at the same time and click on

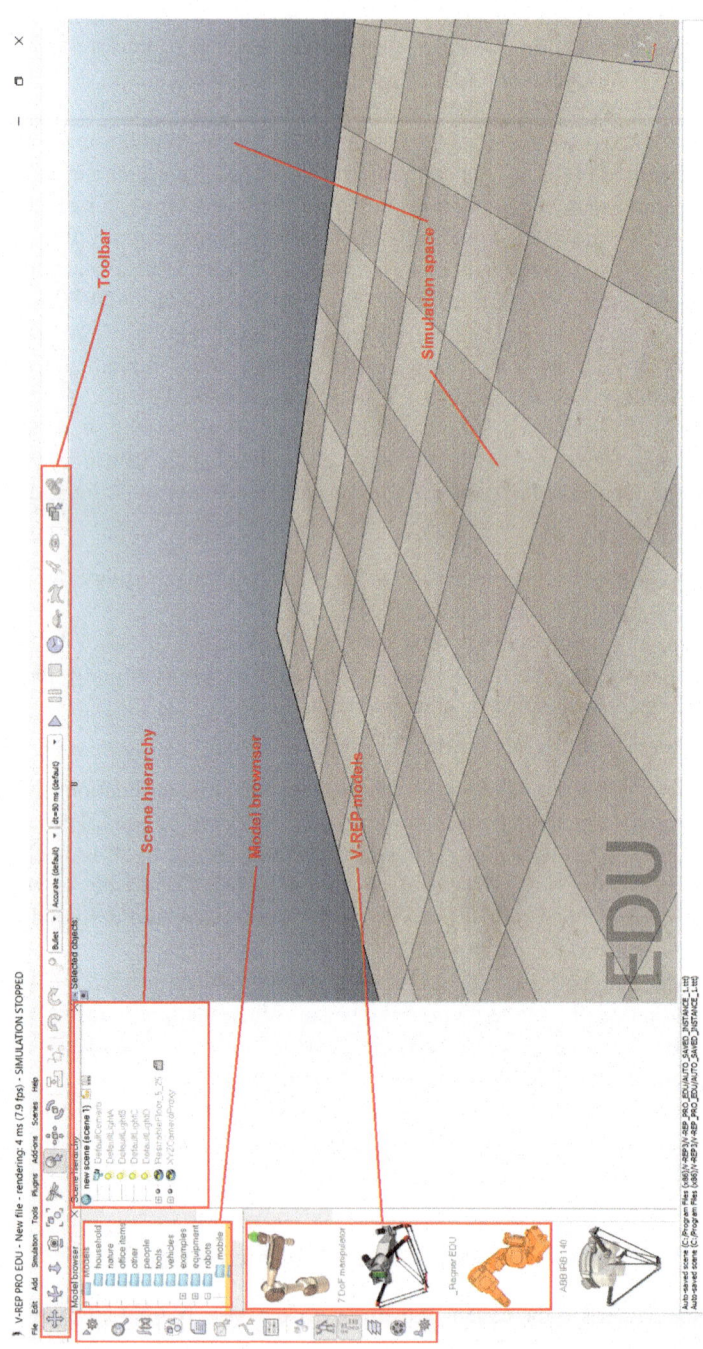

Fig. 2 V-REP window

Edit –> Make last selected object parents

The camera has a perspective angle of 120 degrees and a resolution of 1024 × 1024 pixels. The camera pointing to the ground provides a mirror vision of the workspace quadrants, which must be correct. This orientation can be achieved by rotating the camera to a −90 degree angle along its own Z axis. This will provide correct quadrant position data for the AR-TAG script, and a signal correction on the control script will complete the solution of this small problem.

The next step is to build or choose the ground robot used in the simulation. in this work, the authors built a simple robot to provide the quadricopter–robot cooperative task elements. This robot is available from the chapter on-line repository, so its construction will not be described here, because it is simple and several similar models can be found in the V-REP tutorials. The use of other V-REP robot models is also an interesting way to test the performance of the control architecture.

To finish the inclusion of all the necessary objects into the workspace, we must add an AR-TAG image to the top of each BOB robot. These tag images are generated by the AR-TAG Alvar application, as explained next. The image will be associated with a texture on a plane shape. To add the plane shape to the scene, click on

Add –> Primitive shape –> Plane

Set the plane dimensions to 10 × 10 cm. After that, select the plane object on the Scene Hierarchy menu and click on the Quick Texture button to choose the tag as a texture for the plane.

This step finishes the composition of the workspace and its elements. Figure 3 shows the composed scene finished.

The reference axis details of each scene component is shown on Fig. 4.

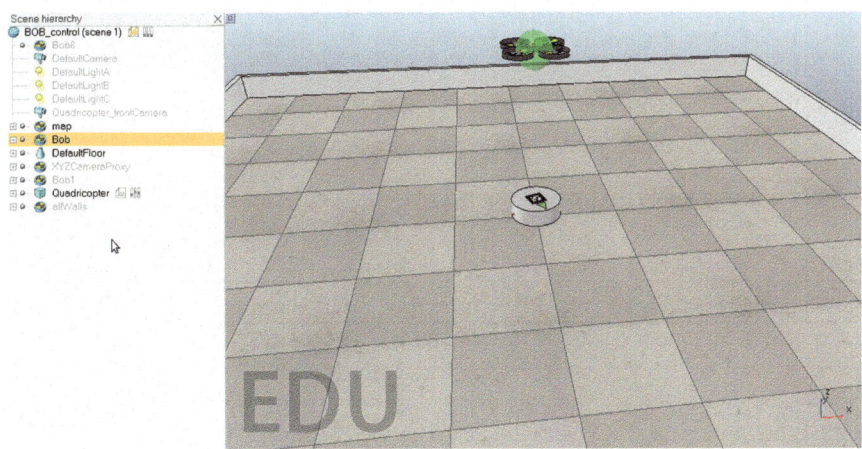

Fig. 3 V-REP complete scene

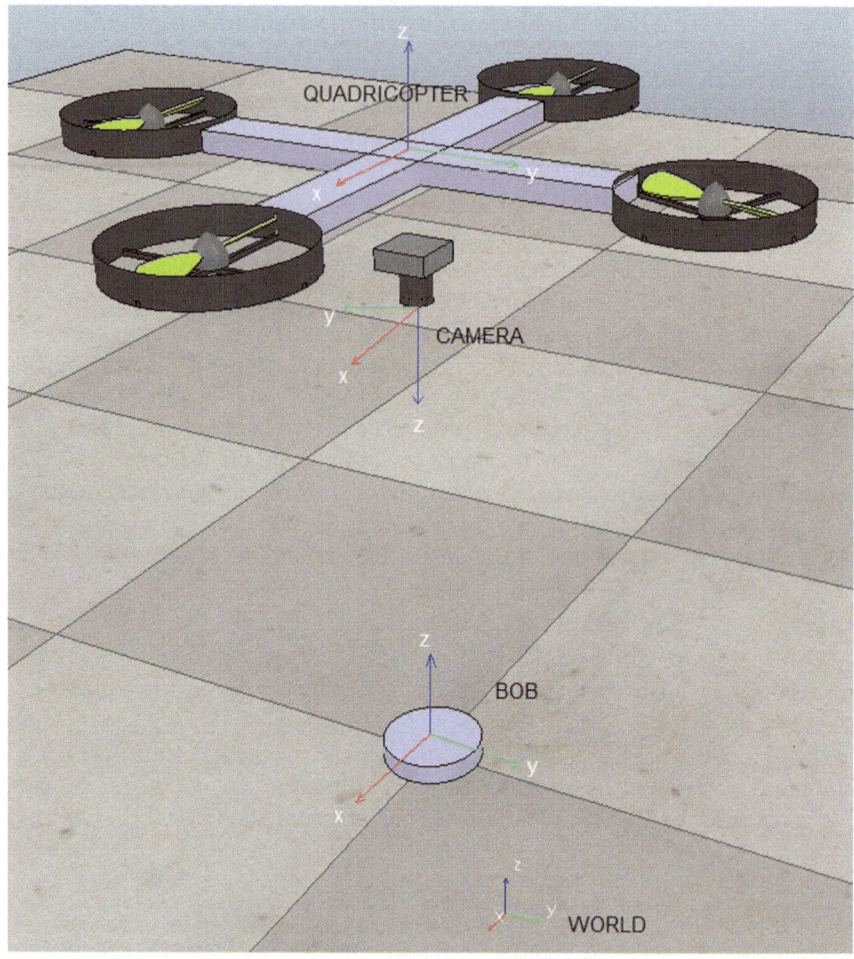

Fig. 4 V-REP component reference axis details

4 General Description of Application Architecture

The proposed application uses position data published to the AR-TAG ROS node to provide control information to a C++ code running on the simulation PC. This code performs all the control actions to lead the BOB robot from an initial position to a desired final set point relative to the quadcopter frame. All of BOB's moves will follow the quadcopter, so when the first changes its position, the other will move to achieve the previously set position defined by the code.

The proposed architecture is composed of five elements on the simulation:

```
1  Quadcopter
2  Global Camera
3  AR-Track_Alvar Package
4  PID Control Script
5  BOB ground robot
```

The quadcopter acts only as a transporter to move the Global Camera within the environment. All the ROS information exchanges are performed as shown in Fig. 5. The Global Camera receives visual information from the V-REP core and sends it to the AR_Track_Alvar package, where the position information of the robot tag is extracted. This information is published to the visualization_marker node and read by the PID script, where the velocity correction calculations are performed. After that, the BOB robot node reads the new velocity information and sends these values to the BOB Lua velocity control script on V-REP. This process continues in a closed loop until the robot reaches its final position, set by the PID control script.

4.1 PID C++ Software Description

The PID software architecture is simple. A C++ software running on a PC receives the values of distance and orientation error from V-REP at each new simulation step and calculates the velocity correction for robot wheels, keeping the displacement inside the tolerance values. Figure 6 shows the PID control loop architecture. Because it is a classical PID loop, the C++ code will not be described here in detail. The complete commented code is available from the on-line book repository [17].

In this example, using the BOB robot, the tolerance values were set to 0.05 rad for the orientation final error and 0.02 m for the final position error. The chosen gains are $P = 0.5$, $I = 0.01$, and $D = 0.01$. For other robots models, it will be necessary to change the PID gains to provide proper working of the control loop. This can be done by using the traditional PID tuning techniques or by using a trial and error approach, changing the values in the code, and running the simulation until the correct values are found.

The estimated position from the AR-TAG is published to the *ar_pose_marker* node. The function subCallback receives the data on the variables *position_x, position_y, position_z, and yaw_angle*.

```
1
2  void subCallback(const visualization_msgs::Marker::
        ConstPtr& marker)
3  {
4
5  pose.orientation.x = (float)marker->id;
6  pose.position.x = marker->pose.position.x;
7  pose.position.y = marker->pose.position.y;
```

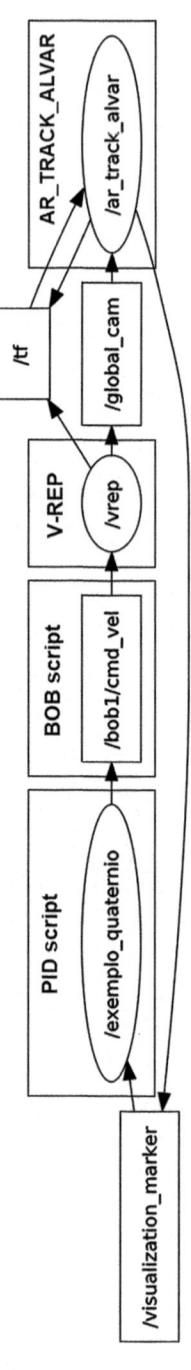

Fig. 5 ROS nodes and topics graphic

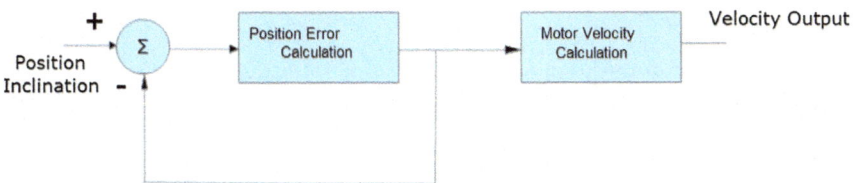

Fig. 6 PID script control blocks

```
8   pose.position.z = marker->pose.position.z;
9
10  // Selects the first AR-Tag if more than one is
        present on the workspace
11  if(marker->id ==1)
12  {
13
14  position_y = -(marker->pose.position.x);
15  position_x = -(marker->pose.position.y);
16  position_z = (marker-->pose.position.z)
17  tf::poseMsgToTF(marker->pose, pose2);
18  yaw_angle = -(tf::getYaw(pose2.getRotation()));
```

In the code above, the *position_x* variable receives *pose.position.y* and the *position_y* variable receives *pose.position.x*. A negative signal multiplication on each variable is done too. This is necessary to correct for quadrant exchanges that occur because the camera is pointed at the ground. This is a small problem found when using a camera fixed to the base of the quadcopter.

The control loop first rotates BOB's front to the set point direction. When BOB reaches the angle tolerance, the displacement loop begins. The PID displacement code controls the distance of the set point and BOB's orientation during until it does not achieve the position tolerance. After the tolerance is achieved, the loop completes and the code stops running. For another displacement, the quadcopter must move to another position and the code starts again.

On V-REP, a simple script is used to subscribe the C++ PID topic and to send to BOB's motor velocity signals. The part of the code responsible for this is as follows:

```
1
2   if (sim_call_type==sim_childscriptcall_actuation) then
3
4   -- Put your main ACTUATION code here
5
6   simExtROS_enableSubscriber('/bob/cmd_vel',1,
        simros_strcmd_set_twist_command,-1,-1,'
        cmd_vel_subdata1')
7   local packedData=simGetStringSignal('cmd_vel_subdata1'
        )
8   if (packedData) then
9   local twistData=simUnpackFloats(packedData)
```

```
10   vx=twistData[1]
11   w=twistData[6]
12   end
13
14   -- Base_controller --
15   r = 1.0000e-01 -- (m) wheel radius
16   L = 0.25
17
18   Vright = - ((w*L)/(2*r)) + (vx/r)
19   Vleft =   ((w*L)/(2*r)) + (vx/r)
20   simSetJointTargetVelocity(LeftmotorHandle,Vleft)
21   simSetJointTargetVelocity(RightmotorHandle,Vright)
22   end
```

Because the quadcopter script was not modified, no comments about it are necessary.

4.2   Description and Use of the AR-TAG Alvar ROS Package

After describing all the objects for simulation and the control scripts for the robot, it is necessary to explain the AR-TAG Alvar package. For this example, the *ar_track_alvar* ROS package was chosen. This package is easy to use and offers several interesting tools:

- It can generating AR-TAGs with variable sizes, image resolution, and data encoding.
- It can identify individual AR-TAGs using cameras or the Kinect sensor.
- It can identify a set of tags as one, avoiding use a group of tags on a body to better pose estimation.

The complete description of the installation and use of this package can be found on the ROS Wiki [6] and is a simple process. Let us explain the necessary steps for this application example.

The installation can be proceeded by the following command line:

$ sudo apt-get install ros-indigo-ar-track-alvar

This will install the package on the Linux OS. After installation, the tags creation tool can be started by the following:

rosrun ar_track_alvar createMarker

This will open a command line configuration menu to enable the creation of the desired tags with the specific information and details necessary. The tags can also be a copy of the link presented in the text, which shows the ready-to-use tag images, if it is convenient. The tags in these files are PNG images, as depicted in Fig. 7.

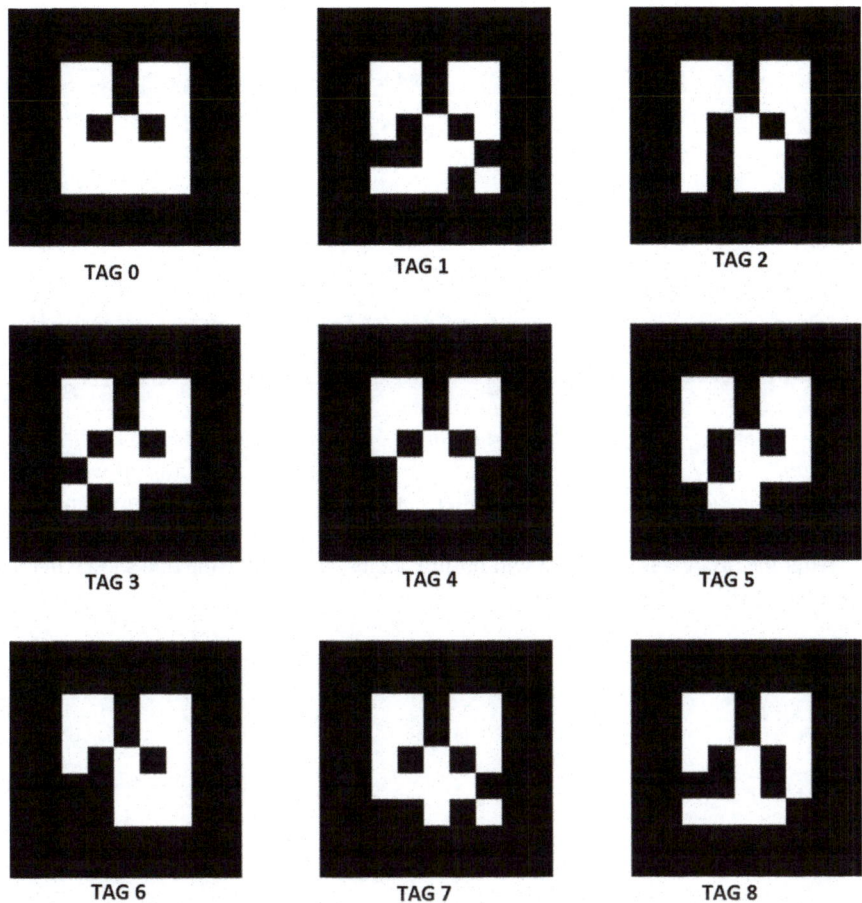

Fig. 7 The first eight tags images of tags alvar

To run properly, the file with the files *package.xml*, *bundle_0_1_2_3_4_5_6.xml*, and *tags_bundle.launch* must be copied to the *catkin_ws* directory. After that, the package can be started by

> roslaunch tags_bundle.launch

The file *tags_bundle.launch* provides information about the ROS topic, which publishes the images to be processed by the package among other information. The content of the file for this application example is as follows:

```
1
2  <launch>
3  <arg name="marker_size" default="10" />
4  <arg name="max_new_marker_error" default="0.05" />
```

```
5   <arg name="max_track_error" default="0.1" />
6   <arg name="cam_image_topic" default="/global_cam" />
7   <arg name="cam_info_topic" default="/global_cam_info"
        />
8   <arg name="output_frame" default="/global_cam" />
9   <arg name="bundle_files" default="$(find
        example_tags_alvar)/bundle_0_1_2_3_4_5_6.xml" />
10  <node name="ar_track_alvar" pkg="ar_track_alvar" type=
        "findMarkerBundlesNoKinect" respawn="false" output=
        "screen" args="$(arg marker_size) $(arg
        max_new_marker_error) $(arg max_track_error) $(arg
        cam_image_topic) $(arg cam_info_topic) $(arg
        output_frame) $(arg bundle_files)" />
11  </launch>
```

In the above code, the line $< argname = "cam\_image\_topic" default = "/global\_cam"/ >$ provides information about the source camera from where the tag images will be received. For another camera name, this information must be changed.

After the process is started, it will prompt a message set like the one shown in the following list:

```
1   started roslaunch server http://drone:37614/
2
3   SUMMARY
4   ========
5
6   PARAMETERS
7   * /rosdistro: indigo
8   * /rosversion: 1.11.21
9
10  NODES
11  /
12  ar_track_alvar (ar_track_alvar/
        findMarkerBundlesNoKinect)
13
14  ROS_MASTER_URI=http://localhost:11311
15
16  core service [/rosout] found
17  process[ar_track_alvar-1]: started with pid [9071]
18  [ INFO] [1505238017.797022505]: Subscribing to info
        topic
19  [ INFO] [1505238018.810070890]: Subscribing to image
        topic
```

This means that the package is running properly.

The package will publish information on two topics:

• /ar_pose_marker: This topic provides the position data (X, Y, Z, and angle) for each tag found by the camera.

- /visualization_marker: This topic publishes information to provide an RVIS (ROS visualization) of all the tags found within the environment. The tags are presented on RVIS as red and green squares on a table, depending on whether the tag is "master" or "common."

The ar_pose_marker message for the tag 0 presented in the simulation is as follows:

```
header:
seq: 2331
stamp:
secs: 0
nsecs:              0
frame_id: ''
markers:
header:
seq: 0
stamp:
secs: 1505262414
nsecs: 490692749
frame_id: /global_cam
id: 0
confidence: 0
pose:
header:
seq: 0
stamp:
secs: 0
nsecs:              0
frame_id: ''
pose:
position:
x: -0.00456776831643
y: -0.00132465785673
z: 0.176076750741
orientation:
x: -0.506383158998
y: 0.489746201693
z: 0.493990549846
w: 0.509605819113
```

The message gives a lot of information, but for our proposes the position and orientation data are important. The C++ code explained before receives the evaluated position and orientation of an individual tag and uses it to execute the control loop. The orientation must be calculated by a quaternion transformation to provide adequate data for the robot orientation, as done with the code.

5 Application Example and Experiments

A simple application task example was created to performs some practical evaluation of the proposed architecture. The task consists of making the BOB robot follow the quadcopter and reach its base projection point on the ground for each new displacement of the aircraft.

To run the application, first start the *ROSCORE* on a prompt. With ROS running, start the V-REP software. On Linux, this can be done on a prompt, typing *./vrep.sh* on the software installation directory. The following INFO messages must appear, showing that the ROS plugin was properly started on V-REP:

```
1  Plugin 'Ros': loading...
2  Plugin 'Ros': load succeeded.
3  Plugin 'RosInterface': loading...
4  Plugin 'RosInterface': load succeeded.
5  Plugin 'RosSkeleton': loading...
6  Plugin 'RosSkeleton': load succeeded.
```

Load the simulation scene and start the simulation by clicking on "play." After V-REP starts properly, run the AR_TAGS_ALVAR package, running the launch script as explained before:

roslaunch tags_bundle.launch

The last step is to start the BOB PID program. Type on a terminal

rosrun "directory" BOB

where "directory" is the file path pointing to the program.

The BOB robot must start moving to reach the programmed set point relative to the quadcopter, as explained below. First, the robot will rotate to align with the pointing direction and then it will start a movement on a straight line, decreasing its velocity as it gets closer to the set final position. When the robot arrives at the point, it rotates one more time to align with the with the forward-facing direction of the quadcopter.

The ground scanning task can be done by moving the quadcopter in small steps (≤ 2 m each time) and restarting the PID program to perform a new round of displacement.

The description given so far shows how to assemble, to configure, and to execute all the necessary components and codes to allow all the applications to run together. Up to now, no technical evaluation of the proposed architecture has been described. Now it is necessary to explain some results of the experiments to demonstrate the obtained performance of the system. A group of simple experiments was executed to verify the performance and to evaluate some important data that allows the correct working of the cooperative system. The experiments provide answers for some important questions, such as optimum quadcopter height for correct position estimation, error estimations with the robot and quadcopter moving, and PID control performance.

All these experimental results show the performance of the proposed example only, but even the simplistic information given here provides a good perception of how the proposed architecture will work on more complex applications.

5.1 Pose and Orientation Estimation for Different Heights Using the Visual Tags

The first experiment has the objective of evaluating the accuracy of position and orientation data estimation provided by the camera fixed onto the mobile quadcopter when it is static at different heights. This is important for determining at which heights the data readings begin to exhibit high variations, which can lead poor performance of the control script.

To perform this evaluation, five tags were placed on the ground within the quadcopter camera visual range, as shown in Fig. 8. The quadcopter was positioned, for the first evaluation, at a height of 1 m and a group of 100 pose estimations was stored. The process was repeated for heights of 1.25, 1.50, 1.75, 2.00, 2.25, 2.50, and 2.75 m. Figures 9 and 10 shows the main estimation error for each height test.

Control of all the small robots demands an accurate position reading to achieve good performance. This is the major difficulty found using visual tags in the proposed application. In this case, the readings will provide less error when the quadcopter stays at lower positions, as expected, but the height must be high enough to allow an adequate camera vision area for the scanning process. If the captured area becomes too small, the displacements of the robot will be small too and a practical scan task

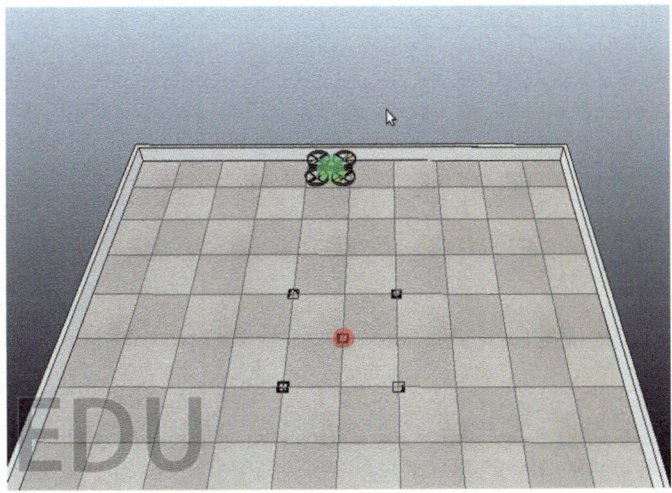

Fig. 8 Position estimation on V-REP

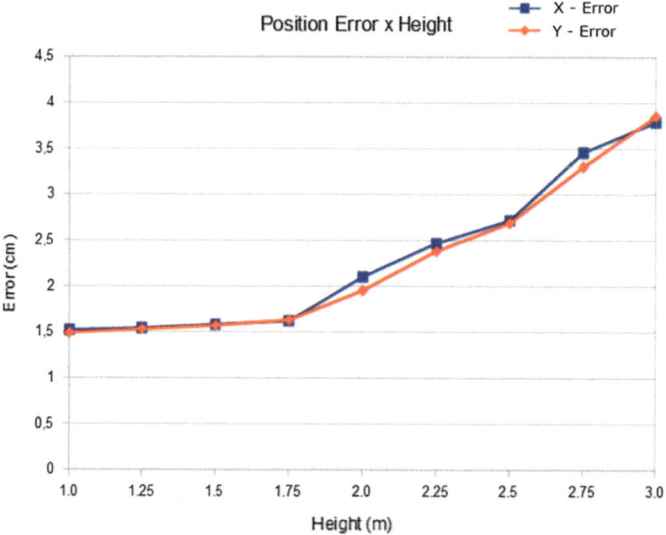

Fig. 9 Positioin estimation error for different heights

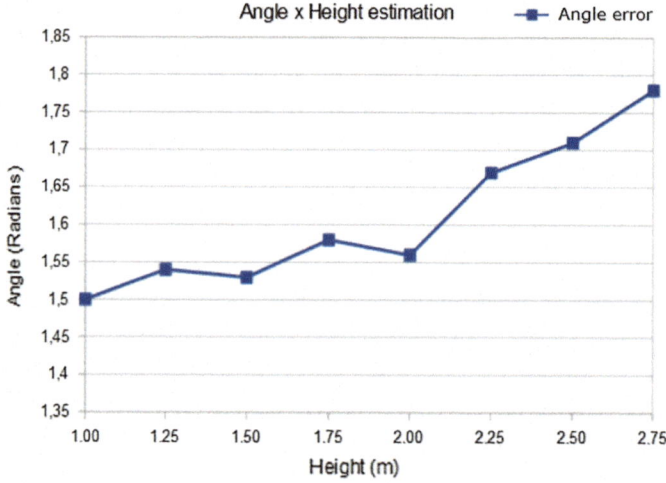

Fig. 10 Angle estimation error for different heights

will not be technically viable. For this application, the optimal height found was 2 m, where the position error does not generate loss of PID control and the visual area is adequate for creating a scan area of 2×2.5 m. If the size of the AR-TAG increases, this height probably will be greater.

The visual range area perceived in this application by the camera increases by a constant factor for each additional meter in height of the quadcopter. For example, at a 3 m height, the range will be 3×3.75 m, at 4 m it will be 4×5, successively. Figure 11 shows the height versus area relation.

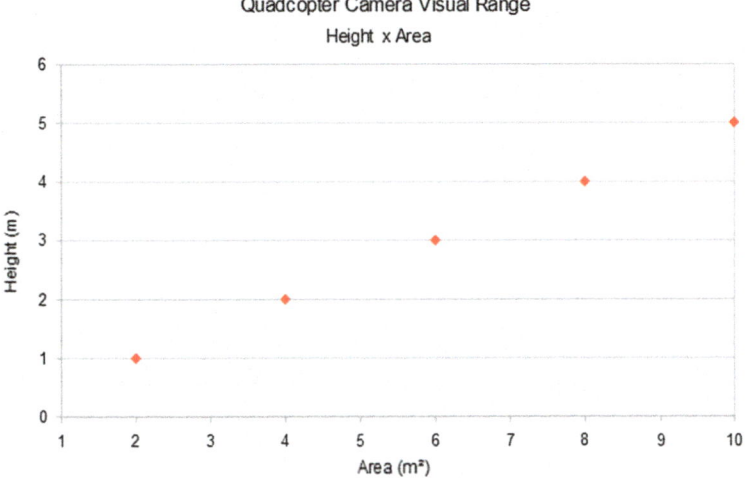

Fig. 11 Camera visual range area for different heights

The objective of this experimental evaluation is to provide some initial performance information about the proposed architecture. As the chapter objective is to be a tutorial, there is no intention of conducting a detailed scientific analysis of the schema. This kind of analysis can be performed on complex applications for each chosen architecture, components, and environment, depending on the proposed objective evaluated.

6 Final Remarks

In this tutorial, we described the use of visual tags and a drone's camera to estimate the position and orientation of a small group of mobile robots and to provide information for a PID controller. This kind of architecture has interesting applications to cooperative multirobot tasks, improvement of indoor robot positions, and visual control of flying robots, for example.

To demonstrate the viability of the purposed architecture some simulations were conducted on V-REP software, yielding good results. This software is a powerful tool for robotic system simulations, and it is not widely known within scientific communities as other simulators such as GAZEBO, which is one of the reasons for its choice in this work. Similar simulations can be performed on GAZEBO with some small changes, and it would be an interesting exercise to evaluate the results found in this work.

A real-world application is not difficult or expensive to implement using cheap quadcopters with ROS interface such as AR-DRONE and small Arduino robots for example. The advantage of this kind of architecture is the possibility of good indoor positioning for mobile cooperative robot systems and the simplicity of the hardware and firmware necessities.

The scripts and programs used in this application can be easily modified to make automatic displacements of the quadcopter and the ground robot and to build new and more complextask architectures. The objective of this tutorial is to offer the basic tools to allow an initial starting point using this kind of schema on cooperative robotic applications. The example application was a simple one to offer a better understanding of all the steps and tools.

Acknowledgements This work was funded by the CNPq - Conselho Nacional de Desenvolvimento Científico e Tecnológico - "National Counsel of Technological and Scientific Development", to which the authors are grateful for the support given to the research.

References

1. R. Federation, *RoboCupSoccer - Small Size* (2016), http://www.robocup.org/leagues/7
2. A. Community, *ARToolKit Documentation* (2017), https://www.artoolkit.org/documentation/
3. H. Kato, M. Billinghurst, I. Poupyrev, *ARToolKit version 2.33: A software library for Augmented Reality Applications. Manual.* (2000), 44, http://www.hitl.washington.edu/research/shared_space/download/
4. ROS.org, Ros package for ar-toolkit - ar_tools (2017), http://wiki.ros.org/ar_tools
5. V. T. R. C. of Finland, *Augmented Reality / 3D Tracking* (2017), http://virtual.vtt.fi/virtual/proj2/multimedia/
6. ROS.org, Ros package for alvar - ar_track_alvar (2017), http://wiki.ros.org/ar_track_alvar
7. U. de Córdoba, *ArUco: a minimal library for Augmented Reality applications based on OpenCV* (2017), http://www.uco.es/investiga/grupos/ava/node/26
8. ROS.org, Ros package for aruco -rar_sys (2017), http://wiki.ros.org/ar_sys
9. A. R. Laboratory, *AprilTag: A robust and flexible visual fiducial system* (2011), https://april.eecs.umich.edu/papers/details.php?name=olson2011tags
10. ROS.org, Ros package for april tags - apriltags_ros (2017), http://wiki.ros.org/apriltags_ros
11. M. Fiala, ARTag, a fiducial marker system using digital techniques, in *Proceedings of the IEEE Computer Society Conference on Computer Vision and Pattern Recognition* vol. 2 (2005), pp. 590–596. https://doi.org/10.1109/CVPR.2005.74
12. K. Okuyama, T. Kawasaki, V. Kroumov, Localization and position correction for mobile robot using artificial visual landmarks, in *Proceedings of the 2011 International Conference on Advanced Mechatronic Systems* (2011), pp. 414–418. https://doi.org/10.1504/IJAMECHS.2012.048395
13. L. George, A. Mazel, Humanoid robot indoor navigation based on 2D bar codes: application to the NAO robot, in *13th IEEE-RAS International Conference on Humanoid Robots (Humanoids)* (2013), pp. 329–335. https://doi.org/10.1109/HUMANOIDS.2013.7029995
14. T. H. Shuyuan Wang, *ROS-Gazebo Supported Platform for Tag-in-Loop Indoor Localization of Quadrocopter.* Intelligent Autonomous Systems 14. IAS 2016. Advances in Intelligent Systems and Computing, vol. 531 (2016)
15. R. Limosani, A. Manzi, L. Fiorini, F. Cavallo, P. Dario, Enabling global robot navigation based on a cloud robotics approach. Int. J. Soc. Robot. **8**(3), 371–380 (2016). https://doi.org/10.1007/s12369-016-0349-8
16. C. Robotics, *Virtual Robot Experimentation Platform* (2017), http://www.coppeliarobotics.com/
17. A. Cantieri, *Book chapter's online reporitory* (2018), https://sourceforge.net/projects/rosbook-2018/files/?source=navbar

Alvaro Rogério Cantieri has been an Associate Professor at the Federal Institute of Paraná (IFPR) since 2010. He obtained his undergraduate degree and Master's degree in electronic engineering at the Federal University of Paraná 1994 and 2000, respectively. He is currently studying for his PhD in electrical engineering and industrial informatics at Federal University of Technology - Paraná (Brazil). He started his teaching career in 1998 in the basic technical formation course of the Polytechnical Institute of Parana (Parana, Brazil) and worked as a Commercial Director of the RovTec Engineering Company, focusing on electronic systems development. His research interests include autonomous multirotor aircraft, image processing, and communications systems.

André Schneider de Oliveira obtained his undergraduate degree in computing engineering at the Universidade of Itajaí Valley (2004), Master's degree in mechanical engineering from Federal de Santa Catarina University (2007), and PhD degree in Automation Systems from Santa Catarina University (2011). He works as an Associate Professor at the Federal University of Technology - Paraná (Brazil). His research interests include robotics, automation, and mechatronics, mainly navigation systems, control systems, and autonomous systems.

Marco Aurélio Wehrmeister received his PhD degree in computer science from the Federal University of Rio Grande do Sul (Brazil) and the University of Paderborn (Germany) in 2009 (double-degree). In 2009, he worked as a Lecturer and Postdoctoral Researcher for the Federal University of Santa Catarina (Brazil). From 2010 to 2013, he worked as tenure-track Professor with the Department of Computer Science of the Santa Catarina State University (Brazil). Since 2013, he has worked as a tenure-track Professor in the Department of Informatics of the Federal University of Technology - Paraná (UTFPR, Brazil). From 2014 to 2016, he was Head of the MSc course on Applied Computing at UTFPR.

João Alberto Fabro is an Associate Professor at the Federal University of Technology - Parana (UTFPR), where he has worked since 2008. From 1998 to 2007, he was with the State University of West-Parana. He has an undergraduate degree in informatics from the Federal University of Paraná (1994), a Master's degree in computing and electrical engineering from Campinas State University (1996), a PhD degree in electrical engineering and industrial informatics from UTFPR (2003) and recently became a Postdoctoral Researcher at the Faculty of Engineering, University of Porto, Portugal (2014). Has experience in computer science, especially computational intelligence, and is actively researching the following subjects: computational intelligence (neural networks, evolutionary computing, and fuzzy systems) and autonomous mobile robotics. Since 2009, has participated in several robotics competitions in Brazil, Latin America, and the World Robocup with both soccer robots and service robots.

Marlon de Oliveira Vaz obtained his undergraduate degree in computer science from the Pontifical Catholic University (PUCPR -1998) and Master's degree in mechanical engineering from PUCPR (2003). He is now a Teacher at the Federal Institute of Parana and pursuing a PhD in electrical and computer engineering at the Federal University of Technology – Parana. He works mainly in the following research areas: graphical computing, image processing, and educational robotics.

Ronniernm Frates Rohrich obtained his undergraduate degree in Electric Engineering at Federal University of Technology - Parana (UTFPR) and Master's degree in Electric Engineering from Federal University of Paraná (UFPR). Is teacher on Electronic Department at Federal University of Technology - Parana (UTFPR).

Magnus Eduardo Goulart obtained his undergraduated degree on languages at Tuiuti University of Paraná - Paraná, and Master's degree at languages at Federal University of Paraná (UFPR). Is professor at Institute of Parana (Parana, Brazil).

Guilherme Hideki is a graduate student in computer engineering at Federal University of Technology - Parana (UTFPR).

Part III
Navigation, Motion Planning and Control

EXOTica: An Extensible Optimization Toolset for Prototyping and Benchmarking Motion Planning and Control

Vladimir Ivan, Yiming Yang, Wolfgang Merkt, Michael P. Camilleri and Sethu Vijayakumar

Abstract In this research chapter, we will present a software toolbox called EXOTica that is aimed at rapidly prototyping and benchmarking algorithms for motion synthesis. We will first introduce the framework and describe the components that make it possible to easily define motion planning problems and implement algorithms that solve them. We will walk you through the existing problem definitions and solvers that we used in our research, and provide you with a starting point for developing your own motion planning solutions. The modular architecture of EXOTica makes it easy to extend and apply to unique problems in research and in industry. Furthermore, it allows us to run extensive benchmarks and create comparisons to support case studies and to generate results for scientific publications. We demonstrate the research done using EXOTica on benchmarking sampling-based motion planning algorithms, using alternate state representations, and integration of EXOTica into a shared autonomy system. EXOTica is an open-source project implemented within ROS and it is continuously integrated and tested with ROS Indigo and Kinetic. The source code is available at https://github.com/ipab-slmc/exotica and the documentation including tutorials, download and installation instructions are available at https://ipab-slmc.github.io/exotica.

V. Ivan (✉) · Y. Yang · W. Merkt · M. P. Camilleri · S. Vijayakumar
School of Informatics, The University of Edinburgh, 10 Crichton Street,
EH8 9AB, Edinburgh, UK
e-mail: v.ivan@ed.ac.uk
URL: https://www.ed.ac.uk/informatics

Y. Yang
e-mail: yiming.yang@ed.ac.uk
URL: https://www.ed.ac.uk/informatics

W. Merkt
e-mail: wolfgang.merkt@ed.ac.uk
URL: https://www.ed.ac.uk/informatics

M. P. Camilleri
e-mail: michael.p.camilleri@ed.ac.uk
URL: https://www.ed.ac.uk/informatics

S. Vijayakumar
e-mail: sethu.vijayakumar@ed.ac.uk
URL: https://www.ed.ac.uk/informatics

© Springer International Publishing AG, part of Springer Nature 2019
A. Koubaa (ed.), *Robot Operating System (ROS)*, Studies in Computational
Intelligence 778, https://doi.org/10.1007/978-3-319-91590-6_7

Keywords Motion planning · Algorithm prototyping · Benchmarking
Optimization

1 Introduction

The ROS community has developed several packages for solving motion planning
problems such as pick-and-place (MoveIt! [1]), navigation [2], and reactive obstacle
avoidance [3]. These tools are easily accessible to the end-user via standardized
interfaces but developing such algorithms takes considerable effort as these interfaces
were not designed for prototyping motion planning algorithms. In this chapter, we
present EXOTica, a framework of software tools designed for development and
evaluation of motion synthesis algorithms within ROS. We will describe how to
rapidly prototype new motion solvers that exploit a common problem definition and
structure which facilitates benchmarking through modularity and encapsulation. We
will refer to several core concepts in robotics and motion planning throughout this
chapter. These topics are well presented in robotics textbooks such as [4] and [5]. This
background material will help you to understand the area of research that motivated
development of EXOTica.

Our motivation to begin this work stems from the need to either implement new
tools or to rely on existing software often designed for solving a problem other than
the one we intended to study. The need to implement and test new ideas rapidly led
us to the specification of a library that is modular and generic while providing useful
tools for motion planning. A guiding principle hereby is to remove implementation-
specific bias when prototyping and comparing algorithms, and hitherto create a
library of solvers and problem formulations.

In this chapter, we will use a well-known algorithm as an example in order to
explain how to use and extend the core components of EXOTica to explore novel
formulations. Consider a robot arm mounted to a workbench as shown in Fig. 1. The
arm consists of several revolute joints actuated by servo motors moving the links of
the robot body. A gripper may be attached to the final link. The task is to compute
a single configuration of the robot arm which will place the gripper at a desired
grasping position—i.e. our example will follow the implementation of an inverse
kinematics solver. Once the problem and motion solver have been implemented, we
can compute the robot configuration using EXOTica with the following code:

```cpp
#include <exotica/Exotica.h>
using namespace exotica;

int main(int argc, char **argv)
{
  MotionSolver_ptr solver = XMLLoader::loadSolver("{
      exotica_examples}/resources/configs/example.xml");
  Eigen::MatrixXd solution;
  solver->solve(solution);
}
```

Fig. 1 Example task of computing the inverse kinematics for a fixed base robot arm. The base frame and the target end effector frame are visualised

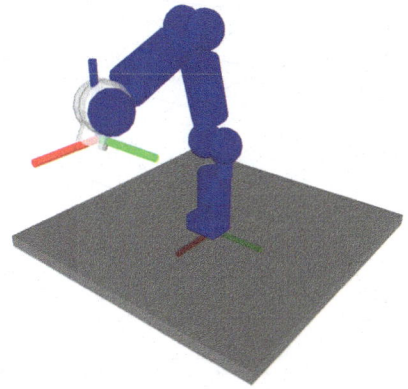

This snippet shows how little code is required to run a motion planning experiment. We load a motion solver and a problem definition from an example configuration file located in the `exotica_examples` package, allocate the output variable, and solve the problem using three lines of code.[1] What this snippet does not show is the definition of the planning problem, the implementation of the algorithm and an array of other tools available in EXOTica. This code and most of the rest of this chapter will focus on motion planning. However, we view motion planning and control as two approaches to solving the same motion synthesis problem at different scales. For example, the problem in Fig. 1 could be viewed as an end-pose motion planning problem as well as operational space control, when executed in a loop. This allows us to formulate complex control problems as re-planning and vice versa. EXOTica provides the tools to implement such systems.

To motivate and explain the EXOTica software framework, we focus on how it can be used in research and prototyping. The first part of this chapter will describe how problems and solvers are defined, and it will provide an overview of the library. The second part of the chapter will demonstrate how EXOTica has been used to aid motion planning research and it will help you to understand how EXOTica may be useful for your research and development.

2 Overview

Prototyping of novel motion planning algorithms relies on defining mathematical models of the robotic system and its environment. To aid this process, EXOTica provides several abstractions and generic interfaces that are used as components

---

[1] This and other basic examples with detailed explanation of each line of code, as well as instruction how to download, compile and run them are available in our online documentation at https://ipab-slmc.github.io/exotica/Installation.html.

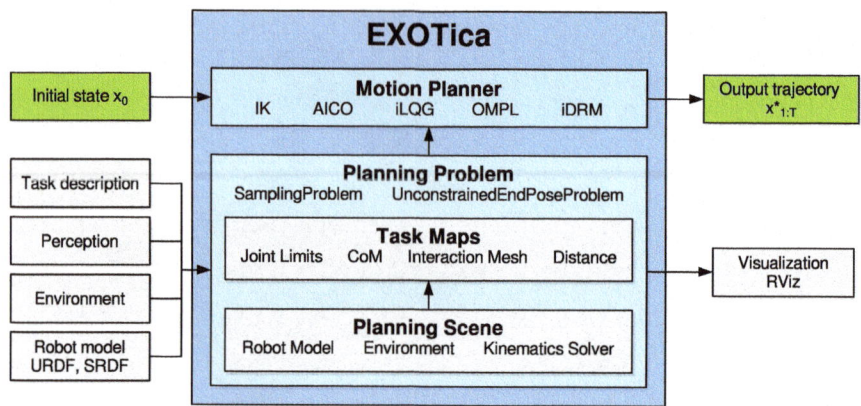

Fig. 2 The core concept of EXOTica highlighting the interplay of the planning scene, problem solver, and efficient batch kinematics solver. Problem and task definitions are generic across solvers and can be loaded easily from configuration files as dynamic plug-ins without the need to recompile. Furthermore, the same problem can be solved using a variety of baseline algorithms provided by EXOTica for benchmarking and comparison

for building algorithms. Figure 2 shows the three components central to algorithm design in EXOTica: (1) a *planning scene*, providing tools to describe the state of the robot and the environment, (2) a *planning problem* formally defining the task, and (3) a *motion solver*. These abstractions allow us to separate problem definitions from solvers. In particular, motion solvers implement algorithms such as *AICO* [6] and *RRTConnect* [7]. These implementations may perform optimization, randomized sampling, or any other computation which requires a very specific problem formulation.

How the problem is formulated is fully contained within the definition of a *planning problem*. Each algorithm solves exactly one type of motion planning problem while one type of problem may be compatible with multiple solvers. As a result, several algorithms can be benchmarked on the exact same problem. When benchmarking two algorithms that are compatible with different types of problems, the problems have to be converted explicitly. This is a useful feature that makes it easy to track differences between problem formulations that are intended to describe the same task.

All planning problems use the *task maps* as components to build cost functions, constraints, or validity checking criteria. Task maps perform useful computations such as forward kinematics, center-of-mass position calculation, and joint limit violation error computation. To further support the extensibility of EXOTica, the motion solvers and the task maps are loaded into EXOTica as plug-ins. As such, they can be developed separately and loaded on demand. One such example is the plug-in which wraps the sampling-based algorithms implemented in the OMPL library [8].

Figure 2 also shows the *planning scene* which separates the computation of kinematics from the computation of task related quantities.

3   System Model

To synthesize motion, we describe the system consisting of the robot and its environment using a mathematical model. This system model may be kinematic or it may include dynamic properties and constraints. EXOTica uses the system model to evaluate the state using tools implemented inside the *planning scene*. The diagram in Fig. 2 shows the *planning scene* as a part of the planning problem where it performs several computations required for evaluating the problem.

3.1 Planning Scene

The *planning scene* implements the tools for updating and managing the robot model and the environment. The robot model is represented by a kinematic tree which stores both the kinematic and dynamic properties of the robot, e.g. link masses and shapes, joint definitions, etc. The environment is a collection of additional models that are not part of the robot tree but that may interact with the robot. The environment may contain reference frames, other simplified models (geometric shapes), and real sensor data based representations such as pointclouds and OctoMaps [9]. The planning scene implements algorithms for managing the objects in the environment (e.g. adding/removing obstacles) as well as computing forward kinematics and forward dynamics.

The system is parametrized by a set of variables that correspond to controllable elements, e.g. the robot joints. The full state of the system is described using these variables and we will refer to it as the *robot state*. In some cases, only a subset of the robot state is controlled. We call this subset the *joint group*. Analogous to the MoveIt! [1] definition of a move group, a joint group is a selection of controlled variables used for planning or control. From now on, whenever we refer to a joint state, we are referring to the state of the joint group.

The system model may be kinematic, kino-dynamic,[2] or fully dynamic. The robot state is then described by joint positions, joint positions and velocities, or full system dynamics respectively. The system dynamics may be provided via a physics simulator but this is outside of the scope of this chapter. We will only consider the kinematic model for simplicity.

The system model is implemented as a tree structure mimicking the structure implemented in the KDL library [10]. Figure 3 illustrates the kinematic tree of a planar robot arm. Every node in the tree has one parent and possibly multiple children. The node defines a spatial transformation from the tip frame of the parent node to its own tip frame. Every node consists of a position offset of the joint, a joint transformation, and a tip frame transformation (see the KDL documentation [10]). The joint transformation is constant for fixed joints. The transformations of all joints

---

[2]Here, we define a kino-dynamic model as one which captures joint positions (kinematics) and joint velocities.

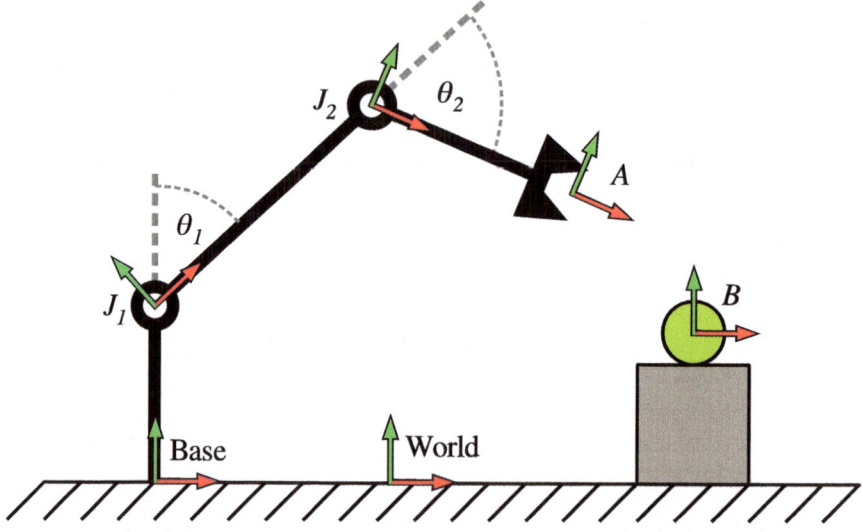

Fig. 3 The planning scene stores the kinematic tree composed of the robot model and the environment. The diagram shows a robot model which has two revolute joints J_1 and J_2 defined by joint angles θ_1 and θ_2 respectively, a base frame and an end effector frame A. A grasping target is located at frame B. The root of the tree is at the world frame. The grasping task can exploit the relative transformation M_A^B

that belong to the controlled *joint group* are updated based on the joint state. During the update, the local transformation of the node is updated and the transformation of the tip w.r.t. the world frame is accumulated. The nodes of the tree are updated in a topological order (from the root to the leafs). This ensures that the tip frame of the parent node is always updated before its children.

The EXOTica *Scene* implements a method for publishing the frames to RViz [11] using tf [12] for debugging purposes. These frames can be visualized using the *tf* and the *RobotModel* plug-ins.[3]

The system model provides an interface to answer kinematic queries. A query can be submitted to the *Scene*, requesting arbitrary frame transformations. Each requested frame has the following format:

- Name of the tip frame (Frame A)
- Offset of the tip frame
- Name of the base frame (Frame B)
- Offset the of base frame

Figure 3 illustrates an example scene. Any existing frame can be used to define a base or a tip frame of a relative transformation. The response to the query will then contain a transformation of the tip frame with respect to the base frame. If an offset is specified, each respective frame will be redefined to include the offset. If

---

[3]Use the tf prefix /exotica to visualize the robot model.

a base frame is not specified, the world frame will be used by default. Since all transformations of the tree nodes w.r.t. the world frame have been computed during the update, the query computation only adds the tip frame to the inverted base frame[4] $M_A^B = M_B^{world-1} M_A^{world}$. The *Scene* has been designed to answer a large number of requests in batches. While some smaller problems, such as simple kinematic chains, may be more costly to update, larger kinematic trees with a large number of leaf nodes are handled more efficiently by simply iterating over the requested frames.

The system model also computes derivatives of the spatial frames w.r.t. the control variables. These are computed as geometric Jacobians (J) and Jacobian derivatives ($\dot{J}$). The Jacobian has six rows and a number of columns corresponding to the number of controlled joints. Each column represents a spatial velocity in form of a *twist*. The twist $^B t_A^i$ describes the linear and angular rate of motion of the tip frame A w.r.t. the joint frame i expressed in the base frame B. We use the notation with the *expressed in frame* in the left superscript. Using the twist representation allows us to correctly compute spatial transformations using the Lie group algebra [13].

The kinematic tree represents the robot kinematic model and the objects in the environment. The robot model can be loaded from a pair of MoveIt! compatible URDF and SRDF files. The URDF file specifies the robot kinematics, joint transformations and range of motion, frame locations, mass properties and collision shapes. The SRDF file specifies the base of the robot (fixed, mobile, or floating), joint groups, and collision pairs. The robot configuration created for MoveIt! is fully compatible with EXOTica. The *Scene* also implements an interface to populate the environment with collision objects from MoveIt! planning scene messages and from MoveIt! generated text files storing the scene objects. The *Scene* may load additional basic shape primitives, meshes, or OctoMaps.

In order to perform collision checking, a *CollisionScene* can be loaded as a plug-in into a *Scene*. This allows for different implementations of collision checking algorithms to be used as required and does not tie EXOTica to a particular collision checking library. For instance, by default, EXOTica ships with two *CollisionScene* implementations using the FCL library—one based on the stable FCL version also used in MoveIt! and one tracking the development revision of FCL. The *Collision-Scene* plug-ins may hereby implement solely binary collision checking, or additional contact information such as signed distance, contact (or nearest) points, as well as contact point normals. This information is captured and exposed in a so-called *CollisionProxy*.

Referring back to the example inverse kinematics problem, the planning scene consists of the kinematics of the robot with a base link rigidly attached to the world frame. We choose to use a simplified version following the DH parameters of the KUKA LWR3 arm which we load from a pair of URDF and SRDF files. This robot has seven revolute joints. The joint group will consist of all seven joints as we intend to control all of them. We will not be performing collision checking in this experiment.

---

[4]Notation: the subscript and superscript denote tip and base frames respectively. M_A^B reads: transformation of frame A w.r.t. frame B.

The *planning scene* is initialized from an EXOTica XML configuration file. The XML file contains the following lines related to the setup of the *planning scene*:

```
<PlanningScene>
  <Scene>
    <JointGroup>arm</JointGroup>
    <URDF>{exotica_examples}/resources/robots/
        lwr_simplified.urdf</URDF>
    <SRDF>{exotica_examples}/resources/robots/
        lwr_simplified.srdf</SRDF>
  </Scene>
</PlanningScene>
```

where the joint group parameter selects a joint group defined in the SRDF file by name. The robot model is loaded from the URDF and SRDF files specified here. When the paths are not specified, EXOTica attempts to load the robot model from the `robot_description` ROS parameter by default. EXOTica additionally allows to set ROS parameters for the planning robot description from specified file paths if desired.

The system model provides access to some generic tools for computing kinematic and dynamic properties of the system. These tools have been designed for performing calculations for solving a wide variety of motion planning problems. The system modeling tools are generic but they can be ultimately replaced with a more specific set of kinematics and dynamics solvers in the final deployment of the algorithm. This is, however, outside of the scope of EXOTica.

4 Problem Definition

EXOTica was designed for prototyping and benchmarking motion synthesis algorithms. The main objective of our framework is to provide tools for constructing problems and prototyping solvers with ease. To do so, we first separate the definition of the problem from the implementation of the solver. Each problem consists of several standardized components which we refer to as *task maps*.

4.1 Task Maps

The core element of every problem defined within EXOTica is the function mapping from the configuration space (i.e. the problem state which captures the model state, a set of controlled and uncontrolled variables, and the state of the environment) to a task space. We call this function a *task map*. For example, a task map computes the center-of-mass of the robot in the world frame. A task map is a mapping from the configuration space to an arbitrary task space. The task space is, in fact, defined by the output of this function. Several commonly used task maps are implemented within EXOTica.

Joint position task map computes the difference between the current joint configuration and a reference joint configuration:

$$\Phi_{\text{Ref}}(x) = x - x_{\text{ref}}, \tag{1}$$

where x is state vector of the joint configuration and x_{ref} is the reference configuration.[5] The whole state vector x may be used or a subset of joints may be selected. This feature is useful for constraining only some of the joints, e.g. constraining the back joints of a humanoid robot while performing a manipulation task. The Jacobian and Jacobian derivative are identity matrices.

Joint limits task map assigns a cost for violating joint limits. The joint limits are loaded from the robot model. The mapping is calculated as:

$$\Phi_{\text{Bound}}(x) = \begin{cases} x - x_{\min} - \epsilon, & \text{if } x < x_{\min} + \epsilon \\ x - x_{\max} + \epsilon, & \text{if } x > x_{\max} - \epsilon , \\ 0, & \text{otherwise} \end{cases} \tag{2}$$

where $x_{\min}$ and $x_{\max}$ are lower and upper joint limits respectively, and $\epsilon \geq 0$ is a safety margin. The Jacobian and Jacobian derivative are identity matrices.

End-effector frame task map captures the relative transformation between the base frame B and the tip frame A:

$$\Phi_{\text{EffFrame}}(x) = M_A^B, \tag{3}$$

where $M_A^B \in SE(3)$ is computed using the system model using the *Scene*. We use the *task space vector* data structure (described later in this section) to handle storage and operations on spatial frames. The Jacobian of this task map is the geometric Jacobian computed by the *Scene*.

End-effector position captures the translation of the relative frame transformation:

$$\Phi_{\text{EffPos}}(x) = P_A^B, \tag{4}$$

where P_A^B is translational part of M_A^B. The Jacobian of this task consists of the rows of the geometric Jacobian corresponding to the translation of the frame.

End-effector orientation captures the rotation of the relative frame transformation:

$$\Phi_{\text{EffRot}}(x) = R_A^B, \tag{5}$$

---

[5]We use notation x for scalar values, x for vectors, X for matrices, and X for vectorized matrices.

where $R_A^B \in SO(3)$ is rotational part of M_A^B. Similarly to the *end-effector frame* task map, the storage and the operations on the resulting $SO(3)$ space are implemented within the *task space vector*. The Jacobian of this task consists of the rows of the geometric Jacobian corresponding to the rotation of the frame.

End-effector distance computes the Euclidean distance between the base and tip frames:

$$\Phi_{\text{Dist}}(x) = \| P_A^B \|. \tag{6}$$

The resulting task map has the same function as the *end-effector position* map. The output, however, is a scalar distance.

Center-of-mass task map computes the center-of-mass of all of the robot links defined in the system model:

$$\Phi_{\text{CoM}}(x) = \sum_i (P_{\text{CoM}_i}^{\text{world}} m_i), \tag{7}$$

where $P_{\text{CoM}_i}^{\text{world}}$ is the position of the center-of-mass of the ith link w.r.t. the world frame, and m_i is mass of the ith body. The Jacobian is computed using the chain rule. This task map can also be initialized to compute the projection of the center-of-mass on the xy-plane. In this case, the z-component is removed.

Collision spheres task map provides a differentiable collision distance metric. The collision shapes are approximated by spheres. Each sphere is attached to the kinematic structure of the robot or to the environment. Each sphere is then assigned a collision group, e.g. $i \in \mathcal{G}$. Spheres within the same group do not collide with each other, while spheres from different groups do. The collision cost is computed as:

$$\Phi_{\text{CSphere}}(x) = \sum_{i,j}^G \frac{1}{1 + e^{5\epsilon(\| P_i^{\text{world}} - P_j^{\text{world}} \| - r_i - r_j)}}, \tag{8}$$

where i, j are indices of spheres from different collision groups, ϵ is a precision parameter, P_i^{world} and P_j^{world} are positions of the centers of the spheres, and r_i, r_j are the radii of the spheres. The sigmoid function raises from 0 to 1, with the steepest slope at the point where the two spheres collide. Far objects contribute small amount of error while colliding objects produce relatively large amounts of error. The precision parameter can be used to adjust the fall-off of the error function, e.g. a precision factor of 10^3 will result in negligible error when the spheres are further than 10^{-3} m apart. The constant multiplier of 5 in Eq. (8) was chosen to achieve this fall-off profile.

In our example, we use the **end-effector position** task map. The task space is therefore $\Phi_{\text{EffPos}}(x) \in \mathbb{R}^3$. The task map is loaded from the XML file. The following lines of the XML configuration file correspond to the task map definition:

```
<Maps>
  <EffPosition Name="Position">
    <EndEffector>
    <Frame Link="lwr_arm_7_link" BaseOffset="0.5 0 0.5
       0 0 0 1"/>
    </EndEffector>
  </EffPosition>
</Maps>
```

where the only parameter of the task map is a single relative spatial frame. This frame defines the translation of the seventh robot link relative to the coordinates $(0.5, 0, 0.5)$ in the world frame.[6] This example is only intended to compute inverse kinematics, we have therefore chosen to only minimize the end-effector position error. However, an arbitrary number of cost terms can be added by adding multiple task maps to this problem definition. For instance, we could easily add another task map to constrain the orientation of the end-effector.

The output of a task map is a representation of the robot's state in the task space. Most task spaces are $\mathbb{R}^n$. As such, they can be stored and handled as vectors of real numbers. However, some task maps output configurations in the $SO(3)$ or the $SE(3)$ space. In this case, Lie group algebra [13] has to be used to correctly compute the additions and subtractions in the task space. The *task space vector* implements operations on task spaces. The *task space vector* is a data structure that keeps track of $SO(3)$ sub-groups within the stored vector. The operations on this vector then implement the Lie group algebra. For example, a spatial frame may be stored as a transformation matrix $M_A^B \in \mathbb{R}^{4 \times 4}$. This matrix will be stored in the *task space vector*. Performing addition and subtraction on the vector will then be incorrect. The correct transformation is performed by a matrix multiplication. The *task space vector* keeps track of transformations stored within its structure and applies the correct operations on them. Furthermore, the result of subtraction is always a geometric twist, e.g. $M_A^B - M_C^B = {}^B t_A^C$. This makes it possible to multiply the result of this operation with a geometric Jacobian, producing a geometrically correct relative transformation. This feature has been used in the implementation of the inverse kinematics solver [14] and the AICO solver [15]. Additionally, a $SO(3)$ rotation can be represented and stored in different ways, e.g. as a unit quaternion $R_Q \in \mathbb{R}^4$ where $\|R_Q\| = 1$, Euler angles $R_{ZYZ}, R_{ZYX}, R_{RPY} \in \mathbb{R}^3$, angle-axis representation $R_A \in \mathbb{R}^3$ where $\|R_A\| = \theta$, rotation matrix $R \in \mathbb{R}^{3 \times 3}$, etc. We handle these representations implicitly. Each sub-group of the *task space vector* stores the size and type of representation that was used. The operations on the vector first convert the task space coordinates into a rotation matrix representation, then the correct spatial operation is applied and a twist is computed. As a result the input and output dimension may vary, i.e. subtraction of two rotations represented as rotation matrices is a function $f(R_1, R_2) : \mathbb{R}^9 \to \mathbb{R}^3$. The result is the angular velocity component of the twist. The *task space vector* is composed by concatenating outputs of multiple task maps. Each task map specifies

---

[6]If no frame is specified, world frame is assumed by default. If a relative offset is not specified, an identity transformation offset is assumed.

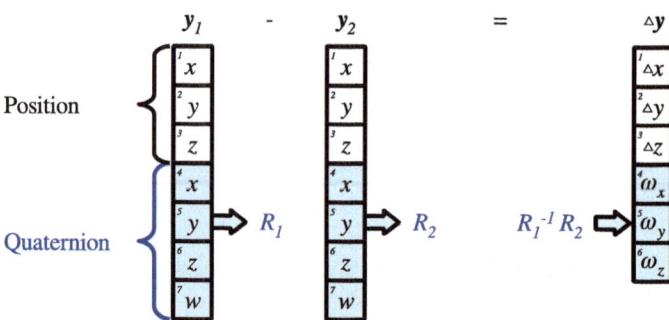

Fig. 4 Task space vector data packing combining three position coordinates x, y, $z \in \mathbb{R}$ and a sub-vector containing a $SO(3)$ rotation represented as a unit quaternion. The subtraction calculation of two task space vectors $\mathbf{y}_1$ and $\mathbf{y}_2$ first converts the quaternions into rotation matrices R_1 and R_2 and performs the rotation operation $R_2^{-1} R_1$. The result is then converted into angular velocities ω_x, ω_y, ω_z and packed into the output vector $\triangle y$. Notice that the dimensionality of $\triangle y \in \mathbb{R}^6$ and $\mathbf{y}_1, \mathbf{y}_2 \in \mathbb{R}^7$ are different

if its output contains any components that have to be handled using the Lie group algebra (see Fig. 4).

The output of a single task map is a segment of the *task space vector*. The input of a task map is the states of the robot model and environment as well as the arbitrary number of frame transformations required for the calculations. These are computed using the *planning scene*. The task map implements the mapping within its `update` method. This method has 3 different overloads depending on what order of derivative is requested: (a) no derivative (e.g. in sampling), (b) first-order derivatives (e.g. Jacobian used in gradient descent), and (c) second-order derivatives. Not all overloads have to be defined, i.e. a collision checking task map may only detect collisions but it will not provide any gradients (derivatives). We exploit this for fast collision checking for sampling-based solvers [5].

The task map will update the task space vector and its derivatives when the solver requires it. These updates are normally triggered by the solver and they do not have to be called manually. This also ensures that the *task space vector* is updated correctly. The collection of task maps is therefore central to formally defining motion planning problems. How the output of the task map is used then depends on the type of the planning problem.

4.2 Planning Problems

A *planning problem* within EXOTica represents a specific formulation of a motion planning problem. Since every formulation has very specific advantages for a particular type of application, the formulations may vary significantly. To provide a unified framework, we identify several categories of common features of different types of problems.

Depending on how the system is modeled, we distinguish: (a) kinematic, (b) kino-dynamic, and (c) dynamic systems. We then categorize the problem based on the *state representation* required by these types of systems: position (x), position and velocity $(x, \dot{x})$, and the full dynamic state $(x, \dot{x}, \ddot{x}, \tau, F)$ where the variables denote positions, velocities, accelerations, joint torques, and external forces respectively. We then distinguish between *planning spaces*: (a) configuration space, and (b) task space (e.g. end-effector position and orientation). These categories define how the state of the the system is stored. This affects both memory layout and the format of the input to the solver (e.g. the start state has to include joint positions and velocities).

Furthermore, we categorize the problem based on the type of the output. The *output type* may include: (a) single configuration (e.g. output of a inverse kinematics solver), (b) a time-indexed trajectory (e.g. output of trajectory optimization), or (c) non-time indexed trajectory (e.g. output of a sampling-based solver). Other types and subtypes of problems do exist. A time-indexed trajectory may have fixed or variable number of time steps or a variable timing between steps. The second category is related to the output type with respect to the controls. Here we consider control paradigms such as position control, velocity control, and torque control. The output of a problem may therefore consist of various combinations of position, velocity, acceleration, and torque trajectories. We refer to this as the *control type*.

Finally, we consider types of common problem formulations, for instance:

An **unconstrained quadratic cost minimization problem** w.r.t. metric Q as presented in [14]:

$$\operatorname*{argmin}_{x}(f(x)^{\top}Qf(x)). \tag{9}$$

A **linear programming problem**:

$$\operatorname*{argmin}_{x}(Qx + c) \tag{10}$$

$$\text{s.t. } Ax \leq b, \tag{11}$$

$$Bx = b. \tag{12}$$

A **quadratic programming problem with linear constraints**, e.g. used by the authors of [16]:

$$\operatorname*{argmin}_{x}(x^{\top}Qx + c^{\top}x)$$

$$\text{s.t. } Ax \leq b, \tag{13}$$

$$Bx = b. \tag{14}$$

A **generic non-linear programming problem**, e.g. as used in [17]:

Table 1 Naming convention of motion planning problems based on problem type categories. Gray name components are omitted for brevity

Planning space	Problem type	State representation	Output type
[CSpace]	[Unconstrained]	[Kinematic]	[EndPose]
[CSpace]	[Unconstrained]	[Kinematic]	[TimeIndexed]
[CSpace]	[Sampling]	[Kinematic]	[NonIndexed]
[CSpace]	[NLP]	[Dynamic]	[TimeIndexed]

$$\operatorname*{argmin}_{x} \| f(x) \|^2$$

$$\text{s.t. } g(x) \le 0, \tag{15}$$

$$h(x) = 0. \tag{16}$$

A **mixed-integer non-linear programming problem with constraints**, such as MIQCQP presented in [18]:

$$\operatorname*{argmin}_{x} \| f(x, i) \|^2$$

$$\text{s.t. } g(x, i) \le 0, \tag{17}$$

$$h(x, i) = 0, \tag{18}$$

$$i \in \mathbb{N}. \tag{19}$$

A **sampling problem**[7]: which is common across several algorithms presented in [5].

$$\operatorname*{arg}_{x} f(x) = \text{True}. \tag{20}$$

A **mixed sampling and optimization problem**, e.g. as presented in [19]:

$$\operatorname*{argmin}_{x} \| f(x) \|^2$$

$$\text{s.t. } g(x) = \text{True}. \tag{21}$$

These are just some of the commonly used formulations of motion planning problems used across literature. EXOTica provides the tools and a generic structure for implementing these types of problem formulations.

EXOTica uses a problem naming system based on this categorization. The names are constructed based on the four main categories: *planning space*, *problem type*,

---

[7]We refer to a problem with binary variables as a sampling problem because randomized or another type of sampling is required to solve them. These types of problems often cannot be solved by numerical optimization because their constraints are not differentiable.

state representation and the *output type*. Table 1 shows how the name is constructed. To achieve brevity, each category has a default type. The default types are *configuration space*, *sampling*, *kinematic*, *non-time indexed trajectory* respectively for each category in Table 1. When the problem falls within the default type for a category, this type is omitted from the name. For example, a problem of type `SamplingProblem` is referring to a configuration space sampling problem using a kinematic robot model and returning a non-time indexed trajectory. Similarly, `NLPDynamicTimeIndexedTorqueControlledProblem` is an example of a fully specified problem corresponding to the one defined in [20]. Three sample problem types which are implemented within EXOTica are highlighted here.

Unconstrained End-Pose Problem defines a problem minimizing the quadratic cost using a kinematic system. The state is represented by joint configurations and the output is a single configuration that minimizes the cost defined by Eq. (9). The cost function is composed of weighted components:

$$f(\boldsymbol{x}) = \sum_i \rho_i \|\Phi_i(\boldsymbol{x}) - \boldsymbol{y}_i^*\|, \tag{22}$$

where $\Phi_i(\boldsymbol{x})$ is the mapping function of the ith task map as defined in Sect. 4.1, $\boldsymbol{y}_i^*$ is the reference or goal in the task space, and ρ_i is the relative weighting of the task. By definition, This problem provides the first derivative of the quadratic cost function. The derivatives of the task terms are provided by the *task map*. Additionally, configuration space weighting W is specified. This allows us to scale the cost of moving each joint (or control variable) individually. Optionally, a nominal pose $\boldsymbol{x}_{\text{nominal}}$ is provided as a reference often used to minimize secondary cost in case of redundancies. This type of problem can be used for solving inverse kinematics problems as proposed in [14].

The example inverse kinematics problem is using this formulation. The full problem definition contains the definition of the *planning scene* and the *task map* as discussed previously. We also set the configuration space weighting W. The problem is then fully defined using the following XML string:

```
<UnconstrainedEndPoseProblem Name="MyProblem">
  <PlanningScene>
    <Scene>
      <JointGroup>arm</JointGroup>
      <URDF>{exotica_examples}/resources/robots/
          lwr_simplified.urdf</URDF>
      <SRDF>{exotica_examples}/resources/robots/
          lwr_simplified.srdf</SRDF>
    </Scene>
  </PlanningScene>
  <Maps>
    <EffPosition Name="Position">
      <EndEffector>
          <Frame Link="lwr_arm_7_link" BaseOffset="0.5 0
              0.5 0 0 0 1"/>
      </EndEffector>
    </EffPosition>
```

```
  </Maps>
  <W> 7 6 5 4 3 2 1 </W>
  <StartState>0 0 0 0 0 0 0</StartState>
  <NominalState>0 0 0 0 0 0 0</NominalState>
</UnconstrainedEndPoseProblem>
```

The weighting W is set to reduce movement of the joints closer to the root (root joint weight 7 to tip joint weight 1). We set the start and the nominal configuration, or state, to a zero vector. This problem is now complete and we can use it to compute the robot configuration which moves the top of the robot to coordinates $(0.5, 0, 0.5)$. EXOTica provides several useful tools that make defining problems using XML more versatile. All strings are automatically parsed as the required data types. Furthermore, file paths containing curly bracket macros will be replaced with catkin package paths, e.g. `{exotica_examples}` will get replaced with the absolute path to the EXOTica package.[8]

Unconstrained Time-Indexed Problem defines a problem minimizing the quadratic cost over a trajectory using a kinematic model of the system. The trajectory is uniformly discretized in time. The time step duration Δt and number of time steps T are specified. The system transition from one time step to another is parametrized by the covariance of the state transition error W (analogous to the weighting W used in the Unconstrained End-Pose Problem) and the covariance to the control error H. These parameters define the evolution of a stochastic system approximated by a linear model over time. See [21] for more details about this model. The cost function is then defined as:

$$f(\boldsymbol{x}) = \sum_t \sum_i \rho_{i,t} \|\Phi_{i,t}(\boldsymbol{x}) - \boldsymbol{y}_{i,t}^*\|, \tag{23}$$

where $t \in (1, \ldots, T)$ is the time index. This type of problem is suitable for use with iLQG-like (iterative linear-quadratic Gaussian [22]) solvers and with the approximate inference control (AICO) algorithm [15].

Sampling Problem defines a class of kinematic problems that do not require any cost function. Each state is evaluated for validity but not for quality as described in Eq. (20). This type of problem also requires a goal configuration $\boldsymbol{x}^*$. The objective of the solvers is to compute a valid trajectory from the start state to the goal state. The validity of each state is checked by applying a threshold ϵ on the output of the task map: $\rho_i(\Phi_i(\boldsymbol{x}) - \boldsymbol{y}_i^*) < \epsilon$. The output trajectory is not indexed on time and the number of configurations may vary between solutions. This type of planning problem is used with sampling-based motion solvers, such as RRT and PRM [5].

The updating of the task maps is always handled by the planning problem. This ensures that all the storage of the task-related properties is consistent and timely. Each problem class is also the storage container for all of the task related data and parameters. For example, the `UnconstrainedEndPoseProblem` stores the task weights ρ, the task map outputs $\Phi(\boldsymbol{x})$ and goals $\boldsymbol{y}^*$ in form of a *task space vector*, and the Jacobians of each task map $J = \frac{\partial \Phi(\boldsymbol{x})}{\partial \boldsymbol{x}}$. Since each problem has a different

---

[8]This feature combines regular expressions with the rospack library to parse catkin package paths.

formulation and structure, how the data is stored may vary. However, each problem has to fully contain all the data and parameters required by the solvers. This ensures modularity and makes it possible to benchmark different solvers on the same set of problems.

5 Motion Solvers

The structure of a planning problem within EXOTica allows us to formally define an interface for solving specific types of problems. The motion solver then takes the problem instance as input and computes the solution. How this computation is performed depends entirely on the implementation of the solver. EXOTica offers several built-in solvers.

In our example, we created a system model and an unconstrained end-pose problem that uses this system model. We will now use our implementation of the inverse kinematics solver to compute the solution.

The inverse kinematics solver implements the regularized, dampened Jacobian pseudo-inverse iterative algorithm described in [14]. This algorithm minimizes the cost function defined in Eq. (22) by iteratively improving the current solution until convergence as described in Algorithm 1.

Algorithm 1 IK solver

Require: C, α
1: $(x_0, x_{\text{nominal}}, W, max\_iter, \epsilon) \leftarrow$ GETPROBLEM
2: $x \leftarrow x_0$
3: $iter \leftarrow 0$
4: **repeat**
5: $\quad (\Phi, J) \leftarrow$ UPDATEPROBLEM(x)
6: $\quad J^\dagger \leftarrow W^{-1} J^{\mathsf{T}} (J W^{-1} J^{\mathsf{T}} + C)^{-1}$
7: $\quad \triangle x \leftarrow \alpha \left[J^\dagger (f(x) - y^*) + (I - J^\dagger J)(x_{\text{nominal}} - x) \right]$
8: $\quad x \leftarrow x + \triangle x$
9: $\quad iter \leftarrow iter + 1$
10: **until** $\| \triangle x \| < \epsilon$ and $iter < max\_iter$
11: **return** x

In Algorithm 1 x is the state of the robot, α is the convergence rate, $J = SJ(x)$ is the concatenated Jacobian matrix multiplied by a diagonal matrix $S = \text{diag}((\rho_1, \rho_2, \ldots))$ constructed from the task map weights ρ, Φ is the concatenated task space vector, I is the identity matrix, x_{nominal} is the nominal configuration that is achieved as a secondary goal in case of redundancies, W is the configuration space cost metric or weighting, and C is the regularization. The solver iterates until convergence ($\| \triangle x \| < \epsilon$) or until the maximum number of iterations is reached. This algorithm implements a local cost minimization method that requires access to the cost function $f(x)$ and its first derivative $J(x) = \frac{\partial f(x)}{\partial x}$. The cost function and its derivative are provided by the unconstrained end-pose problem in line 5. Parameters C and α are properties of the motion solver. Parameters x_0, x_{nominal}, W, max_iter and ϵ are properties of the planning problem and they are extracted using the GET-PROBLEM method. The output of this solver is a single robot configuration solving the inverse kinematics problem.

EXOTica was designed for development of new motion planning algorithms. We provide two more solvers that can be used for benchmarks. The AICO solver is the implementation of the **A**pproximate **I**nference **CO**ntrol algorithm presented in [6]. This algorithm performs trajectory optimization on problems with a quadratic approximation of the cost, linear approximation of the system model, uniformly discretized time axis, and no hard constraints. The complete algorithm, including the pseudo code, is described in [6]. The implementation of AICO within EXOTica uses the unconstrained time-indexed problem to calculate and store all task related properties. The OMPL solver is a wrapper exposing several sampling-based algorithms implemented within the OPEN MOTION PLANNING LIBRARY [8].

6 Python Wrapper

The utility of the EXOTica framework is in fast prototyping of motion synthesis algorithms. The modular structure of EXOTica provides tools for creating new problem and solver types which can now be evaluated and compared with competing algorithms. Setting up tests and benchmarks is straightforward as all parameters of the algorithm are clearly exposed. To make this process even more versatile, we provide a Python wrapper for EXOTica. The following code shows how to initialize and solve our example inverse kinematics problem using Python:

```python
import pyexotica as exo
solver = exo.Setup.loadSolver('{exotica_examples}/
    resources/configs/example.xml')
print(solver.solve())
```

The EXOTica Python wrapper is intended for instantiating problems and solvers to provide a high level interface to the EXOTica tools. This interface is suitable for creating planning services, benchmarks, and unit tests. All core classes and methods are exposed.

7 Applications

The rest of this chapter will provide examples of how the different elements of EXOTica are leveraged for prototyping and evaluating new algorithms.

7.1 Algorithm Benchmarking

The modularity of EXOTica enables us to create benchmarks comparing a variety of problems and solvers. In [23], we construct such benchmark to evaluate several sampling-based algorithms implemented inside the Open Motion Planning Library [8] (OMPL) on a set of reaching problems on a humanoid robot. These algorithms were primarily designed for solving navigation problems and motion planning for fixed base robot arms. However, we have applied these methods to planning whole-body trajectories for manipulation in cluttered environments using humanoid robots.

Valid trajectories for humanoid robots can only contain states that are collision-free while they also have to satisfy additional constraints such as center-of-mass position, foot placement, and sometimes torso orientation. Generating collision-free samples is straightforward by using random sample generators and standard collision checking libraries. However, the additional constraints create a manifold in the unconstrained configuration space. Generating samples which lie on this manifold without having to discard a majority of them in the process is non-trivial. A sampling bias has to be introduced to increase the probability of generating correctly constrained samples. In our approach, a whole-body inverse kinematic solver is employed to produce the constrained samples. We formulate the inverse kinematics problems as a non-linear program (NLP):

$$\operatorname*{argmin}_{\boldsymbol{x}} \|\boldsymbol{x} - \boldsymbol{x}_{\text{nominal}}\|_Q^2,$$
$$\text{s.t. } \boldsymbol{b}_l \leq \boldsymbol{x} \leq \boldsymbol{b}_u,$$
$$c_i(\boldsymbol{x}) \leq 0, i \in \mathcal{C} \tag{24}$$

where $\|\boldsymbol{x} - \boldsymbol{x}_{\text{nominal}}\|_Q^2$ is the squared deviation from the nominal pose $\boldsymbol{x}_{\text{nominal}}$ with respect to a configuration space metric Q. The system is subject to lower and upper bound constraints $\boldsymbol{b}_l$ and $\boldsymbol{b}_u$, and a set of non-linear constraints $\mathcal{C}$. The solver is described in [16]. We will call this solver using the following routine $IK(\boldsymbol{x}_0, \boldsymbol{x}_{\text{nominal}}, \mathcal{C})$, where $\boldsymbol{x}_0$ is start state, $\boldsymbol{x}_{\text{nominal}}$ is the nominal state and $\mathcal{C}$ is the set of constraints.

The majority of algorithms implemented in OMPL perform three basic steps: (1) sample a random state, (2) perform steering, (to compute a *near state* that is close to the random state according to some metric), (3) append the *near state* to the solution if it satisfies all constraints. To preserve compatibility with these algorithms, we augment steps 1 and 2. In step 1, we sample a random unconstrained configuration and return the constrained sample computed using the inverse kinematics (see routine `sampleUniform` in Algorithm 2). In the second step, we then compute the constrained near state using the `sampleUniformNear` routine. The steering function then uses an interpolation routine to check if a path from the current state to the near state is viable. We have augmented the interpolation routine as shown in Algorithm 2 In EXOTica, a new problem type is defined called the `ConstrainedSamplingProblem`. The constraints defined within this problem

Algorithm 2 Humanoid Configuration Space Sampling-based Planning

---

sampleUniform()
1: *succeed* = False
2: **while not** *succeed* **do**
3:     $\bar{\mathbf{x}}_{rand} = RandomConfiguration()$
4:     $\mathbf{x}_{rand}, succeed = IK(\bar{\mathbf{x}}_{rand}, \bar{\mathbf{x}}_{rand}, \mathcal{C})$
    return $\mathbf{x}_{rand}$

---

sampleUniformNear($\mathbf{x}_{near}, d$)
1: *succeed* = False
2: **while not** *succeed* **do**
3:     $a = 0$
4:     **while not** *succeed* **do**
5:         $\bar{\mathbf{x}}_{rand} = RandomNear(\mathbf{x}_{near}, d)$
6:         $\mathcal{C}_{extended} = \mathcal{C}$
7:         $\mathcal{C}_{extended} \leftarrow \|\mathbf{x}_{rand} - \bar{\mathbf{x}}_{rand}\|_Q < a$
8:         $(\mathbf{x}_{rand}, succeed) = IK(\bar{\mathbf{x}}_{rand}, \mathbf{x}_{near}, \mathcal{C}_{extended})$
9:         Increase a
10:     **if** $distance(\mathbf{x}_{rand}, \mathbf{x}_{near}) > d$ **then**
11:         *succeed* = False
    return $\mathbf{x}_{rand}$

---

interpolate($\mathbf{x}_a, \mathbf{x}_b, d$)
1: $\bar{\mathbf{x}}_{int} = InterpolateConfigurationSpace(\mathbf{x}_a, \mathbf{x}_b, d)$
2: *succeed* = False
3: $a = 0$
4: **while not** *succeed* **do**
5:     $\mathcal{C}_{extended} = \mathcal{C}$
6:     $\mathcal{C}_{extended} \leftarrow \|\mathbf{x}_{int} - \bar{\mathbf{x}}_{int}\|_Q < a$
7:     $(\mathbf{x}_{int}, succeed) = IK(\bar{\mathbf{x}}_{int}, \mathbf{x}_a, \mathcal{C}_{extended})$
8:     Increase a
    return $\mathbf{x}_{int}$

---

are passed to the *IK* solver when a constrained solution is required. The motion solver then instantiates the OMPL algorithm overriding the default sampling and and interpolation methods with the ones defined in Algorithm 2.

The second category of problems perform the exploration in the task space. Specifically, the space of positions and orientations of the gripper is used as the state space. We call this space the *end-effector space*. The pose of the gripper does not uniquely describe the configuration of the whole robot. Each gripper position is therefore associated with a full robot configuration to avoid redundancies in the representation. This type of planning problem is using a low dimensional state representation but the connectivity of states depends on the configuration of the whole robot. Therefore, each valid state lies on a manifold in the end-effector space. As a result we formulate a very similar problem to the constrained sampling problem in the configuration space. To implement this, we augmented the solver by replacing the sampling and steering steps of the algorithm with the inverse kinematics solver as described in Eq. (24). Additionally, we use the forward kinematics $\Phi(\mathbf{x})$ to compute

---

Algorithm 3 Humanoid End-Effector Space Sampling-based Planning

---

sampleUniform()
1: $succeed = $ False
2: **while not** $succeed$ **do**
3:    $\bar{\mathbf{y}}_{rand} = RandomSE3()$
4:    $\mathcal{C}_{extended} = \mathcal{C}$
5:    $\mathcal{C}_{extended} \leftarrow \|\bar{\mathbf{y}}_{rand} - \Phi(\mathbf{x}_{rand})\| \leq 0$
6:    $\mathbf{x}_{rand}, succeed = IK(\bar{\mathbf{x}}_{rand}, \bar{\mathbf{x}}_{rand}, \mathcal{C}_{extended})$
7: $\mathbf{y}_{rand} = \Phi(\mathbf{x}_{rand})$
   return $\mathbf{y}_{rand}, \mathbf{x}_{rand}$

---

sampleUniformNear($\mathbf{y}_{near}, d$)
1: $succeed = $ False
2: **while not** $succeed$ **do**
3:    $\bar{\mathbf{y}}_{rand} = RandomNearSE3(\mathbf{y}_{near}, d)$
4:    $\mathcal{C}_{extended} = \mathcal{C}$
5:    $\mathcal{C}_{extended} \leftarrow \|\bar{\mathbf{y}}_{rand} - \Phi(\mathbf{x}_{rand})\| \leq 0$
6:    $\mathbf{x}_{rand}, succeed = IK(\mathbf{x}_{rand}, \mathbf{x}_{near}, \mathcal{C}_{extended})$
7: $\mathbf{y}_{rand} = \bar{\mathbf{y}}_{rand}$
   return $\mathbf{y}_{rand}, \mathbf{x}_{rand}$

---

interpolate($\mathbf{y}_a, \mathbf{y}_b, d$)
1: $\bar{\mathbf{y}}_{int} = InterpolateSE3(\mathbf{y}_a, \mathbf{y}_b, d)$
2: $succeed = $ False
3: $b = 0$
4: **while not** $succeed$ **do**
5:    $\mathcal{C}_{extended} = \mathcal{C}$
6:    $\mathcal{C}_{extended} \leftarrow \|\bar{\mathbf{y}}_{int} - \Phi(\mathbf{x}_{int})\| < b$
7:    $\mathbf{x}_{int}, succeed = IK(\mathbf{x}_a, \mathbf{x}_a, \mathcal{C}_{extended})$
8:    Increase b
9: $\mathbf{y}_{int} = \Phi(\mathbf{x}_{int})$
   return $\mathbf{y}_{int}, \mathbf{x}_{int}$

---

the gripper pose corresponding to a whole robot configuration x. Algorithm 3 show the modifications to Algorithm 2 required to perform planning in the end-effector space. We have implemented this type of problem in EXOTica and we call it the `ConstrainedEndEffectorSamplingProblem`.

With the problem formulation in place, we created several environments containing obstacles such as desks and shelves. We have also defined grasping targets placed on top of the horizontal surfaces. Each environment was then tested with two humanoid robot models: NASA Valkyrie and Boston Dynamics Atlas. Figure 5 shows several scenarios used in the benchmark. We have performed 100 trials in each environment with each robot. We have collected data on computation time, task accuracy, and constraint violation. The results ultimately show that existing (off the shelf) sampling based motion solvers can be used with humanoid robots. The RRT-Connect algorithm outperforms all other methods when planning in configuration

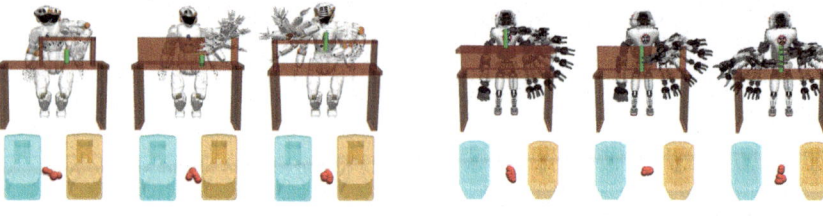

(a) Reaching motion on the NASA
Valkyrie robot.

(b) Reaching motion on the Boston
Dynamics Atlas robot.

Fig. 5 Collision-free whole-body motion generated in different environments with different robot models. The corresponding CoM trajectories are illustrated in the second row (red dots). The benchmark is designed so that one can easily switch to new robot platforms without extensive integration effort

space, with planning times averaging at 3.45 s in the most challenging test scenario. We present the complete benchmark and the analysis in [23].

The benchmark described in this section was aimed at a specific aspect of sampling-based motion solvers. EXOTica enabled us to perform this analysis while maximizing the reuse of existing code. The problem and solver formulations described in this chapter are suitable for creating benchmarks evaluating various aspects of motion planning. In the next section, we explore and compare different problem formulations and how EXOTica can be used to validate that a particular formulation is suitable for solving a very specific task.

7.2 Dynamic Motion Adaptation and Replanning in Alternate Spaces

Adapting motion to dynamic changes is an active area of research. Dynamic environments cause existing plans to become invalid, and it is often necessary to adapt an exiting motion or to re-plan the motion entirely. This process can be computationally expensive when the problem is non-linear and contains local minima. However, a highly non-linear and non-convex problem can sometimes be simplified if we choose an *alternate task space*. A task space introduces a new metric or even a new topology. As a result, motion that is very complex in the configuration may be as simple as a linear interpolation in the appropriate task space. Figure 6 illustrates the effect of using an alternate mapping. EXOTica allows us to define such new task spaces.

In [24], we implemented a novel task map we call the *interaction mesh*. To construct the interaction mesh, we define vertices, points attached to the kinematic structure of the robot or the scene. We then define edges between the vertices to create a mesh. Each edge represents a spatial relationship that has to be preserved. In [24], the mesh is fully connecting all of the vertices. We then compute the Laplace coordinate

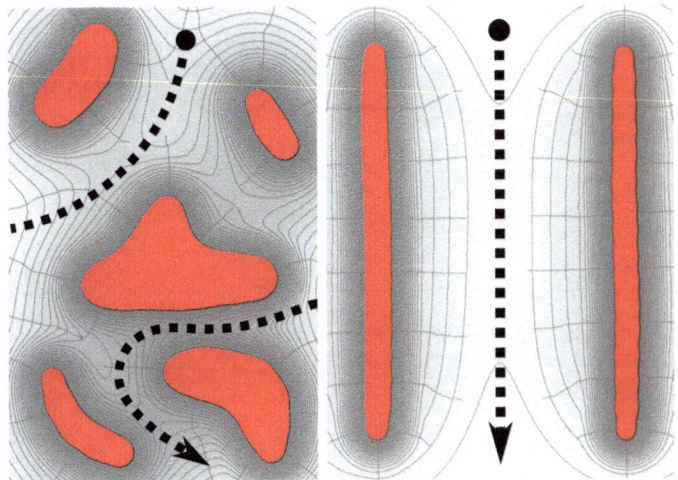

Fig. 6 A complex trajectory in the configuration space (left) becomes a linear trajectory in an alternate space (right)

L_G of each vertex p as

$$L_G(p) = p - \sum_{r \in \partial_G p} \frac{r\, w_{pr}}{\sum_{s \in \partial_G p} w_{ps}},$$

$$w_{pr} = \frac{W_{pr}}{|r - p|}, w_{ps} = \frac{W_{ps}}{|s - p|}, \tag{25}$$

where $\partial_G p$ is the neighbourhood of p in the mesh G, and w_{pr} is a weight which we made inversely proportional to the distance of vertices p, r and multiplied by the edge importance weighting[9] W_{pr}. The weights are then normalized over the neighbouring nodes $s \in \partial_G p$. The position of the vertices p, r and s is computed using kinematics (e.g. $p = \Phi(x)$). The Jacobian of the interaction mesh can be derived using the chain rule.

In EXOTica, we define a task map which implements Eq. (25). This task maps the vertex positions into the space of their Laplace coordinates. We call this the interaction mesh space. In the experiments presented in [24], we recorded a reference trajectory in the interaction mesh space. We have then formulated an unconstrained end-pose problem as defined in Sect. 4.2 to minimize the error in this space. We employed the inverse kinematics solver described in Sect. 5 in a control loop to track the dynamically moving targets in real-time. We have closed the feedback loop with motion tracking data streamed in real-time. Figure 7 shows the timelapse of this experiment.

---

[9]The edge importance weight matrix allows us to further parametrize the representation. For example, high weighting between the target and the end-effector allows us to perform accurate reaching and grasping.

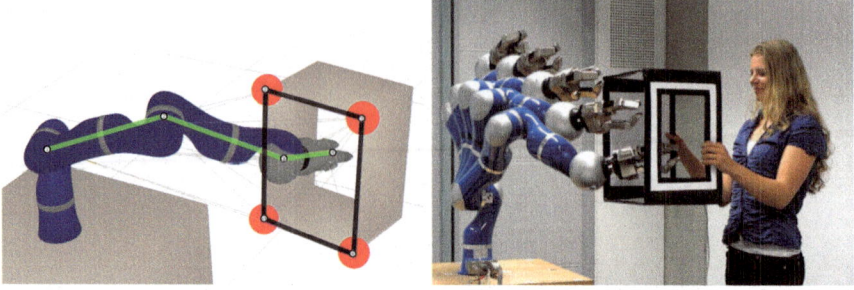

Fig. 7 The vertices and edges of the interaction mesh attached to the kinematic structure of the robot and parts of the dynamically moving obstacle (left), and the timelapse of the optimised motion (right)

Applying local optimisation methods in alternate spaces is a powerful tool. We have achieved fast and robust motion transfer from a human demonstrator to a robot using the interaction mesh and a motion capture suit (see Fig. 8a). This principle has been further studied in [25], where we introduce a more flexible variant of the interaction mesh. We call this alternate representation the *distance mesh*. Distance mesh captures the spatial relationships over edges, rather than the vertices of the mesh. This allows us to encode obstacle avoidance and target reaching behavior more directly. We have demonstrated this technique by performing a welding task, maintaining the contact of the welding tool with the work piece, while reactively avoiding moving obstacles (see Fig. 8b). This experiment also uses the unconstrained end-pose problem formulation and the inverse kinematics solver described in this chapter.

In [24], we also explore a class of representations based on topology of the work space. These alternate spaces abstract away the geometry of the space and capture topological metrics such as winding (see Fig. 8d), wrapping, and enclosing volumes of space. We have applied these metrics to reaching and grasping problems. Each of these representations introduce a space with a topology very different to the topology of the configuration space. However, each of these spaces is still implemented as a task map, and as such, it can be used with any solver within EXOTica. This work was then extended in [26] to optimize area coverage when moving robot links around an object of interest, such as 3D scanning and painting (see Fig. 8c).

The concept of encapsulation of the task map makes it straight forward to define new task spaces. Constructing experiments to evaluate the utility of each space therefore requires only very minimal implementation. This makes EXOTica ideal for rapid prototyping of new task spaces and defining very specialized problems. Once a suitable planning problem is identified, EXOTica can be used to implement a motion planning service that can be integrated with sensing and user interface modules to solve more complex tasks. We will describe this in the next section.

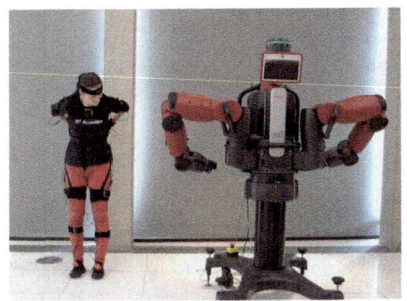

(a) Motion transfer using interaction mesh.

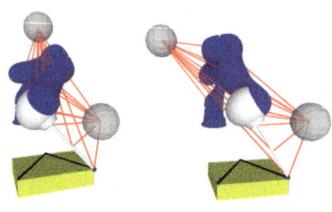

(b) Reactive obstacle avoidance using distance mesh.

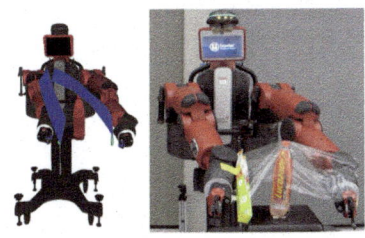

(c) Wrapping an item using spatio-temporal flux representation.

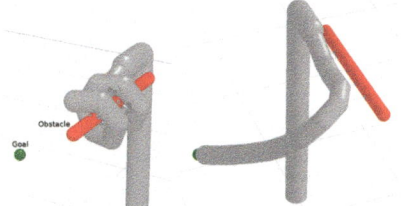

(d) Untangling a multi-link robot using the writhe representation.

Fig. 8 Examples of implementations of different task spaces using the EXOTica framework

7.3 Robust Shared Autonomy with Continuous Scene Monitoring

The level of autonomy of a robot is directly correlated with the predictability of the task and environment. Robots executing a predictable task in a predictable environment such as in a factory setting can work fully autonomously, while for field robots where the environment changes unpredictably, teleoperation is still the accepted gold standard with the human operators as the decision makers and the robot acting as their extension. We focus on a hybrid approach called *shared autonomy*. It is often perceived as a middle ground, combining autonomous sequences requested by the operator and executed by the robot. The operator provides high-level decision making reducing cognitive load and improving the reliability of the systems.

The work in [27] illustrates a complete framework and system application for robust shared autonomy which builds extensively on EXOTica. This systems is composed of four core components:

(1) A *mapping* module acquiring, filtering, and fusing several sources of real sensor data and using it to continuously update the EXOTica planning and collision scene.

(2) An implementation of the *Inverse Dynamic Reachability Map* (iDRM) presented in [28] in EXOTica and integration of the iDRM-based inverse kinematics solver with sampling based motion planning.

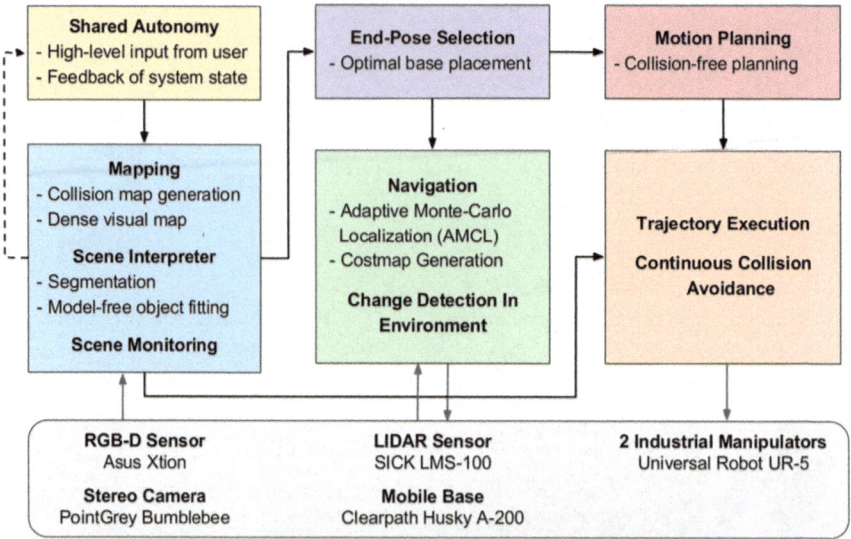

Fig. 9 Overview of the mapping, motion planning, scene monitoring and user interface components with existing ROS-based navigation, control, sensor driver packages

(3) A *user interface* for synthesizing EXOTica planning problems, where the user provides constraint sets for both inverse kinematics and motion planning.

(4) A *scene monitoring* service for detecting and assessing dynamic changes in the environment in real-time and updating the map and the EXOTica planning scene.

Figure 9 shows an overview of the system.

The details of the mapping and user interface components provide updates for the EXOTica planning scene. The sensor data is being processed and filtered to produce an OctoMap that is used for collision avoidance. The details of this process are described in [27]. Other collision objects such as CAD models of known structures can be inserted into the EXOTica scene directly. The most up-to-date snapshot of the scene is then used for motion planning.

The operator interface also relies on the scene monitoring and mapping components to provide the user with a virtual view of the robot environment. The user then specifies high-level goals, such as an object to grasp. The user interface component will process this input and create a set of goals that may contain a grasping target, a navigation target, and motion plan to pick-up or place an item. We compute these goals using EXOTica.

Given a grasping target location from the user, we construct an end-pose planning problem. Since the target location is known but the base of the robot is not fixed, we invert the problem. We compute the location of the base of the robot using the inverse kinematics solver. Here we use the reaching target as a virtual fixed base and we compute an optimal location of the robot base. The problem is formulated as `iDRMEndPoseProblem`. We exploit the iDRM structure to compute collision-free

Fig. 10 (Left) The continuous scene monitoring continuously integrates fused and filtered sensor data in an OctoMap and reasons about changes: Here, a human reaches into the robot working space crossing the planned trajectory (key samples in blue), and the robot halts execution (robot state in red). (Right) Candidate robot base locations computed using iDRM (red and green cubes colored by reachability) and a the optimal robot pose for grasping the red box selected by the user

inverse kinematics for the robot base as proposed in [28, 29]. We have implemented a dedicated solver for this type of problems. The output of the solver is a collision-free robot configuration that we use as a goal for sampling-based motion planning. Additionally, we use the computed robot base location as a target for navigation. The motion planning is then performed using the OMPL interface we have described in Sects. 5 and 7.1.

The motion planning component is implemented as a ROS action server. This creates a clean interface with the rest of the system. We have performed several experiments with the integrated system. Figure 10 shows snapshots of the system performing a pick-and-place task on a bi-manual robot with a mobile base.

8 EXOTica Installation Instructions

The up to date installation instructions are available at:

https://ipab-slmc.github.io/exotica/Installation.html

Example code is available within the `exotica_examples` package. We provide several other examples solving different types of problems as well as applications and tests within this package.

9 Conclusion

The utility of ROS lies in the modularity of its packages. With this in mind, EXOTica was developed as a tool to prototype, evaluate, and rapidly deploy novel motion synthesis algorithms. Existing motion planning tools such as MoveIt! provide mature algorithms and well defined but fairly restrictive interfaces. EXOTica complements this with a more generic and open architecture that allows for rapid prototyping and benchmarking of algorithms at a much earlier stage of development.

References

1. S. Chitta, I. Sucan, S. Cousins, Moveit! [ros topics]. IEEE Robot. Autom. Mag. (RAM) **19**(1), 18–19 (2012)
2. D.V. Lu, M. Ferguson, E. Marder-Eppstein, ROS Navigation Stack, https://github.com/ros-planning/navigation
3. J.-L. Blanco-Claraco, Reactive navigation for 2D robots using MRPT navigation algorithms, http://wiki.ros.org/mrpt_reactivenav2d
4. B. Siciliano, O. Khatib, *Springer Handbook of Robotics* (Springer, Berlin, 2008)
5. S.M. LaValle, *Planning Algorithms* (Cambridge University Press, Cambridge, 2006), http://planning.cs.uiuc.edu/
6. M. Toussaint, Robot trajectory optimization using approximate inference, in *Proceedings of the 26th Annual International Conference on Machine Learning (ICML) ICML '09* (ACM, USA, 2009), pp. 1049–1056
7. J.J. Kuffner, S.M. LaValle, Rrt-connect: an efficient approach to single-query path planning, in *Proceedings of IEEE International Conference on Robotics and Automation (ICRA)* vol. 2 (2000), pp. 995–1001
8. I.A. Şucan, M. Moll, L.E. Kavraki, The open motion planning library. IEEE Robot. Autom. Mag. (RAM) **19**(4), 72–82 (2012), http://ompl.kavrakilab.org
9. A. Hornung, K.M. Wurm, M. Bennewitz, C. Stachniss, W. Burgard, Octomap: an efficient probabilistic 3D mapping framework based on octrees. Auton. Robots (2013), http://octomap.github.com
10. R. Smits, KDL: Kinematics and Dynamics Library, http://www.orocos.org/kdl
11. D. Hershberger, D. Gossow, J. Faust, RViz: 3D visualization tool for ROS, http://wiki.ros.org/rviz
12. T. Foote, E. Marder-Eppstein, W. Meeussen, TF2: transform library for ROS, http://wiki.ros.org/tf2
13. J.-L. Blanco, A tutorial on se(3) transformation parameterizations and on-manifold optimization, Technical report, University of Malaga (2010)
14. Y. Nakamura, H. Hanafusa, Inverse kinematic solutions with singularity robustness for robot manipulator control. ASME Trans. J. Dyn. Syst. Meas. Control **108**, 163–171 (1986)
15. K. Rawlik, M. Toussaint, S. Vijayakumar, On stochastic optimal control and reinforcement learning by approximate inference (extended abstract), in *Proceedings of the 23rd International Joint Conference on Artificial Intelligence (IJCAI)* (AAAI Press, 2013), pp. 3052–3056
16. H. Dai, A. Valenzuela, R. Tedrake, Whole-body motion planning with centroidal dynamics and full kinematics, in *Proceedings of IEEE-RAS International Conference on Humanoid Robots (Humanoids)* (2014), pp. 295–302
17. S. Bradley, A. Hax, T. Magnanti, *Applied Mathematical Programming* (Addison-Wesley, USA, 1977)
18. R. Deits, R. Tedrake, Footstep planning on uneven terrain with mixed-integer convex optimization, in *Proceedings of IEEE-RAS International Conference on Humanoid Robots (Humanoids)* (2014), pp. 279–286
19. B.D. Luders, S. Karaman, J.P. How, Robust sampling-based motion planning with asymptotic optimality guarantees, in *Proceedings of the AIAA Guidance, Navigation, and Control (GNC) Conference*, American Institute of Aeronautics and Astronautics (2013)

20. M. Hutter, H. Sommer, C. Gehring, M. Hoepflinger, M. Bloesch, R. Siegwart, Quadrupedal locomotion using hierarchical operational space control. Int. J. Robot. Res. (IJRR) **33**(8), 1047–1062 (2014)
21. M. Toussaint, A tutorial on Newton methods for constrained trajectory optimization and relations to SLAM, Gaussian Process smoothing, optimal control, and probabilistic inference, in *Geometric and Numerical Foundations of Movements*, ed. by J.-P. Laumond (Springer, Berlin, 2017)
22. E. Todorov, W. Li, A generalized iterative LQG method for locally-optimal feedback control of constrained nonlinear stochastic systems, in *Proceedings of the American Control Conference* (2005), pp. 300–306
23. Y. Yang, V. Ivan, W. Merkt, and S. Vijayakumar, "Scaling Sampling-based Motion Planning to Humanoid Robots," in *Proceedings of IEEE International Conference on Robotics and Biomimetics (ROBIO)*, pp. 1448–1454, 2016
24. V. Ivan, D. Zarubin, M. Toussaint, T. Komura, S. Vijayakumar, Topology-based representations for motion planning and generalization in dynamic environments with interactions. Int. J. Robot. Res. (IJRR) **32**, 1151–1163 (2013)
25. Y. Yang, V. Ivan, S. Vijayakumar, Real-time motion adaptation using relative distance space representation, in *Proceedings of IEEE International Conference on Advanced Robotics (ICAR)* (2015), pp. 21–27
26. V. Ivan, S. Vijayakumar, Space-time area coverage control for robot motion synthesis, in *Proceedings of IEEE International Conference on Advanced Robotics (ICAR)* 2015, pp. 207–212
27. W. Merkt, Y. Yang, T. Stouraitis, C.E. Mower, M. Fallon, S. Vijayakumar, Robust shared autonomy for mobile manipulation with continuous scene monitoring, in *Proceedings of IEEE International Conference on Automation Science and Engineering (CASE)* (2017), pp. 130–137
28. Y. Yang, V. Ivan, Z. Li, M. Fallon, S. Vijayakumar, iDRM: humanoid motion planning with realtime end-pose selection in complex environments, in *Proceedings of IEEE-RAS 16th International Conference on Humanoid Robots (Humanoids)* (2016), pp. 271–278
29. Y. Yang, W. Merkt, H. Ferrolho, V. Ivan, S. Vijayakumar, Efficient humanoid motion planning on uneven terrain using paired forward-inverse dynamic reachability maps. IEEE Robot. Autom. Lett. (RA-L) **2**(4), 2279–2286 (2017)

Vladimir Ivan received his Ph.D. on the topic of Motion synthesis in topology-based representations at the University of Edinburgh where he is currently working as a Research Associate in the School of Informatics. He has previously received a M.Sc. in Artificial Intelligence specializing in Intelligent Robotics at the University of Edinburgh and a B.Sc. in AI and Robotics from the University of Bedfordshire. Vladimir has published over 20 peer-reviewed papers in top level conferences and journals. He contributed to several UK and EU funded academic research projects as well as industry-led projects with partners within EU and Japan and a collaboration with NASA-JSC.

Yiming Yang completed his Ph.D. in Robotics from the University of Edinburgh, in December 2017. His main areas of research interest are robot planning and control algorithms, human-robot interaction, collaborative robots, shared autonomy and machine learning for robotics. He is currently working as a Research Associate at the School of Informatics, University of Edinburgh, where he has been involved in several research projects in robotics, such as the NASA Valkyrie humanoid project, Hitachi Logistics Robotics, collaborative mobile robot manipulation is complex scenes, etc. Yiming has published over ten peer-reviewed articles in international journals, conferences and workshops.

Wolfgang Merkt is a Ph.D. student in Robotics at the University of Edinburgh, with a focus on robust online replanning in high-dimensional and real-world applications through efficient precomputation, machine learning, and storage. Wolfgang obtained an M.Sc. in Robotics from the

University of Edinburgh focusing on motion planning, perception, and shared autonomy, and a B.Eng. in Mechanical Engineering with Management with a thesis on human-to-humanoid motion transfer using topological representations and machine learning. He has been involved in several research projects in robotics, such as the NASA Valkyrie humanoid project, collaborative mobile robot manipulation in complex scenes, and previously worked in service robotics and on industrial robotics applications for small and medium enterprises (co-bots).

Michael P. Camilleri is a Doctoral Student in the CDT in Data Science at the University of Edinburgh, investigating Time-Series Modelling for mouse behavior. His interests are in probabilistic machine learning, Bayesian statistics and Deep Learning. Michael obtained an M.Sc. in Artificial Intelligence at the University of Edinburgh, focusing on the motion planning for the ATLAS robot for the DARPA Robotics Challenge 2013. Following this, he was employed with the same University, and was the lead developer of the original iterations of EXOTica. He subsequently spent time working in industry and academia, in various software engineering, management and lecturing positions.

Sethu Vijayakumar Professor FRSE holds a Personal Chair in Robotics within the School of Informatics at the University of Edinburgh and is the Director of the Edinburgh Centre for Robotics. Since 2007, he holds the Senior Research Fellowship of the Royal Academy of Engineering, co-funded by Microsoft Research and is also an Adjunct Faculty of the University of Southern California (USC), Los Angeles and a Visiting Research Scientist at the ATR Computational Neuroscience Labs, Kyoto-Japan. He has a Ph.D. (1998) in Computer Science and Engineering from the Tokyo Institute of Technology. His research interests include planning and control of anthropomorphic robotic systems, statistical machine learning, human motor control and rehabilitation robotics and he has published over 170 refereed papers in journals and conferences in these areas.

Online Trajectory Optimization and Navigation in Dynamic Environments in ROS

Franz Albers, Christoph Rösmann, Frank Hoffmann
and Torsten Bertram

Abstract This tutorial chapter provides a comprehensive step-by-step guide on the setup of the navigation stack and the *teb_local_planner* package for mobile robot navigation in dynamic environments. The *teb_local_planner* explicitly considers dynamic obstacles and their predicted motions to plan an optimal collision-free trajectory. The chapter introduces a novel plugin to the *costmap_converter* ROS package which supports the detection and motion estimation of moving objects from the local costmap. This tutorial covers the theoretical foundations of the obstacle detection and trajectory optimization in dynamic scenarios. The presentation is designated for ROS Kinetic and Lunar and both packages will be maintained in future ROS distributions.

Keywords Navigation · Local planning · Online trajectory optimization
Dynamic obstacles · Dynamic environment · Obstacle tracking

1 Introduction

In the context of service robotics and autonomous transportation systems mobile robots are required to safely navigate environments populated with humans and other robots. On this occasion, universally applicable motion planning strategies are of utmost importance in mobile robot applications. Online planning is favored over offline approaches as it responds to dynamic environments, map inconsistencies or robot motion uncertainty. Furthermore, online trajectory optimization conciliates

F. Albers (✉) · C. Rösmann · F. Hoffmann · T. Bertram
Institute of Control Theory and Systems Engineering, TU Dortmund University,
44227 Dortmund, Germany
e-mail: franz.albers@tu-dortmund.de

C. Rösmann
e-mail: christoph.roesmann@tu-dortmund.de

F. Hoffmann
e-mail: frank.hoffmann@tu-dortmund.de

T. Bertram
e-mail: torsten.bertram@tu-dortmund.de

© Springer International Publishing AG, part of Springer Nature 2019
A. Koubaa (ed.), *Robot Operating System (ROS)*, Studies in Computational
Intelligence 778, https://doi.org/10.1007/978-3-319-91590-6_8

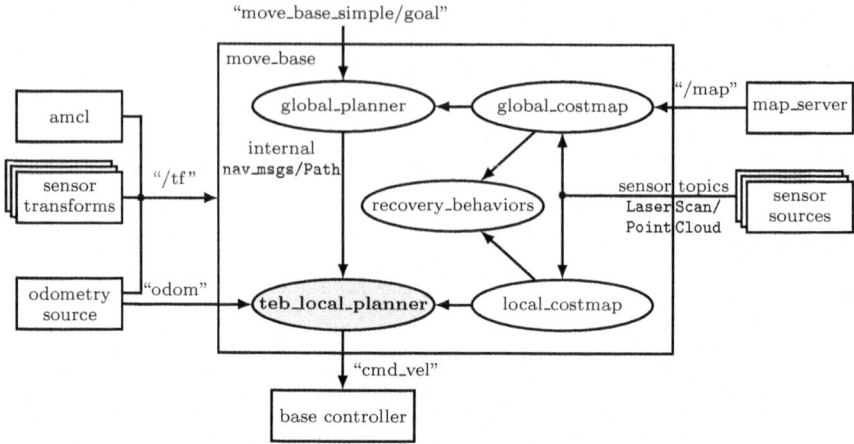

Fig. 1 Overview of the ROS navigation stack including the *teb_local_planner*

among partially conflicting objectives such as control effort, path fidelity, overall
path length or transition time.

The navigation stack[1] [12] along with its plugins for environment representations
(in terms of occupancy grids respectively costmaps) and local and global planners
constitutes a widely established framework for mobile robot navigation in the Robot
Operating System (ROS). Figure 1 shows an overview of the navigation stack setup[2]
employed in the course of this chapter.

A *global_costmap* is generated based on a known map of the environment
(published by the *map_server*). Additionally, sensor readings which are classified
as observations of static obstacles are incorporated into the *global_costmap*. The
global_planner computes an initial path to the goal based on the *global_costmap*.
This path does neither consider any time information nor obstacles which are not
represented in the *global_costmap* (for example because they are dynamic or simply
were not present at the time of mapping). In order to consider these obstacles as well,
the *local_costmap* is generated from fused readings of the robot's sensors. Taking
into account the *local_costmap*, the *local_planner* optimizes the initial plan defined
by the *global_planner* and publishes the appropriate velocities to the underlying *base
controller*. In contrast to the standard architecture of the navigation stack, the default
local_planner is replaced by the gray highlighted *teb_local_planner* plugin in this
chapter. The robot localizes itself using a combination of odometry (motion estimation relative to the robot's starting position based on data from its drive system) and
Adaptive Monte Carlo Localization (amcl) [22]. Transformations between multiple
coordinate frames are provided by the *tf* package.[3]

---

[1] *ROS navigation*, http://wiki.ros.org/navigation.

[2] Adopted from the *move_base* wiki page, http://wiki.ros.org/move_base.

[3] *tf*, http://wiki.ros.org/tf.

The current implementation of the navigation stack assumes a quasi-static environment and neither predicts nor considers the motion of dynamic obstacles (such as humans or other robots) explicitly. The costmap representation only provides a static view of the current environment and lacks the temporal evolution of grid occupancy. Consequently, planners are unable to benefit from the knowledge of a moving obstacle's estimated velocity and heading. In order to achieve robust navigation in dynamic environments, ROS navigation implements so-called inflation in which static occupied costmap cells are inflated by an exponentially decreasing cost decay rate. Thus, the robot plans a more pessimistic trajectory and maintains a larger separation from obstacles than actually required by collision avoidance.

The authors developed a package for dynamic obstacle detection and tracking based on the two-dimensional costmap of the ROS navigation stack. The approach rests upon an algorithm for foreground detection in the rolling window of the costmap that discriminates between occupied cells attributed to moving objects and the static *background*. The approach compensates the robot's ego-motion to obtain an unbiased estimate of obstacle velocities w.r.t. a global frame. The foreground cells are clustered into a set of obstacles for which individual model-based filters (Kalman-Filters) are applied for the ongoing state estimation of the obstacle motion. The node publishes the set of current dynamic obstacles in terms of their estimated location, footprint (shape), translational velocity vector and its uncertainty at every sampling interval. Local trajectory planners utilize the estimated motion of obstacles to plan the future collision-free robot trajectory ahead of time. For that purpose the interface for local planners should not only consider the current costmap but also its temporal evolution due to obstacle motion.

The previous volume of the book *Robot Operating System - The Complete Reference* includes a tutorial chapter on kinodynamic motion planning with Timed-Elastic-Bands (TEB) [18]. The package *teb_local_planner*[4] implements a local planner plugin for the ROS navigation stack. The underlying TEB approach efficiently optimizes the robot trajectory w.r.t. (kino-)dynamic constraints and non-holonomic kinematics of differential-drive, car-like or omnidirectional mobile robots while explicitly incorporating temporal information in order to reach the goal pose in minimal time [16, 17]. With its recent update, the *teb_local_planner* explicitly considers dynamic obstacles based on the estimates provided by the tracker for local planning. Multiple questions and inquiries in (*ROS Answers*) and personal feedback to the authors clearly indicate a broad interest of the community to support dynamic obstacles in ROS mobile robot navigation.

This chapter covers the following topics:

- The current state of the art for mobile robot path planning is briefly summarized in Sect. 2 with a focus on currently available *local_planner* plugins for the ROS navigation stack
- The algorithmic background for local costmap conversion is explained in Sect. 3
- Section 4 introduces the novel `CostmapToDynamicObstacles` plugin for the *costmap_converter* along with an intuitive example setup in Sect. 4.4

---

[4]*teb_local_planner*, http://wiki.ros.org/teb_local_planner.

- Section 5 presents the theoretical foundations of the TEB trajectory optimization methods
- Section 6 discusses the *teb_local_planner* along with a basic test node for spatio-temporal trajectory optimization (Sect. 7)
- Finally, Sect. 8 explains the setup of a navigation task in environments with moving obstacles.

2 Related Work

Collision-free locomotion is a fundamental skill for mobile robots. Especially in dynamic environments, online planning is preferred over offline approaches due to its immediate response to alterations in the vicinity of the robot. The *Dynamic Window Approach* (DWA) constitutes a well-known online trajectory planning approach [4]. It rests upon a dynamic window in the control input space from which admissible velocities for the robot are sampled in each time step. The search space is restricted to collision-free velocities considering the dynamics of the robot. For each sample, a short-term prediction of the future motion is simulated and evaluated w.r.t. a cost function (including a distance measure to the goal and obstacle avoidance terms) by assuming constant control inputs. The DWA was extended by Seder and Petrović for navigation in dynamic environments [21]. The *Trajectory Rollout* approach operates in a similar manner as the original DWA, but rather samples a set of achievable velocities over the entire forward simulation period [6]. In the context of car-like robots, [15] restricts the search space of rotational velocities to the set of feasible solution. However, due to the assumption of constant velocities in the prediction, motion reversals which are required for car-like robots to navigate in confined spaces are not explicitly considered during planning.

Fiorini and Shiller [3] present another approach called *Velocity Obstacles* to velocity-based sampling in the state space. The search space is restricted to admissible velocities that do not collide with dynamic obstacles within the prediction horizon. Fulgenzi et al. extended this method in order to account for uncertainty in the obstacle's motion by means of a Bayesian Occupancy Filter [5].

The *Elastic-Band* approach deforms a path to the goal by applying an internal contraction force resulting in the shortest path and external repulsive forces radiating from the obstacles to receive a collision-free path [23]. However, this approach does not incorporate time information. Hence, the robot's kinodynamic constraints are not considered explicitly and a dedicated path following controller is required. The *2-Step-Trajectory-Deformer* approach incorporates time information in a subsequent planning stage [10]. Resulting trajectories are feasible for holonomic robots with kinodynamic constraints. An extension to non-holonomic robots is presented by Delsart and Fraichard [2]. Gu et al. [7] present a multi-state planning approach which utilizes an optimization-free *Elastic-Band* to generate paths followed by a speed planning stage for car-like robots.

Likewise, the TEB approach augments the *Elastic-Band* method with time information in order to generate time-optimal trajectories [16]. The approach was recently extended to parallel trajectory planning in spatially distinctive topologies [17] and to car-like robot kinematics [19]. The TEB is efficiently integrated with state feedback to repeatedly refine the trajectory w.r.t. disturbances and changing environments. Desired velocities are directly extracted from the planned trajectory.

A variety of the previously presented approaches (DWA, *Trajectory Rollout*, *Elastic-Band* and the *Timed-Elastic-Band*) are provided as local planner plugin in the ROS navigation stack [12]. However, the TEB approach presented in this chapter is currently the only local planner plugin that explicitly incorporates the estimated future motions of dynamic obstacles into trajectory optimization.

The *costmap_converter* package presented in this chapter estimates the velocity of dynamic obstacles according to a constant velocity model. More sophisticated obstacle tracking approaches like the social forces model [11] take cooperative joint motions among a group of agents into account. These models are currently not included, but they might be implemented in potential future package versions.

3   Costmap-Based Obstacle Velocity Estimation

In the ROS navigation stack (see Fig. 1) the costmap represents an occupancy grid. The status of each cell is either free (0), occupied (254), or unknown (255). Costmaps are updated at a specific rate. Unfortunately, the costmap does not include historical data from which velocity and heading of moving obstacles could be inferred. Therefore, the authors implemented a tracker and velocity estimation based on the history of the costmap. The ego-motion is compensated by transforming the observed obstacle velocities from the sensor frame to the global map frame w.r.t. the robot's odometry.

Typical tracking algorithms operate with range measurements but make particular assumptions about the sensor characteristics. Our tracking scheme merely relies on the costmap with already aggregated sensor data, which facilitates its integration into the navigation stack, but arguably sacrifices tracking accuracy and resolution. The tracker is universally applicable as it requires no configuration or adaptation to different types of sensors since data fusion is already accomplished at the costmap stage. The tracker is implemented as a plugin to the existing *costmap_converter* package which provides plugins that convert the static costmap to geometric primitives like lines or polygons. The presented method converts only dynamic obstacles to geometric primitives augmented with estimated location and velocity. Static obstacles are not processed and remain point-shaped.

The foreground detection algorithm extracts dynamic obstacles from the local costmap by subtracting the outputs of two running average filters resulting in a bandpass filter. The binary map labels the cells of moving foreground objects as true, and static and obstacle-free regions as false (see Fig. 2). The centroids and contours of these obstacles are determined by a blob detector from computer vision. The second step is concerned with clustering and tracking connected and coherent cells to individual obstacles and to estimate their location and velocity.

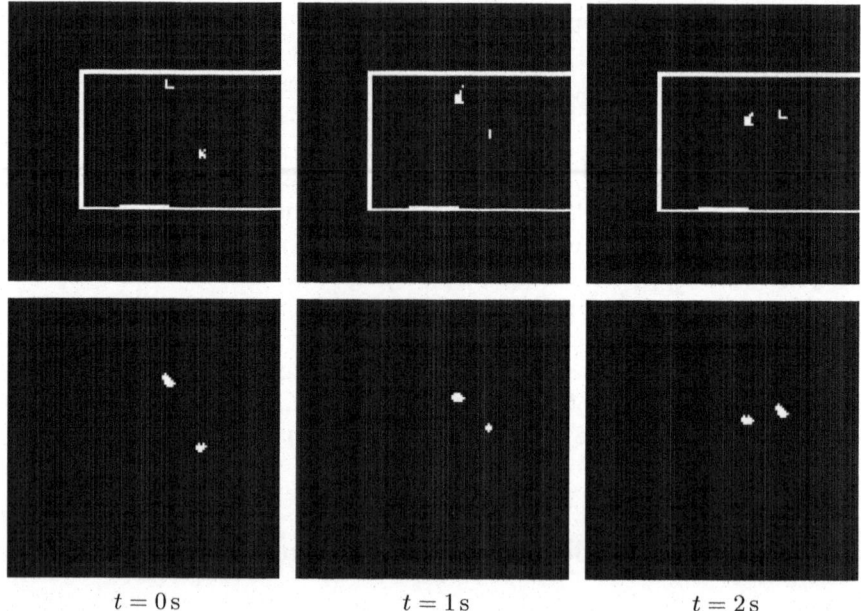

$$t = 0\,\mathrm{s} \qquad\qquad t = 1\,\mathrm{s} \qquad\qquad t = 2\,\mathrm{s}$$

Fig. 2 Evolution of the costmap (upper row) and the corresponding detected dynamic obstacles (lower row) with two dynamic obstacles perceived by a static observer

3.1 Dynamic Obstacle Detection

Though the approach also works when the ego robot is moving, the discussion of the theoretical background assumes that the observing robot and therefore the position of the local costmap w.r.t. to the global frame remains static.

The foreground detection operates with a slow and a fast *running average filter*. These filters are applied to each cell in the costmap. For a single cell these filters are described by:

$$P_f(t + 1) = ((1 - \alpha_f)\, P_f(t) + \alpha_f\, C(t)) \tag{1}$$

$$P_s(t + 1) = ((1 - \alpha_s)\, P_s(t) + \alpha_s\, C(t)) \tag{2}$$

$P_f(t)$ and $P_s(t)$ represent the output of the fast and the slow running average filter at time t, respectively. The gains α_f and α_s define the effect of the current costmap $C(t)$ on P and comply with:

$$0 \leq \alpha_s < \alpha_f \leq 1 \tag{3}$$

For the detection of particular large objects, which form blocks of cells in the local costmap, the Eqs. 1 and 2 are extended by a term that captures the running average filter of the cells nearest neighbors (NN). β denotes the ratio between the contribution of the central cell filter and the effect of the neighboring cells to $P_f(t)$ and $P_s(t)$.

$$P_f(t+1) = \beta\left((1 - \alpha_f)\, P_f(t) + \alpha_f\, C(t)\right) + \frac{(1 - \beta)}{8} \sum_{i \in \mathrm{NN}} P_{f,i}(t) \qquad (4)$$

$$P_s(t+1) = \beta\left((1 - \alpha_s)\, P_s(t) + \alpha_s\, C(t)\right) + \frac{(1 - \beta)}{8} \sum_{i \in \mathrm{NN}} P_{s,i}(t) \qquad (5)$$

These filters identify those cells that are occupied by moving obstacles if they comply with two criteria that filter out high and low frequency noise. The fast filter has to exhibit an activation that exceeds a threshold c_1:

$$P_f(t) > c_1 \qquad (6)$$

In addition, the difference between the fast and the slow filter has to exceed a threshold c_2 in order to eliminate quasi-static obstacles with low frequency noise.

$$P_f(t) - P_s(t) > c_2 \qquad (7)$$

Figure 3 shows the the filtered signal in the local costmap C of the fast filter P_f and slow filter P_s in case of a dynamic obstacle that traverses the cell over a period of two seconds together with the activation thresholds c_1 and c_2.

The thresholding operations generate a binary map that labels dynamic obstacles as ones, whereas free space and static obstacles are labeled as zeros. The sequence of erosion and dilation on the binary map results in a closing operation which reduces noise in the foreground map.

A slightly modified version of OpenCV's `SimpleBlobDetector` extracts obstacle centroids and contours from the binary map. These obstacle features provide the input to the algorithm for simultaneous tracking and velocity estimation of multiple obstacles.

3.2 Dynamic Obstacle Tracking

The centroid of dynamic obstacles progresses with each costmap update and subsequent foreground detection. The assignment of blobs in the current map to obstacle tracks constitutes a data association problem. In order to disambiguate and track multiple objects over time, the current obstacles are matched with the corresponding tracks of previous obstacles. A new track is generated whenever a novel obstacle emerges that is not tracked yet. Tracks that are not assigned to current objects in the

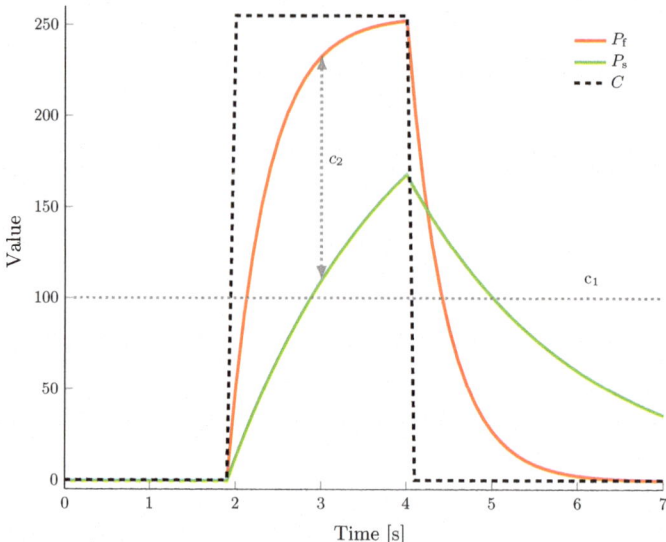

Fig. 3 Slow and fast filter responses

foreground frame are temporarily maintained. The track is removed if it is no longer confirmed by object detections over an extended period of time.

The assignment problem is solved by the so-called Hungarian algorithm, which was originally introduced by [8]. The algorithm efficiently solves weighted assignment problems by minimizing the total Euclidean distance between the tracks and the current set of obstacle centroids.

A Kalman filter estimates the current velocity of tracked obstacles assuming a first order constant velocity model. The constant velocity model sufficiently captures the prevalent motion patterns of humans and robots in indoor environments for the designated spatio-temporal horizon of motion planning.

4 Costmap_Converter ROS Package

This section introduces the technical aspects of the *costmap_converter* ROS package. It handles the conversion of dynamic obstacles extracted from the local costmap (in terms of a `nav_msgs/OccupancyGrid` message) into polygons augmented with velocity and heading information. In addition, the package provides various plugins to convert connected regions in the local costmap to geometric primitives without estimating their velocities. The presentation here focuses on the `CostmapToDynamicObstacles` plugin. The ROS wiki[5] documents the addi-

---

[5]*costmap_converter*, http://wiki.ros.org/costmap_converter.

tional plugins. The utilized plugin can be selected prior to run-time via the parameter `costmap_converter_plugin`. The `CostmapToDynamicObstacles` plugin pursues a two-step approach with the initial conversion and velocity estimation of dynamic obstacles and the subsequent conversion of static obstacles by means of an additional static plugin. The parameter `static_costmap_converter_plugin` specifies the employed plugin for static costmap conversion.

The *costmap_converter* package is available for ROS Kinetic and Lunar. The package defines a pluginlib[6] interface and is primarily intended for direct embedding in the source code. Some applications might prefer a dedicated subscriber `standalone_converter` node to process `nav_msgs/OccupancyGrid` messages.

4.1 Prerequisites and Installation

It is assumed that the reader is accustomed to basic ROS concepts such as navigating the filesystem, creating and building packages, as well as dealing with *rviz*[7], launch files, topics, parameters and *yaml* files discussed in the common ROS beginner tutorials. Familiarity with the concepts and components of ROS navigation such as local and global costmaps and local and global planners (*move_base* node), coordinate transforms, odometry, and localization is expected.

In the following, terminal commands are indicated by a leading $-sign. The *costmap_converter* package is installed from the official ROS repositories by invoking:

```
$ sudo apt-get install ros-kinetic-costmap-converter
```

More recent, albeit experimental, versions of the *costmap_converter* package can be obtained and compiled from source:

```
$ cd ~/catkin_ws/src
$ git clone https://github.com/rst-tu-dortmund/
    costmap_converter
$ cd ../
$ rosdep install --from-paths src --ignore-src --rosdistro
    kinetic -y
$ catkin_make
```

---

[6]*pluginlib*, http://wiki.ros.org/pluginlib.

[7]*rviz*, http://wiki.ros.org/rviz.

The user-created *catkin* workspace is assumed to be located at ~/catkin_ws.

Currently, the *costmap_converter* does not handle incremental costmap updates published to the grid_updates topic. In order to use this package, the full costmap has to be published in every update cycle. Incremental updates can be deactivated by enabling the *costmap_2d*[8] parameter always_send_full_costmap. In the tutorial scenarios introduced in this chapter, this parameter is already defined in the *costmap_common_params.yaml* file.

4.2 Obstacle Messages

Converted obstacles are published as the novel message type ObstacleMsg specifically designed for the purpose of publishing obstacles in conjunction with their velocities. The compact definition of an ObstacleMsg is pictured below.

```
   # Special types:
 2 # Polygon with 1 vertex: Point obstacle
   # Polygon with 2 vertices: Line obstacle
 4 # Polygon with more than 2 vertices: First and last points are
       assumed to be connected

 6 std_msgs/Header header

 8 # Obstacle footprint (polygon descriptions)
   geometry_msgs/Polygon polygon
10
   # Obstacle ID
12 # Specify IDs in order to provide (temporal) relationships
   # between obstacles among multiple messages.
14 int64 id

16 # Individual orientation (centroid)
   geometry_msgs/Quaternion orientation
18
   # Individual velocities (centroid)
20 geometry_msgs/TwistWithCovariance velocities
```

An obstacle is represented by a polygon, i.e. array of vertices. Polygons with a single vertex refer to point obstacles and polygons with two vertices denote line obstacles. In case of true polygons with more than two vertices, the first and last points are assumed to be connected in order to close the polygon perimeter. Additionally, the message provides the velocities, orientation and id of obstacles. The orientation and velocity of an obstacle are specified w. r. t. their respective centroid.

---

[8]*costmap_2d*, http://wiki.ros.org/costmap_2d.

`ObstacleMsgs` are grouped to an `ObstacleArrayMsg` which can include not merely the converted dynamic obstacles, but also the static background of the costmap in terms of point-shaped obstacles.

4.3 Parameters

Most of the package parameters originate from OpenCV's *SimpleBlobDetector*, for which the reader is referred to the OpenCV documentation.[9] This section explains the additional parameters of the *costmap_converter* plugin for dynamic obstacle conversion.

`alpha_fast` (α_f)
Adaption rate of the fast running average filter in Eq. (4). A higher rate indicates a higher confidence in the most recent costmap. `alpha_fast` has to be larger than `alpha_slow` (see Eq. (3)).

`alpha_slow` (α_s)
Adaption rate of the slow running average filter in Eq. (5). A higher rate indicates a higher confidence in the most recent costmap. `alpha_slow` has to be smaller than `alpha_fast` (see Eq. (3)).

`beta` (β)
Ratio, of the contribution to the running filterscenter cell relative to the neighboring cells (see Eqs. (4) and (5)).

`min_occupancy_probability` (c_1)
Threshold of the fast filter for classification of a cell as foreground.

`min_sep_between_slow_and_fast_filter` (c_2)
Threshold of minimal difference between the slow and fast running average filters for classification of the cell as foreground.

`max_occupancy_neighbors`
Maximum mean value of the 8-neighborhood for classification of the cell as foreground.

`morph_size`
Size of the structuring element (circle) used for the closing operation applied to the binary map after foreground detection.

---

[9]OpenCV *SimpleBlobDetector*,
http://docs.opencv.org/3.3.0/d0/d7a/classcv_1_1SimpleBlobDetector.html.

`dist_thresh`
Maximum Euclidean distance between obstacles and tracks to be considered for matching in the assignment problem.

`max_allowed_skipped_frames`
Maximum number of frames for which a dynamic obstacle is tracked without confirmation in the current foreground map.

`max_trace_length`
Maximum number of points representing in the object trace.

`publish_static_obstacles`
Include obstacles from the static background cells. By default, static costmap cells are subsequently converted to polygons by the `CostmapToPolygonsDBSMCCH` plugin for static costmap conversion.

4.4 Prototype Scenario for Obstacle Velocity Estimation

This section introduces a minimal *stage*[10] simulation setup with costmap conversion respectively velocity estimation for dynamic obstacles. The simulation setup consists of an observing robot and a single dynamic obstacle moving in a simple square environment. A costmap conversion scenario including navigation by means of the *teb_local_planner* is discussed in Sect. 8.

Stage is a fast and lightweight mobile robot simulator. Although this tutorial chapter refers to *stage*, the demo code is equally applicable to other simulation environments such as *gazebo*.[11] In case *Stage* is not yet installed along with the full ROS distribution package invoke:

```
$ sudo apt-get install ros-kinetic-stage-ros
```

Clone (or download and unzip) the *teb_local_planner_tutorials* package[12] for this tutorial:

```
$ cd ~/catkin_ws/src
$ git clone -b rosbook_volume3 https://github.com/rst-tu-
    dortmund/teb_local_planner_tutorials.git
```

---

[10] *stage_ros*, http://wiki.ros.org/stage_ros.

[11] *gazebo_ros_pkgs,* http://wiki.ros.org/gazebo_ros_pkgs.

[12] *teb_local_planner_tutorials*,   https://github.com/rst-tu-dortmund/teb_local_planner_tutorials/tree/rosbook_volume3.

The simulated environment in *stage* consists of a static map and other agents such as robots or (dynamic) obstacles. The additional agents are included in a separate definition file for the observing robot. Observe the robot and sensor definitions for *stage* defined in *myRobot.inc* located in the subfolder *stage*. The following sensor and robot definitions are utilized in the course of this tutorial:

```
define mylaser ranger
(
   sensor
   (
      # just for demonstration purposes
      range [ 0.1 25 ] # minimum and maximum range
      fov 360.0 # field of view
      samples 1920 # number of samples
   )
   size [ 0.06 0.15 0.03 ]
)

define myrobot position
(
   size [ 0.25 0.25 0.4 ] # (x,y,z)
   localization "gps" # exact localization
   gui_nose 1 # draw nose on the model showing the heading
   drive "diff" # diff-drive
   color "red" # red model
   mylaser(pose [ -0.1 0.0 -0.11 0.0 ]) # add mylaser sensor
)
```

The first code block defines a range sensor named `mylaser` with a complete field of view (`fov 360.0`). The second block defines a robot which utilizes the previously defined ranging sensor. The dynamic obstacle model is defined in the *myObstacle.inc* file in a similar way:

```
define myobstacle position
(
   localization "gps" # exact localization
   size [ 0.25 0.25 0.4 ] # (x,y,z)
   gui_nose 1 # draw nose on the model showing the heading
   drive "omni" # omni-directional movement possible
   color "blue" # blue model
)
```

The *stage* environment along with robots and obstacles is defined in the file *emptyBox.world*:

```
   ## include our robot and obstacle definitions
 2 include "robots/myRobot.inc"
   include "robots/myObstacle.inc"

 4
   ## Simulation settings
 6 resolution 0.02
   interval_sim 100 # simulation timestep in milliseconds

 8
   ## Load a static map
10 model
   (
12   name "emptyBox"
     bitmap "../maps/emptyBox.png"
14   size [ 6.0 6.0 2.0 ]
     pose [ 0.0 0.0. 0.0 0.0]
16   laser_return 1
     color "gray30"
18 )

20 # throw in a robot and an obstacle
   myrobot
22 (
     pose [ -2.0 0.0 0.0 -90.0 ] # initial pose (x,y,z,beta[deg])
24   name "myRobot"
   )

26
   myobstacle
28 (
     pose [ 0.0 1.0 0.0 0.0 ] # initial pose (x,y,z,beta[deg])
30   name "myObstacle"
   )
```

The robot definition files are included in lines 2–3. General simulation settings are defined in lines 6–7. Line 10–18 define the map model as an empty square box with an edge length of 6 m. The observing robot and the dynamic obstacle are defined from line 21 onward. Next, we will inspect the *costmap_conversion.launch* file step by step which is located in the *launch* subfolder of the package.

```
   <!-- ************* Stage Simulator **************** -->
 2 <node pkg="stage_ros" type="stageros" name="stageros" args="$(find
       teb_local_planner_tutorials)/stage/emptyBox.world">
     <remap from="/robot_0/base_scan" to="/robot_0/scan"/>
 4 </node>

 6 <!-- ****************** Maps ********************* -->
   <node name="map_server" pkg="map_server" type="map_server" args="$(find
       teb_local_planner_tutorials)/maps/emptyBox.yaml" output="screen">
 8   <param name="frame_id" value="map"/>
   </node>
```

These commands start the *stageros* and *map_server* nodes and load the *emptyBox* map.

```
<!-- ******* Localization ******** -->
2 <!-- See stage world file for initial poses -->
<node pkg="tf" type="static_transform_publisher" name="perfect_loc_robot
    " args="-2 0 0 -1.570796 0 0 /map robot_0/odom 100" />
4 <node pkg="tf" type="static_transform_publisher" name="
    perfect_loc_obstacle" args="0 1 0 0 0 0 /map robot_1/odom 100" />
```

In this section `static_transform_publishers` are launched publishing the transformations to the perfect robot and obstacle locations for localization.

```
<!-- ************* Navigation Ego Robot ************ -->
2 <group ns="robot_0">
    <param name="tf_prefix" value="robot_0"/>

4

    <node pkg="move_base" type="move_base" respawn="false" name="
    move_base" output="screen">
6   <rosparam file="$(find teb_local_planner_tutorials)/cfg/diff_drive/
    costmap_common_params.yaml" command="load" ns="global_costmap" />
    <rosparam file="$(find teb_local_planner_tutorials)/cfg/diff_drive/
    costmap_common_params.yaml" command="load" ns="local_costmap" />
8   <rosparam file="$(find teb_local_planner_tutorials)/cfg/diff_drive/
    local_costmap_params.yaml" command="load" />
    <rosparam file="$(find teb_local_planner_tutorials)/cfg/diff_drive/
    global_costmap_params.yaml" command="load" />
10  <rosparam file="$(find teb_local_planner_tutorials)/cfg/diff_drive/
    teb_local_planner_params.yaml" command="load" />

12  <param name="base_local_planner" value="teb_local_planner/
    TebLocalPlannerROS" />
    <param name="controller_frequency" value="5.0" />
14  <param name="controller_patience" value="15.0" />
    <remap from="map" to="/map"/>
16  </node>
    </group>
```

This code brings up the ROS navigation stack in the namespace of *robot_0*. Since the navigation stack contains the *costmap_2d* package, it is required for costmap conversion. Various parameter files for the local and global costmap are loaded. The *teb_local_planner* is utilized as local planner, though other local planners can be used for costmap conversion.

```
   <!-- *********** Costmap conversion *********** -->
 2 <node name="standalone_converter" pkg="costmap_converter" type="
      standalone_converter" output="screen">
   <param name="converter_plugin" value="
      costmap_converter::CostmapToDynamicObstacles" />
 4   <param name="costmap_topic" value="/robot_0/move_base/local_costmap/
      costmap" />
   <param name="odom_topic" value="/robot_0/odom" />
 6 </node>
```

These commands launch the *costmap_converter* standalone node with the previously introduced CostmapToDynamicObstacles plugin. The odom_topic is required in order to compensate a robots ego motion while estimating obstacle velocities. The *costmap_converter* subscribes to the costmap_2d and publishes an ObstacleArrayMsg with an array of ObstacleMsgs, each containing estimated obstacle velocities and shapes.

```
   <!-- ****************** Obstacles ****************** -->
 2 <group ns="robot_1">
   <param name="tf_prefix" value="robot_1"/>
 4   <node name="Mover" pkg="teb_local_planner_tutorials" type="
      move_obstacle.py" output="screen"/>
   <node name="visualize_obstacle_velocity_profile" pkg="
      teb_local_planner_tutorials" type="
      visualize_obstacle_velocity_profile.py" output="screen" />
 6 </group>
```

Two distinctive nodes are launched in the namespace of *robot_1*. The script *move_obstacle.py* actuates the obstacle (by publishing a cmd_vel message to the obstacles namespace) with a constant velocity by sampling a velocity in the opposite direction in case of an encounter with the walls. The collision with a wall is detected from the base_pose_ground_truth topic in the obstacles namespace. Additionally, a node which plots both the estimated and ground truth velocities of an obstacle is started by the *visualize_obstacle_velocity_profile.py* script.

```
   <!-- *************** Visualisation *************** -->
 2 <node name="rviz" pkg="rviz" type="rviz" args="-d $(find
      teb_local_planner_tutorials)/cfg/rviz_navigation_cc.rviz">
   <remap from="/move_base_simple/goal" to="/robot_0/move_base_simple/
      goal" />
 4 </node>
```

This section starts up *rviz* and loads a predefined configuration file. Launch the simulation by invoking:

```
roslaunch teb_local_planner_tutorials costmap_conversion.launch
```

stage and *rviz* pop up and display the simulated environment. The estimated footprint of the dynamic obstacle is indicated by a green polygon in *rviz*. Estimated and ground truth velocities are visualized in a live plot. In case of very slow obstacle velocities the *costmap_converter* might consider the obstacle as static and therefore lose track of the object. Try to customize the obstacle detection by changing the parameters of the *costmap_converter* using the *rqt_reconfigure* tool:

```
$ rosrun rqt_reconfigure rqt_reconfigure
```

It is recommended to invoke costmap conversion in a separate thread since the conversion of dynamic obstacles is a time critical task for safe navigation. The computational effort for one cycle of obstacle detection in the local costmap, velocity estimation and the publishing of these informations is depicted in Fig. 4. Regardless of the number of tracked obstacles, the median for one conversion cycle is located at around 1 ms. These measurements were taken using an Intel Core i5-6500 processor along with 8GB RAM. As the default costmap conversion rate is 5 Hz, costmap conversion can also be performed on less powerful hardware.

In the following, the estimated velocities provide the basis for optimal spatiotemporal trajectory planning by the *teb_local_planner*. Costmap conversion and velocity estimation do not depend on the *teb_local_planner* package and might be incorporated standalone into custom applications.

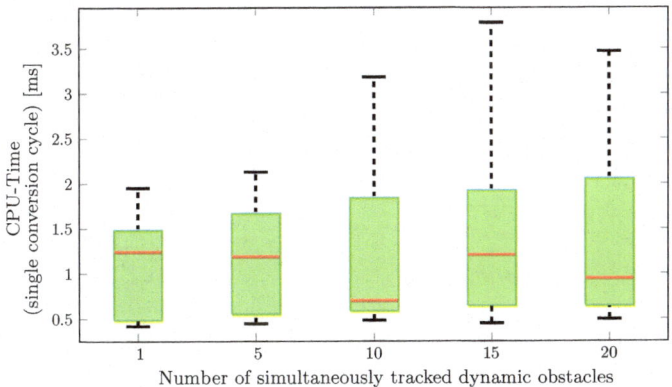

Fig. 4 Computational performance of the *costmap_converter*

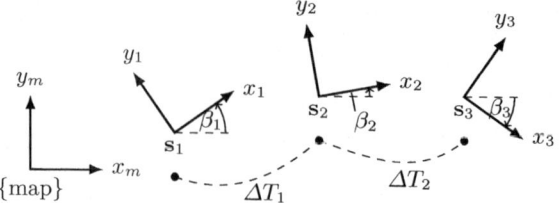

Fig. 5 Discretized trajectory with $n = 3$ poses

5 Theoretical Foundations of TEB

This section introduces the fundamental concepts of the TEB trajectory planning approach. It discusses the theoretical foundations for the successful integration and customization of the *teb_local_planner* in mobile robot applications. A more detailed description of online trajectory optimization with Timed-Elastic-Bands is given in [17].

The tutorial chapter on kinodynamic motion planning with Timed-Elastic-Bands published in the second volume of this book [18] discusses the theoretical foundations of the TEB approach. Even though the main features of TEB remain the same, for the sake of a self-contained presentation the basic operation of TEB optimization is explained in summary. The main modification is the extension from a quasi-static environment to a dynamic world that explicitly takes the future poses of moving obstacles into account for planning an optimal spatio-temporal trajectory.

5.1 Trajectory Representation and Optimization

A discretized trajectory $\mathbf{b}$ is defined by an ordered sequence of robot poses $\mathbf{s}_k = [x_k, y_k, \beta_k]^\mathsf{T} \in \mathbb{R}^2 \times S^1$ with $k = 1, 2, \ldots, N$ and time stamps $\Delta T_k \in \mathbb{R}_{>0}$ with $k = 1, 2, \ldots, N - 1$.

$$\mathbf{b} = \left[\mathbf{s}_1, \Delta T_1, \mathbf{s}_2, \Delta T_2, \ldots, \Delta T_{N-1}, \mathbf{s}_N \right]^\mathsf{T} \tag{8}$$

ΔT_k denotes the transition time between two consecutive poses $\mathbf{s}_k$ and $\mathbf{s}_{k+1}$, respectively. Figure 5 depicts an example trajectory with three poses. The reference frame of the trajectory representation is denoted as *map*-frame.[13]

The optimal trajectory $\mathbf{b}^*$ is obtained by minimizing a cost function which captures partially conflicting objectives and constraints of motion planning. These objectives include energy consumption, path length, the total transition time, or a weighted

---

[13]Conventions for names of common coordinate frames in ROS are listed at http://www.ros.org/reps/rep-0105.html.

combination of the above. Admissible solutions are restricted to a feasible set by penalizing trajectories, which do not comply with the kinodynamic constraints of the mobile robot.

The TEB optimization problem is defined as an aggregated nonlinear least-squares cost function, which considers conflicting sets of objectives $\mathcal{J}$ and penalties $\mathcal{P}$, each weighted by a factor σ_i:

$$\mathbf{b}^* = \underset{\mathbf{b} \backslash \{\mathbf{s}_1, \mathbf{s}_N\}}{\mathrm{argmin}} \sum_i \sigma_i f_i^2(\mathbf{b}) , \quad i \in \{\mathcal{J}, \mathcal{P}\} \tag{9}$$

The notation $\mathbf{b} \backslash \{\mathbf{s}_1, \mathbf{s}_N\}$ implies that neither the start pose $\mathbf{s}_1 = \mathbf{s}_s$ nor the goal pose $\mathbf{s}_N = \mathbf{s}_g$ are subject to optimization. During optimization the trajectory is clipped at the current robot pose $\mathbf{s}_s$ and the desired goal pose $\mathbf{s}_g$.

In order to account for the dynamic environment and to refine the trajectory during runtime, a model predictive control scheme is applied. Thus, the optimization problem (9) is solved repeatedly in each sampling interval[14] with respect to the current robot pose and velocity. The current robot pose and velocity are provided by a localization scheme. In compliance with the basic concept of model predictive control [13], during each time step only the first control action of the computed trajectory is commanded to the robot. The *teb_local_planner* pursues a warm-start approach, hence the optimal trajectory of the previous time interval serves as initial solution for the subsequent optimization problem.

In the navigation stack, the *base controller* interface typically subscribes to a *cmd_vel* message (see Fig. 1) composed of translational and angular velocities. These components can easily be extracted from the optimal trajectory $\mathbf{b}^*$ by investigating finite differences both on the position and orientation part. As car-like robots often require the steering angle rather than the angular velocity, the steering angle can be calculated from the turn rate and the car-like robot's kinematic model, e.g. refer to [19].

The TEB optimization problem is mapped onto a hyper-graph in which vertices correspond to the poses $\mathbf{s}_k$ and time intervals ΔT_k that form the solution vector and the (hyper)-edges denote the cost terms f_i that set up the nonlinear program. In addition, fixed vertices not subject to optimization include start and goal pose ($\mathbf{s}_1$ or $\mathbf{s}_N$), obstacle positions $\mathcal{O}_i$, or other static parameters. The prefix *hyper* indicates that an edge connects an arbitrary number of vertices. An edge connecting various parameters represents a cost term, that is dependent on these parameters.

The resulting hyper-graph is efficiently solved by the *g2o-framework*[15] [9]. The framework exploits the sparse structure of the system matrix by using the Levenberg–Marquardt Algorithm. The sparse structure emerges from a formulation that expresses relationships in the solution by soft rather than hard constraints. The computational efficiency of the algorithm benefits from the sparse structure in the

---

[14]The sampling interval can be adjusted by means of the parameter `controller_frequency` provided by the *move_base* node of the navigation stack.

[15]*libg2o*, http://wiki.ros.org/libg2o.

Cholesky-decomposition step. The hyper-graph formulation comes with the additional advantage of modularity which allows the seamless integration of additional constraints and objectives.

Soft constraints tolerate a certain amount of violation at the price of an abrupt increase in the cost function. Let $\mathcal{B}$ denote the entire set of potential trajectories such that $\mathbf{b} \in \mathcal{B}$. An inequality constraint $g_i(\mathbf{b}) \geq a$ with $g_i : \mathcal{B} \to \mathbb{R}$ is approximated by a positive semi-definite penalty function which captures the amount of constraint violation:

$$f_i(\mathbf{b}) = \max\{0, -g_i(\mathbf{b}) + a + \epsilon\} \quad \forall i \in \mathcal{P} \tag{10}$$

The parameter ϵ adds a margin to the lower bound a of the inequality constraint such that the cost merely vanishes for $g_i(\mathbf{b}) \geq a + \epsilon$. The theory of penalty optimization methods [14] postulates that the weights of the individual penalty terms should tend towards infinity in order to comply with the truly optimal solution. Unfortunately, large weights result in a numerically ill-conditioned optimization problem for which the underlying solver does not converge properly. For this reason, the TEB approach approximates the optimal trajectory with finite weights in order to achieve a computationally more efficient solution.

The TEB approach employs multiple cost terms f_i for trajectory optimization. For example, limited velocities and accelerations, compliance with non-holonomic kinematics or transition to the goal pose in minimal time are considered. Section 6.3 summarizes the currently implemented cost terms of the optimization problem (9). The TEB cost structure is extended by a novel distinctive penalty term for dynamic obstacles that complements the previous penalty term for static obstacles. The following section considers and analyzes different options for the configuration and parametrization of the cost term for dynamic obstacles.

5.2 Transition Time Estimation for Dynamic Obstacles

Collision avoidance demands a minimal separation of the robot and obstacle poses. The spatio-temporal distance between an obstacle and a pose $\mathbf{s}_k$ along the trajectory is calculated based on the predicted poses of dynamic obstacles at time Δt_k rather than their current location. The robot-obstacle distance calculations consider the polygonal footprints of the robot and the obstacles. Distances between the robot and obstacles are bounded from below by the minimal separation (a) and an additional tolerance value ϵ. Hence, these distances can be directly inserted into the penalty function for inequality constraints (10) along with parameters a and ϵ. The penalty term itself is used in the optimization problem (9) for each penalty and each objective term listed in Sect. 6.3.

Under the assumption that dynamic obstacles $\mathcal{O}_i$ maintain their current (estimated) speed $\hat{v}$ and orientation the constant velocity model predicts the future obstacle poses $\hat{P}(\mathcal{O}_i,)$.

$$\hat{P}(\mathcal{O}_i, t) = P(\mathcal{O}_i, t_0) + \Delta t_k \cdot \hat{v}(\mathcal{O}_i) \tag{11}$$

The current obstacle position $P(\mathcal{O}_i, t_0)$ and the estimated velocity $\hat{v}(\mathcal{O}_i)$ are provided by the *costmap_converter* package. The predicted obstacle pose depends on the robot's transition time Δt_k between its current and its kth future pose.

This transition time amounts to the accumulated time steps $\sum_{i=1}^{k} \Delta T_i$ up to pose $\mathbf{s}_k$. However, the dependency of the total transition time Δt_k on all previous time intervals ΔT_i with $i \leq k$ compromises the sparsity of the system matrix which causes a degradation of computationally efficiency.

In order to preserve computational efficiency an improved strategy exploits the iterative nature of the online trajectory optimization. Changes of the environment and the underlying nonlinear program between two consecutive control cycles are rather small. Thus it is valid to approximate the true total transition time Δt_k by summation of the time intervals $\Delta T_k'$ in the previous optimization step and consider the Δt_k as constant within the obstacle pose prediction step. This approximation maintains the sparse structure of the system matrix and causes a significant improvement in performance.

5.3 Planning in Distinctive Spatio-Temporal Topologies

The previously introduced TEB approach is subject to local optimization. Instead of finding the globally optimal solution, the optimized trajectory might get stuck in local minima due to the presence of obstacles. Identifying these local minima coincides with exploring and analyzing distinctive topologies between start and goal poses. For instance, the robot might circumnavigate an obstacle either on the left- or right-hand side. In case of moving obstacles the concept of a traversal to left and right are augmented by a traversal before or after passage of a dynamic obstacle. Therefore the new trajectory optimization not only considers the spatial topology of the trajectory but also its temporal dimension.

The approach rests upon two theorems of electromagnetism, the Biot–Savart and Ampere's law. It defines a new equivalence relation in order to distinguish among trajectories of distinctive topologies in the three-dimensional x-y-t-space. Notice, the additional temporal dimension that extends the mere spatial analysis of the previous TEB trajectory optimization [18]. The TEB ROS implementation explores and optimizes multiple trajectories in distinctive spatio-temporal topologies in parallel and selects the best candidate trajectory at each sampling interval. However, the theory of this method is beyond the scope of this tutorial. For a detailed description of this approach, the reader is referred to [1].

6 teb_local_planner ROS Package

This section provides an overview of the *teb_local_planner* ROS package which employs the previously described TEB approach for online trajectory optimization. Since the prerequisites and basics of the *teb_local_planner* did not change since the release of the last volume of this book, some fundamental parts from [18] are only briefly revisited.

6.1 Prerequisites and Installation

To install and configure the *teb_local_planner* package for a particular application, comply with the following limitations and prerequisites:

- Although current online trajectory optimization approaches feature mature computational efficiency, their application still requires substantial CPU resources. Depending on the desired trajectory length respectively resolution as well as the number of considered obstacles, common desktop computers or modern notebooks usually cope with the computational burden. However, older and embedded systems might not be capable to perform trajectory optimization at a reasonable rate.
- Results and discussions on stability and optimality properties for online trajectory optimization schemes are widespread in the literature, especially in the field of model predictive control. However, since these results are often theoretical and the planner is confronted with e.g., sensor and actuator uncertainty and dynamic environments in real applications, finding a feasible and stable trajectory in every conceivable scenario is not guaranteed. Especially due to noisy velocity estimations for dynamic obstacles, planned trajectories may oscillate. In order to generate a feasible trajectory, the planner detects and resolves failures by post-introspection of the optimized trajectory.
- Even though the presentation focuses on differential-drive robots, the package currently supports car-like and omnidirectional robots as well. For a detailed tutorial on the setup and configuration of car-like robots, the reader is referred to [18].
- Officially supported ROS distributions are Kinetic and Lunar. Legacy versions of the planner without support for dynamic obstacles are also available in Indigo and Jade. Support of future distributions is expected. The package is released for both default and ARM architectures.

Similar to the installation of the *costmap_converter* package, the *teb_local_planner* is installed from the official ROS repositories by invoking:

```
$ sudo apt-get install ros-kinetic-teb-local-planner
```

As before, the distribution name `kinetic` should be adapted to match the currently installed distribution. As an alternative, compile the most recent versions of the *teb_local_planner* from source:

```
$ cd ~/catkin_ws/src
$ git clone https://github.com/rst-tu-dortmund/
    teb_local_planner.git  --branch kinetic-devel
$ cd ../
$ rosdep install --from-paths src --ignore-src --rosdistro
    kinetic -y
$ catkin_make
```

6.2 Integration with ROS Navigation

As a plugin for the ROS navigation stack, the *teb_local_planner* package replaces the navigation stack's default local planner (refer to Fig. 1). The global planner of the *move_base* package plans a global path to the goal according to a global costmap which corresponds to a known map of the environment and serves as initial solution for the local planner. However, the global costmap does not support dynamic obstacles. Therefore, robocentric sensor readings are fused with the global costmap in order to calculate a local costmap. The *teb_local_planner* computes a feasible trajectory corresponding to the local costmap and publishes the associated velocity commands. The robot is localized with respect to the global map by means of the *amcl* node which employs an adaptive monte carlo localization algorithm which compensates for the accumulated odometric error.

The package wiki page[16] contains the complete list of *teb_local_planner* parameters which are nested in the relative namespace of the *move_base* node, e.g. `/move_base/TebLocalPlannerROS/param_name`. Assuming a running instance of the *teb_local_planner*, parameters are configured at runtime by launching the *rqt_reconfigure* GUI:

```
$ rosrun rqt_reconfigure rqt_reconfigure
```

6.3 Included Cost Terms: Objectives and Penalties

The *teb_local_planner* optimizes the planned trajectory respectively current control commands by minimizing a designated cost function (9) composed of objective and penalty terms which are approximated by a penalty function (10). For a detailed explanation of the mathematics behind these cost terms the reader is referred to [17]. Currently implemented cost terms f_i, their respective weights σ_i and related ROS

---

[16]teb_local_planner, http://wiki.ros.org/teb_local_planner.

parameters are summarized below according to [18]. A distinction is made between alternative cost functions for static and dynamic obstacles.

Minimal Total Transition Time (Objective)
Description: Minimizes the total transition time to the goal in order to find a time-optimal solution.
Weight parameter: `weight_optimaltime`
Related parameters: `selection_alternative_time_cost`

Via-points (Objective)
Description: Minimizes the distance to specific via-points which define attractors for the trajectory.
Weight parameter: `weight_viapoint`
Related parameters: `global_plan_viapoint_sep`

Compliance with Non-holonomic Kinematics (Objective)
Description: Enforces the geometric constraint for non-holonomic robots that requires two consecutive poses s_k and s_{k+1} to be located on a common arc of constant curvature. Actually the compliance is an equality constraint ensured by a large weight rather than an objective.
Weight parameter: `weight_kinematics_turning_radius`
Related parameters: `min_turning_radius`

Limiting Translational Velocity (Penalty)
Description: Constrains the translational velocity v_k to the interval $[-v_{back}, v_{max}]$. v_k is computed for each time interval ΔT_k with s_k, s_{k+1} by means of finite differences.
Weight parameter: `weight_max_vel_x`
Related parameters: `max_vel_x` (v_{max}), `max_vel_x_backwards` (v_{back})

Limiting Angular Velocity (Penalty)
Description: Constrains the angular velocity to $|\omega_k| \leq \omega_{max}$ by means of finite differences.
Weight parameter: `weight_max_vel_theta`
Related parameters: `max_vel_theta` ω_{max}

Limiting Translational Acceleration (Penalty)
Description: Constrains the translational acceleration to $|a_k| \leq a_{max}$ by means of finite differences.
Weight parameter: `weight_acc_lim_x`
Related parameters: `acc_lim_x` (a_{max})

Limiting Angular Acceleration (Penalty)
Description: Constrains the angular acceleration to $|\dot{\omega}_k| \leq \dot{\omega}_{max}$ by means of finite differences.

Weight parameter: `weight_acc_lim_theta`
Related parameters: `acc_lim_theta` ($\dot{\omega}_{max}$)

Limiting the Minimum Turning Radius (Penalty)
Description: This penalty enforces a minimum turning radius designated for car-like robots with limited steering angles. Differential drive robots are able to turn in place with a turning radius $r_{min} = 0$.
Weight parameter: `weight_kinematics_turning_radius`
Related parameters: `min_turning_radius` (r_{min})

Penalizing Backward Motions (Penalty)
Description: This cost term reflects a bias for forward motions, even though small weights still allow backward motions.
Weight parameter: `weight_kinematics_forward_drive`

Limiting Distance to Static Obstacles (Penalty)
Description: Enforces a minimum separation d_{min} of poses along the planned trajectory to a static obstacle (incorporates the robot footprint). In addition, obstacle inflation considers a buffer zone (larger than d_{min} in order to take effect) around an obstacle.
Weight parameters: `weight_obstacle`, `weight_obstacle_inflation`
Related parameters: `min_obstacle_dist` (d_{min}), `inflation_dist`

Limiting Distance to Predicted Positions of Dynamic Obstacles (Penalty)
Description: Enforces a minimum separation d_{min} of poses along the planned trajectory to the dynamic obstacle pose predicted according to a constant velocity model. The distance refers to the obstacle pose at the corresponding time step according to closest separation of the robot footprint and the most recent estimated obstacle shape. Again, obstacle inflation accounts for a buffer zone. The obstacle motion prediction is disabled by the parameter `include_dynamic_obstacles` which causes all obstacles to be considered as static.
Weight parameters: `weight_dynamic_obstacle`,
                    `weight_dynamic_obstacle_inflation`
Related parameters: `min_obstacle_distance` (d_{min}),
                    `dynamic_obstacle_inflation_dist`,
                    `include_dynamic_obstacles`

7   Testing Spatio-Temporal Trajectory Optimization

The *teb_local_planner* package includes a basic test node (*test_optim_node*) for testing and analysis of the trajectory optimization between a fixed start and goal pose. It supports parameter configurations and performance validation on the target hardware. Obstacles are represented by interactive markers[17] and can be animated with the *rviz* GUI.

The *teb_local_planner* package provides a launch file in order to start up the *test_optim_node* along with a preconfigured *rviz* node:

```
$ roslaunch teb_local_planner test_optim_node.launch
```

Rviz shows the planned trajectories and obstacles. The *teb_local_planner* plans multiple trajectories in parallel and selects the optimal trajectory indicated by red arrows. Select the menu button *interact* to move the obstacles around and observe the optimization and reconfiguration of the planned trajectories. Trajectory optimization is customized at runtime with:

```
$ rosrun rqt_reconfigure rqt_reconfigure
```

Select the *test_optim_node* from the list of available nodes and enable dynamic obstacles with the parameter `include_dynamic_obstacles`. Furthermore, set the parameter `visualize_with_time_as_z_axis_scale` to some positive number, e.g. 0.2 (refer to Fig. 6). The temporal evolution of the trajectory and obstacle configuration (indicated by red line markers) is displayed along the z-axis according to the scaling factor. Currently, the obstacles are static and maintain their pose.

Obstacles in the *test_optim_node* subscribe to individual topics, which specify their velocities in the *map*-frame. The velocity of an obstacle is defined by publishing a `geometry_msgs/Twist` message to the corresponding topic:

```
$ rostopic pub --once /test_optim_node/obstacle_0/cmd_vel
    geometry_msgs/Twist  '{linear:  {x: 0.2, y: 0.3, z: 0.0},
    angular: {x: 0.0, y: 0.0, z: 0.0}}'
```

In order to define the velocities of the remaining obstacles, replace the respective obstacle number. Admissible topics can be listed with:

```
$ rostopic list /test_optim_node/ | grep obstacle_
```

Only the translational x- and y-components of the message are considered. The planned trajectory avoids dynamic obstacles according to the predictions of their future positions. The current prediction operates with the ground-truth velocities provided by the message published previously. In case of estimated obstacle velocities, these predictions are obviously less accurate, especially for remote poses. This causes the planned trajectories to oscillate in particular as the uncertainties of pose estimates increase with the transition time.

---

[17]*Interactive markers*, http://wiki.ros.org/interactive_markers.

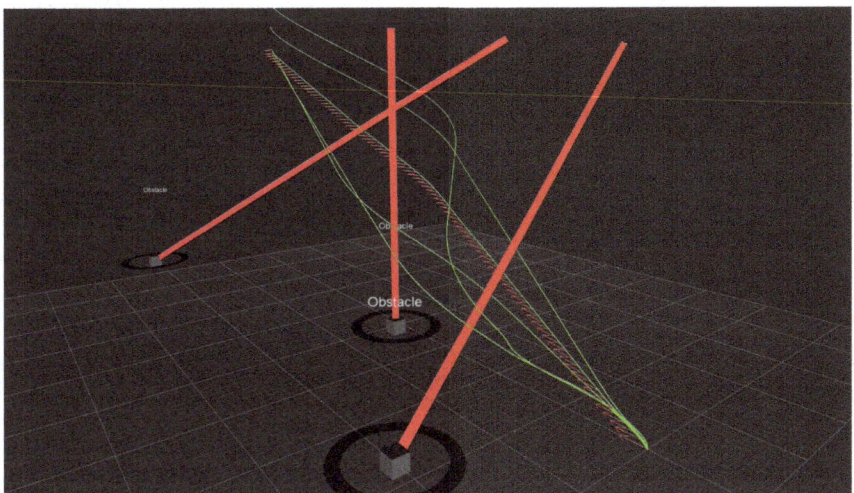

Fig. 6 Spatio-temporal obstacle avoidance in *rviz*

Switch back to the *rqt_reconfigure* GUI and customize the optimization with different parameter settings. Adjust parameters one at a time, as some parameters significantly influence the optimization performance. In case of insufficient performance on your target system, either decrease the parameters `no_inner_iterations` or `no_outer_iterations` to reduce the number of executed inner respectively outer iterations per TEB optimization step. As an alternative, increase the reference time step `dt_ref` slightly to obtain a coarser trajectory with fewer poses. These workarounds may deteriorate the optimality of the planned trajectory.

8 Obstacle Motion Predictive Planning

This section introduces a simple albeit challenging scenario for local path planning algorithms in a highly dynamic environment and illustrates the benefits of spatio-temporal trajectory planning. The scenario mimics an arcade game in which the robot traverses a corridor while evading intruders. Figure 7 illustrates the obstacles moving back and forth across the corridor along parallel paths with a constant velocity. Note, that the intruder motion proceeds blindly in a purely open loop manner. The obstacle velocities are estimated from the costmap conversion and do not reflect ground truth velocities. The analysis compares the static *teb_local_planner* with its dynamic extension.

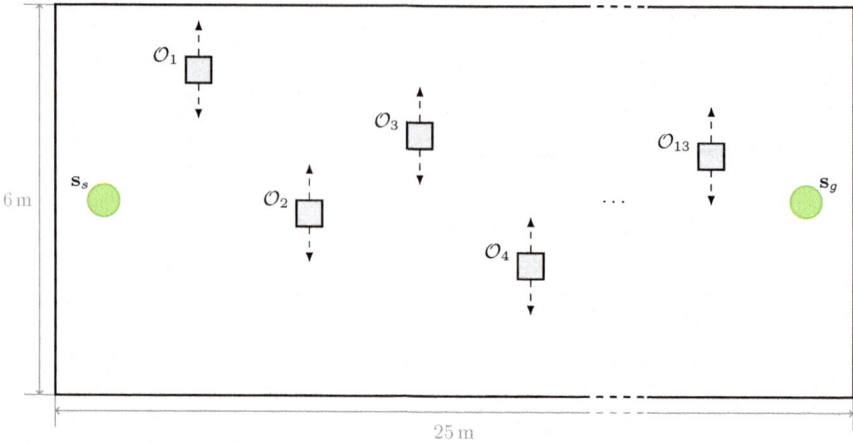

Fig. 7 Setup of the corridor scenario

In the following details of the launch file parameters and configuration are explained.

```
<!-- ************** Global Parameters ************** -->
2 <param name="/use_sim_time" value="true"/>
```

The global parameter `use_sim_time` enables ROS to operate with a simulated clock rather than the CPU system time.

```
<!-- ************** Stage Simulator ***************** -->
2 <node pkg="stage_ros" type="stageros" name="stageros" args="$(find
     dynamic_obstacle_test)/stage/corridor.world">
   <remap from="/robot_0/base_scan" to="/robot_0/scan"/>
4 </node>
```

This command starts up the *stage* node and loads the environment configuration from *corridor.world* file.

```
<!-- ******************* Maps ********************* -->
2 <node name="map_server" pkg="map_server" type="map_server" args="$(find
     dynamic_obstacle_test)/maps/corridor.yaml" output="screen">
   <param name="frame_id" value="map"/>
4 </node>
```

This code segment starts the *map_server* node with the parameters defined in the *corridor.yaml* file.

```
<!-- ************* Navigation ROBOT 0 ************* -->
<group ns="robot_0">
  <param name="tf_prefix" value="robot_0"/>

  <node pkg="tf" type="static_transform_publisher" name="
    link1_broadcaster" args="2 3 0 0 0 1 /map /robot_0/odom 100" />

  <node pkg="move_base" type="move_base" respawn="false" name="
    move_base" output="screen">
    <rosparam file="$(find teb_local_planner_tutorials)/cfg/diff_drive/
      costmap_common_params.yaml" command="load" ns="global_costmap" />
    <rosparam file="$(find teb_local_planner_tutorials)/cfg/diff_drive/
      costmap_common_params.yaml" command="load" ns="local_costmap" />
    <rosparam file="$(find teb_local_planner_tutorials)/cfg/diff_drive/
      local_costmap_params.yaml" command="load" />
    <rosparam file="$(find teb_local_planner_tutorials)/cfg/diff_drive/
      global_costmap_params.yaml" command="load" />
    <rosparam file="$(find teb_local_planner_tutorials)/cfg/diff_drive/
      teb_local_planner_params.yaml" command="load" />

    <!-- Here we load our costmap conversion settings -->
    <!-- If you uncomment the following line, disable the
      ground_truth_obstacles node at the bottom of this script! -->
    <!-- rosparam file="$(find teb_local_planner_tutorials)/cfg/
      diff_drive/costmap_converter_params.yaml" command="load" /-->

    <param name="TebLocalPlannerROS/include_costmap_obstacles" value="
      False" />
    <param name="TebLocalPlannerROS/include_dynamic_obstacles" value="
      True" />

    <param name="base_global_planner" value="navfn/NavfnROS" />
    <!--param name="base_global_planner" value="global_planner/
      GlobalPlanner" />
    <param name="planner_frequency" value="1.0" />
    <param name="planner_patience" value="5.0" /-->

    <param name="base_local_planner" value="teb_local_planner/
      TebLocalPlannerROS" />
    <param name="controller_frequency" value="5.0" />
    <param name="controller_patience" value="15.0" />
    <remap from="map" to="/map"/>
  </node>
</group>
```

This code initializes the navigation stack and launches the *move_base* node in the namespace of `robot_0` (which corresponds to the observing robot). Several parameters regarding the local and global costmap are loaded. The utilized plugin for the *costmap_converter* is specified in the *costmap_converter_params.yaml* file. The *teb_local_planner* plugin replaces the default local planner.

```
<!-- ****************** Obstacles ******************** -->
2 <group ns="robot_1">
    <param name="tf_prefix" value="robot_1"/>
4   <node name="Mover" pkg="dynamic_obstacle_test" type="move_obstacle.py"
        output="screen"/>
    <param name="pos_ub" value="5.0" />
6   <param name="pos_lb" value="1.0" />
    <param name="vel_y" value="0.3" />
8 </group>
```

Intruder obstacles are set in motion by a `cmd_vel` message published by the *Mover* node in the obstacle's namespace. The code is replicated for each obstacle with its respective namespace. The *Mover* node in *move_obstacle.py* subscribes to the `base_pose_ground_truth` topic published by *stage*. Each obstacle moves back and forth between `pos_lb` and `pos_ub` with the velocity `vel_y`. In case no velocity `vel_y` is specified, a random velocity is sampled at each turnaround.

```
<!-- **************** Visualisation **************** -->
2 <node name="rviz" pkg="rviz" type="rviz" args="-d $(find
    teb_local_planner_tutorials)/cfg/rviz_navigation_cc.rviz">
    <remap from="/move_base_simple/goal" to="/robot_0/move_base_simple/
        goal" />
4 </node>
```

This code launches the visualization tool *rviz* and loads a configuration file which contains some predefined *rviz* displays, e.g. the robots footprint.

```
<!-- ************ Ground Truth Obstacles ************ -->
2 <node name="ground_truth_obstacles" pkg="dynamic_obstacle_test" type="
    publish_ground_truth_obstacles.py" output="screen" />
```

This code segment launches the node that publishes the ground truth velocities of the obstacles. The node subscribes to the `base_pose_ground_truth` topic and publishes an `ObstacleArrayMsg` further processed by the *teb_local_planner*. Ground truth velocities of the obstacles are only utilized for comparison with the ideal optimal collision avoidance maneuvers. Launch the simulation:

```
$ roslaunch teb_local_planner_tutorials corridor_scenario.
    launch
```

rviz and *stage* show the simulation setup. The dynamic obstacles move across the corridor. By default, dynamic obstacles are not explicitly considered during the trajectory optimization. In order to monitor the reference trajectory planning without motion prediction for dynamic obstacles, publish a navigation goal at the end of the corridor using the '2D Nav Goal' button in *rviz*. Observe the robots behavior when obstacles cross its way. The robot responds only until the obstacle is located directly in front of the robot. The static perspective severely compromises the planning due to the mismatch of the static environment and its true status within the evolving scenario. As a consequence, the robot merely relies on reactive control and exhibits detours. In many cases, the planned trajectory lacks robustness frequently causing collisions with the intruders. Close the simulation (*Ctrl+C*).

Relaunch the simulation and enable obstacle motion predictive planning in the *rqt_reconfigure* tool with the parameter `use_dynamic_obstacles`. Additionally, the temporal dimension of the planned trajectory can be visualized as *z*-axis in *rviz* by means of the parameter `visualize_with_time_as_z_axis_scale`. Again, publish a navigation goal at the end of the corridor and observe the planned trajectories. The robot avoids the obstacles ahead of time which not only significantly reduces the total transition time to the goal but also results in less frequent collisions. Customize trajectory planning with different optimization weights and adapt the minimal distance to dynamic obstacles (`min_obstacle_distace`).

Activate costmap conversion by uncommenting the line loading the costmap conversion settings in the *corridor_scenario.launch* file:

```
<rosparam file="$(find teb_local_planner_tutorials)/cfg/diff_drive/
    costmap_converter_params.yaml" command="load" />
```

Disable the node that publishes ground truth data about the obstacles from the simulation setup by removing the following lines from the launch file *corridor_scenario.launch*:

```
<!-- *********** Ground Truth Obstacles *********** -->
<node name="ground_truth_obstacles" pkg="dynamic_obstacle_test" type="
    publish_ground_truth_obstacles.py" output="screen" />
```

Relaunch the simulation, enable `use_dynamic_obstacles` in the *rqt_reconfigure* tool, and publish a navigation goal at the end of the corridor using the '2D Nav Goal' button. Footprints of the observed obstacles are visualized as red polygons. Notice, the slight oscillations of the planned trajectory attributed to the uncertainty of obstacle velocity estimations. Nevertheless, the robot avoids most obstacles merely based on their predicted movements.

The enhanced trajectory planning of the *teb_local_planner* in dynamic environments due to the incorporation of estimated obstacle velocities is confirmed in simulations. Figure 8 depicts a comparison of the total transition times between the legacy and the current version of the *teb_local_planner*. Compared to the default configuration, these results were recorded with a slightly increased minimum distance to obstacles in order to increase the influence of the rather small obstacles. In the introduced corridor scenario with random initial obstacle positions, the mean transition time can be reduced by more than 20 s. Additionally, the prediction of obstacle movement results in a more robust trajectory which is shown by the reduced collision probability (see Table 1). Note, that the robot does not collide actively, but rather passively due to too optimistic trajectory planning and the blind motion of obstacles. Collisions occurring with the current version of the *teb_local_planner* are mostly the result of abrupt changes in direction performed by dynamic obstacles at corridor walls. In these cases the planned trajectory avoids the current motion of an obstacle, but does not account for the sudden change in the obstacles movement direction.

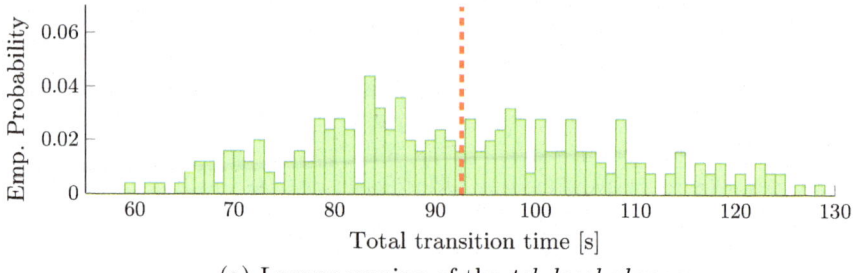

(a) Legacy version of the *teb_local_planner*

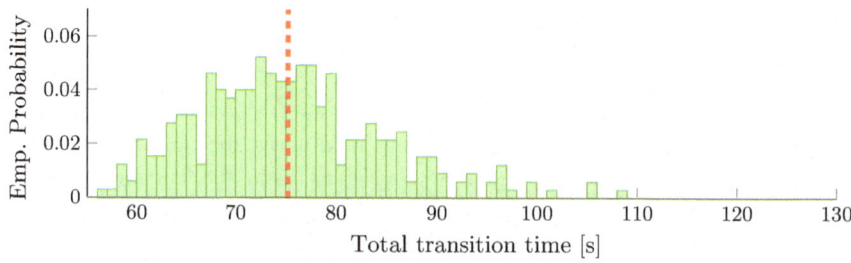

(b) *teb_local_planner* with dynamic obstacle support

Fig. 8 Comparison of total transition times

Table 1 Comparison of total transition times and collision probabilities

	Legacy version	Current version
Number of simulations	368	470
Number of collisions	117	96
Emp. collision probability	0,318	0,205
Emp. mean transition time [s]	92,64	71,00
Standard deviation [s]	15,71	8,41

9 Conclusion

This tutorial chapter provides a comprehensive step-by-step guide for the setup of the *teb_local_planner* and *costmap_converter* ROS packages for mobile robot navigation with explicit consideration of dynamic obstacles. The literature contains several advanced approaches for optimal trajectory planning with motion prediction of dynamic obstacles. However, these algorithms are currently not compatible with the static local costmap representation of the ROS navigation stack. The proposed costmap conversion exploits the established architecture of the navigation stack and estimates velocities, orientations, and shapes of dynamic obstacles only based on information on the local costmap. These informations are incorporated into the Timed-Elastic-Band approach to calculate optimal spatio-temporal trajectories in the presence of dynamic obstacles. Besides its ability to include dynamic obstacles explicitly into trajectory optimization, to our best knowledge, the *teb_local_planner* is currently the only local planner plugin for the ROS navigation stack, which is capable of trajectory planning for car-like robots without the use of an additional low-level controller.

Future work addresses the automatic tuning of cost function weights in order to improve the overall performance of the *teb_local_planner*. Furthermore, a benchmark suite for the comparative analysis of ROS local planners would be a valuable addition to ROS in order to support users in their decision for the local planner most suitable for their particular mobile robot applications.

References

1. S. Bhattacharya, M. Likhachev, V. Kumar, Identification and representation of homotopy classes of trajectories for search-based path planning in 3D. in *Proceedings of Robotics: Science and Systems* (2011)
2. V. Delsart, T. Fraichard, Reactive trajectory deformation to navigate dynamic environments. in *European Robotics Symposium* (2008), pp. 233–241
3. P. Fiorini, Z. Shiller, Motion planning in dynamic environments using velocity obstacles. Int. J. Robot. Res. **17**(7), 760–772 (1998)
4. D. Fox, W. Burgard, S. Thrun, The dynamic window approach to collision avoidance. IEEE Robot. Autom. Magaz. **4**(1), 23–33 (1997)
5. C. Fulgenzi, A. Spalanzani, C. Laugier, Dynamic obstacle avoidance in uncertain environment combining PVOs and occupancy grid. in *IEEE International Conference on Robotics and Automation (ICRA)* (2007)
6. B. Gerkey, K. Konolige, Planning and control in unstructured terrain. in *Proceedings of the ICRA Workshop on Path Planning on Costmaps* (2008)
7. T. Gu, J. Atwood, C. Dong, J. M. Dolan, J.-W. Lee, Tunable and stable real-time trajectory planning for urban autonomous driving. in *IEEE International Conference on Intelligent Robots and Systems (IROS)* (2015), pp. 250–256
8. H.W. Kuhn, The Hungarian method for the assignment problem. Nav. Res. Logist. Q. **2**, 83–97 (1955)
9. R. Kümmerle, G. Grisetti, H. Strasdat, K. Konolige, W. Burgard, G2o: a general framework for graph optimization. in *IEEE International Conference on Robotics and Automation (ICRA)* (2011), pp. 3607–3613
10. H. Kurniawati, T. Fraichard, From path to trajectory deformation. in *IEEE/RSJ International Conference on Intelligent Robots and Systems (RISO)* (2007), pp. 159–164
11. M. Luber, A. Stork, G.D. Tipaldi, K.O. Arras, People tracking with human motion predictions from social forces. in *IEEE International Conference on Robotics and Automation (ICRA)* (2010), pp. 464–469
12. E. Marder-Eppstein, E. Berger, T. Foote, B. Gerkey, K. Konolige, The office marathon: robust navigation in an indoor office environment. in *IEEE International Conference on Robotics and Automation (ICRA)* (2010)
13. M. Morari, J.H. Lee, Model predictive control: past, present and future. Comput. Chem. Eng. **23**(4–5), 667–682 (1999)
14. J. Nocedal, S.J. Wright, *Numerical Optimization*, Operations Research (Springer, New York, 1999)
15. K. Rebai, O. Azouaoui, M. Benmami, A. Larabi, Car-like robot navigation at high speed. in *IEEE International Conference on Robotics and Biomimetics (ROBIO)* (2007), pp. 2053–2057
16. C. Rösmann, W. Feiten, T. Wösch, F. Hoffmann, T. Bertram, Trajectory modification considering dynamic constraints of autonomous robots. in *7th German Conference on Robotics (ROBOTIK)* (2012), pp. 74–79

17. C. Rösmann, F. Hoffmann, T. Bertram, Integrated online trajectory planning and optimization in distinctive topologies. Robot. Auton. Syst. **88**, 142–153 (2017)
18. C. Rösmann, F. Hoffmann, T. Bertram, Online trajectory planning in ROS under kinodynamic constraints with timed-elastic-bands, *Robot Operating System (ROS) - The Complete Reference 2*, vol. 707, Studies in Computational Intelligence (Springer International Publishing, 2017)
19. C. Rösmann, F. Hoffmann, T. Bertram, Kinodynamic trajectory optimization and control for car-like robots. in *IEEE/RSJ International Conference on Intelligent Robots and Systems (IROS)* (2017), pp. 5681–5686
20. C. Rösmann, M. Oeljeklaus, F. Hoffmann, T. Bertram, Online trajectory prediction and planning for social robot navigation. in *IEEE International Conference on Advanced Intelligent Mechatronics (AIM)* (2017), pp. 1255–1260
21. M. Seder, I. Petrović, Dynamic window based approach to mobile robot motion control in the presence of moving obstacles. in *IEEE International Conference on Robotics and Automation (ICRA)* (2007), pp. 1986–1991
22. S. Thrun, D. Fox, W. Burgard, F. Dellaert, Robust Monte Carlo localization for mobile robots. Artif. Intell. **128**, 99–141 (2001)
23. S. Quinlan, O. Khatib, Elastic bands: connecting path planning and control. in *IEEE International Conference on Robotics and Automation (ICRA)* (1993), pp. 802–807

Franz Albers received his B.Sc. and M.Sc. degree in electrical engineering and computer science from the Technische Universität Dortmund, Germany, in 2015 and 2017, respectively. He is currently working as a Dr.-Ing. candidate at the Institute of Control Theory and Systems Engineering, Technische Universität Dortmund, Germany. His research interests include trajectory planning and automotive systems.

Christoph Rösmann was born in Münster, Germany, on December 8, 1988. He received the B.Sc. and M.Sc. degree in electrical engineering and information technology from the Technische Unversität Dortmund, Germany, in 2011 and 2013, respectively. He is currently working towards the Dr.-Ing. degree at the Institute of Control Theory and Systems Engineering, Technische Universität Dortmund, Germany. His research interests include nonlinear model predictive control, mobile robot navigation and fast optimization techniques.

Frank Hoffmann received the Diploma and Dr. rer. nat. degrees in physics from the Christian-Albrechts University of Kiel, Germany. He was a postdoctoral Researcher at the University of California, Berkeley from 1996–1999. From 2000 to 2003, he was a lecturer in computer science at the Royal Institute of Technology, Stockholm, Sweden. He is currently a Professor at TU Dortmund and affiliated with the Institute of Control Theory and Systems Engineering. His research interests are in the areas of robotics, computer vision, computational intelligence, and control system design.

Torsten Bertram received the Dipl.-Ing. and Dr.-Ing. degrees in mechanical engineering from the Gerhard Mercator Universität Duisburg, Germany, in 1990 and 1995, respectively. In 1990, he joined the Gerhard Mercator Universität Duisburg, Germany, in the Department of Mechanical Engineering, as a Research Associate. During 1995–1998, he was a Subject Specialist with the Corporate Research Division, Bosch Group, Stuttgart, Germany, In 1998, he returned to Gerhard Mercator Universität Duisburg as an Assistant Professor. In 2002, he became a Professor with the Department of Mechanical Engineering, Technische Universität Ilmenau, Germany, and, since 2005, he has been a member of the Department of Electrical Engineering and Information Technology, Technische Universität Dortmund, Germany, as a Professor of systems and control engineering and he is head of the Institute of Control Theory and Systems Engineering. His research fields are control theory and computational intelligence and their application to mechatronics, service robotics, and automotive systems.

A Backstepping Non-smooth Controller for ROS-Based Differential-Drive Mobile Robots

Walter Fetter Lages

Abstract This chapter presents a non-linear controller for a mobile robot based on feedback linearization, non-smooth feedback and backstepping. The stability and convergence of the controller to the reference pose is proved by using the Lyapunov theory and the Barbalat Lemma. The controller design is based on a robot model considering its kinematics and dynamics, and hence the control inputs are the torques applied on the wheels. Contrariwise to most available implementation of controllers in the Robot Operating System, which implements a set of single input, single output controllers using the proportional + integral + derivative control law, here a truly multi-input, multi-output non-linear controller is considered. Results showing the effectiveness of the proposed controller for the setting point and the trajectory tracking problems were obtained by using the Gazebo robot simulator and Rviz.

Keywords Backstepping · Multi-input multi-output · Non-linear controller
Differential drive · Mobile robot

1 Introduction

This chapter proposes a control law for a differential-drive mobile robot which considers its dynamic model and not only the usual kinematic model. This control law is based on feedback linearization to handle the dynamics of the robot, non-smooth feedback to cope with the non-holonomicity of the kinematic model and backstepping to consider the cascade structure of the complete model (considering the kinematics and the dynamics effects) of the mobile robot.

The proposed controller is implemented on the Robot Operating System (ROS). Despite ROS being a widely used framework nowadays, its documentation and examples covering low-level controllers are poor and almost all implementations are based

W. F. Lages (✉)
Universidade Federal do Rio Grande do Sul, Av. Osvaldo Aranha, 103,
Porto Alegre, RS 90035-190, Brazil
e-mail: fetter@ece.ufrgs.br
URL: http://www.ece.ufrgs.br/fetter

© Springer International Publishing AG, part of Springer Nature 2019
A. Koubaa (ed.), *Robot Operating System (ROS)*, Studies in Computational
Intelligence 778, https://doi.org/10.1007/978-3-319-91590-6_9

on single input, single output (SISO) controllers using the classical proportional + integral + derivative (PID) control law. Most references and textbooks on ROS [9, 21, 23, 26] do not even cover the implementation of controllers. On the other hand, robots are, in general, non-linear multi-input, multi-output systems (MIMO), for which the use of independent PID controllers for each degree of freedom (DoF) is not adequate if high performance is desired.

As the controller proposed in this work is non-linear and MIMO, its implementation in ROS is generic and representative of any controller and can be used as example for further implementation of any other control law (either linear or non-linear, SISO or MIMO, etc.), while the controllers available from standard ROS packages, such as ros_controllers are only representatives of SISO controllers. To capitalize on the generality of the implementation of the proposed controller, the inner working of the real-time loop of ROS is explained in detail, motivating the further implementation of advanced controllers in ROS such as in [18], where a computed torque controller is implemented to control a biped robot.

To verify the controller performance, the mobile robot is described in the Unified Robot Description Format (URDF) [21] and simulated by using the Gazebo simulator [11]. It is important to note that the robot model used for control purposes was derived independently from the model used by Gazebo to simulate the mobile robot. Furthermore, the robot is described in URDF in greater detail than the model used for control purposes considers. Hence, there is some model mismatch between the model used for simulation and the model used for control, as would be the case with a real robot, showing the robustness of the proposed approach. The stability and convergence of the closed-loop system is proved by using Lyapunov stability analysis and the Barbalat Lemma [25].

Similar concepts are used in the lyap_control ROS package [32]. However, its implementation is based on a ROS node and topics are used to communicate with sensors and actuators. This architecture is not real-time safe, as opposite to a controler based on the ros_control package as proposed here. This problem is even worse because a known drawback of the controller used in lyap_control is that it requires very small sampling periods, which are difficult to achieve without a proper real-time system. Furthermore, it requires the DoFs of the system to be the same as the number of its inputs, which does not hold for a differntial-drive mobile robot.

2   Background

In this section the background required to understand the implementation of the controller is presented. Section 2.1 presents the details of the model of a differential-drive mobile robot, including its dynamics and kinematics. That model is the base for the design of a backstepping controller presented in Sect. 2.2. The implemantion of that controller requires information about the robot pose, which for the purpose of this chapter is computed by the odometry procedure described in Sect. 2.3. Of

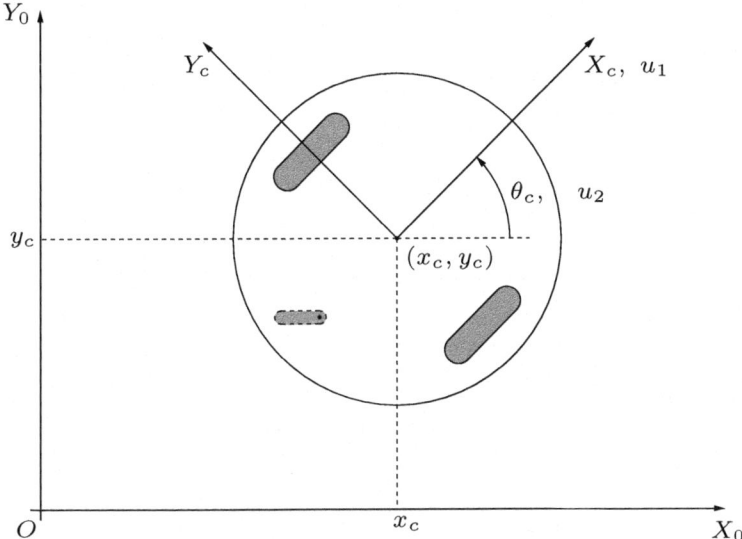

Fig. 1 Coordinate systems

course, as any odometry-only based pose estimation, it is subject to drift and for more ambitious applications, a proper pose estimation with sensor fusion should be used. Nonetheless, the procedure described in Sect. 2.3 can be used as the odometry sensor for that more sophisticated sensor fusion based pose estimation.

2.1 Mobile Robot Model

The model of the mobile robot used for control purposes is described in this section. Figure 1 shows the coordinate systems used to describe the mobile robot model, where X_c and Y_c are the axes of the coordinate system attached to the robot and X_0 and Y_0 form the inertial coordinate system. The pose (position and orientation) of the robot is represented by $x = \begin{bmatrix} x_c & y_c & \theta_c \end{bmatrix}^T$.

The dynamic model of the TWIL mobile robot used in this work can be obtained based on the Lagrange-Euler formulation [6] and is given by [14]:

$$\begin{aligned} \dot{x} &= B(x)u \\ \dot{u} &= f(u) + G\tau \end{aligned} \tag{1}$$

where u is the vector of the linear and angular velocities of the robot and τ is the vector of input torques on the wheels. $B(x)$ is a matrix whose structure depends on the kinematic (geometric) properties of the robot, $f(u)$ and G depend on the

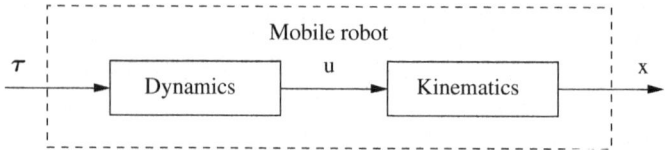

Fig. 2 Cascade between dynamics and the kinematic model

kinematic and dynamic (mass and inertia) parameters of the robot and for the TWIL mobile robot are given by:

$$\boldsymbol{B}(\boldsymbol{x}) = \begin{bmatrix} \cos\theta_c & 0 \\ \sin\theta_c & 0 \\ 0 & 1 \end{bmatrix} \tag{2}$$

$$\boldsymbol{f}(\boldsymbol{u}) = \begin{bmatrix} 0 & f_{12} \\ f_{21} & 0 \end{bmatrix} \begin{bmatrix} u_1 u_2 \\ u_2^2 \end{bmatrix} = \boldsymbol{F} \begin{bmatrix} u_1 u_2 \\ u_2^2 \end{bmatrix} \tag{3}$$

$$\boldsymbol{G} = \begin{bmatrix} g_{11} & g_{12} \\ g_{21} & g_{21} \end{bmatrix} \tag{4}$$

where $f_{12} = 0.08444\,\text{m/rad}^2$, $f_{21} = 3.7706\,\text{m}^{-1}$, $g_{11} = g_{12} = 2.6468\,\text{kg}^{-1}\text{m}^{-1}$ and $g_{21} = -g_{22} = -16.0840\,\text{kg}^{-1}\text{m}^{-2}$ are constants depending only on the geometric and inertia parameters of the robot.

Note that the dynamic model of the robot is a cascade between its kinematic model (the first expression of (1)), which considers velocities as inputs, and its dynamics (the second expression of (1)), which considers torques as inputs, as shown in Fig. 2.

2.2 Control of the Mobile Robot

Differential-drive mobile robots are nonholonomic systems [6]. An important general statement on the control of nonholonomic systems has been made by [4], who has shown that it is not possible to asymptotically stabilize the system at an arbitrary point through a time-invariant, smooth state feedback law. In spite of it, the system is controllable [2].

Ways around Brockett's conditions for asymptotic stability are time-variant control [8, 24, 27, 31], non-smooth control [2, 7, 29] and hybrid control laws [17]. Most of those control laws considers only the kinematics of the mobile robot and hence only the first equation of (1), neglecting the dynamics effects described by the second equation of (1). Nonetheless, due to the cascading structure of (1), it is possible to use a control law designed to control only the kinematics of the mobile robot to design a control law for the whole model (1) in a procedure called backstepping [12].

Here, for the kinematic part of the robot model, a set of possible input signals is obtained by using a non-smooth coordinate transform. A general way of designing control laws for nonholonomic systems through non-smooth coordinate transform was presented by [2]. Here, we consider a non-smooth coordinate transform similar to [15], which was already implemented in ROS [14].

Before the design of the non-smooth control law and the backstepping, two other steps are performed: a feedback linearization [10, 28] to further simplify the model of the mobile robot and a coordinate change in order to enable the steering of the robot to any reference pose and not only to the origin.

Feedback Linearization By using the feedback [14]:

$$\tau = G^{-1} (v - f(u)) \tag{5}$$

where v is a new input vector, it is possible linearize the second expression of (1) to obtain:

$$\dot{x} = B(x)u$$
$$\dot{u} = v \tag{6}$$

Note that the model (6) still has a cascade structure, but with a much simpler dynamics with accelerations instead of torque as input. That cascade structure enables the design of a backstepping controller.

Offset to Origin In the following sections, a controller for (6) is designed under the assumption that the robot should converge to the origin. However, it is interesting to stabilize the robot at any pose $x_r = \begin{bmatrix} x_{cr} & y_{cr} & \theta_{cr} \end{bmatrix}^T$. This can be done by a coordinate change [15] given by (see Fig. 3):

$$\bar{x} = \begin{bmatrix} \cos\theta_{cr} & \sin\theta_{cr} & 0 \\ -\sin\theta_{cr} & \cos\theta_{cr} & 0 \\ 0 & 0 & 1 \end{bmatrix} (x - x_r) \tag{7}$$

Non-Smooth Control In this section, a non-smooth control law for stabilizing the kinematics of the mobile robot is presented without details. See [14] for the details of how this control law is obtained. Also, see [16] for coordinate changes for details on control laws for mobile robot not based on differential-drive kinematics.

By considering a coordinate change [3], which is similar to a change to polar coordinates,

$$e = \sqrt{\bar{x}_1^2 + \bar{x}_2^2} \tag{8}$$

$$\psi = \text{atan2}(\bar{x}_2, \bar{x}_1) \tag{9}$$

$$\alpha = \bar{x}_3 - \psi \tag{10}$$

$$\eta_1 = u_1 \tag{11}$$

$$\eta_2 = u_2 \tag{12}$$

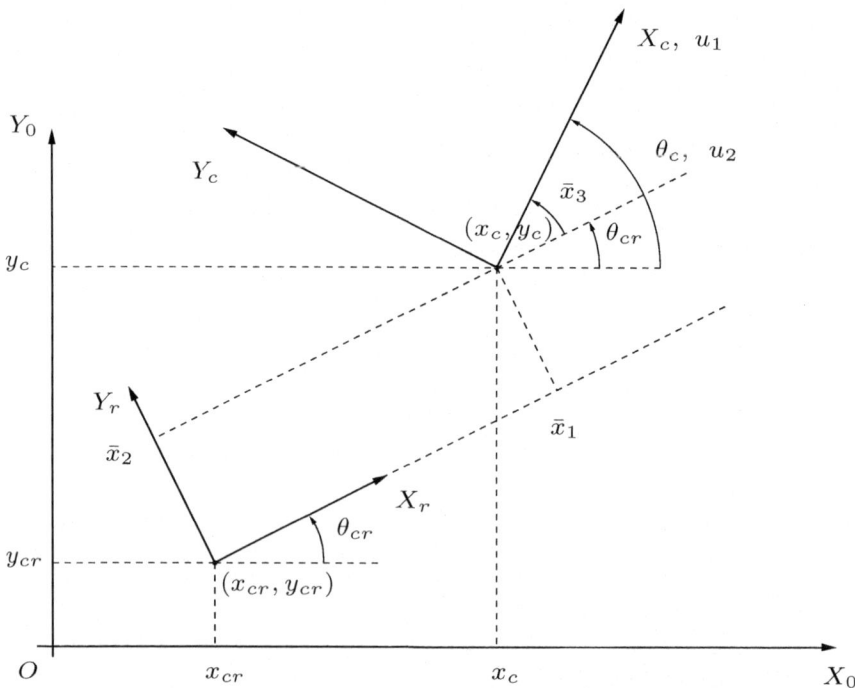

Fig. 3 Robot coordinates with respect to the reference frame

the first expression of system model (6) can be rewritten as:

$$\dot{e} = \cos \alpha \eta_1$$
$$\dot{\psi} = \frac{\sin \alpha}{e} \eta_1 \tag{13}$$
$$\dot{\alpha} = -\frac{\sin \alpha}{e} \eta_1 + \eta_2.$$

which is forced to converge to the origin by the following input signal [1]:

$$\eta_1 = -\gamma_1 e \cos \alpha \tag{14}$$

$$\eta_2 = -\gamma_2 \alpha - \gamma_1 \cos \alpha \sin \alpha + \gamma_1 \frac{\lambda_3}{\lambda_2} \cos \alpha \frac{\sin \alpha}{\alpha} \psi \tag{15}$$

with $\lambda_i > 0$ and $\gamma_i > 0$.

Backstepping Although (14)–(15) are able to stabilize the first equation of (6), they can not stabilize (6) as a whole since its input is $\boldsymbol{v}$ and not $\boldsymbol{u}$. Note, however that (6) can be seen as a cascade between two subsystems and in this case, it is possible to apply a backstepping procedure [12] to $\boldsymbol{u}$ in order to obtain an expression for $\boldsymbol{v}$.

By applying the transforms (7)–(10) to (6) it is possible to write:

$$\dot{e} = \cos \alpha u_1 \tag{16}$$

$$\dot{\psi} = \frac{\sin \alpha}{e} u_1 \tag{17}$$

$$\dot{\alpha} = -\frac{\sin \alpha}{e} u_1 + u_2 \tag{18}$$

$$\dot{u}_1 = v_1 \tag{19}$$

$$\dot{u}_2 = v_2 \tag{20}$$

Then, by adding $\cos \alpha (\eta_1 - \eta_1)$ to (16), $\frac{\sin \alpha}{e}(\eta_1 - \eta_1)$ to (17) and $-\frac{\sin \alpha}{e}(\eta_1 - \eta_1) + (\eta_2 - \eta_2)$ to (18) and rearranging it is possible to write:

$$
\begin{aligned}
\dot{e} &= \cos \alpha \eta_1 + \cos \alpha (u_1 - \eta_1) \\
\dot{\psi} &= \frac{\sin \alpha}{e} \eta_1 + \frac{\sin \alpha}{e}(u_1 - \eta_1) \\
\dot{\alpha} &= -\frac{\sin \alpha}{e} \eta_1 + \eta_2 - \frac{\sin \alpha}{e}(u_1 - \eta_1) + (u_2 - \eta_2) \\
\dot{u}_1 &= v_1 \\
\dot{u}_2 &= v_2
\end{aligned}
\tag{21}
$$

and by defining:

$$e_1 \triangleq u_1 - \eta_1 \tag{22}$$

$$e_2 \triangleq u_2 - \eta_2 \tag{23}$$

$$\bar{v}_1 \triangleq v_1 - \dot{\eta}_1 \tag{24}$$

$$\bar{v}_2 \triangleq v_2 - \dot{\eta}_2 \tag{25}$$

results in:

$$
\begin{aligned}
\dot{e} &= \cos \alpha \eta_1 + \cos \alpha e_1 \\
\dot{\psi} &= \frac{\sin \alpha}{e} \eta_1 + \frac{\sin \alpha}{e} e_1 \\
\dot{\alpha} &= -\frac{\sin \alpha}{e} \eta_1 + \eta_2 - \frac{\sin \alpha}{e} e_1 + e_2 \\
\dot{e}_1 &= \bar{v}_1 \\
\dot{e}_2 &= \bar{v}_2
\end{aligned}
\tag{26}
$$

Then, by replacing η_1 and η_2 from (14) and (15):

$$
\begin{aligned}
\dot{e} &= -\gamma_1 e \cos^2 \alpha + \cos \alpha e_1 \\
\dot{\psi} &= -\gamma_1 \sin \alpha \cos \alpha + \frac{\sin \alpha}{e} e_1
\end{aligned}
$$

$$\dot{\alpha} = -\gamma_2\alpha + \gamma_1\frac{\lambda_3}{\lambda_2}\cos\alpha\frac{\sin\alpha}{\alpha}\psi - \frac{\sin\alpha}{e}e_1 + e_2$$

$$\dot{e}_1 = \bar{v}_1$$

$$\dot{e}_2 = \bar{v}_2 \tag{27}$$

Let the following candidate to Lyapunov function:

$$V_1 = \frac{1}{2}\left(\lambda_1 e^2 + \lambda_2\alpha^2 + \lambda_3\psi^2 + \lambda_4 e_1^2 + \lambda_5 e_2^2\right) \tag{28}$$

which, by differentiating with respect to time and replacing the system equations from (27) gives:

$$\dot{V}_1 = -\gamma_1\lambda_1 e^2\cos^2\alpha - \gamma_2\lambda_2\alpha^2 + \lambda_1 e\cos\alpha e_1$$
$$- \lambda_2\alpha\frac{\sin\alpha}{e}e_1 + \lambda_3\psi\frac{\sin\alpha}{e}e_1$$
$$+ \lambda_4 e_1\bar{v}_1 + \lambda_2\alpha e_2 + \lambda_5 e_2\bar{v}_2 \tag{29}$$

Then, by choosing:

$$\bar{v}_1 \triangleq -\gamma_3 e_1 - \frac{\lambda_1}{\lambda_4}e\cos\alpha + \frac{\lambda_2}{\lambda_4}\alpha\frac{\sin\alpha}{e} - \frac{\lambda_3}{\lambda_4}\psi\frac{\sin\alpha}{e} \tag{30}$$

$$\bar{v}_2 \triangleq -\gamma_4 e_2 - \frac{\lambda_2}{\lambda_5}\alpha \tag{31}$$

results in:

$$\dot{V}_1 = -\gamma_1\lambda_1 e^2\cos^2\alpha - \gamma_2\lambda_2\alpha^2 - \gamma_3\lambda_4 e_1^2 - \gamma_4\lambda_5 e_2^2 \leq 0 \tag{32}$$

which proves that V_1 is indeed a Lyapunov function for the system (27). Then, it follows that e, α, e_1 and e_2 are bounded, which implies that $\ddot{V}_1$ is bounded as well, which ensures that $\dot{V}_1$ is uniformly continuous. Furthermore, since $\dot{V}_1$ is uniformly continuous, it follows from the Barbalat lemma [25, 28] that $\dot{V}_1 \rightarrow 0$, which implies that $e \rightarrow 0$, $\alpha \rightarrow 0$, $e_1 \rightarrow 0$ and $e_2 \rightarrow 0$. It remains to prove that ϕ converges to zero. By applying the Barbalat's lemma to $\dot{\alpha}$ it follows that $\dot{\alpha} \rightarrow 0$ in (27), which implies that $\psi \rightarrow 0$.

The control law for the system (6) is then, from (24)–(25) given by:

$$v_1 = \bar{v}_1 - \dot{\eta}_1 \tag{33}$$

$$v_2 = \bar{v}_2 - \dot{\eta}_2 \tag{34}$$

A block diagram of the proposed control schema is presented in Fig. 4.

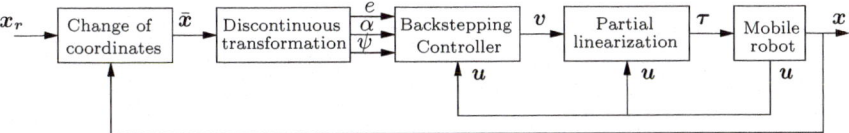

Fig. 4 Block diagram of the proposed controller

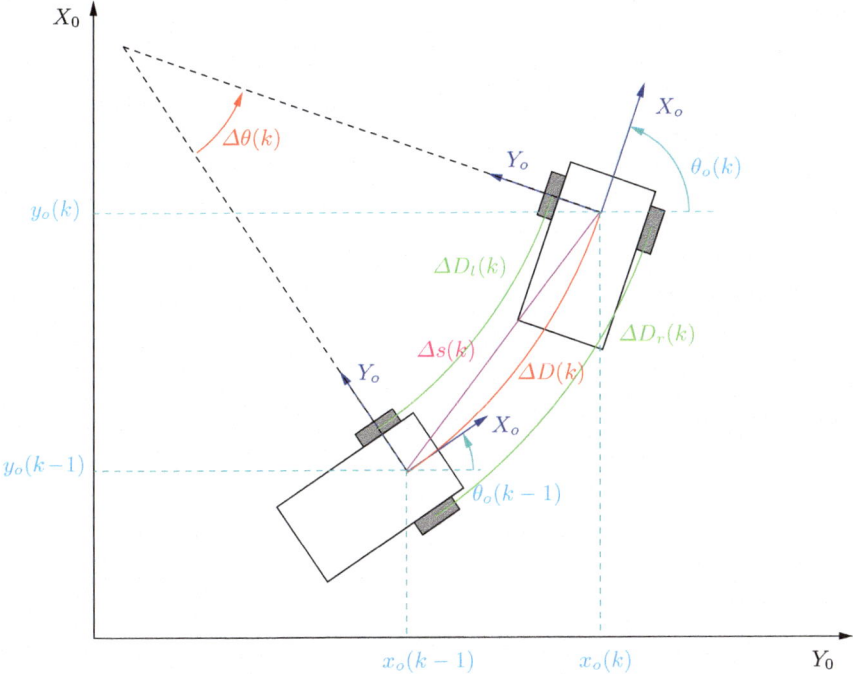

Fig. 5 Odometry supposing an arc of circumference trajectory

2.3 Odometry

For a differential-drive mobile robot, all factible trajectories are composed of arcs of circumferences. The straight line is just a circumference with infinite radius. Hence, it makes more sense to compute the odometry by supposing an arc of circumference between adjacent sampling points. This is more precise than supposing an straight line between sampling points as it is usually done. Therefore, as shown in Fig. 5, the the pose estimate $\hat{x}_o$ is given by [22]:

$$\hat{x}_o(k) = \hat{x}_o(k-1) + \Delta \hat{x}_o(k) \tag{35}$$

with

$$
\Delta\hat{x}_o(k) = \begin{bmatrix} \Delta s(k)\cos\left(\theta(k) + \frac{\Delta\theta(k)}{2}\right) \\ \Delta s(k)\sin\left(\theta(k) + \frac{\Delta\theta(k)}{2}\right) \\ \Delta\theta(k) \end{bmatrix} \tag{36}
$$

$$
\Delta s(k) = \Delta D(k)\frac{\sin\left(\frac{\Delta\theta(k)}{2}\right)}{\frac{\Delta\theta(k)}{2}} \tag{37}
$$

and

$$
\begin{bmatrix} \Delta D(k) \\ \Delta\theta(k) \end{bmatrix} = \begin{bmatrix} \frac{1}{2}\left(\Delta\varphi_r(k)r_r + \Delta\varphi_l(k)r_l\right) \\ \frac{1}{2b}\left(\Delta\varphi_r(k)r_r - \Delta\varphi_l(k)r_l\right) \end{bmatrix} = \Delta U(k) \tag{38}
$$

where $\Delta\varphi_r(k)$ and $\Delta\varphi_l(k)$ are the angular displacement of the right and left wheels, respectively, r_r and r_l are the radii of the right and left wheels, respectively and $2b$ is the axial distance between the right and left wheels..

Note that usually odometry is computed by assuming a straight line trajectory between two successive points, while in (35)–(38) an arc of circumference is assumed, which is the actual path of the robot if there is no error in kinematics parameters and the wheels do not slip.

3 ROS Setup

This section describes the installation of some packages useful for the implementation of the backstepping controller described in Sect. 2. Some of them are not present in a standard installation of ROS and should be installed. Also, some custom packages with our implementation of the TWIL robot model and the backstepping controller should be installed. Most of those packages are already described in detail in [13] and/or in [14] and hence, it will not be repeated here. Just the installation instructions for those packages will be given.

3.1 Setting up a Catkin Workspace

The packages to be installed for implementing ROS controllers assume an existing catkin workspace. If it does not exist, it can be created with the following commands (assuming a ROS Indigo version):

```
1  source /opt/ros/indigo/setup.bash
2  mkdir -p ~/catkin_ws/src
3  cd ~/catkin_ws/src
```

```
4  catkin_init_workspace
5  cd ~/catkin_ws
6  catkin_make
7  source ~/catkin_ws/devel/setup.bash
```

3.2 *ros_control*

The ros_control meta-package includes a set of packages to implement generic controllers. It is not included in the standard ROS desktop installation, hence it should be installed. On Ubuntu, it can be installed from Debian packages with the command:

```
1  sudo apt-get install ros-indigo-ros-control
```

3.3 *ros_controllers*

This meta-package implements a set of controllers to be used in ROS. In particular, the joint_state_controller controller, is used to obtain the output of the robot system. This meta-package is not included in the standard ROS desktop installation, hence it should be installed. On Ubuntu, it can be installed from Debian packages with the command:

```
1  sudo apt-get install ros-indigo-ros-controllers
```

3.4 *gazebo_ros_pkgs*

This is a collection of ROS packages for integrating the ros_control controller architecture with the Gazebo simulator [11] and is not included in the standard ROS desktop installation, hence it should be installed. On Ubuntu, it can be installed from Debian packages with the command:

```
1  sudo apt-get install ros-indigo-gazebo-ros-pkgs ros-indigo-gazebo-ros-control
```

3.5 *twil*

This is a meta-package with the package with the description of the TWIL robot. It contains an URDF description of the TWIL mobile robot. More specifically it includes the following packages:

`twil_description:` URDF description of the TWIL mobile robot.

`twil_ident:` identification of the dynamic parameters of the TWIL mobile robot. This package is not used here and was described in [14].

The `twil` meta-package can be downloaded and installed in the ROS catkin workspace with the commands:

```
1  cd ~/catkin_ws/src
2  wget http://www.ece.ufrgs.br/twil/indigo-twil-20180303.tgz
3  tar -xzf indigo-twil-20180303.tgz
4  cd ~/catkin_ws
5  catkin_make
6  source ~/catkin_ws/devel/setup.bash
```

3.6 *arc_odometry*

This package implements an odometry procedure supposing that the robot trajectory between two adjacent sampling points is an arc of circunference, as described in Sect. 2.3. Its implementation is described in Sect. 5.2 and it can be downloaded and installed in the ROS catkin workspace with the commands:

```
1  cd ~/catkin_ws/src
2  wget http://www.ece.ufrgs.br/ros-pkgs/indigo-arc-odometry-20180303.tgz
3  tar -xzf indigo-arc-odometry-20180303.tgz
4  cd ~/catkin_ws
5  catkin_make
6  source ~/catkin_ws/devel/setup.bash
```

3.7 *pose2d_trajectories*

This package implements simple trajectory generators for testing the controllers. It can be downloaded and installed in the ROS catkin workspace with the commands:

```
1  cd ~/catkin_ws/src
2  wget http://www.ece.ufrgs.br/ros-pkgs/indigo-pose2d-trajectories-20180303.tgz
3  tar -xzf indigo-pose2d-trajectories-20180303.tgz
4  cd ~/catkin_ws
5  catkin_make
6  source ~/catkin_ws/devel/setup.bash
```

3.8 *nonsmooth_backstep_controller*

This packake has the implementation of the backstepping controller. The background for the theory of operation of the controller is described in Sect. 2, while its implementation is discussed in Sect. 5. It can be downloaded and installed in the ROS catkin workspace with the commands:

```
cd ~/catkin_ws/src
wget http://www.ece.ufrgs.br/ros-pkgs/indigo-nonsmooth-backstep-controller\
-20180303.tgz
tar -xzf indigo-nonsmooth-backstep-controller-20180303.tgz
cd ~/catkin_ws
catkin_make
source ~/catkin_ws/devel/setup.bash
```

4 Testing the Installed Packages

A simple test for the installation of the packages described in Sect. 3 is performed here.

The installation of the ROS packages can be done by loading the TWIL model in Gazebo and launching the controller with the commands:

```
source /opt/ros/indigo/setup.bash
source ~/catkin_ws/devel/setup.bash
roslaunch nonsmooth_backstep_controller gazebo.launch
```

The robot should appear in Gazebo as shown in Fig. 6.

Then, start the simulation by clicking in the play button in the Gazebo panel, open a new terminal and issue the following commands to move the robot.

```
source /opt/ros/indigo/setup.bash
source ~/catkin_ws/devel/setup.bash
rosrun nonsmooth_backstep_controller pose_step.sh 5 0 0
```

The `pose_step.sh` script publishes the setting point pose for the controller. In this case, $x_c = 5$, $y_c = 0$ and $\theta = 0$. Hence the robot should move to a point 5 m ahead of its current position.

If everything is right, the TWIL robot should move for some seconds and then stop, as shown in Fig. 7.

Note that the trajectory of the robot is not necessary a straight line. The trajectory is implicit determined by the controller and its parameters. If a specific trajectory to reach a final point is desired, then, the reference to the controller should not be the final point itself, but a sequence of points produced by a trajectory generator. See the example in Sect. 6 for an 8 trajectory.

By checking the `/nonsmooth_backstep_controller/status` topic with the command:

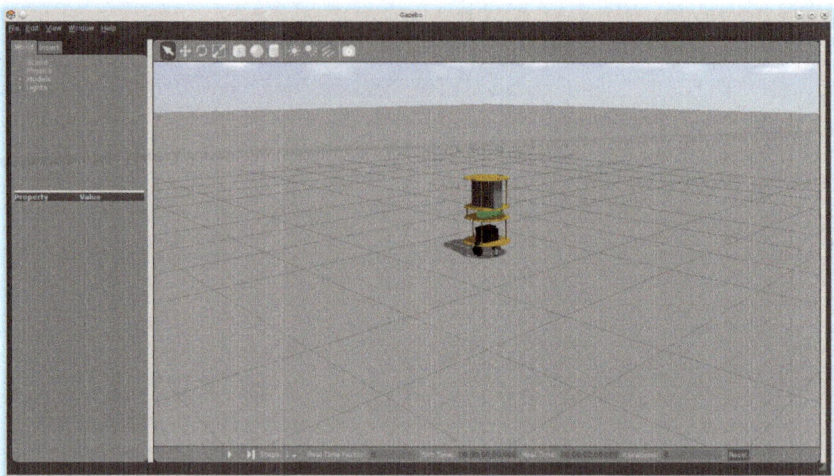

Fig. 6 TWIL mobile Robot in Gazebo

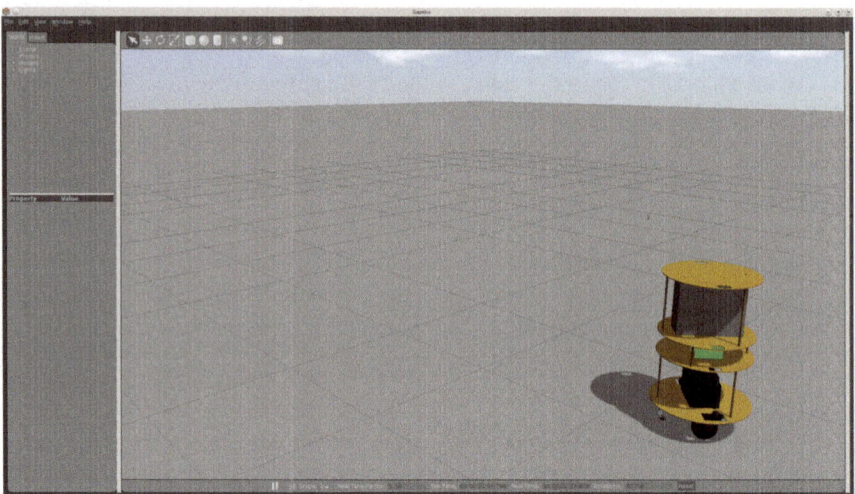

Fig. 7 Gazebo with TWIL robot after a step test motion

```
rostopic echo /nonsmooth_backstep_controller/status
```

it is possible to verify that the robot pose (the process_value field in the topic) converges to the desired pose.

In this simulation, the TWIL mobile robot is driven by the nonsmooth backstepping controller, which receives a message of the geometry_msgs/Pose2D type and forces the robot to converge to the 2D pose it describes.

The `pose_step.sh` is a script with an example of how to publish the reference for the nonsmooth backstepping controller, which is message of the `geometry_msgs/Pose2D` type. The script just publishes the required values by using the `rostopic` command. In a real application, that references would be generated by a planning package, such as MoveIt! [30] or a robot navigation package, such as the Navigation Stack [19, 20].

5 Implementation of the Backstepping Controller in ROS

This section describes the details of the implementation of the controller described in Sect. 2.2 and some other ROS packages which are used by the examples.

5.1 The `twil_description` Package

The `twil_description` package has the URDF description of the TWIL mobile robot and is covered in detail in [14]. Hence, it will not be discussed further here.

5.2 The `arc_odometry` Package

The `arc_odometry` package, shown if Fig. 8 implements the odometry procedure described in Sect. 2.3. Th package contains a library, called `libarc_odometry.so` and a ROS node, called `odometry_publisher`. The purpose of the library is to enable the reuse of the code in the implementation of other packages, while the node enables the standalone use of the arc odometry with already existing packages.

The library is implemented in the `diff_odometry.h` and `diff_odometry.cpp` files, which define the `DiffOdometry` class, shown in Listing 1. The `update()` function computes the odometry, as described by (35)–(38). For convenience of use, there are two overloaded versions of the function: one taking as argument the angular displacement on each wheel and another one taking as arguments a vector with both angular displacements. The other functions are used to get the pose and velocity of the robot as a vector or as individual components and to set the parameters of the odometry and initial pose.

Listing 1 DiffOdometry class

```
1    class DiffOdometry
2    {
3        public:
4        DiffOdometry(double wheelBase,std::vector<double> wheelRadius);
5        ~DiffOdometry(void);
6
7        void update(long leftDisp,double rightDisp,const ros::Duration &
8            duration);
9        void update(const Eigen::Vector2d &deltaPhi,const ros::Duration &
10           duration);
11       double x(void) const {return x_[0];}
12       double y(void) const {return x_[1];}
13       double heading(void) const {return x_[2];}
14       void getPose(Eigen::Vector3d &x) const {x=x_;}
15       double linear(void) const {return u_[0];}
16       double angular(void) const {return u_[1];}
17       void getVelocity(Eigen::Vector2d &u) const {u=u_;}
18       void setParams(double wheelBase,std::vector<double> wheelRadius);
19       void setPose(const Eigen::Vector3d &x) {x_=x;}
20
21       private:
22       std::vector<double>wheelRadius_;
23       double wheelBase_;
24
25       Eigen::Vector2d u_;
26       Eigen::Vector3d x_;
27    };
28 }
```

By using the library, the `odometry_publisher` ROS node is implemented
in the `odometry_publisher.cpp` file. It subscribes to the `joint_states`
topic to obtain the joint positions, which are used to compute the joint displacements
and the odometry and publishes the `odom` and `/tf` topics, as shown in Fig. 9.

Fig. 8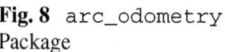
Package

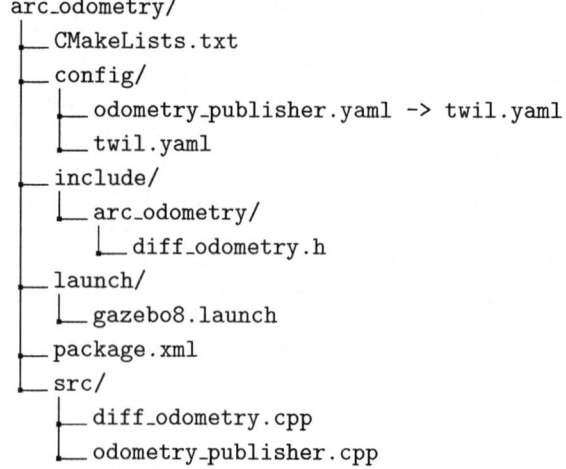

Fig. 9 Topics used by the odometry_publisher

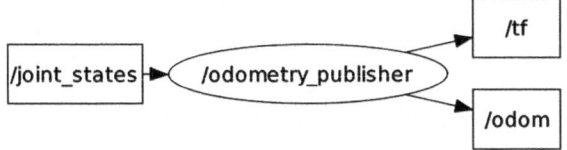

The parameters for the odometry_publisher ROS nodes are configured in the config/odometry_publisher.yaml file. In the default package it is a symbolic link to the twil.yaml, which is configured with the values for the TWIL robot.

In this chapter, the odometry_publisher node is not used. The implementation of the non-smooth backstepping controller uses directly the library as it makes simpler to synchronize dht pose and velocity estimateswot the controller updates.

5.3 The pose2d_trajectories Package

The purpose of the pose2d_trajectories package is to implement nodes for generating the messages with the reference trajectories to be followed by the controllers implemented in the nonsmooth_backstep_controller package. It is not a full featured trajectory generator, but just a collection of simple trajectory publishers to be used to test the controllers. The reference for the non-smooth backstepping controller is a geometry_msg:Pose2D message but the Rviz tool, in its default configuration, can not show this type of message, but can show geometry_msgs:PoseStamped, this package also implements a node to convert geometry_msgs:Pose2D messages in geometry_msgs:PoseStamped messages.

5.4 The nosmooth_backstep_controller Package

The nonsmooth_backstep_controller package, shown in Fig. 10, implements the controller proposed in Sect. 2.2. The files in the config directory specify the parameters for the controller, such as the joints of the robot associated to the controller, its gains and sampling rate. The files in the src directory are the implementation of the controller described in Sect. 2.2 (the NonSmoothBackstep Controller class), while the include directory holds the files with the declaration of that class. The nonsmooth_backstep_controller_plugins.xml file specifies that the class implementing the controller is a pluginsfor the ROS controller manager. The files in the launch directory are used to launch (i.e. load) the

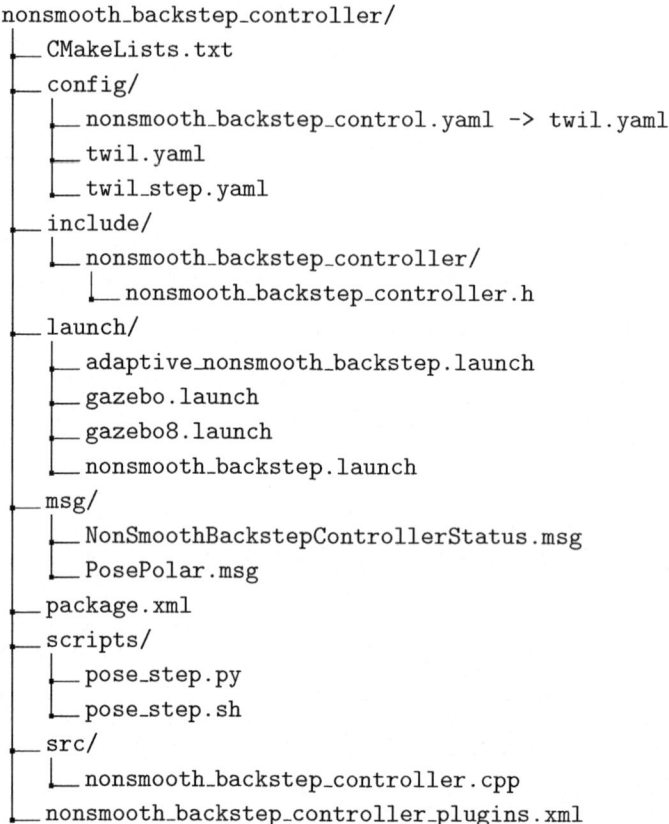

Fig. 10 nonsmooth_backstep_controller package

controller with the respective configuration files. The messages defined in the msg directory are used for publishing the status of the nonsmooth backstepping controller.

The effort_controllers/NonSmoothBackstepController is the implementation of the controller described in Sect. 2.2. It is a MIMO controller, which receives the desired pose of the robot as its reference and computes the torques to apply to each wheel of the robot. Figure 11 shows the topics subscribed and published by the controller.

The reference to the controller is received through the /nonsmooth_backstep _controller/command topic, and actuates the robot wheels by calling some member functions of the JointHandle class. See [13] for details.

Three other topics are also published by the controller: the /nonsmooth_ backstep_controller/status topic, which exposes many internal variables of the controller, the /nonsmooth_backstep_controller/odom topic, where the odometry data computed by the controller for its internal use is published and the /tf topic, where the transformation of the base frame of the mobile

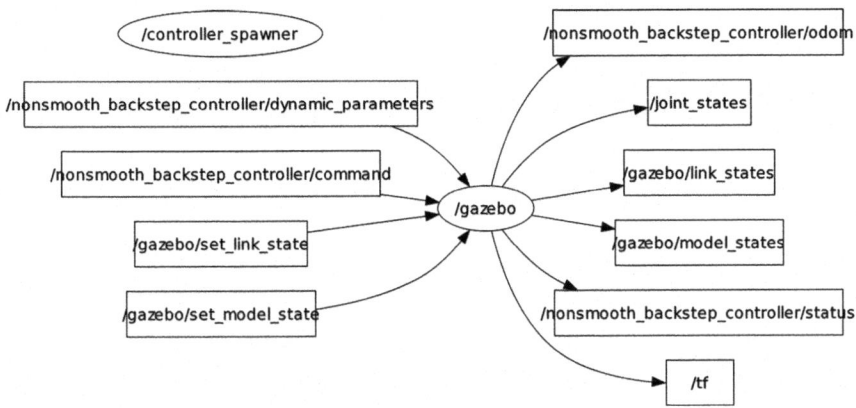

Fig. 11 Topics used by the nonsmooth backstepping controller

robot with respect to the `odom` frame is published. The name of the base frame of the mobile robot is configured in the `.yaml` file in the `config` directory.

The `/nonsmooth_backstep_controller/status` topic can be used for debugging and tuning of the controller parameters while the `/nonsmooth_backstep_controller/odom` topic can be useful for other nodes in the system estimating the pose of the robot through data fusion.

The `/nonsmooth_backstep_controller/dynamic_parameters` topic is used to receive updates to the parameters of the dynamic model (3, 4), in an adaptive controller similar to the one described in details in [13], but this feature is not used here.

Listing 2 shows the `NonSmoothBackstepController` class. It is a typical controller using the `EffortJointInterface`. The public members are the constructor and the destructor of the class and the functions called to load the controller (`init()`), to start it (`starting()`) and to update the controller at each sampling time (`update()`).

Listing 2 `NonsmoothBackstepController` class

```
1   class NonSmoothBackstepController: public controller_interface::
2       Controller<hardware_interface::EffortJointInterface>
3   {
4       public:
5       NonSmoothBackstepController(void);
6       ~NonSmoothBackstepController(void);
7
8       bool init(hardware_interface::EffortJointInterface *robot,
9               ros::NodeHandle &n);
10      void starting(const ros::Time& time);
11      void update(const ros::Time& time,const ros::Duration& duration);
12
13      private:
14      ros::NodeHandle node_;
15      hardware_interface::EffortJointInterface *robot_;
16      std::vector<hardware_interface::JointHandle> joints_;
```

```
17
18          boost::scoped_ptr<realtime_tools::RealtimePublisher
19              <nonsmooth_backstep_controller::NonSmoothBackstepControllerStatus>
20              >  status_publisher_ ;
21
22          boost::shared_ptr<realtime_tools::RealtimePublisher
23              <nav_msgs::Odometry> > odom_publisher_;
24          boost::shared_ptr<realtime_tools::RealtimePublisher
25              <tf::tfMessage> > tf_odom_publisher_;
26
27          ros::Subscriber sub_command_;
28          ros::Subscriber sub_parameters_;
29
30          Eigen::Matrix2d Ginv_;
31          Eigen::Matrix2d F_;
32
33          arc_odometry::DiffOdometry odom_;
34
35          Eigen::Vector3d xRef_;
36
37          Eigen::Vector2d eta_;
38
39          double time_step_;
40          ros::Time lastSamplingTime_;
41
42          Eigen::Vector2d phi_;
43
44          std::vector<double> lambda_;
45          std::vector<double> gamma_;
46
47          void commandCB(const geometry_msgs::Pose2D::ConstPtr &command);
48          void parametersCB(const std_msgs::Float64MultiArray::ConstPtr &command)
            ;
49      };
```

The class has many handles which are private members: for the node, for the EffortJointInterface and for the joints, handles for publishing and subscribing topics and many internal variables and callback functions for the subscribed topics. All those are the usual for any ROS node. However, the handles for publishing require special attention: They are not the usual ROS publisher handles. These are not real-time safe and then can not be used from the update loop of a real-time controller. To overcome this problem, there is the realtime_tools package, which implements wrappers for the usual ROS publishers. The wrapped version of the publisher can be used in a realtime loop. The wrapper implements this feature by creating non-real-time threads to handle the message publishing.

The implementation of the constructor and destructor functions is trivial. they just call the constructors of member objects and their destructors, respectively. The init() and starting() functions, called on controller load and starting are simple, as well. They just initialize variables and load controller parameters from the parameter server.

More interesting things are done in the update() function, shown in Listing 3. This function is called by the controller manager at a sampling rate defined in the ⟨controlPeriod⟩ tag in the URDF description of the robot. See details in the `twil_description` package. The controller manager uses this same rate for all enable controller in the system. However, while ay be good for some controllers, it may be inadequate for others. Hence, if a specific controller should be run at a slower sampling rate, it should perform a downsampling to its required sampling rate. This is done by letting its `update()` function to implement a downsampling. The actual sampling period of the nonsmooth backstepping controller is configured by the `time_step` parameter and loaded in the parameter server though the `.yaml` file in the `config` directory.

Listing 3 `NonsmoothBackstepController::update()` function

```
 1    void NonSmoothBackstepController::update(const ros::Time& time,
 2            const ros::Duration& duration)
 3    {
 4        ros::Duration dt=time-lastSamplingTime_;
 5
 6        if(fabs(dt.toSec()-time_step_) > time_step_/20) return;
 7        lastSamplingTime_=time;
 8
 9        // Incremental encoders sense angular displacement and
10        // not velocity
11        // phi[0] is the left wheel angular displacement
12        // phi[1] is the right wheel angular displacement
13        Eigen::Vector2d deltaPhi=-phi_;
14        for(unsigned int i=0;i < joints_.size();i++)
15        {
16            phi_[i]=joints_[i].getPosition();
17        }
18        deltaPhi+=phi_;
19
20        odom_.update(deltaPhi,dt);
21
22        Eigen::Vector3d x;
23        odom_.getPose(x);
24
25        Eigen::Vector2d u;
26        odom_.getVelocity(u);
27
28        // Change of coordinates
29        Eigen::Matrix3d R;
30        R << cos(xRef_[2]), sin(xRef_[2]), 0.0,
31              -sin(xRef_[2]), cos(xRef_[2]), 0.0,
32              0.0, 0.0, 1.0;
33        Eigen::Vector3d xBar=R*(x-xRef_);
34
35        // Discontinuous transformation
36        double e=sqrt(sqr(xBar[0])+sqr(xBar[1]));
37        double psi=atan2(xBar[1],xBar[0]);
38        double alpha=xBar[2]-psi;
39
40        // deta=(eta(k)-eta(k-1)/dt
41        Eigen::Vector2d deta=-eta_;
42
43        // Backstepping
44        eta_[0]=-gamma_[0]*e*cos(alpha);
45
46        if(fabs(alpha ) > DBL_EPSILON) eta_[1]=-gamma_[1]*alpha
47                -gamma_[0]*sin(alpha)*cos(alpha)+gamma_[0]*lambda_[2]*psi*
```

```
48                      sin(alpha)/
49                  lambda_[1]/alpha*cos(alpha);
50          else eta_[1]=gamma_[0]*lambda_[2]*psi/lambda_[1];
51          deta+=eta_;
52          deta/=dt.toSec();
53
54          Eigen::Vector2d eb=u-eta_;
55
56          Eigen::Vector2d vBar;
57          if(fabs(e) > DBL_EPSILON) vBar[0]=-gamma_[2]*eb[0]
58                  -lambda_[0]/lambda_[3]*e*cos(alpha)
59                  +lambda_[1]/lambda_[3]*alpha*sin(alpha)/e
60                  -lambda_[2]/lambda_[3]*psi*sin(alpha)/e;
61          else vBar[0]=-gamma_[2]*eb[0]-lambda_[0]/lambda_[3]*e*cos(alpha);
62          vBar[1]=-gamma_[3]*eb[1]-lambda_[1]/lambda_[4]*alpha;
63
64          Eigen::Vector2d v=vBar+deta;
65
66          // Linearization
67          Eigen::Vector2d uf(u[0]*u[1],sqr(u[1]));
68          Eigen::Vector2d torque=Ginv_*(v-F_*uf);
69
70          // Apply torques
71          for(unsigned int i=0;i < joints_.size();i++)
72          {
73                  joints_[i].setCommand(torque[i]);
74          }
75
76          if(status_publisher_ && status_publisher_->trylock())
77          {
78                  status_publisher_->msg_.header.stamp=time;
79
80                  status_publisher_->msg_.set_point.x=xRef_[0];
81                  status_publisher_->msg_.set_point.y=xRef_[1];
82                  status_publisher_->msg_.set_point.theta=xRef_[2];
83
84                  status_publisher_->msg_.process_value.x=x[0];
85                  status_publisher_->msg_.process_value.y=x[1];
86                  status_publisher_->msg_.process_value.theta=x[2];
87
88                  status_publisher_->msg_.process_value_dot.x=u[0]*cos(x[2]);
89                  status_publisher_->msg_.process_value_dot.y=u[0]*sin(x[2]);
90                  status_publisher_->msg_.process_value_dot.theta=u[1];
91
92                  status_publisher_->msg_.error.x=xRef_[0]-x[0];
93                  status_publisher_->msg_.error.y=xRef_[1]-x[1];
94                  status_publisher_->msg_.error.theta=xRef_[2]-x[2];
95
96                  status_publisher_->msg_.time_step=dt.toSec();
97
98                  for(int i=0;i < torque.size();i++)
99                      status_publisher_->msg_.command[i]=torque[i];
100
101                 for(int i=0;i < lambda_.size();i++)
102                     status_publisher_->msg_.lambda[i]=lambda_[i];
103
104                 for(int i=0;i < gamma_.size();i++)
105                     status_publisher_->msg_.gamma[i]=gamma_[i];
```

```
106              status_publisher_->msg_.polar_error.range=e;
107              status_publisher_->msg_.polar_error.angle=psi;
108              status_publisher_->msg_.polar_error.orientation=alpha;
109
110              for(int i=0;i < eta_.size();i++)
111                  status_publisher_->msg_.backstep_set_point[i]=eta_[i];
112
113              for(int i=0;i < deta.size();i++)
114                  status_publisher_->msg_.backstep_set_point_dot[i]=deta[i];
115
116              for(int i=0;i < u.size();i++)
117                  status_publisher_->msg_.backstep_process_value[i]=u[i];
118
119              for(int i=0;i < eb.size();i++)
120                  status_publisher_->msg_.backstep_error[i]=eb[i];
121
122              for(int i=0;i < vBar.size();i++)
123                  status_publisher_->msg_.backstep_command[i]=vBar[i];
124
125              for(int i=0;i < v.size();i++)
126                  status_publisher_->msg_.linear_dynamics_command[i]=v[i];
127
128              status_publisher_->unlockAndPublish();
129          }
130
131      if(odom_publisher_ && odom_publisher_->trylock())
132      {
133              odom_publisher_->msg_.header.stamp=time;
134
135              odom_publisher_->msg_.pose.pose.position.x=x[0];
136              odom_publisher_->msg_.pose.pose.position.y=x[1];
137              odom_publisher_->msg_.pose.pose.orientation.z=sin(x[2]/2);
138              odom_publisher_->msg_.pose.pose.orientation.w=cos(x[2]/2);
139
140              odom_publisher_->msg_.twist.twist.linear.x=u[0]*cos(x[2]);
141              odom_publisher_->msg_.twist.twist.linear.y=u[0]*sin(x[2]);
142              odom_publisher_->msg_.twist.twist.angular.z=u[1];
143
144              odom_publisher_->unlockAndPublish();
145      }
146
147      if(tf_odom_publisher_ && tf_odom_publisher_->trylock())
148      {
149          geometry_msgs::TransformStamped &odom_frame=
150                  tf_odom_publisher_->msg_.transforms[0];
151          odom_frame.header.stamp=time;
152          odom_frame.transform.translation.x=x[0];
153          odom_frame.transform.translation.y=x[1];
154          odom_frame.transform.rotation.z=sin(x[2]/2);
155          odom_frame.transform.rotation.w=cos(x[2]/2);
156
157              tf_odom_publisher_->unlockAndPublish();
158      }
159  }
160
```

Here, the backstepping controller should run with a sampling period of 10 ms, as configured by the time_step parameter in the .yaml file in the config directory. If the update() function is called in a shorter interval, it just returns.

Then, the incremental encoders mounted at the robot wheels are read with a call to the `getPosition()` function. Note that incremental encoders read angular displacement (and not velocity), which divided by the interval of time measured from the former read gives the average velocity within the interval. If the `getVelocity()` function were used instead, the result would be the instantaneous velocity at the sampling time, which, for the purpose of computing the odometry is less appropriate than the average velocity due to the noise. The `arc_odometry` library (see Sect. 5.2) is used at this point to compute the robot pose and velocity.

In the sequence, the offset to origin, described in Sect. 2.2 and the non-smooth coordinate change (8)–(10) are computed. The reference for the controller, used in the offset to origin is received through the `command` topic, as seen in Fig. 11. It is a `geometry_msgs/Pose2D` message. Then, the backstepping control is computed, resulting in the virtual input for the linearized system, v.

With the virtual input for the linearized system, it is possible to compute the torques to be applied to the robot, by using (5). The values for the parameters G and F (used to compute $f(u)$) were identified through experiments with the actual robot, following the method described in [14], and are used to initialize the associated variables in the `init()` function of the controller. The default values are read from the `.yaml` file in the `config` directory, but they can be changed by publishing their values and the associated covariances to the `dynamic_parameters` topic, as can be seen in Fig. 11. That is useful for implementing adaptive control versions of the controller.

The torque is then applied to the robot by calling the `setCommand()` function for each joint, concluding the control cycle. The remainder of the `update()` function is devoted to publish the information produced by the controller. However, as the controller code should be real-time safe, the publishers are not the usual ROS publishers, but wrapped real-time versions, as discussed in Sect. 5.4, which implies that the publication should be protected by mutexes [5]. This is done by calling the function `trylock()` before accessing the data fields of the message and then publishing it using the `unlockAndPublish()` function, instead of the usual function for publishing messages used in ROS nodes.

6   Results

This section presents the results of simulations of the TWIL robot in Gazebo using the backstepping controller, proposed in previous sections. Two types of references are used. First, the setting point problem, also known as the parking problem is considered. In this problem, a given target pose is applied to controler input and the robot should converge to this pose. This type of reference is used to evaluate the controller transient reponse. Note that a trajectory to drive the robot from initial pose to the target pose is not specified and is implicit determined by the controller parameters (λ_i and γ_i), but the controller ensures the convergence to the specified position and orientation.

If a specific trajectory is desired, then the trajetory tracking problem should be considered. In this problem, the robot should converge to a reference trajectory and follow it. Here, it is important to note that the actual trajectory of the robot is subject to the non-holonomic constraints. For a differential-drive robot, that means that its orientation should be aligned with the tangent of its trajectory in $X_0 \times Y_0$ plane. And this may be different from the orientation specified in the reference trajectory, even for a well planed factible trajectory, due to noise. Then, the controller attempt to force to robot to follow the reference position and the infactible orientation may produce some very unnatural trajectory following behavior. The solution to this problem is to tune the controller parameters (λ_i and γ_i) to permit a larger error in orientation. Then, the robot would be able to follow the reference position and align itself with the tangent to the actual trajectory, while allowing a larger orientation error if necessary.

From the discussion above, it should be clear that the set of controller parameters, in particular those related to orientation (λ_2 and λ_3) which are good for the setting point problem are not good for the trajectory tracking problem and vice-versa, as in the first problem the orientation should converge to the reference, and in the second problem some error should be allowed to enable the robot to align with the tangent to the actual trajectory. Here, the values were chosen as $\lambda_1 = 200.0$, $\lambda_2 = 6.0$, $\lambda_3 = 6.0$, $\lambda_4 = 500.0$, $\lambda_5 = 10000.0$, $\gamma_1 = 10$, $\gamma_2 = 1.0$, $\gamma_3 = 10.0$ and $\gamma_4 = 50.0$, which give an average performance for both cases. Of course, if better convergence of the orientation in the setting point problem is desired, the values of λ_2 and λ_3 can be increased, but then the preformance in the trajectory tracking would deteriorate.

6.1 Setting Point

For the setting point tests, the controller is loaded with the command:

```
roslaunch nonsmooth_backstep_controller gazebo.launch
```

and then, the `pose_step.sh` script can be used to supply a setting point to the controller, as done in Sect. 4,

The Gazebo simulator is started in paused mode so that the initial condition for the simulation can be checked. By pressing the play button, the simulation starts.

Step in X_0 Direction In this case, the reference is set to $x_c = 1$, $y_c = 0$ and $\theta = 0$, by issuing the command:

```
rosrun nonsmooth_backstep_controller pose_step.sh 1 0 0
```

Figure 12 shows the Cartesian position of the robot while following the step reference in X_0 direction. In this figure it is possible to see the offset at the initial point of the trajectory and the robot convergence to the trajectory. Note that the tracking error is small and in part due to the pose estimation based on odometry.

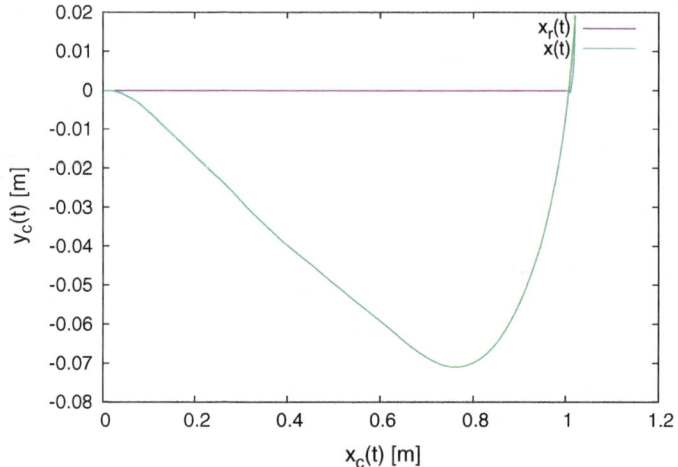

Fig. 12 Cartesian position $y_c(t) \times x_c(t)$ for a step reference in X_0 direction

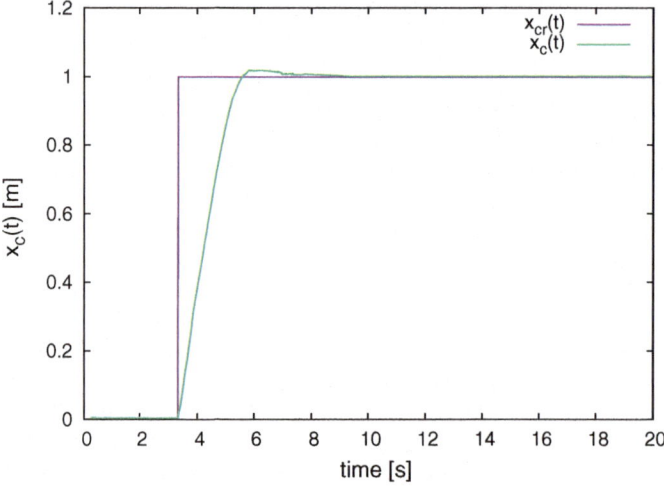

Fig. 13 Cartesian position $x_c(t) \times t$ for a step reference in X_0 direction

Figures 13, 14 and 15 show the reference pose and the robot pose over the time. Those figures show the transitories of $x_c(t)$, $y_c(t)$ and $\theta_c(t)$ while they converge to the reference values.

The torque applied in the wheels to follow the trajectory is shown in Fig. 16, while Fig. 17 shows the intermediate variables $\boldsymbol{\eta}$ and $\mathbf{u}$. Note that components of $\mathbf{u}$ converge to the respective components of $\boldsymbol{\eta}$, as enforced by the backstepping procedure. Figure 18 shows the backstepping errors e_1 and e_2.

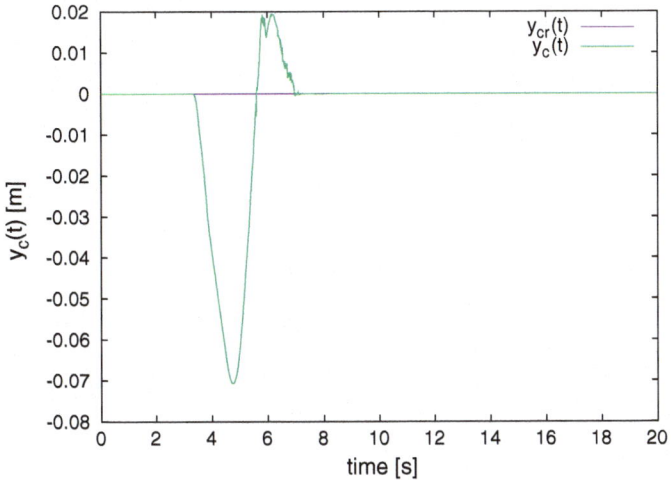

Fig. 14 Cartesian position $y_c(t) \times t$ for a step reference in X_0 direction

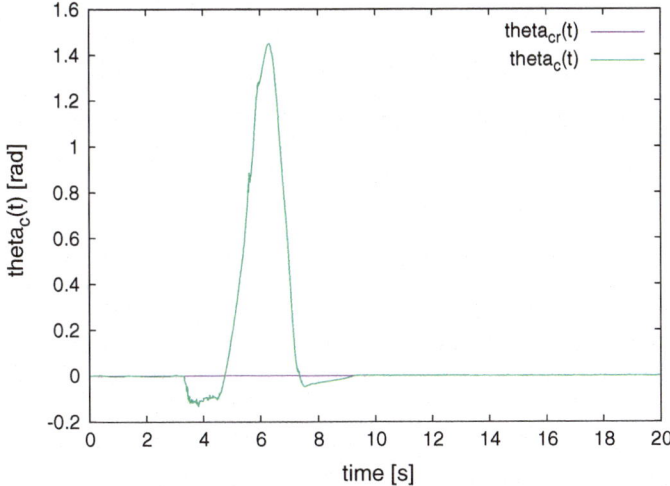

Fig. 15 Cartesian orientation $\theta_c(t) \times t$ for a step reference in X_0 direction

Step in Y_0 Direction In this case, the reference is set to $x_c = 0$, $y_c = 1$ and $\theta = 0$, by issuing the command:

```
1  rosrun nonsmooth_backstep_controller pose_step.sh 0 1 0
```

Note that, as the robot can not move instantaneously in the Y_0 direction, this is a difficult motion for any controller.

Figure 19 shows the Cartesian position of the robot while following the step reference in Y_0 direction. In this figure it is possible to see the offset at the initial

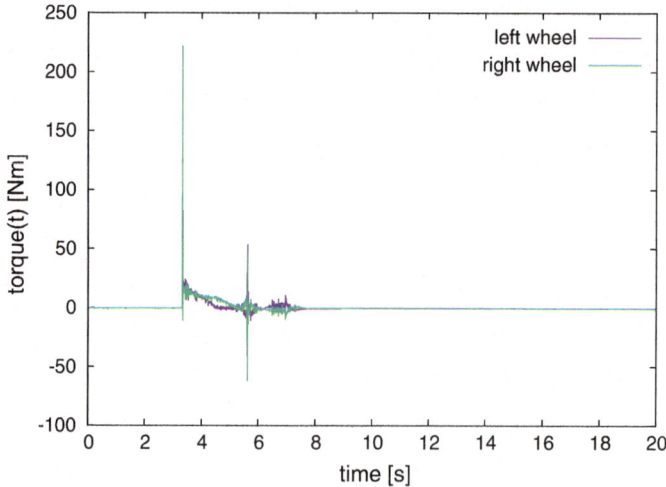

Fig. 16 Torque applied to wheels for a step reference in X_0 direction

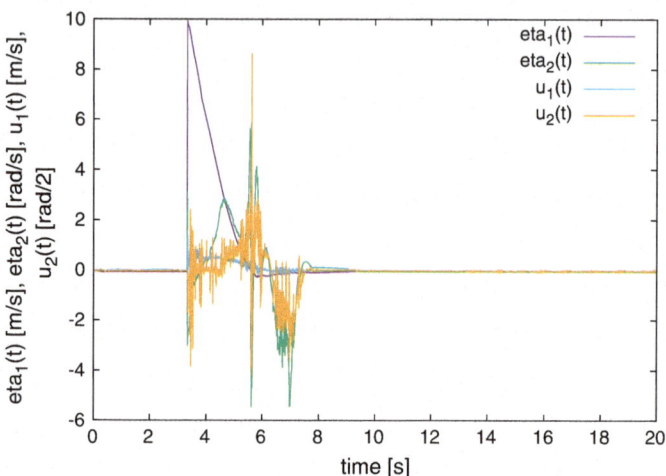

Fig. 17 Backstepping reference $\eta(t)$ and $\mathbf{u}(t)$ for a step reference in X_0 direction

point of the trajectory and the robot convergence to the trajectory. Note that the tracking error is small and in part due to the odometry based pose estimation.

Figures 20, 21 and 22 show the reference pose and the robot pose over the time. Those figures show the transitories of $x_c(t)$, $y_c(t)$ and $\theta_c(t)$ while they converge to the reference values.

The torque applied in the wheels to follow the trajectory is shown in Fig. 23, while Fig. 24 shows the intermediate variables η and $\mathbf{u}$. Note that components of $\mathbf{u}$ converge to the respective components of η, as enforced by the backstepping procedure. Figure 25 shows the backstepping errors e_1 and e_2.

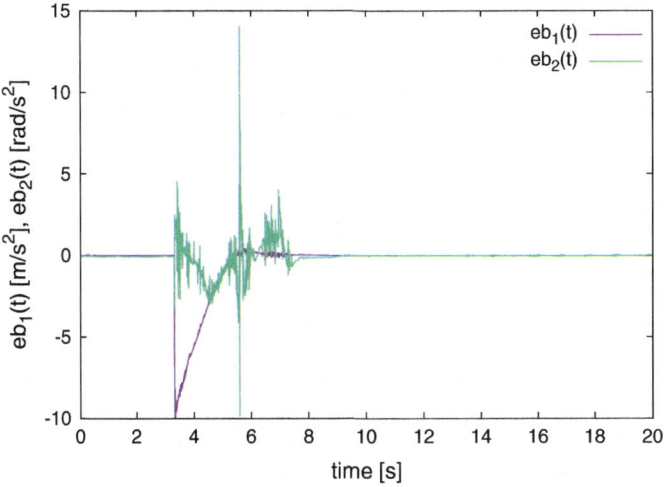

Fig. 18 Backstepping errors $e_1(t)$ and $e_2(t)$ for a step reference in X_0 direction

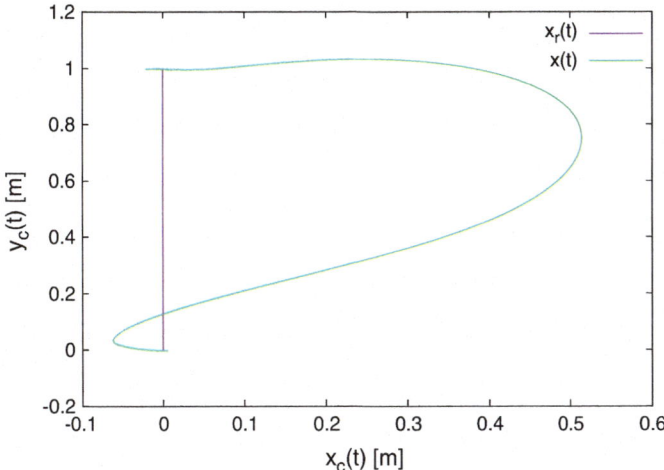

Fig. 19 Cartesian position $y_c(t) \times x_c(t)$ for a step reference in Y_0 direction

6.2 Trajectory Tracking

A launch file is provided to launch a test scenario where the nonsmooth backstepping controller is used to follow a trajectory in for of an 8. This launch files, launches the backsteping controller, a tajectory generator, the Gazebo simulator and a visualization in Rviz. It can be launched with the command:

```
roslaunch nonsmooth_backstep_controller gazebo8.launch
```

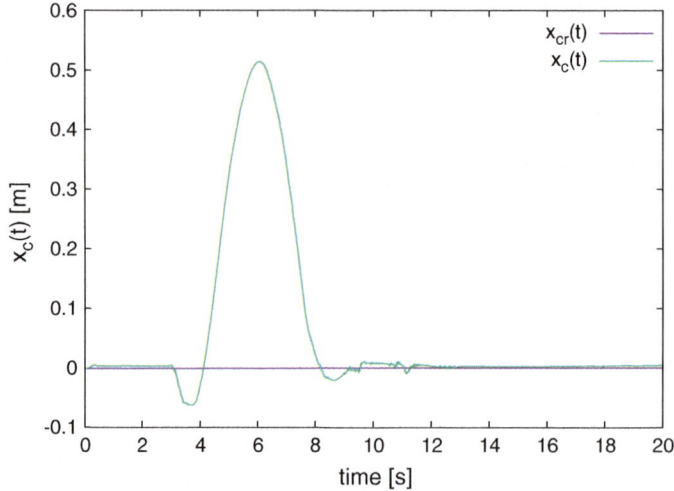

Fig. 20 Cartesian position $x_c(t) \times t$ for a step reference in Y_0 direction

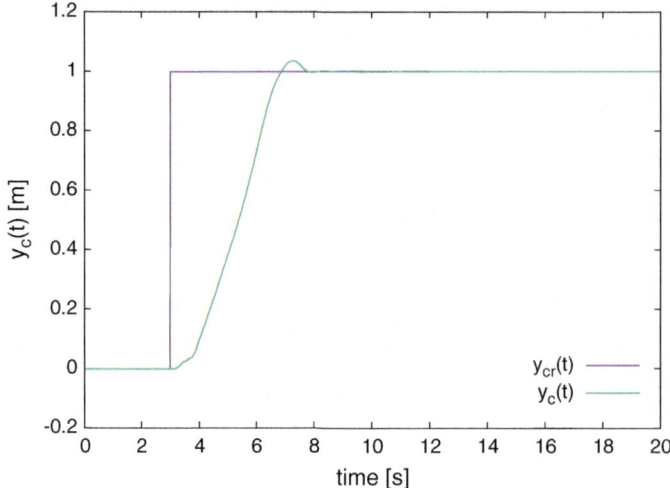

Fig. 21 Cartesian position $y_c(t) \times t$ for a step reference in Y_0 direction

Both, the Gazebo simulator and the Rviz visualization are launched because it is more convenient to see reference trajectory and the trajectory performed by the robot in Rviz, while the simulation in Gazebo is more realistic.

The Gazebo simulator is started in paused mode so that the initial condition for the simulation can be checked. By pressing the play button, the simulation starts.

Figure 26 shows this initial condition in Rviz. The blue arrow shows the reference trajectory point and the read arrow shows the current pose of the TWII robot. Note

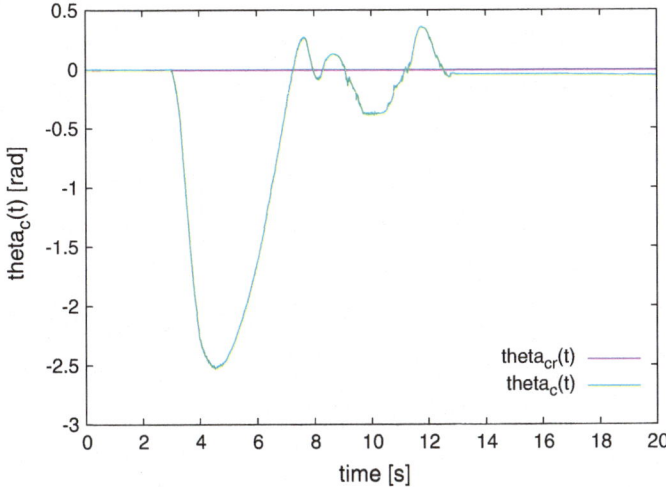

Fig. 22 Cartesian orientation $\theta_c(t) \times t$ for a step reference in Y_0 direction

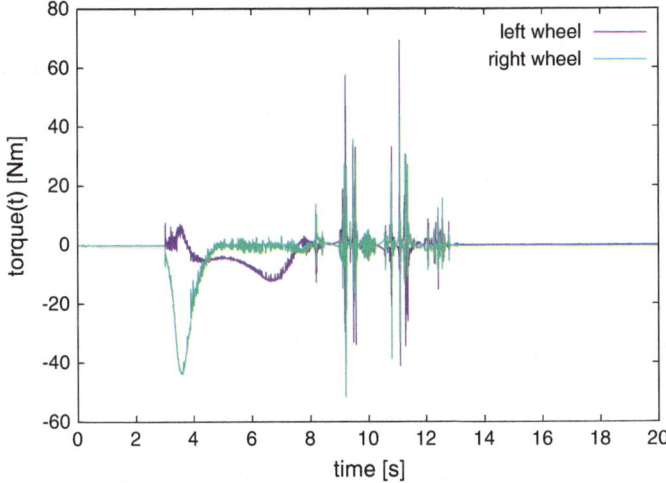

Fig. 23 Torque applied to wheels for a step reference in Y_0 direction

that at the start of the simulation, the current pose of the robot $(0, 0, 0)$ and the initial pose of the reference trajectory $(0, -0.5, 0)$ are not the same. Therefore, the robot should converge to the trajectory and then follow it.

Figure 27 shows the situation after 100 s. The read arrows show the history of poses of the TWIL robot, while the blue arrow shows the next pose to be followed. Note that in the start, the robot robot moves in order to converge to the trajectory and that is a difficult motion as this type of robot can not move sideways. The controller should then move away from the trajectory in order to be able to correct the offset in

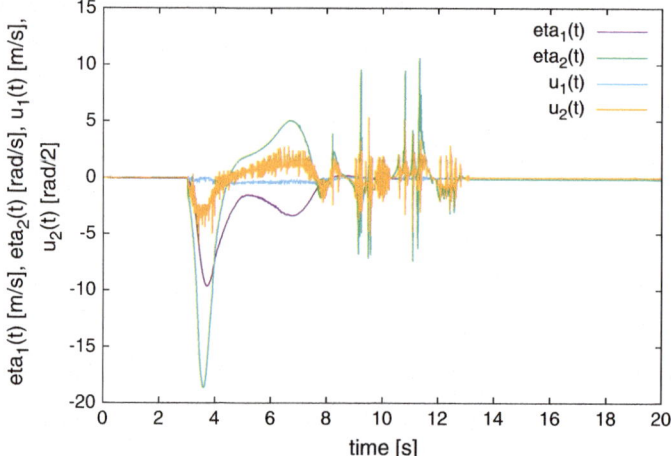

Fig. 24 Backstepping reference $\eta(t)$ and $\mathbf{u}(t)$ for a step reference

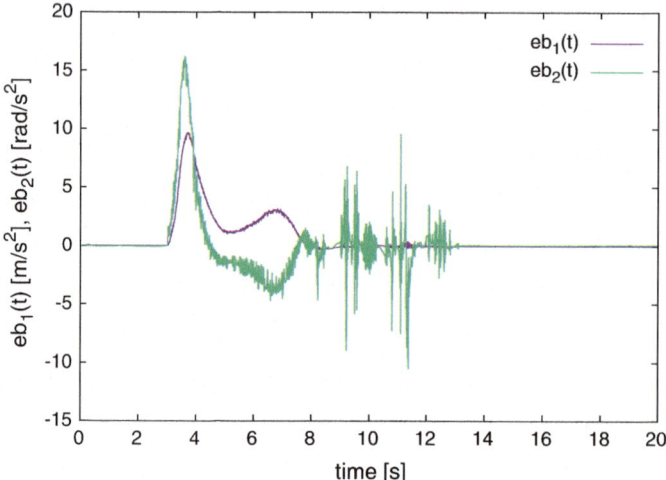

Fig. 25 Backstepping errors $e_1(t)$ and $e_2(t)$ for a step reference in Y_0 direction

Y direction an then correct the orientation to track the trajectory. As those objectives are conflicting, the motion of the robot seems to be erratic, but it finally converge to the reference trajectory.

Figure 28 shows the computation graph for the node launched by the launch file. Note that the controller itself is not represented by a node since it is a plugin. The /nonsmooth_backstep_controller/command topic receives the reference and the /tf topic publishes the pose of the robot.

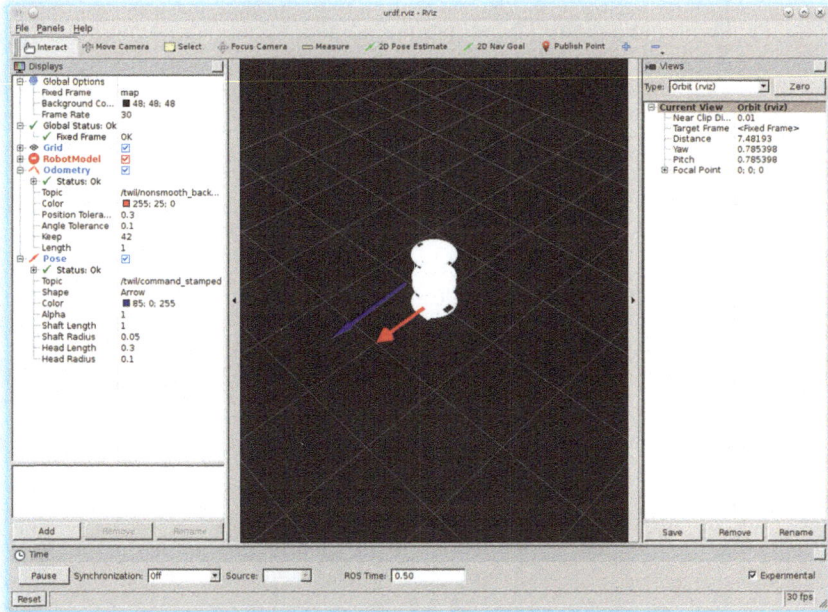

Fig. 26 Reference trajectory pose and Twil initial pose in Rviz

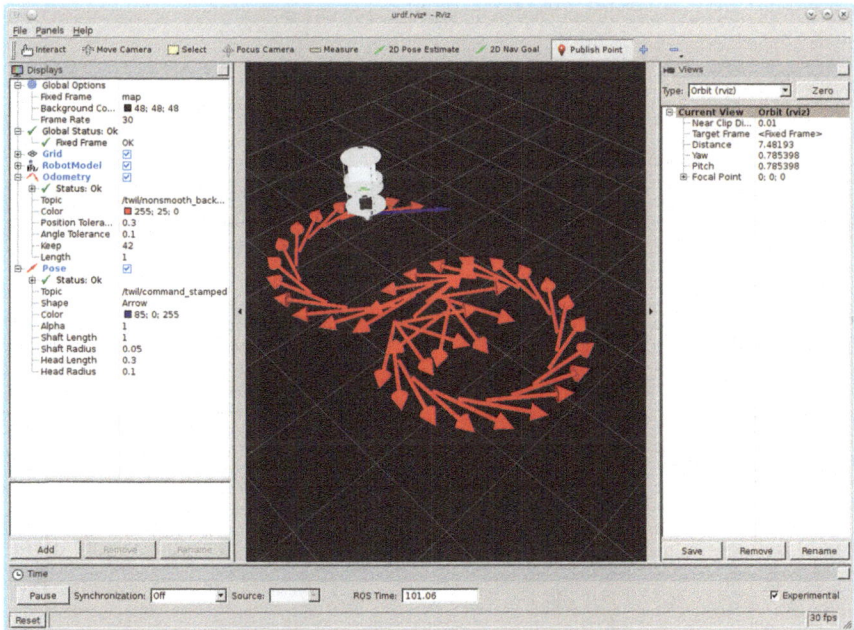

Fig. 27 Reference trajectory pose and Twil pose in Rviz after 100 s

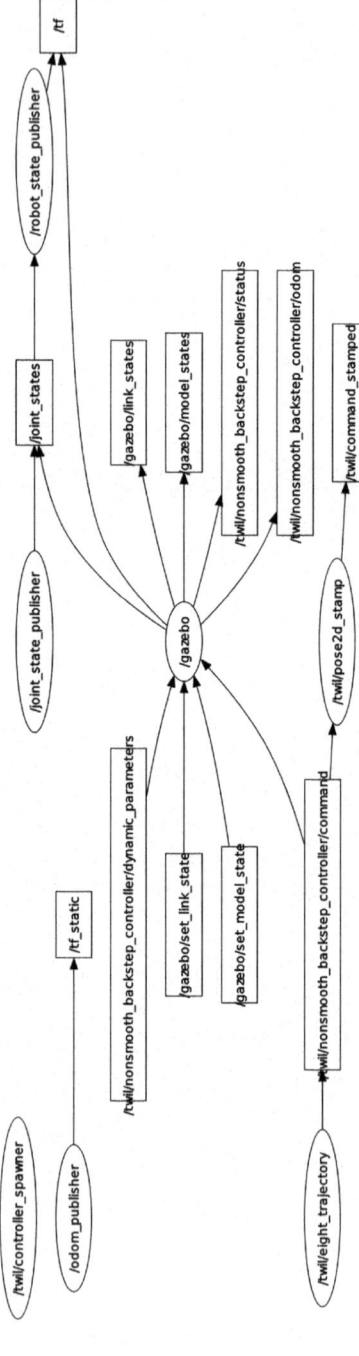

Fig. 28 Computation graph for the backstepping controller with trajectories displayed in Rviz

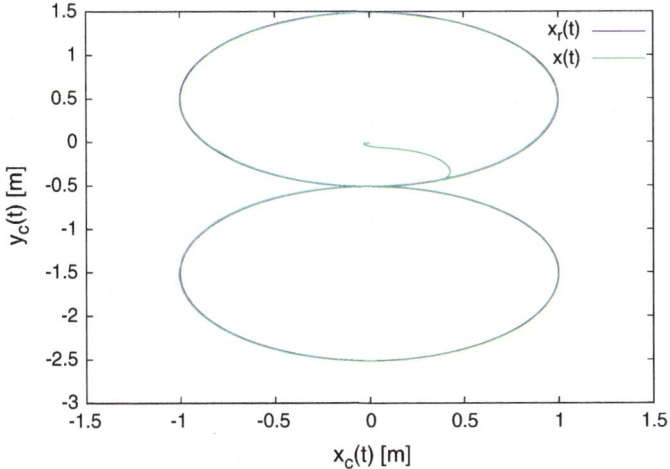

Fig. 29 Cartesian position $y_c(t) \times x_c(t)$ with the backstepping controller

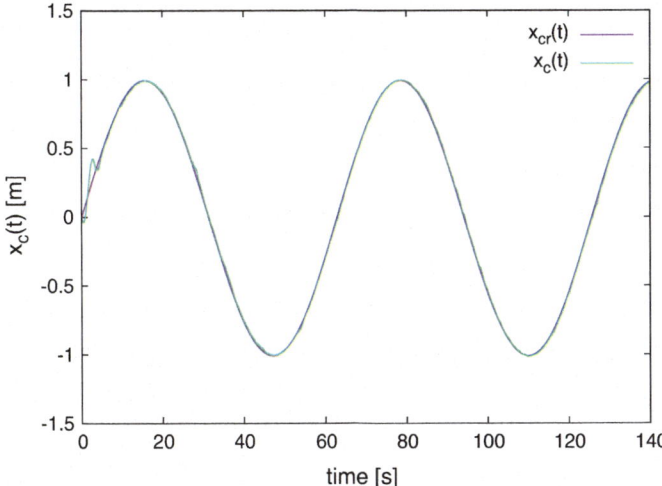

Fig. 30 Cartesian position $x_c(t) \times t$ with the backstepping controller

Figure 29 shows the Cartesian position of the robot while following the 8 reference trajectory. In this figure it is possible to see the offset at the initial point of the trajectory and the robot convergence to the trajectory. Note that the tracking error is small and in part due to the odometry based pose estimation.

Figures 30, 31 and 32 show the reference pose and the robot pose over the time. It can be seen the initial transitory converging to the reference and then its following with a very small error.

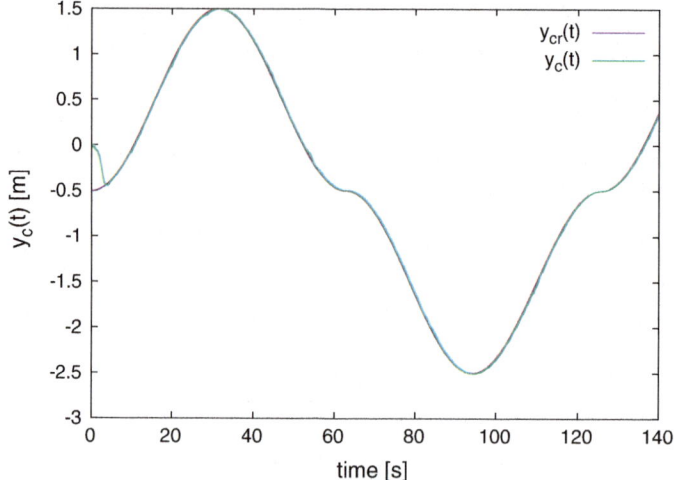

Fig. 31 Cartesian position $y_c(t) \times t$ with the backstepping controller

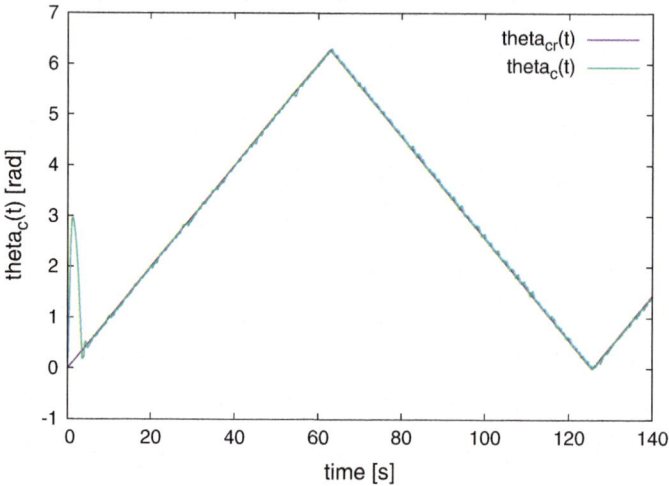

Fig. 32 Cartesian orientation $\theta_c(t) \times t$ with the backstepping controller

The torque applied in the wheels to follow the trajectory is shown in Fig. 33, while Fig. 34 shows the intermediate variables $\boldsymbol{\eta}$ and $\mathbf{u}$. Note that components of $\mathbf{u}$ converge to the respective components of $\boldsymbol{\eta}$, as enforced by the backstepping procedure. Figure 35 shows the backstepping errors e_1 and e_2.

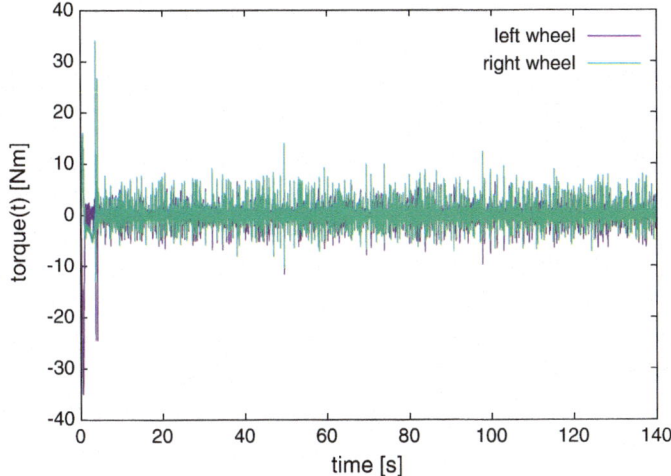

Fig. 33 Torque applied to wheels with the backstepping controller

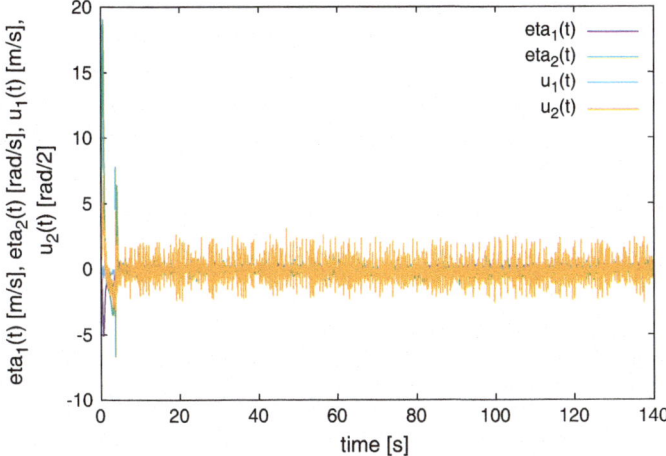

Fig. 34 Backstepping reference $\eta(t)$ and $\mathbf{u}(t)$

7 Conclusion

This work presented a controller for a mobile robot which considers the dynamic model of robot. This controller uses feedback linearization to compensate for the part of the mobile robot model related to its dynamics and then a non-smooth coordinate transform to enable the development of a Lyapunov-based non-linear control law. Then, backstepping is used to compute the appropriate inputs for the linearizing controller. The convergence of the system to the reference is proved by using the

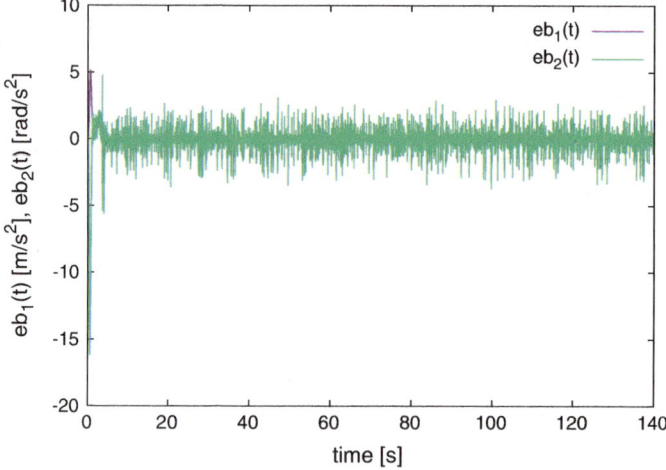

Fig. 35 Backstepping errors $e_1(t)$ and $e_2(t)$

Barbalat lemma. The proposed controller was implemented in ROS as a MIMO non-linear controller, which is an important departure from the traditional low-level ROS controllers which consider a SISO system using PID controllers. The performance of the controller for the setting point and the trajectory tracking problems were shown using the Gazebo simulator and Rviz.

References

1. M. Aicardi, G. Casalino, A. Bicchi, A. Balestrino, Closed loop steering of unicycle-like vehicles via lyapunov techniques. IEEE Robot. Autom. Mag. **2**(1), 27–35 (1995)
2. Astolfi, A.: On the stabilization of nonholonomic systems, in *Proceedings of the 33rd IEEE American Conference on Decision and Control* (IEEE Press, Lake Buena Vista, Piscataway, 1994), pp. 3481–3486
3. Barros, T.T.T., Lages, W.F.: Development of a firefighting robot for educational competitions, in *Proceedings of the 3rd International Conference on Robotics in Education* (Prague, Czech Republic, 2012)
4. R.W. Brockett, *New Directions in Applied Mathematics* (Springer, New York, 1982)
5. A. Burns, A. Wellings, *Real-Time Systems and Programming Languages*, vol. 3 (Addison-Wesley, Reading, 2001)
6. G. Campion, G. Bastin, B. D'Andréa-Novel, Structural properties and classification of kinematic and dynamical models of wheeled mobile robots. IEEE Trans. Robot. Autom. **12**(1), 47–62 (1996)
7. Canudas de Wit, C., Sørdalen, O.J.: Exponential stabilization of mobile robots with nonholonomic constraints. IEEE Trans. Autom. Control **37**(11), 1791–1797 (1992)
8. J. Godhavn, O. Egeland, A lyapunov approach to exponential stabilization of nonholonomic systems in power form. IEEE Trans. Autom. Control **42**(7), 1028–1032 (1997)
9. P. Goebel, *ROS by Example* (Lulu, Raleigh, NC, 2013), http://www.lulu.com/shop/r-patrick-goebel/ros-by-example-hydro-volume-1/paperback/product-21460217.html

10. A. Isidori, *Nonlinear Control Systems*, vol. 3 (Springer, Berlin, 1995)
11. Koenig, N., Howard, A.: Design and use paradigms for gazebo, an open-source multi-robot simulator, in *Proceedings of the 2004 IEEE/RSJ International Conference on Intelligent Robots and Systems (IROS 2004)*. vol. 3 (IEEE Press, Sendai, Japan, 2004), pp. 2149–2154
12. P.V. Kokotović, Developments in nonholonomic control problems. IEEE Control Syst. Mag. **12**(3), 7–17 (1992)
13. W.F. Lages, Implementation of real-time joint controllers, in *Robot Operating System (ROS): The Complete Reference (Volume 1)*. Studies in Computational Intelligence, vol. 625, ed. by A. Koubaa (Springer International Publishing, Switzerland, 2016), pp. 671–702
14. W.F. Lages, Parametric identification of the dynamics of mobile robots and its application to the tuning of controllers in ROS, in *Robot Operating System (ROS): The Complete Reference. Studies in Computational Intelligence*, ed. by A. Koubaa, vol. 2 (Springer International Publishing, Cham, Switzerland, 2017), pp. 191–229. https://doi.org/10.1007/978-3-319-54927-9_6
15. W.F. Lages, J.A.V. Alves, Differential-drive mobile robot control using a cloud of particles approach. Int. J. Adv. Robot. Syst. **14**(1) (2017). https://doi.org/10.1177/1729881416680551
16. W.F. Lages, E.M. Hemerly, Smooth time-invariant control of wheeled mobile robots, in *Proceedings of The XIII International Conference on Systems Science* (Technical University of Wrocław, Wrocław, Poland, 1998)
17. P. Lucibello, G. Oriolo, Robust stabilization via iterative state steering with an application to chained-form systems. Automatica **37**(1), 71–79 (2001)
18. E.H. Maciel, R.V.B. Henriques, W.F. Lages, Control of a biped robot using the robot operating system, in *Proceedings of the 6th Workshop on Applied Robotics and Automation* (Sociedade Brasileira de Automática, São Carlos, SP, Brazil, 2014)
19. E. Marder-Eppstein, *Navigation Stack* (2016), http://wiki.ros.org/navigation
20. E. Marder-Eppstein, E. Berger, T. Foote, B. Gerkey, K. Konolige, The office marathon: robust navigation in an indoor office environment, in *2010 IEEE International Conference on Robotics and Automation (ICRA)* (IEEE Press, Anchorage, 2010), pp. 300–307
21. A. Martinez, E. Fernández, *Learning ROS for Robotics Programming* (Packt Publishing, Birmingham, 2013)
22. S. Murata, T. Hirose, On board locating system using real-time image processing for a self-navigating vehicle. IEEE Trans. Indust. Electron. **40**(1), 145–153 (1993)
23. J.M. O'Kane, *A Gentle Introduction to ROS* (CreateSpace Independent Publishing Platform, 2013), http://www.cse.sc.edu/~jokane/agitr/
24. J.B. Pomet, B. Thuilot, G. Bastin, G. Campion, A hybrid strategy for the feedback stabilization of nonholonomic mobile robots, in *Proceedings of the IEEE International Conference on Robotics and Automation* (IEEE Press, Nice, France, 1992), pp. 129–134
25. V.M. Popov, *Hyperstability of Control Systems, Die Grundlehren der matematischen Wissenshaften*, vol. 204 (Springer, Berlin, 1973)
26. M. Quigley, B. Gerkey, K. Conley, J. Faust, T. Foote, J. Leibs, E. Berger, R., Wheeler, A. Ng, ROS: an open-source robot operating system, in *Proceedings of the IEEE International Conference on Robotics and Automation, Workshop on Open Source Robotics* (IEEE Press, Kobe, Japan, 2009)
27. F. Rehman, M. Rafiq, Q. Raza, Time-varying stabilizing feedback control for a sub-class of nonholonomic systems. Eur. J. Sci. Res. **53**(3), 346–358 (2011)
28. J.J.E. Slotine, W. Li, *Applied Nonlinear Control* (Prentice-Hall, Englewood Cliffs, 1991)
29. O.J. Sørdalen, Feedback Control of Nonholonomic Mobile Robots. Thesis (dr. ing.), The Norwegian Institute of Technology, Trondheim, Norway (1993)
30. I.A. Sucan, S. Chitta, MoveIt! (2015), http://moveit.ros.org
31. A.R. Teel, R.M. Murray, G.C. Walsh, Non-holonomic control systems: from steering to stabilization with sinusoids. Int. J. Control **62**(4), 849–870 (1995)
32. A. Zelenak, M. Pryor, Stabilization of nonlinear systems by switched Lyapunov function, in *Proceedings of the ASME 2015 Dynamic Systems and Control Conference* (The American Society of Mechanical Engineers, Columbus, 2015). https://doi.org/10.1115/DSCC2015-9650

Walter Fetter Lages graduated in Electrical Engineering at Pontifícia Universidade Católica do Rio Grande do Sul (PUCRS) in 1989 and received the M.Sc. and D.Sc. degrees in Electronics and Computer Engineering from Instituto Tecnológico de Aeronáutica (ITA) in 1993 and 1998, respectively. From 1993 to 1997 he was an assistant professor at Universidade do Vale do Paraíba (UNI-VAP), from 1997 to 1999 he was an adjoint professor at Fundação Universidade Federal do Rio Grande (FURG). In 2000 he moved to the Universidade Federal do Rio Grande do Sul (UFRGS) where he is currently a full professor. In 2012/2013 he held a PostDoc position at Universität Hamburg. Dr. Lages is a member of IEEE, ACM, the Brazilian Automation Society (SBA) and the Brazilian Computer Society (SBC).

University Rover Challenge: Tutorials and Team Survey

Daniel Snider, Matthew Mirvish, Michal Barcis
and Vatan Aksoy Tezer

Abstract In this tutorial chapter we present a guide to building a robot through 11 tutorials. We prescribe simple software solutions to build a wheeled robot and manipulator arm that can autonomously drive and be remotely controlled. These tutorials are what worked for several teams at the University Rover Challenge 2017 (URC). Certain tutorials provide a quick start guide to using existing Robot Operating System (ROS) tools. Others are new contributions, or explain challenging topics such as wireless communication and robot administration. We also present the results of an original survey of 8 competing teams to gather information about trends in URC's community, which consists of hundreds of university students on over 80 teams. Additional topics include satellite mapping of robot location (mapviz), GPS integration (original code) to autonomous navigation (move_base), and more. We hope to promote collaboration and code reuse.

Keywords Outdoor robot · Arm control · Autonomous navigation
Teleoperation · Panoramas · Image overlay · Wireless · GPS · Robot administration

D. Snider (✉)
Ryerson University, Toronto, ON, Canada
e-mail: danielsnider12@gmail.com

M. Mirvish
Bloor Collegiate Institute, Toronto, ON, Canada
e-mail: matthewmirvish@hotmail.com

M. Barcis
University of Wroclaw, Wroclaw, Poland
e-mail: mbarcis@mbarcis.net

V. A. Tezer
Istanbul Technical University, Istanbul, Turkey
e-mail: vatanaksoytezer@gmail.com

© Springer International Publishing AG, part of Springer Nature 2019
A. Koubaa (ed.), *Robot Operating System (ROS)*, Studies in Computational
Intelligence 778, https://doi.org/10.1007/978-3-319-91590-6_10

1   Introduction

The University Rover Challenge (URC) is an engineering design competition held
in the Utah desert that requires large teams of sometimes 50 or more university students. Students spend a year preparing and building from scratch a teleoperated and
autonomous rover with an articulated arm. This chapter gives an overview of eight
rover designs used at the URC, as well as a deep dive into contributions from three
design teams: Team R3 (Ryerson University), Team Continuum (University of Wroclaw), and Team ITU (Istanbul Technical University). We detail how to build a rover
by piecing together existing code, lowering the challenges for new Robot Operating
System (ROS) [14] users. We include 11 short tutorials, 7 new ROS packages, and an
original survey of 8 teams after they participated in the URC 2017 rover competition.

1.1   Motivation

At URC there is a rule limiting the budget that is allowed to be spent by teams on
their rover to $15,000 USD.[1] Therefore students typically engineer parts and software
themselves rather than buying. This makes ROS's free and open source ecosystem a
natural fit for teams to cut costs and avoid re-engineering common robotics software.
At URC 2017, Team R3 spoke to several teams who are not using ROS but want to,
and others who want to expand their use of it.

Several authors of this chapter joined URC because they are passionate about
hands on learning and ROS. After our URC experience we are even more confident
that ROS is an incredible framework for building advanced robotics quickly with
strong tooling that makes administering ROS robots enjoyable.

Our motivation comes from a passion to share lessons that enable others to build
better robots. Software is eating the world and there is a large positive impact that
can be made in the application of software interacting with the physical world.
Our software is freely shared so it can have the largest unencumbered impact and
usefulness.

1.2   Main Objectives

Aim one is to help ROS users quickly learn new capabilities. Therefore, our contributions are in the form of tutorials. These are made relevant and interesting by
giving them in the context of the URC competition. Readers can better assess the
usefulness of the tutorials by comparing solutions given to the other approaches that
our survey of eight other teams has revealed in Sect. 3. Many sections of this chapter

---

[1]URC 2017 Rules http://tinyurl.com/urc-rules.

give detailed descriptions of software implementations used at the competition by 3 different teams: Team R3 from Ryerson University in Toronto, Canada, and Team Continuum from the University of in Wroclaw, Poland, and the ITU Rover Team from Istanbul Technical University in Turkey. Original ROS packages are documented with examples, installation and usage instructions, and implementation details. At the end of the chapter readers should have a better sense of what goes into building a rover and of the University Rover Challenge (URC) that took place in Utah, 2017.

1.3   Overview of Chapter

Following the introduction and background sections, a wide angle look at rover systems with a survey and two case-studies is presented. Then specific tutorials delve into new packages and implementations mentioned in the case studies and team survey.

Section 2 "Background" provides an explanation of the URC rover competition and some of its rules.

Section 3 "Survey of URC Competing Teams" presents the results of an original survey of 8 teams who competed at URC 2017. It details each team's rover computer setup, ROS packages, control software, and avionics hardware for communication, navigation, and monitoring.

Section 4 "Case Study: Continuum Team" gives a case-study of their rover and what lead them to a second place result at the URC 2017 competition.

Section 5 "Case Study: Team R3" gives a case-study of the ROS software architecture used in Team R3's Rover. It provides the big picture for some of the tutorials in later sections which dive into more detailed explanations.

Section 6 "Tutorial: Autonomous Waypoint Following" details the usage of a new, original ROS package that will queue multiple move_base navigation goals and navigate to them in sequence. This helps URC teams in the autonomous terrain traversal missions.

Section 7 "Tutorial: Image Overlay Scale and Compass" details a new, original ROS package that meets one of the URC requirements to overlay an image of a compass and scale bar on imagery produced by the rover. It is intended to add context of the world around the rover.

Section 8 "Tutorial: A Simple Drive Software Stack" details the usage and technical design of a new, original ROS package that will drive PWM motors given input from a joystick in a fashion known as skid steering. The new package contains Arduino firmware and controls a panning servo so that a teleoperator can look around with a camera while driving.

Section 9 "Tutorial: A Simple Arm Software Stack" details the usage and technical design of a new, original ROS package that will velocity control arm joint motors, a gripper, and camera panning. The new package contains Arduino firmware to control PWM motors and a servo for the camera.

Section 10 "Tutorial: Autonomous Recovery after Lost Communications" details the technical design and usage of a new, original ROS package that uses ping to determine if the robot has lost connection to a remote base station. If the connection is lost then motors will be stopped or an autonomous navigation goal will be issued so as to reach a configurable location.

Section 11 "Tutorial: Stitch Panoramas with Hugin" details the usage and technical design of a new, original ROS package that will create panoramic images using ROS topics. At the URC competition teams must document locations of interest such as geological sites with panoramas.

Section 12 "Tutorial: GPS Navigation Goal" details the usage and technical design of a new, original ROS package that will convert navigation goals given in latitude and longitude GPS coordinates to ROS frame coordinates.

Section 13 "Tutorial: Wireless Communication" gives a detailed explanation of the primary and backup wireless communication setup used between ITU Rover Team's rover and base station for up to 1 km in range.

Section 14 "Tutorial: Autonomous Navigation by Team R3" explains the technical architecture of the autonomous system used at URC 2017 by Team R3 from Ryerson University, Toronto. It is based on the ZED stereo camera, the RTAB-Map ROS package for simultaneous localization and mapping (SLAM), and the move_base navigation ROS package.

Section 15 "Tutorial: MapViz Robot Visualization Tool" presents the MapViz ROS package and illustrates how a top-down, 2D visualization tool with support for satellite imagery can be useful for outdoor mobile robotics and URC. Our original Docker container created to ease the use of MapViz with satellite imagery is also documented.

Section 16 "Tutorial: Effective Robot Administration" discusses a helpful pattern for robot administration that makes use of tmux and tmuxinator to roslaunch many ROS components in separate organized terminal windows. This makes debugging and restarting individual ROS components easier.

Section 17 "Conclusion" ends with the main findings of the chapter and with ideas for further collaboration between URC teams and beyond.

1.4 Prerequisite Skills for Tutorials

The tutorials in this chapter expect the following skills at a basic level.

- ROS basics (such as `roslaunch`)
- Command line basics (such as `bash`)
- Ubuntu basics (such as `apt` package manager)

Fig. 1 A URC competition judge watches as the winning team of 2017, Missouri University of Science and Technology, completes the Equipment Servicing Task

2 Background

2.1 About the University Rover Challenge

The University Rover Challenge is an international robotics competition run annually by The Mars Society. Rovers are built for a simulated Mars environment with challenging missions filling three days of competition. It is held in the summer time at the very hot Mars Desert Research Station, in Utah. There were 35 rovers and more than 500 students from seven countries that competed in the 2017 competition (Fig. 2).[2] The winning team's rover can be seen in Fig. 1.

Rovers must be operated remotely by team members who cannot see the rover or communicate with people in the field: violations are punished by penalty points. Teams must bring and setup their own base station in a provided trailer or shelter and a tall communication mast nearby for wireless communication to the rover. The rover may have to travel up to 1 km and even leave direct line of sight to the wireless communication mast.

The idea is that the rover is on Mars (the Utah desert serves as a substitute) performing scientific experiments and maintenance to a Mars base. An assumption is made that the rovers are being operated by astronauts on or orbiting Mars rather than on Earth and therefore there is no major communication delay.

---

[2]URC 2017 competition score results and standings http://urc.marssociety.org/home/urc-news/americanroverearnsworldstopmarsrovertitle.

Fig. 2 Group photo of URC 2017 finalists at the Mars Desert Research Station in Utah

2.2 University Rover Challenge Tasks

The URC rules detail four tasks.[3] The science cache task involves retrieving and
testing subsurface soil samples without contamination. For this the rover must have
an auger to drill into the hard desert soil. After the science task teams present to a
panel of judges about their scientific findings. The evidence collected by the rover's
cameras, soil collection, and minimum three 3 sensors (e.g. temperature, humidity,
pH) is presented in a way that purports the possibility of water and life on Mars.

The extreme retrieval and delivery task requires teams to search out tools and
marked rocks in the desert and then use an arm on the rover to bring them back to the
base station. The equipment servicing task has teams perform finer manipulations
with their rover arm to start a fake generator. This consists of pouring a fuel canister,
pressing a button, flicking a switch and other manipulation tasks.

In the autonomous task, teams must start with their rover within 2 m of the designated start gate and must autonomously navigate to the finish gate, within 3 m. Teams
are provided with GPS coordinates for each gate and the gates are marked with a
tennis ball elevated 10–50 cm off the ground and are not typically observable from
a long distance. Teams may conduct teleoperated excursions to preview the course
but this will use their time. Total time for this task is 75 min per team and the total
distance of all stages will not exceed 1000 m.

Teams must formally announce to judges when they are entering autonomous
mode and not transmit any commands that would be considered teleoperation,
although they can monitor video and telemetry information sent from the rover.
On-board systems are required to decide when the rover has reached the finish gate.

[3]URC 2017 Rules http://tinyurl.com/urc-rules.

The newer 2018 rules[4] are very similar, but with more difficult manipulation tasks such as typing on a keyboard and a more demanding autonomous traversal challenge that explicitly calls for obstacle avoidance, something that Team R3 had last year and is explained in detail in Sect. 14.

2.3 Planetary Rovers beyond the Competition

Although the University Rover Challenge (URC) competition is a simulation of planetary rovers for Mars, there are significant differences between student built URC rovers and realistic planetary rovers. For example URC teams are limited in budget, manpower and engineering knowledge level. The Martian environment also poses challenging conditions such as radiation, low atmospheric pressure, very low oxygen levels, and a lack of communication and navigation systems found on Earth such as GPS.

The following paragraphs compare the systems of NASA's Curiosity rover [9, 10] and URC rovers. When comparing, the technology difference between the production year of Curiosity (2011) and now (2017) should also be kept in mind.

On Board Computer (OBC) The Curiosity Rover carries redundant 200 MHz BAE RAD750 CPUs, which is a special CPU that is designed to work in high radiation environments and has 256 MB RAM, 2 GB Flash, 256 KB EEPROM. The CPU runs a real time operating system called VxWorks [4, 15]. Interestingly, most of the URC rovers use on board computers that have more features and computing power than the planetary rovers due to the improvement of technology over the years. For example, Team R3 uses a Jetson TX1, running Ubuntu 16.04 which has 4 GB RAM.

Autonomous Navigation Although URC rovers and Curiosity have several common sensors, the lack of GPS and harsh environmental conditions at Mars leads their respective autonomous navigation algorithms to be built and work differently. Both of them use their Internal Measurement Units (IMU) and cameras to navigate, combining the odometry from internal sensors, visual odometry and some custom image processing algorithms to reach their target. The difference is that while URC rovers can rely on their GPS to navigate, Curiosity must rely on its position data from internal sensors only. Also, as Curiosity is navigating on the harsh Mars terrain it decides to navigate through the better terrain using a complex system of image processing algorithms [8].

Cameras Curiosity has 17 cameras that are used for various objectives, such as obstacle avoidance, navigation and science. Some of these cameras are very high resolution due to their scientific intent [8]. URC rovers generally have fewer cameras, such as 2 or 3, and less resolution is available because of cost limitations for teams at the competition.

---

[4]URC 2018 Rules http://tinyurl.com/urc-rules2018.

Wireless Communication Curiosity can communicate directly with the Earth with its X-band (7–12 GHz) communication modules or it can communicate with satellites orbiting Mars, specifically the Mars Reconnaissance Orbiter (MRO) or Mars Odyssey Orbiter, over a 400 MHz UHF link [8]. URC rovers generally prefer a 2.4 GHz UHF link to communicate with their ground stations. This difference is because Curiosity has to communicate with Earth from an average distance of 225 million km, while URC rovers has to communicate with their respective ground stations from a maximum of 1 km.

Power Unlike most planetary rovers which use solar power, Curiosity carries 4.8 kg of radioactive plutonium-238 to provide energy to its instruments for 14 years [8]. On the other hand, URC rovers are generally powered with Li-Po or Li-Ion batteries that usually last for several hours as the maximum mission time is limited and there are breaks between missions, unlike Curiosity's years long mission.

3 Survey of URC Competing Teams

Figure 3 and 4 show eight teams that were surveyed for their rover computer setup, ROS packages, control software, and avionics hardware (for communication, navigation, and monitoring). The team survey results have been edited and condensed for publication.

This survey serves the purpose of providing a broad overview of components and rover development styles before describing in subsequent sections a few detailed implementations provided by the teams who authored this chapter: Team R3 (Ryerson University), Team Continuum (University of Wroclaw), and Team ITU (Istanbul Technical University).

Several trends that emerged from the survey are interesting to note. Of the 8 teams surveyed, teams that used Raspberry Pis or STM microcontrollers all placed better than teams that used Arduinos or Teensy microcontrollers. For teleoperator input, Logitech controllers or joysticks were extremely popular and used by all teams. Teams most often expressed difficulty using IMUs or regretted not testing their rover enough. A wide variety of autonomous systems were experimented with: from custom OpenCV implementations, to using existing vision-based obstacle avoidance software (RTAB-Map), to GPS only approaches.

The winning approach of the Mars Rover Design Team from Missouri University of Science and Technology utilized a large number of custom solutions. Their high quality solutions and extremely comprehensive testing is exemplified in just one case by developing a custom UDP communication software. Dubbed "RoveComm", their communication system can reduce latency and increase video quality and was key to successful teleoperation.

	Mars Rover Design Team	Team Continuum	Cornell Mars Rover	ITU Rover Team	UWRT Robotics	Ryerson Rams Robotics (R3)	SJSU Robotics	Team Anveshak
School Name	Missouri University of Science and Technology	University of Wroclaw, Poland	Cornell University, USA	Istanbul Technical University, Turkey	University of Waterloo, Canada	Ryerson University, Canada	San Jose State University, USA	Indian Institute of Technology, Madras
Final Score (Rank)	403.4 (1)	336.3 (2)	264.1 (11)	243.1 (13)	225.7 (15)	190.9 (21)	164.3 (26)	151.4 (29)
Computers on rover	Raspberry Pi, TIVA-C Connected, MSP-432, Launchpad-C2000	A Banana Pi, 3x Raspberry Pi, 1x Jetson (optionally), multiple STM microcontrollers	A Intel NUC N82E16856102053, and 8x PIC32 MX530F128H microcontroller	A Raspberry Pi 3 with 64gb SD card running Ubuntu 16.04, STM32F103 microcontrollers	A FitPC miniature fanless PC	A Jetson TX1 with 32 GB SD card, Ubuntu 16.04, and 2x Arduino Mega microcontrollers	Odroid XU4, and Teensy 3.2 microcontroller	A Thinkpad T460 laptop running Ubuntu 14.04, and Arduino microcontrollers
Joysticks	Xbox Controller, Logitech Extreme 3D Pro	Logitech Gamepads	Logitech Gamepad F310, Thrustmaster VG T16000M FCS Joystick	2x Logitech Extreme 3D Pro, one for driving and one for the arm	2x Logitech joysticks for the arm, and an Xbox controller for driving	Xbox 360 Controller for driving, Logitech Extreme 3D Pro for arm	Logitech Extreme 3D Pro Joystick	2x Logitech F310 Gamepads, one for telemetry control and one for auger/arm
Cameras	Lorex, Sony EFFIO CCD Superhead	Standard Raspberry Pi cameras and two with wide angle lenses	Logitech HD Laptop Webcam C615, x264 video encoding	5 IP cameras used for security and an Xbox 360 Kinect v1 for image processing and fake laser	2x Pointgrey cameras, 1x USB Camera	ZED depth camera, 2x BL170 degree fisheye cameras	CCD 700TVL Composite video cameras (RunCam Swift 2.0)	SJ-CAM, IP-Camera, and a Logitech webcam. Cameras were interfaced using the "motion" Linux package, though it lags and quality was not great
GPS	MTK 3339	Ublox GPS	USGlobalsat BU-353-S4	Radiolink M8N	Microstrain	Linx FM Series GPS Receiver	UBlox GPS 7	ROS All Sensors Android App
IMU	LSM9DS1	Tried multiple units, nothing really worked	SparkFun SEN-13762, chip: MPU-9250	GY-80	Microstrain	MPU-9250 module, couldn't get it working	BNO055	ROS All Sensors Android App on Moto Play G4 phone
Software Packages	Energia, TI motorware, OpenCV	ROS kinetic with joint_state_controller, rviz, rqt, robot_localization, and more	ROS Kinetic with control-toolbox, dwa-local-planner, gazebo-ros-pkgs, gpsd-client, image-transport-plugins, image-rotate, pid, ros-controllers, spacenav-node, usb-cam, rplidar-ros, and gmapping	ROS Kinetic with packages depthimagetolasers can, huksy_control, move_base, actionlib, cv_bridge, image_transport and more	ROS Indigo with packages socket_canbridge, rosbridge_server, teleop_twist_joy, and more	ROS Kinetic with packages rqt_image_view, rtabmap, move_base, mapviz, joy, rtimulib_ros, zed_ros_wrapper, rgbd_odometry, usb_cam, and nmea_navsat_driver	Custom framework RoverCore-S, RoverCore-F, RoverCore-MC, built in house	ROS Kinetic and Indigo with packages joy, rosserial, amcl, and robot_localization
Autonomous System	OpenCV, Python	Implemented on our own using GPS and distance to the goal. A control PID with some constraints and logic necessary to back up if necessary to leads us to a given point. Goals are set when previous one was reached.	ROS move_base	ROS move_base and as a backup navigation using yaw and gps. Also, a C++ OpenCV tennis ball finding algorithm on top of ROS. We could find and navigate to the tennis ball from 8m.	move_base and robot_localization	ZED depth camera, rtabmap, move_base. We first teleoperate to build a SLAM map and find the tennis ball by human eye, then we go back to the start and set an autonomous goal in the SLAM map.	GPS and drive system, no need for anything else	We had plans of using AMCL and sensor fusion by making use of the existing packages in ROS, but ran out of time.
Arm Control Software	Custom solution in Energia. interfaced with custom control software RED (Rover Engagement Display) at base station	Tried MoveIt but implemented our own	Some experiments with MoveIt inverse kinematics but used forward kinematics at competition	Wrote our own inverse kinematics and simulation in Unity using C#	Wrote our own PWM library for arm motors	We had plans to use MoveIt but due to lack of testing time used velocity control for each joint mapped to a joystick	We wrote firmware into our framework for our Teensy 3.2 MCUs	Open-loop control with commands sent to an Arduino

Fig. 3 Survey of eight rover teams that competed in URC 2017

4 Case Study: Continuum Team

In the following section a case study is presented of the Continuum team (University of Wroclaw) and their rover, Aleph1. It is particularly interesting because they managed to score second place during the URC 2017 and had multiple other successes since they debuted in 2015. Michal Barcis decided to share with us some insights about their rover and his opinions on the competition.

	Mars Rover Design Team	Team Continuum	Cornell Mars Rover	ITU Rover Team	UWRT Robotics	Ryerson Rams Robotics (R3)	SJSU Robotics	Team Anveshak
School Name	Missouri University of Science and Technology	University of Wroclaw, Poland	Cornell University, USA	Istanbul Technical University, Turkey	University of Waterloo, Canada	Ryerson University, Canada	San Jose State University, USA	Indian Institute of Technology, Madras
Final Score (Rank)	403.4 (1)	336.3 (2)	264.1 (11)	243.1 (13)	225.7 (15)	190.9 (21)	164.3 (26)	151.4 (29)
Wireless radios and antennas	Ubiquiti 900MHz, Cloverleaf MIMO antenna on rover and dual polarity yagi at base station	Ubiquiti Bullet	Base station antenna was the Ubiquiti AM-2G15-120, rover antenna was the Super Power Supply B00O7ZEK7S, rover and base transceiver was the Ubiquiti Rocket M2	Microhard pDDL2450 could achieve 1km in non-line of sight with 5 dBi omnidirectional antennas. We also backed up comms except the cameras and the TCP link via a RF link with 433 MHz LoRa module.	2.4 GHz and 900 MHz antennas	Ubiquiti M2 Rockets 2.4GHz 802.11n MIMO paired with TP-Link 2408CL omnidirectional antennas	Ubiquiti Rockets M900 and the directional Ubiquiti Loco M900	TP-Link WA 5210 2.4GHz with included directional antenna
Battery System	LGChem18650HE4 Lithium Ion, 80 set up in custom pack, 10 set in parallel with 8 of those sets in series	Custom LiPo modules	1x MaxAmps 7S LIPO Battery	Tattu 6 cell LiPo 22Ah	1x Tattu 6 cell 22Ah Tattu LiPo	Panasonic NCR18650BD 3.7V 3200mAh Li-Ion 4 batteries in series to achieve 14.8V and 6 in parallel to achieve a 19.2Ah	3x Zippy LiPos 7S with a power board we designed	3x 24V LiPo batteries for drive, 2x 12V LiPo batteries for auger/arm
Wired Communication Protocols	I2C, RS232, RoveComm (Custom UDP)	CAN built-into the bananapi with two networks, one for driving wheels, another for the manipulator	CAN bus for interboard, UART for Intel NUC to microcontroller	I2C for sensors, USB for Raspberry Pi to microcontroller	CAN for most things, USB for drive motor controller, I2C/SPI for sensors	I2C sensors, USB for cameras, USB serial for Arduinos, UART for GPS, PWM for motor controllers	I2C, UART, Bluetooth (RFCOMM), SPI, PWM, PPM	Serial from the main computer to the various Arduinos
Sensor Fusion	Kalman filtering and custom filtering	robot_ localization	robot_ localization	robot_ localization, custom EKF backup in microcontrollers	robot_ localization	robot_ localization, didn't end up using due to IMU issues	None	None
Team Strengths	Manufacturing capabilities and access to programs that allow us to have many custom components on our rover.	Drive and manipulator controls. Also, I think being just 10 people and our motivation a lot. Everyone has important work to do	Modularity	Our wireless communication modules. We never lost control or communication to our rover at the competition.	A lot of different experiences from team members because of our coop program.	Tmux for terminal organization, keeping things simple, team dedication, and keeping it fun.	The absolute passion from each and every member of our team as well as our team manage system.	Dedicated team, always ready to learn new things, not shy of challenges. We made great strides in learning ROS in a matter of 3 months.
Improvements for next year	Fix bugs and flaws we found while at URC 2017 and push the boundaries of innovation as we build a new rover.	More field tests of the whole Rover.	Ease of use: easy way to launch and monitor the entire system. Live sensor diagnostics and robust CV.	I really want to add machine learning for finding the tennis ball from further.	Improve our project management.	Clearly labelled wires and pin outs, avoid USB hubs, and a geologist team member.	Secure bigger budget, start earlier, and update our technologies.	More development time, exploit ROS even more, test things more often, more collaboration with other teams.
Source code	github.com/ mst-mrdt	Inverse kinematics only gist.github.com/ danielsnider/5181ca50 cef0ec8fdea5c11279a 9fdbc	https://drive.google. com/open?id=0B1r9 QYTd8YNrWXNjNm dtcGlwMjQ	github.com/ itu-rover	github.com/ uwrobotics	github.com/ teamr3/ URC	github.com/ kammce/ RoverCore-S	github.com/ Team-Anveshak/ rover-control

Fig. 4 Survey of eight rover teams that competed in URC 2017. Cont.

The differences between team Continuum and other participants will be identified in order to find the key strengths that supported their achievements.

4.1 Recipe for Success

The teams, especially the ones that managed to place themselves in the first ten places during the competition, do not differ very much. Both software and hardware solutions are similar. Many teams also decided to implement programs using ROS.

We will try to identify some features that distinguish the Continuum team and let the reader decide which of them, if any, were the most advantageous.

One of the key differences is the size of the team. On average there were only around 12 members working on the rover during the period between 2014 and 2017. This makes the Continuum one of the smallest groups on the URC. Such an approach has both positive and negative effects: the smaller workforce means each person has more work to do and there is less shared knowledge, but also makes each member more important and increases motivation. Each of the key components in the rover had a person responsible for it.

There is one especially interesting hardware component that the Continuum team decided to do a bit differently than other teams and the team was often asked about. It is the choice of cameras. Aleph1 is equipped with inexpensive Raspberry Pi cameras [2]. Although much better devices in terms of specification are available on the market, those cameras had a big advantage as it was possible to easily integrate and customize them using raspberry pi. Therefore, it was easy for the team to experiment with different configurations and find a compromise between good quality and low latency.

The team was also asked what would be one thing they wished to do differently next time and the answer was always the same: more field tests with all components of the rover. This is also the advice that was often given by other teams and it seems very reasonable. By testing the rover with similar tasks as in the competition, it is possible to identify problems sooner and fix them. It also forces the team to complete the work sooner. Of course, to do that properly, a lot of self-control and good organization of the whole group is necessary.

4.2   Rover Manipulator Arm

In the following section, the robotic arm (or "manipulator") of the Aleph1 rover is described. Team Continuum decided to focus on this particular element, because it is crucial in most of the tasks at URC and at the same time is relatively hard to control.

Before starting the work on the arm controller, the Continuum team decided to conduct a survey of currently available solutions of similar problems in ROS. One of the most promising options was the MoveIt! Motion Planning Framework [13]. Unfortunately, it was not designed with teleoperation in mind and the team was unable to make it perform reasonably well with a goal specified in real time. Therefore, they decided to implement their own solution, tailored for the specific hardware they were using.

The main component that allows the team to control the arm is the inverse kinematics software (python source code[5]). It utilizes the feedback of four relative encoders placed on the joints of the manipulator to provide the operator two important fea-

---

[5]Continuum Inverse kinematics python source: https://gist.github.com/danielsnider/5181ca50ce f0ec8fdea5c11279a9fdbc.

Fig. 5 3D visualization of the Aleph1 rover used by the team during teleoperation

tures: the visualization of the state of the device and the ability to give more intuitive commands to the effector. For example, with this system it is possible to move the gripper up, down, front or back and the speeds of all the motors are automatically adjusted to reach given position.

Another big advantage of the arm system that proved to be very helpful during the competition is that even when the data connection is not good, it is possible to operate the manipulator. As soon as the instruction reaches the rover, the arm will position itself in the correct way deterministically. This would not be true when using an alternative way of controlling robots where the device is performing some action for as long as a button is being pressed and the loss of packets from the control station might change the result of the operation. Therefore, without such a system the operator needs to depend on feedback from cameras which might be delayed or not even available.

In Figs. 5, 6 and 7 the graphical user interface used to control and visualize the state of the manipulator and the whole rover is presented. Figure 8 shows the photo of an actual setup in the base station. The GUI is mainly used to support the operator in collision detection and pose estimation, because the visual feedback from the cameras often was not sufficient. The manipulator could be controlled using a mouse, but usually a Logitech game pad was used.

Team Continuum also implemented a semi-automatic system for picking objects up and for flicking manually operated switches. The system was developed for the European Rover Competition (ERC) 2016 because such functionality provided bonus points. Even though it was not deployed during the URC 2017, we have decided to present it in this section, because it is an interesting example of a relatively simple extension to the already described system, enabling much more complex tasks.

To get additional points during ERC 2016 the team must have positioned the effector at least 20 cm from the object it wanted to pick up or from the two-state switch. Then, the operator should announce he is starting the autonomous mode and

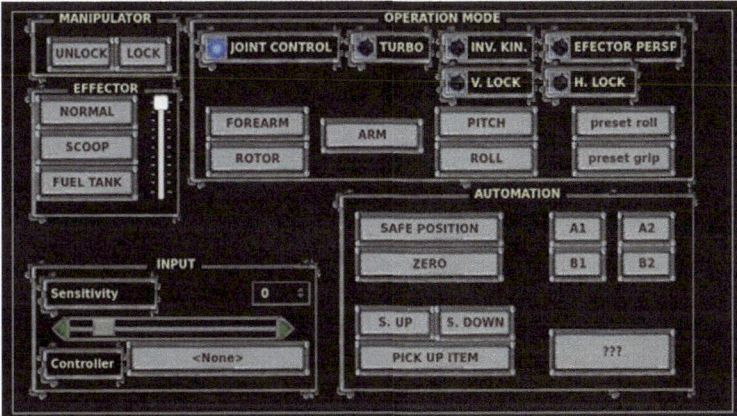

Fig. 6 Team Continuum's GUI used to control the manipulator of the rover

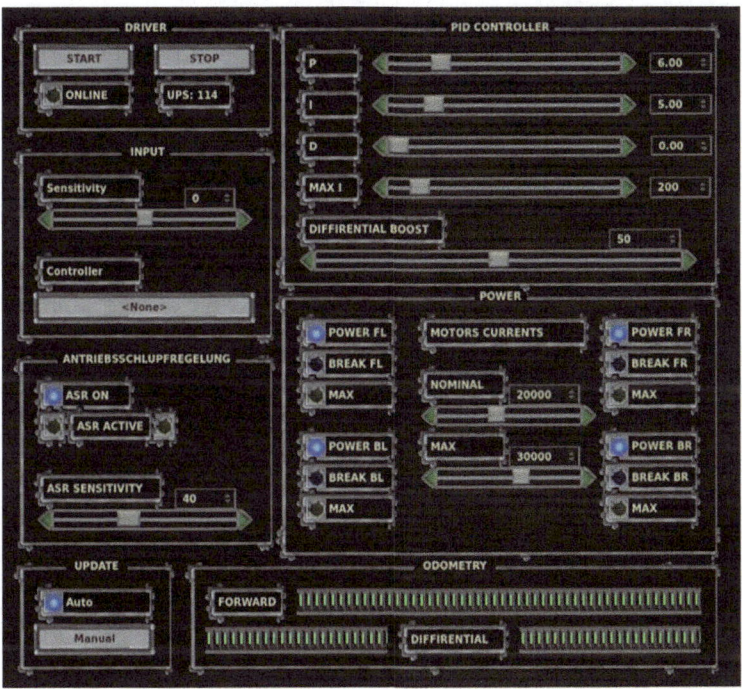

Fig. 7 Team Continuum's GUI used to control the movements of the rover

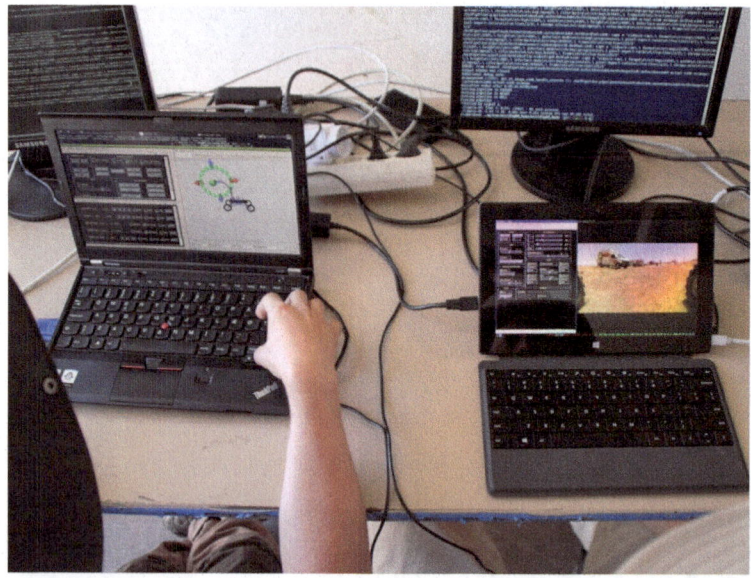

Fig. 8 Team Continuum's base station setup. A screenshot of their rover control GUI can be seen in Fig. 7

put down the controller. Next, the rover should pick up the object or actuate the switch and move back at least 20 cm.

The team decided to implement a simple idea: they wanted to place the effector exactly 20 cm from the object and directly in front of it. Then, using the inverse kinematics system they were able to execute pre-recorded movements in order to complete the task.

The first challenge they faced was how to measure exact distance from objects. They decided to mount two laser pointers on the effector, directed in the direction of each other. They can be seen in multiple scenes of the recently released video from the University of Wroclaw.[6] The two observable dots meet exactly at 20 cm.

The two preprogrammed arm movements were developed as follows:

For flicking switches, the effector went forward for 20 cm and 2.5 cm down simultaneously, then up for 5 cm and back to the initial position (20 cm back and 2.5 cm down). Due to the mechanical construction and the softness of the end of the arm, Aleph1 was able to switch most of the actuators using this technique, both very small and big ones.

For picking objects up the effector goes down 20 cm, then the grip motor engages until the force measurement on this motor crosses a predetermined threshold, then it goes back up 20 cm.

---

[6]A recent video of from the University of Wroclaw of their rover has incredible cinematography: https://www.youtube.com/watch?v=MF8DkKDBXtg.

The arm of the Aleph1 rover is far from the state of the art manipulators. It is not as fast or precise as it could be and the control sometimes is tricky. However, even though the described solutions might seem simple it was still one of the most advanced robotic arms in URC 2017. It is hard to construct a robust and fail-proof manipulator that is mountable on a movable platform and meets the strict mass and costs limits of URC. The Continuum team has proven that this is possible and the capabilities of such a simple platform can be exceptional.

5 Case Study: Team R3

5.1 ROS Environment

At Team R3 (Ryerson University), our Kinetic ROS software built on Ubuntu 16.04 had five main systems: the drive system, the autonomous system, the global positioning system, the visual feedback system, and the odometry system. The full diagram, as seen in Fig. 9, has been made available in Microsoft Visio format.[7]

To learn more about each software component, links to the most relevant documentation are provided.

Autonomous System Team R3's autonomous system consists of the ZED depth camera and the RTAB-Map ROS package for SLAM mapping [6]. The authors have also contributed gps_goal and follow_waypoints ROS packages for added goal setting convenience. For further details about the autonomous system see the tutorial in Sect. 14.

Listing 1 Autonomous system software used in Team R3's rover (numbers refer to Fig. 9).

```
1. zed–ros–wrapper (http://wiki.ros.org/zed–ros–wrapper)
2. follow_waypoints.py (http://wiki.ros.org/follow_waypoints)
3. rgbd_odometry (http://wiki.ros.org/rtabmap_ros#rgbd_odometry)
4. rtabmap (http://wiki.ros.org/rtabmap_ros)
21. gps_goal.py (http://wiki.ros.org/gps_goal)
```

Odometry System At the URC competition teams are surrounded by sandy desert terrain in Utah. As a result of wheel slippage on sand, Team R3 did not use wheel odometry. Instead we focused on fusing IMU and visual odometry into a more reliable position and orientation. However, because our IMU was not working well enough to produce a good fused result, at the competition we did not actually use the ekf_localization ROS nodes of the odometry system [12]. Instead we relied on odometry from the rgbd_odometry ROS node only and this worked well for our autonomous system based on the RTAB-Map package.

---

[7]Team R3's rover software architecture diagram in Microsoft Visio format https://github.com/danielsnider/ros-rover/blob/master/diagrams/Rover_Diagram.vsdx?raw=true.

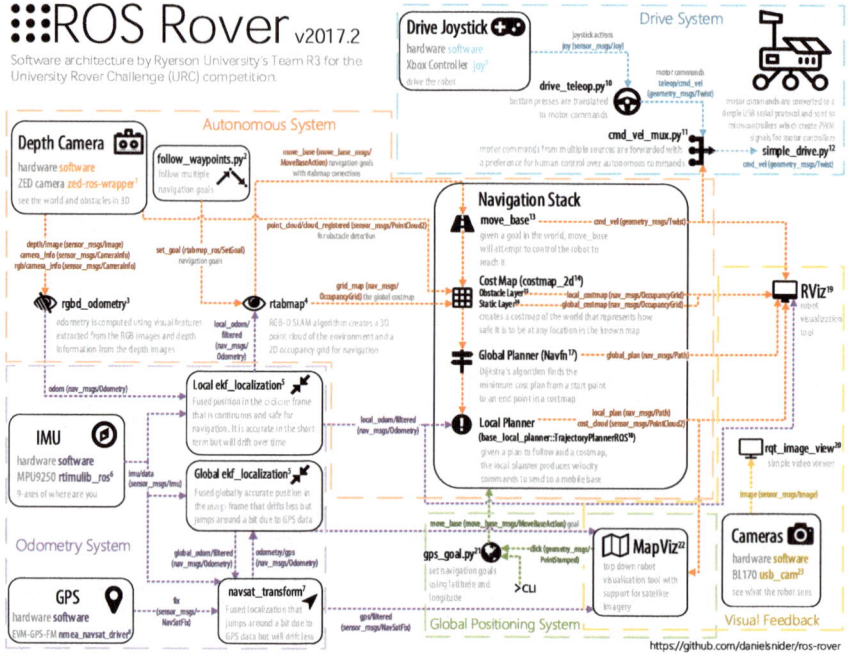

Fig. 9 ROS software architecture of Team R3's rover

Listing 2 Odometry software components used in Team R3's rover (numbers refer to Fig. 9).

```
5.  ekf_localization (http://docs.ros.org/kinetic/api
    /robot_localization/html/)
6.  rtimulib_ros (https://github.com/romainreignier/
    rtimulib_ros)
7.  navsat_transform (http://docs.ros.org/kinetic/api
    /robot_localization/html/)
8.  nmea_navsat_driver (http://wiki.ros.org/
    nmea_navsat_driver)
```

Drive System The drive software created by Team R3 is called simple_drive because it does not produce wheel odometry or transforms. That is left to the autonomous system via SLAM and is more robust in slippery desert environments. The simple_drive package controls the velocity of 6 motors with PWM pulses using an Arduino dedicated to driving. The motors are D.C. motors where the voltage on its terminals is given by the duty-cycle of the PWM signal. For further details of the drive system see the tutorial in Sect. 8.

Listing 3 Drive system software used in Team R3's rover (numbers refer to Fig. 9).

```
9.  joy (http://wiki.ros.org/joy)
10.  drive_teleop.py (http://wiki.ros.org/
     simple_drive#drive_teleop)
11.  cmd_vel_mux.py (http://wiki.ros.org/simple_drive
     #cmd_vel_mux)
12.  simple_drive.py (http://wiki.ros.org/
     simple_drive#simple_drive-1)
```

Navigation Stack The navigation stack used by Team R3 follows the commonly used development patterns of the ROS navigation stack.[8] Our setting choices were inspired by RTAB-Map's tutorial[9] and we share our tips in Sect. 14.

Listing 4 Navigation software stack used in Team R3's rover (numbers refer to Fig. 9).

```
13.  move_base (http://wiki.ros.org/move_base)
14.  Cost Map costmap_2d (http://wiki.ros.org/
     costmap_2d)
15.  Cost Map Obstacle Layer (http://wiki.ros.org/
     costmap_2d/hydro/obstacles)
16.  Cost Map Static Layer (http://wiki.ros.org/
     costmap_2d/hydro/staticmap)
17.  Global Planner Navfn (http://wiki.ros.org/navfn)
18.  Local Planner base_local_planner (http://wiki.
     ros.org/base_local_planner)
```

Visual Feedback To assist in teleoperation of the robot, sensors and navigation plans were visualized in rviz for local information such as point clouds and in mapviz for a global context that includes satellite imagery. The visual feedback software and joystick software were executed remotely on a laptop in the control station. The rest of software was executed on a Jetson TX1 inside the rover.

Listing 5 Visual feedback software used by Team R3 (numbers refer to Fig. 9).

```
19.  RViz (http://wiki.ros.org/rviz)
20.  rqt_image_view (http://wiki.ros.org/
     rqt_image_view)
22.  MapViz (http://wiki.ros.org/mapviz)
23.  usb_cam (http://wiki.ros.org/usb_cam)
```

6 Tutorial: Autonomous Waypoint Following

The following tutorial documents an original ROS package follow_waypoints that will buffer move_base goals until instructed to navigate to them in sequence

---

[8]ROS Navigation stack http://wiki.ros.org/navigation.

[9]RTAB-Map tutorial for the ROS navigation stack http://wiki.ros.org/rtabmap_ros/Tutorials/StereoOutdoorNavigation.

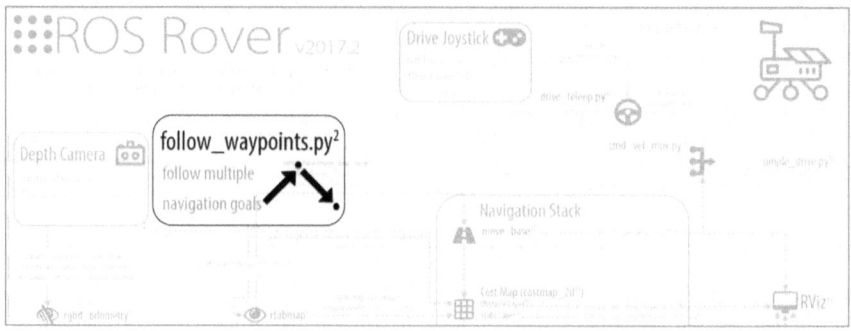

Fig. 10 ROS node follow_waypoints as seen in the larger architecture diagram. See Fig. 9 for the full diagram

(Fig. 10).[10, 11] If you can autonomously navigate from A to B, then you can combine multiple steps of A to B to form more complicated paths and use cases. For example, do you want your rover to take the scenic route? Are you trying to reach your goal and come back? Do you need groceries on the way home from Mars?

Team R3 (Ryerson University) has developed the follow_waypoints ROS package to use actionlib to send the goals to move_base in a robust way. The code structure of follow_waypoints.py is a barebones state machine. For this reason it is easy to add complex behavior controlled by state transitions. For modifying the script to be an easy task, you should learn about the Python state machine library in ROS called SMACH.[12, 13] The state transitions in the script occur in the order GET_PATH (buffers goals into a path list), FOLLOW_PATH, and PATH_COMPLETE and then they repeat.

6.1   Usage in the University Rover Challenge (URC)

A big advantage of waypoint following is that the rover can go to points beyond reach of Wi-Fi. In the autonomous traversal task, Team ITU's rover at one point lost connection but then got it back again when it reached the waypoint.

Other possible uses of waypoint following: To navigate to multiple goals in the autonomous task with a single command (use in combination with GPS goals, see Sect. 12). To search a variety of locations, ideally faster than by teleoperation.

---

[10]Source code for follow_waypoints ROS package https://github.com/danielsnider/follow_waypoints.

[11]Wiki page for follow_waypoints ROS package http://wiki.ros.org/follow_waypoints.

[12]SMACH state machine library for python http://wiki.ros.org/smach.

[13]One alternative to SMACH is py-trees, a behavior tree library http://py-trees.readthedocs.io/en/devel/background.html.

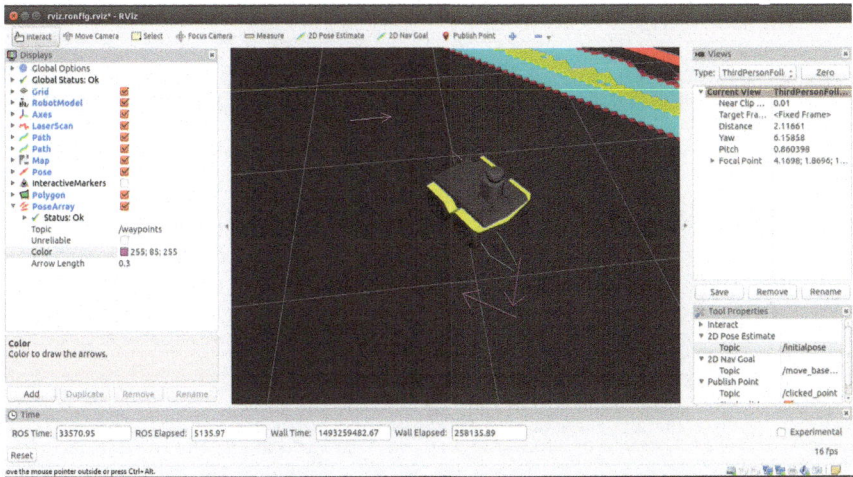

Fig. 11 A simulated Clearpath Jackal robot navigating to one of several waypoints displayed as pink arrows

To allow for human assisted obstacle avoidance where an obstacle is known to fail detection.

6.2   Usage Instructions

1. Install the ROS package:

```
$ roslaunch follow_waypoints follow_waypoints.launch
```

2. Launch the ROS node:

```
$ roslaunch follow_waypoints follow_waypoints.launch
```

3. To set waypoints you can either publish a ROS Pose message to the /initialpose topic directly or use RViz's tool "2D Pose Estimate" to click anywhere. Figure 11 shows the pink arrows representing the current waypoints in RViz. To visualize the waypoints in this way, use the topic /current_waypoints, published by follow_waypoints.py as a PoseAarray type.

4. To initiate waypoint following send a "path ready" message.

```
$ rostopic pub /path_ready std_msgs/Empty -1
```

To cancel the goals do the following. This is the normal move_base command to cancel all goals.

```
$ rostopic pub -1 /move_base/cancel actionlib_msgs/
    GoalID -- {}
```

6.3 Normal Output

When you launch and use the follow_waypoints ROS node you will see the following console output.

```
$ roslaunch follow_waypoints follow_waypoints.py

[INFO] : State machine starting in initial state 'GET_PATH'
    with userdata: ['waypoints']
[INFO] : Waiting to receive waypoints via Pose msg on topic /
    initialpose
[INFO] : To start following waypoints: 'rostopic pub /
    path_ready std_msgs/Empty -1'
[INFO] : To cancel the goal: 'rostopic pub -1 /move_base/cancel
    actionlib_msgs/GoalID -- {}'
[INFO] : Received new waypoint
[INFO] : Received new waypoint
[INFO] : Received path ready message
[INFO] : State machine transitioning 'GET_PATH':'success'-->'
    FOLLOW_PATH'
[INFO] : Executing move_base goal to position (x,y):
    0.0123248100281, -0.0620594024658
[INFO] : Executing move_base goal to position (x,y):
    -0.0924506187439, -0.0527720451355
[INFO] : State machine transitioning 'FOLLOW_PATH':'success
    '-->'PATH_COMPLETE'
[INFO] : ##############################
[INFO] : ##### REACHED FINISH GATE #####
[INFO] : ##############################
[INFO] : State machine transitioning 'PATH_COMPLETE':'success
    '-->'GET_PATH'
[INFO] : Waiting to receive waypoints via Pose msg on topic /
    initialpose
```

Listing 6 Normal console output seen when launching the follow_waypoints ROS node.

7 Tutorial: Image Overlay Scale and Compass

In an effort to add context to imagery recorded by the rover, Team R3 has developed a ROS package image_overlay_compass_and_scale that can add an indication of scale and compass to images and video streams.[14,15] A compass graphic will be embedded into imagery in a way that makes north direction apparent. A scale bar is also added so that the size of objects in images is more easily interpreted (Figs. 12 and 13).

Compass and scale values must be provided using standard ROS Float32 messages. Alternatively, a command interface can be used without ROS.

This tool meets one of the requirements of URC 2017 (in an automated way) and is applied to images of soil sampling sites and scenic panoramas for scientific and geological purposes.

---

[14]Source code for the image_overlay_compass_and_scale ROS package https://github.com/danielsnider/image_overlay_scale_and_compass.

[15]Wiki page for the image_overlay_compass_and_scale ROS package http://wiki.ros.org/image_overlay_scale_and_compass.

Fig. 12 Example of the image_overlay_compass_and_scale ROS package

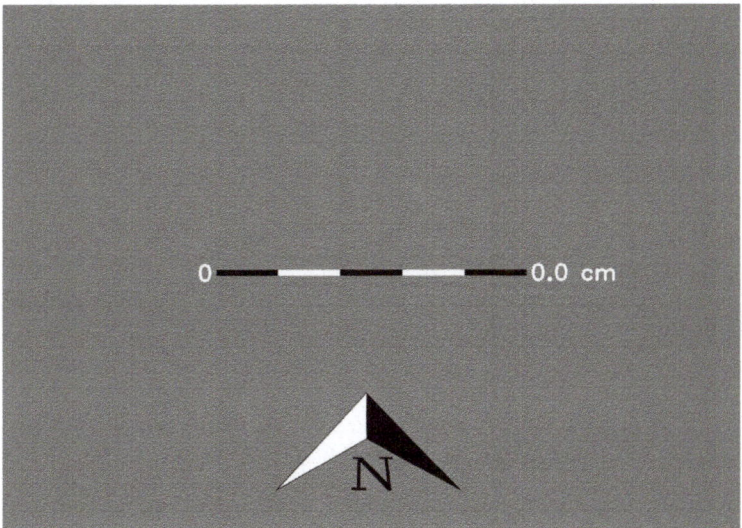

Fig. 13 This is what to expect when nothing is received by the node. This is the default published image

This package uses the OpenCV python library to overlay a compass graphic, the scale bar and dynamic text which is set using a ROS topic [1].

The implementation of overlaying the compass graphic on the input image follows these steps: (1) resize the compass graphic to be 60% the size of the input image's smaller side (whichever is smaller, x or y resolution). (2) Rotate the compass to the degrees specified on the /heading input topic. (3) Warp the compass to make it appear that the arrow is pointing into the image. This assumes that the input image has a view forward facing with the sky in the upper region of the image.

7.1 Usage Instructions

1. Install the ROS package:

```
$ sudo apt-get install ros-kinetic-image-overlay-
   compass-and-scale
```

2. Launch the ROS node:

```
$ roslaunch image_overlay_compass_and_scale overlay.
   launch
```

3. Publish heading and scale values

```
$ rostopic pub /heading std_msgs/Float32 45 # unit
   is degrees
$rostopic pub /scale std_msgs/Float32 133 # unit is
   centimeters
```

4. View resulting image

```
$ rqt_image_view /overlay/compressed
```

7.2 Command Line Interface (CLI)

The image_overlay_compass_and_scale ROS package includes a command line interface to invoke the program once and output an image with the overlayed graphics. You can envoke the tool as seen in Listing 7. Note that rosrun is not needed for this more basic form of execution.

Listing 7 Example usage of the image_overlay_compass_and_scale CLI script to save the image overlay to disk instead of publishing to ROS.

```
$ roscd image_overlay_compass_and_scale
$ ./src/image_overlay_compass_and_scale/image_overlay
    .py --input-image
~/mars.png --heading 45 --scale-text 133 --output-
    file output.png
```

8 Tutorial: A Simple Drive Software Stack

The authors have published a new ROS package, simple_drive, used at the University Rover Challenge (URC) 2017 on Team R3's rover (Figs. 14 and 15).[16,17] It proved simple and effective in desert conditions. The package is simple in the sense that it does not publish TF odometry from wheel encoders because wheels slip very substantially on sand.

The package implements skid steering joystick teleoperation with three drive speeds, dedicated left and right thumbsticks control left and right wheel speeds, control of a single axis panning servo to look around the robot, a cmd_vel multiplexer to support a coexisting autonomous drive system, and Arduino firmware to send PWM

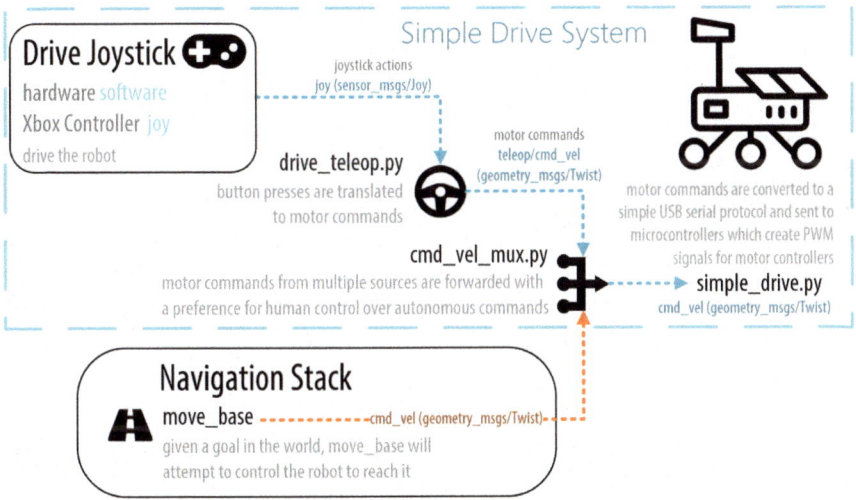

Fig. 14 Architecture of the drive ROS software used by Team R3. See Fig. 9 for a larger diagram. Note that the drive_firmware microcontroller software component is not illustrated but it is useful to note that it communicates with simple_drive.py

[16]Source code for simple_drive ROS package https://github.com/danielsnider/simple_drive.

[17]Wiki page for simple_drive ROS package http://wiki.ros.org/simple_drive.

Fig. 15 Team R3's rover being teleoperated by the simple_drive ROS package

commands to control the speed of drive motors and the position the panning servo. For the sake of simplicity, this package does not do the following: TF publishing of transforms, wheel odometry publishing, PID control loop, no URDF, or integration with ros_control. Though all of these simplifications are normal best practices in sophisticated robots. The simple_drive package gives ROS users the ability to advance their robot more quickly and hopefully to find more time to implement best practices.

This package is divided into four parts: drive_teleop ROS node, cmd_vel_mux ROS node, simple_drive ROS node, drive_firmware Arduino code. In the following sections we will explain the main features and implementation details but for the full ROS API documentation please see the simple_drive online ROS wiki.

8.1 Usage Instructions

1. Install the ROS package:

```
$ sudo apt-get install ros-kinetic-simple-drive
```

2. Install the drive_firmware onto a microcontroller connected to your motors and wheels by PWM. See Sect. 8.5 for detailed instructions. The microcontroller must also be connected to the computer running the simple_drive ROS node by a serial connection (e.g. USB).

3. Launch the three simple_drive ROS nodes separately or together using the included drive.launch file:

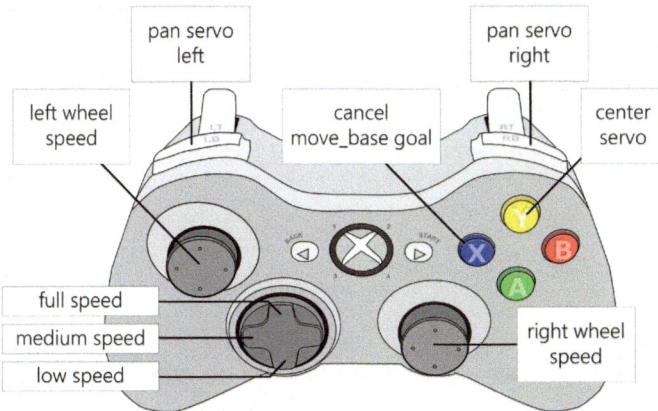

Fig. 16 The Xbox 360 joystick button layout for the simple_drive package (diagram is available in Visio format). Image credit: Microsoft Corporation

```
$ roslaunch simple_drive drive_teleop.launch joy_dev
    :=/dev/input/js0
$ roslaunch simple_drive cmd_vel_mux.launch
$ roslaunch simple_drive
simple_drive.launch serial_dev:=/dev/ttyACM0
# OR all-in-one launch
$ roslaunch simple_drive drive.launch
```

4. Your robot should now be ready to be driven.

8.2   *drive_teleop ROS Node*

The drive_teleop node handles joystick input commands and outputs desired drive speeds to the cmd_vel_mux ROS Node. This node handles joystick inputs in a in skid steering style, also known as diff drive or tank drive where the left joystick thumbstick controls the left wheels and the right thumbstick controls the right wheels.[18] We refer to this layout as tank drive through this section.

More specifically, this node converts sensor_msgs/Joy messages from the joy ROS node into geometry_msgs/Twist messages which represent the desired drive speed. Figure 16 shows that there are programmed buttons to set the drive speed to low, medium, or high speed, look around with a single axis servo, and cancel move_base goals at any moment.

Typically the servo is used to move a camera so that the teleoperator can pan around the surroundings of the robot. The servo's rotation speed (in degrees per button

---

[18]Xbox 360 joystick button layout diagram in Visio format https://github.com/danielsnider/ros-rover/blob/master/diagrams/simple_drive_Xbox_Controller.vsdx?raw=true.

press) can be set using the `servo_pan_speed` ROS parameter. The minimum and maximum angle of servo rotation in degrees can be set using the `servo_pan_min` and `servo_pan_max` ROS parameters respectively.

The button mapping was tested on an Xbox 360 controller and should require little or no modification for similar controllers, if they support a DirectInput mode.

8.3 cmd_vel_mux ROS Node

The cmd_vel_mux node receives movement commands on two sensor_msgs/Twist topics, one for teleoperation and one for autonomous control, typically move_base. Movement commands are multiplexed (i.e. forwarded) to a final topic for robot consumption with a preference for human control over autonomous commands. If any teleoperation movement command is received the cmd_vel_mux node will block autonomous movement commands for a set time defined by the `block_duration` ROS parameter.

8.4 simple_drive ROS Node

The simple_drive node sends commands to motors by communicating with a micro-controller over serial using the protocol defined by the drive_firmware. The simple_drive node listens to geometry_msgs/Twist for motor commands and std_msgs/Float32 for the servo position. The serial device that simple_drive communicates with is set with the `serial_dev` and `baudrate` ROS parameters.

This node is very simple and could be eliminated if your microcontroller supports ROS. For example Arduinos can use rosserial_arduino. However, this node is written in Python so you could more easily add complex functionality in Python and then in your microcontroller do the minimum amount of work necessary, thus allowing for the use of smaller and more lightweight microcontrollers.

8.5 drive_firmware Ardiuno Software

The drive_firmware is software for an Arduino microcontroller and it does not run as a ROS node. The Arduino is assumed to be dedicated to use of the simple_drive package. It does the minimum amount of work possible to receive motor commands from the simple_drive node over a USB serial connection and output voltages to digital PWM output to be received by motor controllers. We use Spark motor controller connected to D.C. motors where the voltage on its terminals is given by the duty-cycle of the PWM signal.

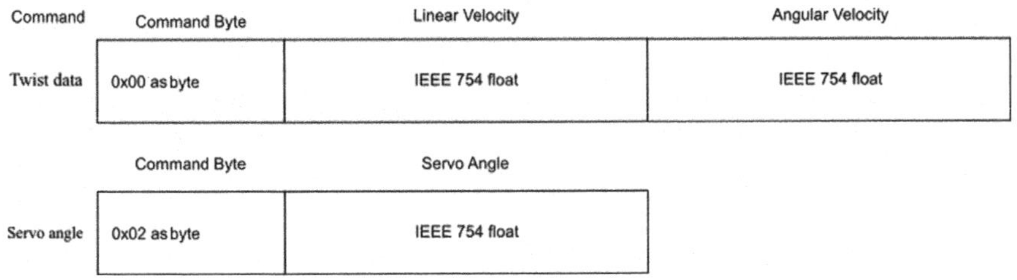

Command	Command Byte	Linear Velocity	Angular Velocity
Twist data	0x00 as byte	IEEE 754 float	IEEE 754 float

	Command Byte	Servo Angle
Servo angle	0x02 as byte	IEEE 754 float

Fig. 17 Diagram of the serial format used by drive_firmware to communicate between microcontroller and an on board computer

We tested on an Arduino Mega 2560 and Arduino Due, however many other boards should work with the same code and setup steps thanks to PlatformIO. You may need to change to change the pin numbers. Note that this software does not stop moving the robot if no messages are received, or if communications are lost.

Serial Communication Protocol The drive_firmware uses a serial protocol that is designed for simplicity rather than integrity of messages. As such it should not be perceived as especially robust. However, it has worked consistently in our experience.

Figure 17 shows how data is encoded over the serial connection. A header command byte is transmitted followed by one or two (depending on the command) IEEE 754 standard binary float values. Linear and angular velocity are expected to be between -1.0 and 1.0, which are linearly scaled to motor duty-cycles by the drive_firmware.

An example packet following this format could be encoded as bytes: 0x00 0x3f 0x80 0x00 0x00 0x00 0x00 0x00 0x00. This would be decoded as a twist data (leading 0x00) with 1.0 for linear velocity (0x3f 0x80 0x00 0x00) and 0.0 for angular velocity (0x00 0x00 0x00 0x00).

Tank to Twist Calculation When left and right joystick inputs are received by the drive_teleop node, representing left and right wheel linear velocities[19] (i.e. skid steering or differential drive), a conversion calculation to a geometry_msgs/Twist with linear and rotational velocities is performed as seen in Eqs. 1 and 2. The parameter b is half of the distance between the rover's wheels in m, V is the linear velocity in m/s in X axis, w is the angular velocity around Z axis in rad/s, V_r is the right wheel linear velocity in m/s and V_l is the left wheel linear velocity in m/s.

$$V = \frac{V_r + V_l}{2} \tag{1}$$

$$w = \frac{V_r - V_l}{2b} \tag{2}$$

---

[19]Wheel linear velocity is meant to be the speed at which distance is travelled and not rpm.

Twist to Tank Calculation When a geometry_msgs/Twist containing linear and rotational velocities is received by the drive_firmware, corresponding linear velocities are calculated for the left and right sides of wheel banks on the vehicle as seen in Eqs. 3 and 4.

$$V_l = V - wb \tag{3}$$

$$V_r = V + wb \tag{4}$$

PlatformIO We deploy the drive_firmware to an Arduino microcontroller using PlatformIO because it allows for a single source code to be deployed to multiple platforms.[20] PlatformIO supports approximately 200 embedded boards and all major development platforms such as Atmel, ARM, STM32 and more.

8.6 Install and configure drive_firmware

The following steps demonstrate how to install and configure drive_firmware component of the simple_drive package. These steps were tested on Ubuntu 16.04.
1. Install PlatformIO[21]:

```
$ sudo python -c "$(curl -fsSL
https://raw.githubusercontent.com/platformio/
    platformio/master/scripts/get-platformio.py)"
# Enable Access to Serial Ports (USB/UART)
$ sudo usermod -a -G
dialout <your username here>
$ curl https://raw.githubusercontent.com/platformio/
    platformio/develop/scripts/99-platformio-udev.
    rules
> /etc/udev/rules.d/99-platformio-udev.rules
# After this file is installed, physically unplug
    and reconnect your board.
$ sudo service udev restart
```

---

[20]PlatformIO is an open source ecosystem for IoT development http://platformio.org/.

[21]More information on how to install PlatformIO is here http://docs.platformio.org/en/latest/installation.html#super-quick-mac-linux.

2. Create a PlatformIO project[22]:

```
$ roscd simple_drive
$ cd ./drive_firmware/
# Find the microcontroller that you have in the list
    of PlatformIO boards
$ pio boards | grep mega2560
# Use the name of your board to initialize your
    project
$ pio init --board megaatmega2560
```

3. Modify the microcontroller pin layout to match wirings to motor controller hardware. First open the file containing the pin settings then change the pin numbers as needed:

```
$ vim src/main.cpp +4

  1 // Pins to Left Wheels
  2 #define pinL1 13
  3 #define pinL2 12
  4 #define pinL3 11
  5 // Pins to Right Wheels
  6 #define pinR1 9
  7 #define pinR2 8
  8 #define pinR3 7
  9 // Pin to the Servo
 10 #define pinServo 5
```

4. Depending on the specs of your motor controllers, modify the PWM settings as needed (values are duty-cycles in microseconds):

```
$ vim src/main.cpp +17

  1 // PWM specs of the Spark motor controller.
    Spark manual:
  2 //      http://www.revrobotics.com/content/docs
    /LK-ATFF-SXAO-UM.pdf
  3 #define sparkMax 1000 // Default full-reverse
    input pulse
  4 #define sparkMin 2000 // Default full-forward
    input pulse
```

5. Deploy the drive_firmware to the microcontroller:

```
$ pio run --target upload
```

6. Your robot is now ready to be driven.

---

[22]More documentation about PlatformIO: http://docs.platformio.org/en/latest/quickstart.html.

9 Tutorial: A Simple Arm Software Stack

In this tutorial the original simple_arm package is presented.[23,24] The simple_arm package is teleoperation software and firmware for an arm with 6° of freedom. Forces input by the operator's joystick motions are converted to individual motor velocities. We simply command motors to move by applying voltages: there is no feedback. We do not control the arm by assigning it a position. This makes the arm more difficult to control but simpler to implement because there is no hardware or software to sense joint positions.

At the University Rover Competition (URC) the rover's manipulator arm is probably the most important component because it is needed for a large portion of point awarding tasks. It is also very complex. Team R3 ran out of development time to integrate position encoders with MoveIt!, so the arm software in simple_arm that went to URC 2017, and that is described here, was very simple yet still effective.

In the science URC mission, the arm was used to drill soil and collect samples. In the equipment servicing tasks the arm was used for unscrewing a cap, pouring a container of liquid, and more. In the extreme delivery and retrieval mission the arm was used to pick up and carry hand tools and small rocks.

As seen in Fig. 18, the arm is controlled by a joystick where each arm motor is controlled by a different button or single axis of motion on the joystick. The simple_arm packages increases the rotational velocity of motors as the joystick is pushed further or twisted more.[25]

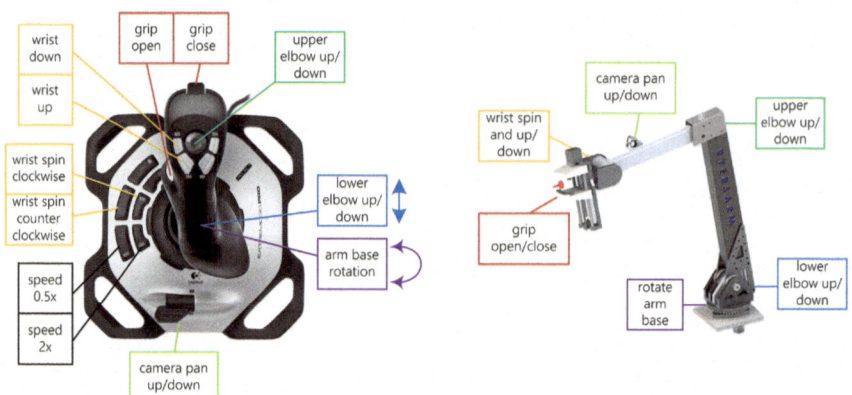

Fig. 18 The Logitech Extreme 3D Pro joystick button layout for the simple_arm package (diagram is also available in Visio format). Image credit: Logitech International

---

[23]Source code for simple_arm ROS package https://github.com/danielsnider/simple_arm.

[24]Wiki page for simple_arm ROS package http://wiki.ros.org/simple_arm.

[25]Logitech Extreme 3D Pro joystick button layout diagram in Visio format https://github.com/danielsnider/ros-rover/blob/master/diagrams/simple_arm_joystick_diagram.vsdx?raw=true.

Features of the simple_arm package include: velocity control of individual arm joint motors, fast and slow motor speed modifier buttons, buttons to open and close a gripper, control of a camera servo (one axis only), simple Arduino firmware to send PWM signals to control the velocity of motors and position of the camera servo. For the sake of simplicity, this package does not implement the following best practices: no tf publishing, no URDF, no joint limits and no integration with ros_control or MoveIt!. The simple_arm package gives ROS users the ability advance their robot more quickly and hopefully to find more time to implement best practices.

9.1 Usage Instructions

1. Install the ROS package:

```
$ sudo apt-get install ros-kinetic-simple-arm
```

2. Launch the ROS node and specify as arguments the joystick device path for controlling the arm and the Arduino to control the arm motors:

```
$ roslaunch simple_arm simple_arm.launch
joystick_serial_dev:=/dev/input/js0
microcontroller_serial_dev:=/dev/ttyACM0
```

In most cases however, the joystick is connected to another computer, such as a teleoperation station. To do this, run the joy ROS node separately over the ROS network.

3. Install the arm_firmware onto a microcontroller as described in Sect. 8.5. The microcontroller must be connected to the arm's motors and to the on board computer running the simple_arm ROS node by a serial connection (ex. USB).

4. Your robot arm is ready to be moved.

9.2 simple_arm ROS Node

The simple_arm node is written in python and simply converts ROS sensor_msgs/Joy messages from the common joy joystick ROS node into serial messages. The serial messages are sent to the arm_firmware on the microcontroller to drive the robot arm. The serial messages follow a simple protocol. Each command is a list of 7 floats, one velocity command for each of the 6 joints and one target angle to control the single axis camera.

The button mapping implemented by the simple_arm node can be seen in Fig. 18 and was tested on a Logitech Extreme 3D Pro joystick. The button layout should only need small modifications if any to work for similar controllers that support a DirectInput mode.

9.3 arm_firmware Arduino Software

The arm_firmware microcontroller code does the minimum amount of work possible to receive motor commands from a USB serial connection and output voltages to digital PWM to be received by motor controllers. It simply receives and fires commands to the lower hardware level with no feedback. We use the Victor SP motor controller to control our D.C. motors where the voltage is given by the duty-cycle of the PWM signal.

We connected an Arduino by USB serial to our robot's on-board computer and dedicated its use to the simple_arm package. We tested on an Arduino Mega 2560, however many other boards should work with the same code and setup steps thanks to PlatformIO. You may need to change the pin numbers. For details on how to install and use PlatformIO with the simple_arm package see the ROS wiki page or Sect. 8.5 which contains very similar instructions.

Please note that this software does not stop moving the robot if no messages are received for certain period of time.

10   Tutorial: Autonomous Recovery after Lost Communications

At the URC competition, Team R3 (Ryerson University) was worried about travelling into a communication deadzone and losing wireless control of our rover from a distant base station. This is one of the challenges put forth by URC competition and is often found in the real world.

We have published a new package, called lost_comms_recovery that will trigger when the robot loses connection to the base station and it will navigate to a configurable home or stop all motors.[26,27] The base station connection check uses ping to a configurable list of IPs. The monitoring loop waits 3 s between checks and by default failure is triggered after 2 consecutive failed pings. Each ping will wait up to one second to receive a response.

While this node tries to add a safety backup system to your robot, it is far from guaranteeing any added safety. What is safer than relying on this package, is using motor control software that sets zero velocity after a certain amount of time not receiving any new commands.

---

[26]Source code for lost_comms_recovery ROS package https://github.com/danielsnider/lost_comms_recovery.

[27]Wiki page for lost_comms_recovery ROS package http://wiki.ros.org/lost_comms_recovery.

10.1 Usage Instructions

1. Install:

```
$ sudo apt-get install ros-kinetic-lost-comms-
    recovery
```

2. Launch:

```
$ roslaunch lost_comms_recovery lost_comms_recovery.
    launch
ips_to_monitor:=192.168.1.2
```

3. Then the following behavior will take place:

If move_base is running, an autonomous recovery navigation will take place. The default position of the recovery goal is the origin (0, 0) of the frame given in the goal_frame_id ROS parameter and the orientation is all 0s by default. This default pose can be overridden if a messaged is published on the recovery_pose topic. If move_base is already navigating to a goal it will not be interrupted and recovery navigation will happen when move_base is idle.

If move_base is not running when communication failure occurs then motors and joysticks are set to zero by publishing a zero geometry_msgs/Twist message and a zero sensor_msgs/Joy message to simulate a joystick returning to a neutral, non-active position.

10.2 Normal Output

When you launch and use the lost_comms_recovery ROS node you will see the following console output.

```
$ roslaunch lost_comms_recovery lost_comms_recovery.
    launch
ips_to_monitor:=192.168.190.136

[INFO] Monitoring base station on IP(s):
    192.168.190.136.
[INFO] Connected to base station.
[INFO] Connected to base station.
...
[ERROR] No connection to base station.
[INFO] Executing move_base goal to position (x,y)
    0.0, 0.0.
[INFO] Initial goal status: PENDING
[INFO] This goal has been accepted by the simple
    action server
[INFO] Final goal status: SUCCEEDED
[INFO] Goal reached.
```

Listing 8 Normal console output seen when launching the lost_comms_recovery ROS node.

11 Tutorial: Stitch Panoramas with Hugin

Hugin is professional software popularly used to create panoramic images by compositing and rectifying multiple still images [11]. We have created a package called hugin_panorama to wrap one high level function of Hugin, the creation of panoramas.[28,29]

In the science task of the URC competition, teams are awarded points if they document locations of scientific interest such as geological and soil sampling sites with panoramas.

Our package uses the Hugin command line tools to compose panoramas in 8 steps according to a well-documented workflow.[30,31] To summarize, it consists of creating a Hugin project file, finding matching feature control points between images, pruning control points with large error distances, finding vertical lines across images to be straightened, doing the straightening and other photometric optimization, optimal cropping, and saving to tiff and compressed png image formats. The compressed panoramic image is published on at output ROS topic.

An example panorama can be seen in Fig. 19. Despite the fact that the panorama was created using low resolution raw images, the competition judges still awarded the it full points.

The hugin_panorama launch implementation[32] makes use of the image_saver[33] node provided by the image_view package. The image_saver node will save all images from a sensor_msgs/Image topic as jpg/png files. The saved images are used as the source image parts when creating the panoramas.

11.1 *Usage Instructions*

1. Install the ROS package:

```
$ sudo apt-get install ros-kinetic-hugin-panorama
    hugin-tools
enblend
```

---

[28]Source code for hugin_panorama ROS package https://github.com/danielsnider/hugin_panorama.

[29]Wiki page for hugin_panorama ROS package http://wiki.ros.org/hugin_panorama.

[30]The Hugin image processing library http://hugin.sourceforge.net/.

[31]Panorama scripting with Hugin http://wiki.panotools.org/Panorama_scripting_in_a_nutshell.

[32]Main launch file of the hugin_panorama package https://github.com/danielsnider/hugin_panorama/blob/master/launch/hugin_panorama.launch.

[33]Documentation for the image_saver ROS node http://wiki.ros.org/image_view#image_view.2BAC8-diamondback.image_saver.

Fig. 19 Panorama example made in Utah at the URC competition using the hugin_panorama ROS package. The graphic overlays were created with our ROS package that is presented in Sect. 7

2. Launch the ROS node:

```
$ roslaunch hugin_panorama hugin_panorama.launch
    image:=/image_topic
```

3. Save individual images for input to the panorama: (order doesn't matter)

```
$ rosservice call /hugin_panorama/image_saver/save
# change angle of camera
$ rosservice call /hugin_panorama/image_saver/save
# repeat as many times as you like...
```

4. Stitch the panorama:

```
$ rosservice call /hugin_panorama/stitch
```

5. View resulting panorama:

```
$ rqt_image_view /hugin_panorama/panorama/compressed
# or open the panorama file
$ roscd hugin_panorama; eog ./images/output.png
```

6. Start again:

```
$ rosservice call /hugin_panorama/reset
```

This command will clear the images waiting to be stitched so you can start collecting images for an entirely new panorama.

11.2   Live Panorama Mode

If you have more than one camera on your robot and you want to stitch images together repetitively in a loop, then use stitch_loop.launch. However, expect a slow frame rate of less than 1 Hz because this package is not optimized for speed.

1. Launch the stitch_loop node:

```
$ roslaunch hugin_panorama stitch_loop.launch image1
    :=/image_topic2 image2:=/image_topic2
```

2. View resulting live panorama:

```
$ rqt_image_view /hugin_panorama/panorama/compressed
```

If you have more than two cameras then the quick fix is to edit the simple python script (rosed hugin_panorama stitch_loop.py) and the launch file (rosed hugin_panorama stitch_loop.launch) to duplicate some parts.

12   Tutorial: GPS Navigation Goal

In the autonomous task of the URC competition, a series of goal locations are given to teams as approximate GPS coordinates. Rovers are expected to autonomously drive to the GPS location and then find and stop near a tennis ball marker. To achieve the GPS navigation requirements of this task, Team R3 has created the gps_goal package.[34, 35] We believe this is the first packaged for ROS solution to convert navigation goals in given in GPS coordinates to ROS frame coordinates. The package uses one known GPS location in the ROS frame to facilitate converting between coordinate systems. Figure 20 shows how this package fits into Team R3's larger ROS software architecture.

The new gps_goal package uses the WGS84 ellipsoid[36] and geographiclib[37] python library to calculate the surface distance between GPS points. WGS84 is the standard coordinate system for GPS and thus the packages configures GeographicLib to use it because it is important for calculating the correct distance between GPS points.

---

[34]Source code for gps_goal ROS package https://github.com/danielsnider/gps_goal.

[35]Wiki page for gps_goal ROS package http://wiki.ros.org/gps_goal.

[36]The World Geodetic System (WGS) 84 is the reference coordinate system used by the Global Positioning System (GPS). WGS84 uses degrees. It consists of a latitudinal axis from -90 to $90°$ and a longitudinal axis from -180 to $180°$. As it is the standard coordinate system for GPS it is also commonly used in robotics. https://en.wikipedia.org/wiki/World_Geodetic_System# A_new_World_Geodetic_System:_WGS_84.

[37]The GeographicLib software library https://geographiclib.sourceforge.io/.

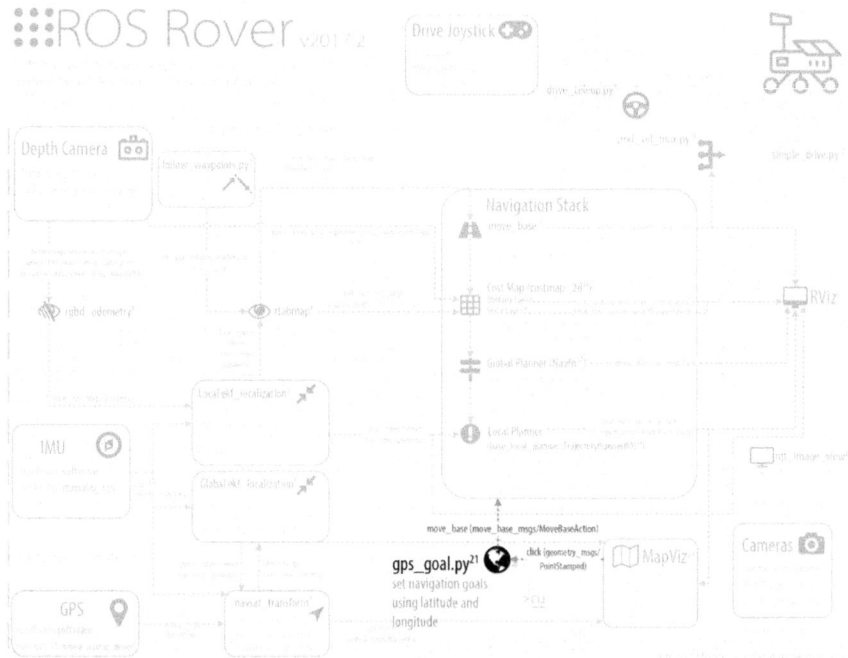

Fig. 20 The gps_goal ROS node seen within Team R3's rover system. See Fig. 9 for the full diagram

The GPS goal can be set using a geometry_msgs/PoseStamped or sensor_msgs/NavSatFix message. The robot's desired yaw, pitch, and roll can be set in a PoseStamped message but when using a NavSatFix they will always be set to 0°.

The goal is calculated in a ROS coordinate frame by comparing the goal GPS location to a known GPS location at the origin (0,0) of a ROS frame given by the local_xy_frame ROS parameter which is typically set to 'world' but can be any ROS frame. This initial origin GPS location is best published using a helper initialize_origin node (see Sect. 12.3 below for more details).

12.1 Usage Instructions

1. Install the ROS package:

```
$ sudo apt-get install ros-kinetic-gps-goal
```

2. Launch the ROS node:

```
$ roslaunch gps_goal gps_goal.launch
```

3. Set a known GPS location using one of the following approaches (a) or (b). The given GPS location will be attached to the origin (0,0) of the ROS frame given by the local_xy_frame ROS parameter. This is used to calculate the distance to the goal.

3a. Use the next GPS coordinate published on a ROS topic (requires package ros-kinetic-swri-transform-util):

```
$ roslaunch gps_goal initialize_origin.launch origin
   :=auto
```

3b. Or set the initial origin manually using a rostopic publish command:

```
$ rostopic pub /local_xy_origin geometry_msgs/
    PoseStamped '{ header: { frame_id: "/map" }, pose
    : { position: { x: 43.658, y: -79.379 } } }' -1
```

4. Set a navigation goal using GPS coordinates set with either a Pose or NavSatFix GPS message.

```
$ rostopic pub /gps_goal_fix sensor_msgs/NavSatFix
    "{latitude:38.42, longitude: -110.79}" -1
OR
$ rostopic pub /gps_goal_pose geometry_msgs/
    PoseStamped '{ header: { frame_id: "/map" }, pose
    : {
position: { x: 43.658, y: -79.379 } } }' -1
```

12.2 Command Line Interface (CLI)

Alternatively, a Command Line Interface (CLI) is available to set GPS navigation goals. When using the CLI interface you can use one of two coordinate formats: either degree, minute, and seconds (DMS) or decimal GPS format. Using command line arguments, users can also set the desired roll, pitch, and yaw final position. You can invoke the gps_goal script once using the Command Line Interface (CLI) with any of the following options.

Listing 9 Example usages of the gps_goal CLI script to set a navigation goal.

```
$ roscd gps_goal
$ ./src/gps_goal/gps_goal.py --lat 43.658 --long
    -79.379
# decimal format
OR
$ ./src/gps_goal/gps_goal.py --lat
43,39,31 --long -79,22,45
# DMS format
```

12.3 *initialize_origin Helper ROS Node*

The initialize_origin node will continuously publish (actually in a latched manner[38]) a geometry_msgs/PoseStamped on the local_xy_origin topic and this is the recommended approach over manually publishing the origin GPS location with rostopic pub. This location is the origin (0,0) of the frame (typically world) given by the local_xy_frame parameter to the initialize_origin node. This location is used to calculate distances for goals. One message on this topic is consumed when the node starts only.

This node is provided by the swri_transform_util package (apt-get install roskinetic-swri-transform-util) and it is often launched as a helper node for MapViz, a top-down robot and world visualization tool that is detailed in Sect. 15. There are two modes for initialize_origin: static or auto.

Static Mode You can hard code a GPS location (useful for testing) for the origin (0,0). In the following example the coordinates for the Mars Desert Research Station (MDRS) are hard coded in initialize_origin.launch and selected on the command line with the option "origin:=MDRS".

```
$ roslaunch gps_goal initialize_origin.launch origin
  :=MDRS
```

Auto Mode When using the "auto" mode, the origin will be to the first GPS fix that it receives on the topic configured in the initialize_origin.launch file.

```
$ roslaunch gps_goal initialize_origin.launch origin
  :=auto
```

Launch example Starting the initialize_origin ROS node can be done in the following way.

Listing 10 An example launch config to start the initialize_origin ROS node.

```
<node pkg="swri_transform_util" type="
   initialize_origin.py"
name="initialize_origin" output="screen">
  <param name="local_xy_frame" value="/world"/>
  <param name="local_xy_origin" value="MDRS"/> <!--
     setting "auto" here will set the origin to the
     first GPS fix that it receives -->
  <remap from="gps" to="gps"/>
  <rosparam param="local_xy_origins">
    [{ name: MDRS,
       latitude: 38.40630,
       longitude: -110.79201,
       altitude: 0.0,
       heading: 0.0}]
  </rosparam>
</node>
```

---

[38]When a connection is latched, the last message published is saved and automatically sent to any future subscribers of that connection.

12.4 Normal Output

When you launch and use the gps_goal ROS node or CLI interface you will see the following console output.

```
$ roscd gps_goal $ ./src/gps_goal/gps_goal.py --lat
   43.658 --long
-79.379

[INFO]: Connecting to move_base...
[INFO]: Connected.
[INFO]: Waiting for a message to initialize the
   origin GPS location...
[INFO]: Received origin: lat 43.642, long -79.380.
[INFO]: Given GPS goal: lat 43.658, long -79.379.
[INFO]: The distance from the origin to the goal is
   97.3 m.
[INFO]: The azimuth from the origin to the goal is
   169.513 degrees.
[INFO]: The translation from the origin to the goal
   is (x,y) 91.3, 13.6 m.
[INFO]: Executing move_base goal to position (x,y)
   91.3, 13.6, with 138.14 degrees yaw.
[INFO]: To cancel the goal: 'rostopic pub -1 /
   move_base/cancel actionlib_msgs/GoalID -- {}'
[INFO]: Inital goal status: PENDING
[INFO]: Final goal status: COMPLETE
```

Listing 11 Normal console output seen when launching the gps_goal ROS node or from the CLI interface.

13 Tutorial: Wireless Communication

At Team ITU (Istanbul Technical University), communication from our base station to our mobile outdoor rover was of utmost importance and received a good amount of development time. Non-line of sight (NLOS) communication is an important issue and often a cause of unsuccessful runs in URC. Although there are various commercial products available claiming to solve the issues of long range and high throughput communication, many of them don't solve the problems as advertised. So a combination of systems and products were tested and used in the competition.[39,40]

The primary system must be capable of delivering a video feed to the ground station for operators to use while piloting the vehicle remotely. This system is generally preferred to be a Wi-Fi network that can multiplex high-res video and data

[39]Team ITU's low level communication source code https://github.com/itu-rover/2016-2017-Sensor-GPS-STM-Codes-/blob/master/STM32/project/mpu_test/Src/main.c.

[40]Team ITU's high level communication source code https://github.com/itu-rover/communication.

Fig. 21 UHF LoRa radio
module for long range
communication. Image
credit: Semtech Corporation

LoRa SX1278
433MHz /-140dBm /3500m

traffic at high speeds. After performing a series of unsuccessful non-line of sight
(NLOS) tests at long distances (400–500 m) with the Ubiquiti Bullet M2, a popular
2.4 GHz product, a less popular Microhard pDDL2450 module was chosen based on
our sponsor's advice. Thankfully, the sponsor had a few spares and donated them to
Team ITU. In tests, this 2.4 GHz module was generally successful in sending useful
data from the vehicle's instruments and 720p video feed compressed with MJPEG
from five cameras at the same time over a 1 km range in NLOS course with 8dBi
omnidirectional antennas. This module is physically connected to the high level on-
board computer (OBC), a Raspberry Pi 3 running Ubuntu 16.04, with a standard
CAT5 Ethernet cable.

Secondly, our radio frequency (RF) backup link was able to pass over the natural
obstacles such as large rocks and hills. This link operates on lower UHF frequencies
and lower baud rates to increase performance needed to send crucial information
about the vehicle's condition to ensure health and function of the rover at extreme
distances (5 km). We consider our rover's heartbeat, GPS position, attitude and current speed to be important data that should be sent through the RF link. Tests were
conducted using 433 MHz LoRa modules (see Fig. 21) on both sides with 3dBi omnidirectional antennas, in 9600 and 115200 baud rates. The tests showed that these
modules have no problem sending the data over a 5 km range in NLOS conditions.
The LoRa modules were wired to the standard RX-TX wires on our STM32F103
microprocessor and uses UART communication. The microprocessor also controls
the driving system, communicates with the sensors, the Raspberry Pi 3 on-board
computer (OBC) (Fig. 22).

To make the LoRa RF link active in the C# language see Listing 12.

Listing 12 Start communication with the LoRa RF module in the C# language.

```
SerialPort _serialPort = new SerialPort("COM1", 115200, Parity.None
    , 8, StopBits.One);
_serialPort.Open();
```

In normal, connected conditions, only the Wi-Fi network system is active and the
RF system is inactive. In these conditions all the processing is done in the Raspberry
Pi 3 on-board computer (OBC). Commands from the human pilot reach the OBC
first and then are distributed to the low level STM microprocessor via a RS232
link. The video feed is active and the pilot can easily drive the rover using the video

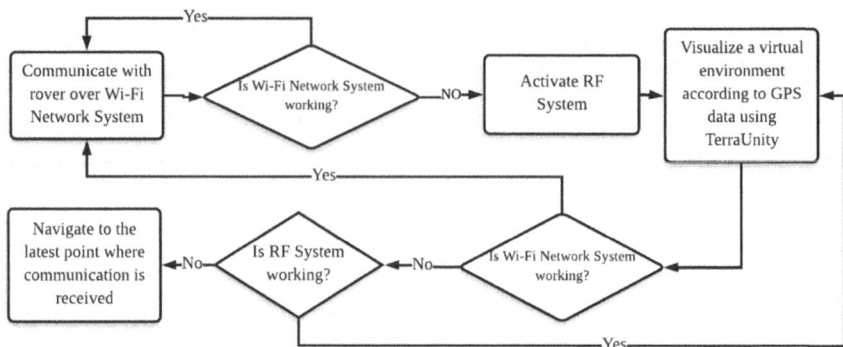

Fig. 22 Flowchart of decision making algorithm used on ITU's rover

images from the cameras. The low band RF system was remarkably reliable although interruptions were encountered at times with the Wi-Fi network.

In the case of losing the Wi-Fi network system or OBC, the video feed will be lost and piloting the vehicle without a video feed is almost impossible. So an innovative solution was implemented to overcome this problem and continue the mission. In such conditions, first the RF link is activated and crucial information and the pilot commands are redirected to this link. Therefore, the pilot directly communicates with the low level processor. To help the pilot visualize the environment a virtual environment around the GPS coordinates is simulated with computer graphics. This visual environment is created using the TerraUnity software. The software creates a one-to-one, colorized, topographical landscape with natural objects loaded from a 3D map database of the location.[41] This way the driver could look at the ground station monitor and see the terrain around the vehicle in the Unity graphics simulation, which was pretty precise in our tests. The software was found to be very successful in creating a realistic environment and it gives a clue to the driver about where the vehicle is and what natural obstacles are around it. Knowing the locations of natural obstacles is essential as the rover has physical limits that prevent it from navigating some terrain. Although the TerraUnity solution is an imperfect representation of the world, it provides a useful avenue to continue the mission in a catastrophic failure scenario.

Finally, in the case where both Wi-Fi and RF communication is lost to the rover, a third backup system is initialized where the rover navigates autonomously to the last GPS point that it communicated successfully with the ground station.

---

[41] TerraUnity computer graphics software used to the visualize the rover at its GPS location in a topographical simulation of earth http://terraunity.com/.

14 Tutorial: Autonomous Navigation by Team R3

In this section we present Team R3's (Ryerson University) autonomous software architecture that was designed for outdoor autonomous driving at the University Rover Competition (URC) 2017. The design uses a stereo camera and SLAM to navigate to a goal autonomously and avoid static obstacles. For a description of the requirements of the autonomous task for URC 2017, please refer back to Sect. 2.2 of this chapter.

In the following subsections we will elaborate on the ZED stereo camera, rbgd_odometry, RTAB-Map SLAM, and move_base.[42] Figure 23 depicts a diagram of Team R3's autonomous software design. To see this diagram within a larger diagram with more of Team R3's rover software components see Sect. 5.

14.1 ZED Depth Camera

The ZED stereo camera[43,44,45] and ROS wrapper software perform excellently for the price of $450 (Fig. 24). With the ZED camera Team R3 was able to avoid obstacles such as rocks and steep cliffs. However, the rover could not move quickly, no faster

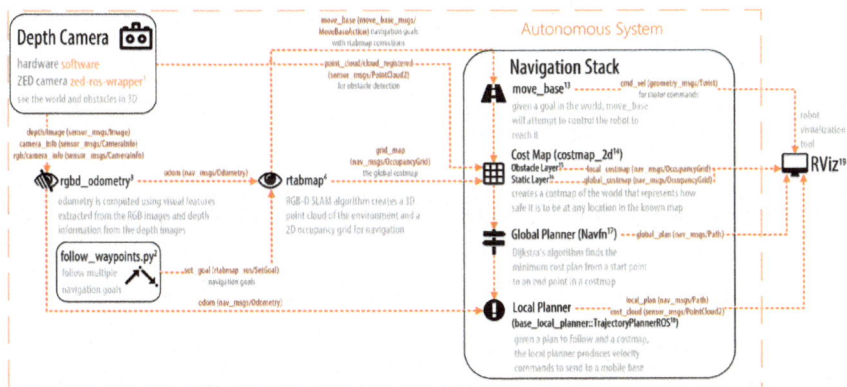

Fig. 23 Team R3's autonomous navigation system used at URC 2017 rover competition. This diagram is also available in Visio format

[42] Team R3's autonomous software architecture diagram in Microsoft Visio format https://github.com/danielsnider/ros-rover/blob/master/diagrams/team_r3_AUTO_Diagram.vsdx?raw=true.

[43] ZED stereo camera technical specs https://www.stereolabs.com/zed/.

[44] More ZED camera documentation https://www.stereolabs.com/documentation/guides/using-zed-with-ros/ZED_node.html.

[45] Team R3's ZED launch file https://github.com/teamr3/URC/blob/master/rosws/src/rover/launch/zed_up.launch.

Fig. 24 The ZED depth camera. Image credit: Stereolabs

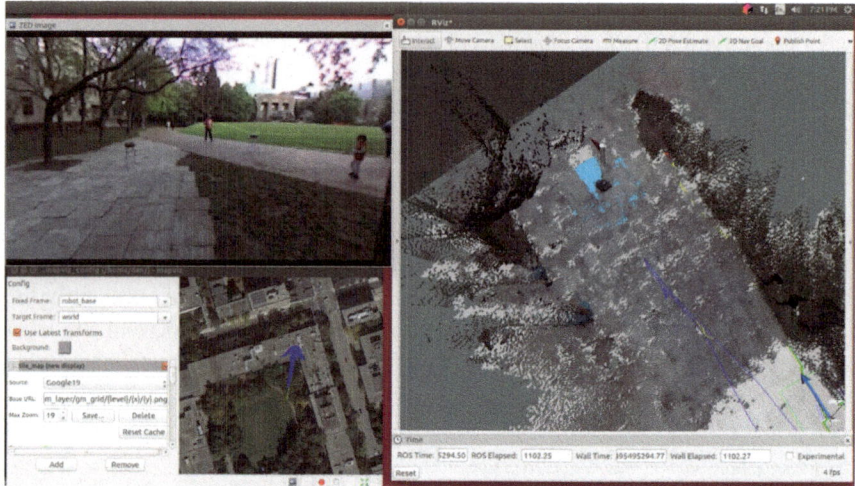

Fig. 25 Screenshot of R3's autonomous system tests with RTAB-Map (video available on YouTube)

than slow-moderate human walking speed, because the performance of our on board computer, a Nvidia Jetson TX1,[46] was fully utilized. It is important to know that a restriction of the ZED camera is that it requires an Nvidia GPU, a dual-core processor, and 4 GB of RAM. All of which the Nvidia Jetson TX1 has.

The ZED camera combined with RTAB-Map for SLAM localization and mapping worked reasonably robustly even in Utah's desert where the ground's feature complexity is low and even with a significant amount of shaking on the pole to which our ZED camera was attached (Figs. 24 and 25).

Here is a tip when using the ZED camera: launch the node with command line arguments so you can more easily find the right balance between performance and resolution. At URC we wanted the lowest latency so we default to VGA resolution,

---

[46]Nvidia Jetson TX1 technical specs https://developer.nvidia.com/embedded/buy/jetson-tx1-devkit.

at 10 FPS, and low depth map quality. Also, note that the ZED camera is designed for outdoor textured surfaces. Indoor floors that are featureless will make testing more difficult. Also when testing indoors, you may use a blinder on top of the camera so that it doesn't see the ceiling as an obstacle.

```
$ roslaunch rover zed_up frame_rate:=30 resolution
  :=2 depth_quality:=3
```

```
1  <launch>
2      <arg name="frame_rate" default="10"/>
3      <arg name="resolution" default="3"/>
4      <arg name="depth_quality" default="1"/>
5      <node output="screen" pkg="zed_wrapper" name="
           zed_node" type="zed_wrapper_node">
6          <param name="frame_rate" value="$(arg
               frame_rate)"/>
7          <!-- Image resolution options: -->
8          <!--    0': HD2K,    1': HD1080,    2': HD720
               , 3': VGA -->
9          <param name="resolution" value="$(arg
               resolution)"/>
10         <!-- Depth map quality options: -->
11         <!--    0': NONE,    1': PERFORMANCE,    2':
               MEDIUM,    3': QUALITY -->
12         <param name="quality" value="$(arg
               depth_quality)"/>
13     </node>
14     <node pkg="image_transport" type="republish"
           name="zed_camera_feed" args="raw in:=rgb/
           image_rect_color out:=rgb_republished"/>
15  </launch>
```

Listing 13 Arguments in a ROS launch file so to allow easy changing of the quality of the ZED stereo camera.

Reduce Bandwidth Used by Video Streams To lower the amount of data on our wireless link, on line 14 we publish the ZED camera as JPEG compressed stills and Theora video streaming using the republish node of the image_transport ROS package.[47] Republish listens on one uncompressed (raw) image topic and republishes JPEG compressed stills and Theora video on different topics.

To lower bandwidth even further you can convert images to greyscale, cutting data usage by 3. Team R3 has a small ROS node for this.[48]

Additionally, you should use the republish node when more than one ROS node is subscribing to a depth or image stream over a wireless connection. Instead you should have one republish node subscribe at the base station, then multiple ROS nodes at the base station can subscribe to the republish node without consuming a lot of wireless bandwidth. This is also referred to as a ROS relay.

---

[47]Republish ROS node documentation http://wiki.ros.org/image_transport#republish.

[48]Python ROS script to reduce bandwidth usage of video streams https://github.com/teamr3/URC/blob/master/rosws/src/rover/src/low_res_stream.py.

Another easy way to reduce bandwidth used by the ZED camera is to downsample its pointcloud using the VoxelGrid nodelet in the pcl_ros ROS package.[49]

14.2   Visual Odometry with rgbd_odometry

The ZED camera does not have a gyroscope or accelerometer in it. It uses visual information for odometry and it is quite good. We found that the rgbd_odometry[50,51] node provided by the RTAB-Map package produces better visual odometry than the standard ZED camera odometry algorithm. Visual odometry was very robust to jitter and shaking as the rover moved over rough terrain, even with our camera on a tall pole which made the shaking extreme. Optimizations to rgbd_odometry used by Team R3 at the URC 2017 rover competition are shown in Listing 14

```
1  <launch>
2      <node output="screen" type="rgbd_odometry" name=
           "zed_odom" pkg="rtabmap_ros">
3          <!-- 2D SLAM makes the position drift less
                over time -->
4          <param name="Reg/Force3DoF" type="string"
                value="true"/>
5          <!-- Change if camera is tilted downwards or
                any non-level pose -->
6          <param name="initial_pose" value="0 0 0 0 0
                0"/>
7
8          <!-- Options to Reduce Resource Usage -->
9          <!-- 0=Frame-to-Map (F2M) 1=Frame-to-Frame (
                F2F) -->
10         <param name="Odom/Strategy" value="1"/>
11         <!-- Correspondences: 0=Features Matching,
                1=Optical Flow -->
12         <param name="Vis/CorType" value="1"/>
13         <!-- maximum features map size, default 2000
                -->
14         <param name="OdomF2M/MaxSize" type="string"
                value="1000"/>
15         <!-- maximum features extracted by image,
                default 1000 -->
16         <param name="Vis/MaxFeatures" type="string"
                value="600"/>
17     </node>
18  </launch>
```

Listing 14   Important settings to optimize the rgbd_odometry ROS node.

---

[49]pcl_ros ROS documentation http://wiki.ros.org/pcl_ros.

[50]rgbd_odometry ROS node documentation http://wiki.ros.org/rtabmap_ros#rgbd_odometry.

[51]Team R3's rgbd_odometry launch file https://github.com/teamr3/URC/blob/master/rosws/src/rover_navigation/launch/rgbd_odometry.launch.

14.3   3D Mapping in ROS with RTAB-Map

Using depth camera data, RTAB-Map[52,53] creates a continuously growing point cloud of the world using simultaneous localization and mapping (SLAM) [5]. Inherent to the SLAM algorithm is pinpointing your own location in the map that you are building as you move. Using this map, RTAB-Map then creates an occupancy grid map [3], which represents free and occupied space, needed to avoid obstacles in the rover's way. RTAB-Map's algorithm has real-time constraints so that when mapping large-scale environments time limits are respected and performance does not degrade [7].

In the launch file seen in Listing 15, lines 3–5 show configurations to reduce noisy detection of obstacles. If you set MaxGroundAngle to 180°, this effectively disables obstacle detection, which can be both useful and dangerous.[54]

RTAB-Map also performs loop closures. Loop closure is the problem of recognizing a previously-visited location and updates the beliefs accordingly.[55] When an image is matched to a previously-visited location, a loop closure is said to have occurred. At this point RTAP-Map will adjust the map to compensate for drift that occurred since the last time the location was visited. Lines 8–10 of listing 15 increase the likelihood of loop closures being detected.

```
1  <launch>
2      <node pkg="rtabmap_ros" name="rtabmap" type="
          rtabmap" output="screen">
3          <!-- Improve obstacle detection -->
4          <param name="Grid/MaxGroundAngle" value="110
              "/> <!-- Maximum angle between point's
              normal to ground's normal to label it as
              ground. Points with higher angle
              difference are considered as obstacles.
              Default: 45 -->
5          <param name="grid_eroded" value="true"/> <
              !-- remove obstacles which touch 3 or
              more empty cells -->
6
7          <!-- Improve loop closure chances -->
8          <param name="RGBD/
              LoopClosureReextractFeatures" type="
              string" value="true"/> <!-- Extract
              features even if there are some already
              in the nodes, more loop closures will be
              accepted. Default: false -->
```

---

[52] RTAB-Map documentation http://wiki.ros.org/rtabmap_ros.

[53] Team R3's launch file for RTAB-Map https://github.com/teamr3/URC/blob/master/rosws/src/rover_navigation/launch/rtabmap.launch.

[54] Video of an autonomous navigation by Team R3 with RTAB-Map and the ZED stereo camera https://www.youtube.com/watch?v=p_1nkSQS8HE.

[55] More information about loop closures https://en.wikipedia.org/wiki/Simultaneous_localization_and_mapping\#Loop_closure.

```
9       <param name="Vis/MinInliers" type="string"
            value="10"/> <!-- Minimum feature
            correspondences to compute/accept the
            transformation. Default: 20 -->
10      <param name="Vis/InlierDistance" type="
            string" value="0.15"/> <!-- Maximum
            distance for feature correspondences.
            Used by 3D->3D estimation approach (the
            default approach). Default: 0.1 -->
11    </node>
12  </launch>
```

Listing 15 Important settings for tuning the RTAB-Map 3D mapping ROS package.

Team R3's main strategy for the autonomous task of URC 2017 was to: 1. Build a SLAM map by teleoperating from the start gate all the way to the tennis ball objective (actually this could be an autonomous navigation attempt by using the GPS location of the tennis ball as the goal), 2. then we would put a flag[56] in RViz to mark where we observed the tennis ball, 3. then we would teleoperate back to the start gate, 4. complete a loop closure to correct for drift, 5. and then use RViz to set an autonomous goal for where we saw the tennis ball.

14.4 *move_base Path Planning*

The ROS navigation stack,[57] also known as move_base,[58] is a collection of components/plugins that are selected and configured by YAML configuration files as seen in Listing 16. For global path planning, the NavFn plugin is used which implements Dijkstra's shortest path algorithm.

```
1  <launch>
2      <node pkg="move_base" type="move_base" name="
           move_base" output="screen" clear_params="true
           ">
3          <rosparam file="$(find rover)/
               costmap_common_params.yaml" command="load
               " ns="global_costmap"/>
4          <rosparam file="$(find rover)/
               costmap_common_params.yaml" command="load
               " ns="local_costmap"/>
5          <rosparam file="$(find rover)/
               local_costmap_params.yaml" command="load"
               />
```

```
6        <rosparam file="$(find rover)/
            global_costmap_params.yaml" command="load
            "/>
7        <rosparam file="$(find rover)/
            base_local_planner_params.yaml" command="
            load"/>
8    </node>
9 </launch>
```

Listing 16 move_base is configured by independent YAML files that configure the subcomponents of move_base.

The most interesting configuration file for move_base is the base_local_planner_params.yaml.[59] Given a path for the robot to follow and a costmap, the base_local_planner[60] produces velocity commands to send to a mobile base. This configuration is where you set minimum and maximum velocities and accelerations for your robot, as well as goal tolerance. Make sure that the minimum velocity multiplied by the sim_period is less than twice the tolerance on a goal. Otherwise, the robot will prefer to rotate in place just outside of range of its target position rather than moving towards the goal.

```
1  TrajectoryPlannerROS:
2    acc_lim_x:  0.5
3    acc_lim_y:  0.5
4    acc_lim_theta: 1.00
5
6    max_vel_x:  0.27
7    min_vel_x:  0.20
8    max_rotational_vel: 0.4
9    min_in_place_vel_theta: 0.27
10   max_vel_theta: 0.1
11   min_vel_theta: -0.1
12   escape_vel: -0.19
13
14   xy_goal_tolerance: 1
15   yaw_goal_tolerance: 1.39626 # 80 degrees
16
17   holonomic_robot: false
18   sim_time: 1.7 # set between 1 and 2. The higher he
            value, the smoother the path (though more
            samples would be required)
```

Listing 17 Important velocity settings in the base_local_planner_params.yaml configuration file of move_base.

---

[59]Team R3's config file for base_local_planner_params.yaml configuration of move_base https://github.com/teamr3/URC/blob/master/rosws/src/rover_navigation/config/base_local_planner_params.yaml.

[60]Documentation for base_local_planner http://wiki.ros.org/base_local_planner.

15   Tutorial: MapViz Robot Visualization Tool

At the URC competition, Team R3 (Ryerson University) improved their situational awareness by using MapViz (Figs. 26 and 27).[61] Mapviz is a ROS-based visualization tool with a plug-in system similar to rviz but focused only on a top down, 2D view of data. Any 3D data is flattened into the 2D view of MapViz. Created by the Southwest Research Institute in Florida for their outdoor autonomous robotics research, it is still under active open source development at the time of writing in December 2017. Using a plugin called tile-map, Google Maps satellite view can be viewed in the MapViz plugin called Tile Map.

The authors have contributed a Docker container[62] to make displaying Google Maps in MapViz as easy as possible. This container runs software called MapProxy which converts from the format of Google Maps API to a standard format called Web Map Tile Service (WMTS) which MapViz Tile Map plugin can display. The authors have set MapProxy's configuration[63] to cache any maps that you load to ~/mapproxy/`cache_data/` so that they are available offline.

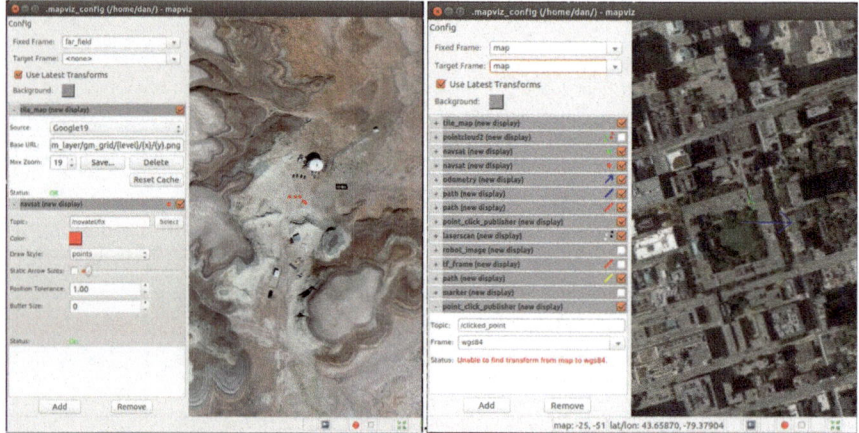

Fig. 26 Screenshots of MapViz ROS visualization tool. Red dots indicate gps coordinates. Many more visualization layers are possible

---

[61]Documentation and source code for MapViz https://github.com/swri-robotics/mapviz.

[62]Docker container for proxying Google Maps toMapViz https://github.com/danielsnider/MapViz-Tile-Map-Google-Maps-Satellite.

[63]MapProxy config file https://github.com/danielsnider/docker-mapproxy-googlemaps/blob/master/mapproxy.yaml.

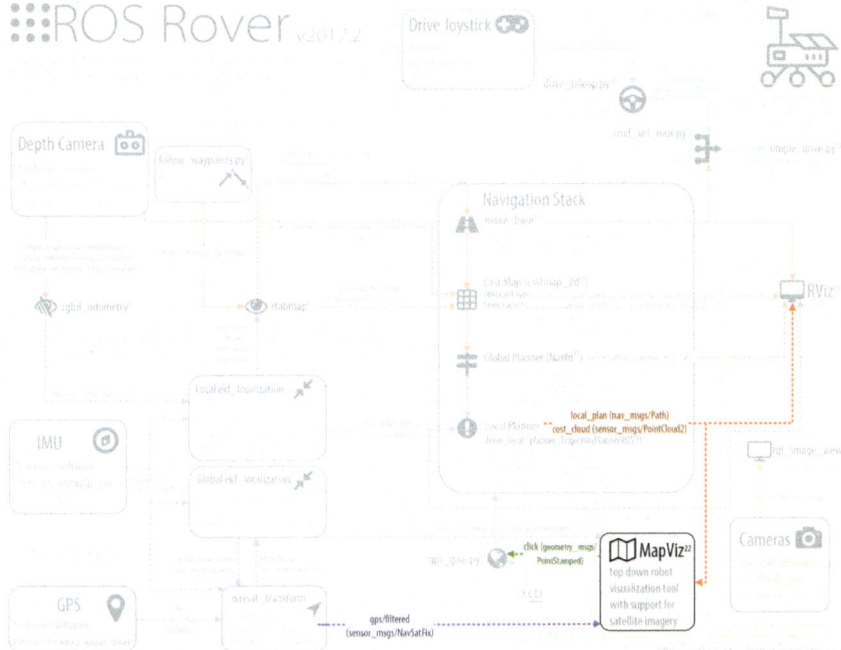

Fig. 27 MapViz ROS node seen within Team R3's rover system. See Fig. 9 for the full diagram

15.1   Usage Instructions

The goal of this tutorial is to install and configure MapViz to display Google Maps.
1. Install mapviz, the plugin extension software, and the plugin for supporting tile maps which is needed to display Google Maps:

```
$ sudo apt-get install ros-kinetic-mapviz ros-
    kinetic-mapviz-plugins ros-kinetic-tile-map
```

2. Launch the MapViz GUI application:

```
$ roslaunch mapviz mapviz.launch
```

3. Use Docker to set up a proxy of the Google Maps API so that it can be cached and received by MapViz in WMTS format. To make this as simple as possible, run the Docker container created by the authors:

```
$ sudo docker run -p 8080:8080 -d -t -v ~/mapproxy:/
    mapproxy danielsnider/mapproxy
```

The -v~/mapproxy:/mapproxy option is a shared volume, a folder that is synced between the Docker container and the host computer. The ~/mapproxy folder needs to be created, though it could be in another location. The -t option allocates a pseudo-tty which gives the program a terminal environment to run in. It is needed for most programs. The -p option sets the Docker port mapping between host and container.

4. Confirm MapProxy is working by browsing to http://127.0.0.1:8080/demo/. The MapProxy logo will be displayed and you can click on "Image-format png" to get an interactive map. Also, test that the first map tile is working by browsing to http://localhost:8080/wmts/gm_layer/gm_grid/0/0/0.png.

5. In the MapViz GUI, click the "Add" button and add a new map_tile display component.

6. In the "Source" dropdown select "Custom WMTS Source...".

7. In the "Base URL:" field enter the following: http://localhost:8080/wmts/gm_layer/gm_grid/{level}/{x}/{y}.png

8. In the "Max Zoom:" field enter 19 and Click "Save...". This will permit MapViz to zoom in on the map 19 times.

Google Maps will now display in MapViz. To set a default location in the world to display at program start up time, you can edit ~/.mapviz_config.

```
$ vim ~/.mapviz_config
# edit the following lines offset_x: 1181506
   offset_y: -992564.2
```

Listing 18 MapViz setting for default viewing location (within a ROS frame) when GUI opens.

16 Tutorial: Effective Robot Administration

In this tutorial, Team R3 (Ryerson University) shares two favorite preferences for making command line administration of ROS robots easier both for the URC competition and any other use.

16.1 *tmux Terminal Multiplexer*

Tmux is a popular linux command line program that can take over one terminal window and organize many terminals into grouped layouts and will continue running when you close the parent window or lose an SSH connection.[64] Multiple people can join a tmux session to share an identical terminal view of a Linux system. Many technology professionals (especially linux and IT professionals) see tmux as essential to their workflow.[65]

Tmux works harmoniously with ROS's modular design. Separate tmux windows can display different ROS components (Fig. 28). Almost any ROS component can

---

[64]Homepage for Tmux the terminal multiplexer https://github.com/tmux/tmux/wiki.

[65]A crash course to learn tmux https://robots.thoughtbot.com/a-tmux-crash-course.

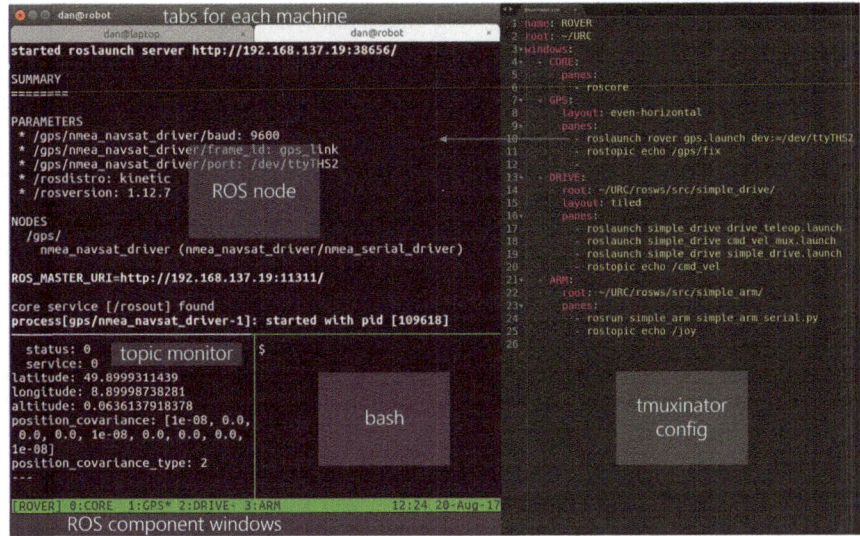

Fig. 28 An annotated example of tmuxinator's usefulness for ROS

be launched, controlled, and debugged using ROS's command line tools. Using tmuxinator, you can codify the launching and debugging commands that you most often use into a repeatable layout.[66]

At URC 2017, Team R3 used tmuxinator to launch all our robot's software systems. There was a section that contained the terminals running the drive software, another section with terminals running the arm software, another for the IMU, for the GPS, for the cameras, etc. All of it in an organized way. Just about every software can be started on the command line using tmuxinator after the rover's computer boots.

The tmuxinator configuration used by Team R3 was split into two sides: the robot config,[67] and the base station config.[68] The robot configuration launches all of the rover's software. The base station configuration launches all of the software needed to visualize and control the robot remotely by the teleoperator.

Using Tmuxinator can be thought of as a quick way to create a very simple user interface to help administer a robot (but it is not a replacement for Rviz and other existing tools). Building a robot GUI as a web interface or desktop application can be useful for some applications, and for novices who are unwilling to learn common command line tools, but such a GUI will require a lot more "plumbing" and "glue code" to create.

---

[66]Tmuxinator is an tool for tmux that lets to write tmux layout configuration files for repeatable layouts https://github.com/tmuxinator/tmuxinator.

[67]The tmuxinator config used by Team R3 to start all the rover software components https://github.com/teamr3/URC/blob/master/.tmuxinator.yml.

[68]The tmuxinator config used by Team R3 to start all the base station software components https://github.com/teamr3/URC/blob/master/devstuff/dan/.tmuxinator.yml.

An implementation imperfection is that tmuxinator starts all of its panes (i.e. terminals) at the same time: running multiple roslaunch instances may try and fail to create multiple masters. The solution we implemented was to run a roscore separately, which has the added benefit of being able to stop and start roslaunches without worrying about which one is running the master. This still has the problem of roslaunches starting before the roscore though, so to solve this naively we have used a small wait time, for example "sleep 3; roslaunch..." in our tmuxinator config.

16.2   ROS Master Helper Script

Team R3 has developed a script to make it a little easier to connect your computer to a remote master that is not on your machine.[69] The script when run will automatically set bash environment variables needed for ROS networking to work in a convenient way. The script will detect if your robot is online using ping (using a static IP for your robot) and set your ROS_MASTER_URI environment variable to point to your robot. If your robot is not online your own computer's IP will be used for your ROS_MASTER_URI, assuming you will do local or simulation development since you are away from your robot. To use this script run `source set_robot_as_ROS_master.sh` or add it to your `~/.bashrc`. Also, the script sets your own machine's ROS_IP environment variable because it is needed in any case for ROS networking.

17   Conclusion

This chapter presented an overview of rover systems through the lenses of the University Rover Competition. Design summaries of 8 URC teams were surveyed and implementation details from 3 URC teams were discussed in a series of tutorials. Several new ROS packages were documented with examples, installation and usage instructions, along with implementation details.

To summarize the main findings in this chapter: Rovers can be built by integrating existing software thanks to the ROS ecosystem. The URC competition is very challenging and students learned a lot by participating. A variety of creative rover designs exist and the best rover teams were the most prepared and practiced.

We hope this chapter spurs greater collaboration between teams. Ideally, the teams of URC will look past the competitive nature of the event and view collaborating and building better robots as the more important goal. Building on a common core frees up time to focus on the hardest parts. Here are a few ways to further collaboration: (1) Contribute to a ROS package or the ROS core. (2) Open an issue, feature request,

---

[69]Team R3's ROS master helper script https://gist.github.com/danielsnider/13aa8c21e4fb12621b7 d8ba59a762e75.

or pull request. (3) Discuss URC on the URC Hub forum. (4) Discuss ROS on their forum. (5) Contribute to a book like this.

Acknowledgements We thank our advisor Professor Michael R. M. Jenkin P. Eng., Professor of Electrical Engineering and Computer Science, York University, NSERC Canadian Field Robotics Network. We also thank and appreciate the contributions of our survey respondents: Khalil Estell from San Jose State University, SJSU Robotics, and Jacob Glueck from Cornell University, Cornell Mars Rover, and Hunter D. Goldstein from Cornell University, Cornell Mars Rover, and Akshit Kumar from Indian Institute of Technology, Madras, Team Anveshak, and Jerry Li from University of Waterloo, UWRT, and Gabe Casciano from Ryerson University, Team R3, and Jonathan Boyson from Missouri University of Science and Technology (Missouri S&T), Mars Rover Design Team.

References

1. G. Bradski, The OpenCV Library. Dr. Dobb's J. Softw. Tools (2000)
2. Camera Module - Raspberry Pi Documentation, https://www.raspberrypi.org/documentation/hardware/camera/. Accessed 23 Dec 2017
3. P. Fankhauser, M. Hutter, A universal grid map library: implementation and use case for rough terrain navigation, in *Robot Operating System (ROS) - The Complete Reference (Volume 1)*, ed. by A. Koubaa (Springer, Berlin, 2016), www.springer.com/de/book/9783319260525. (Chap. 5, ISBN: 978-3-319-26052-5)
4. D. Helmick, M. Bajracharya, M.W. Maimone, Autonomy for mars rovers: past, present, and future. Computer **41**, 44–50 (2008). https://doi.org/10.1109/MC.2008.515. (ISSN: 0018-9162)
5. M. Labbe, F. Michaud, Appearance-based loop closure detection for online large-scale and long-term operation. IEEE Trans. Robot. **29**(3), 734–745 (2013)
6. M. Labbe, F. Michaud, Online global loop closure detection for large-scale multi-session graph-based slam, in *2014 IEEE/RSJ International Conference on Intelligent Robots and Systems (IROS 2014)* (IEEE, 2014), pp. 2661–2666
7. M. Labbé, F. Michaud, Long-term online multi-session graph-based SPLAM with memory management. Auton. Robot. 1–18 (2017)
8. Learn About Me: Curiosity, NASA, https://marsmobile.jpl.nasa.gov/msl/multimedia/interactives/learncuriosity/index-2.html. Accessed 03 Feb 2018
9. Mars Science Laboratory, NASA, https://mars.nasa.gov/msl/. Accessed 28 Dec 2017
10. Mars Science Laboratory - Curiosity, NASA, https://www.nasa.gov/mission_pages/msl/index.html. Accessed 28 Dec 2017
11. S. Montabone, *Beginning Digital Image Processing: Using Free Tools for Photographers* (Apress, 2010). ISBN: 978-1-430-22841-7
12. T. Moore, D. Stouch, A generalized extended Kalman filter implementation for the robot operating system, in *Proceedings of the 13th International Conference on Intelligent Autonomous Systems (IAS-13)* (Springer, Berlin, 2014)
13. MoveIt! Motion Planning Framework, http://moveit.ros.org/. Accessed 24 Dec 2017
14. M. Quigley et al., ROS: an open-source robot operating system, in *ICRA Workshop on Open Source Software* (2009)
15. The Rover's Brains, NASA, https://mars.jpl.nasa.gov/msl/mission/rover/brains/. Accessed 03 Feb 2018

Daniel Snider received a Bachelor of Information Technology (BIT) from the University of Ontario Institute of Technology in Ontario, Canada (2013). He is currently a member of the R3 robotics team at Ryerson University and works as a computer vision software developer at Sick-Kids Research Institute, associated with the University of Toronto. His current research is on the design of modular polyglot frameworks (such as ROS) and scientific workflow systems.

Matthew Mirvish is currently a student at Bloor Collegiate Institute. He is also currently a member of the R3 robotics team at Ryerson University and helps write the code for their rover for URC. He mainly specializes in the autonomous task, but usually ends up helping with anything he can get his hands on. In his spare time he dabbles with microcontrollers as well as creating small computer games.

Michal Barcis received a bachelor's and master's degree in computer science from University of Wroclaw, Poland, in 2015 and 2016, respectively. He is currently pursuing the Ph.D. degree as a researcher with the Alpen-Adria-Universitat Klagenfurt, Austria. Between 2015 and 2017 he was a member of the Continuum Student Research Group at the University of Wroclaw. His research interests include robotics, computer networks and machine learning.

Vatan Aksoy Tezer is currently studying Astronautical Engineering in Istanbul Technical University, Turkey. He was the software sub-team leader of ITU Rover Team during the 2016–2017 semester and worked on autonomous navigation, communication, computer vision and high level control algorithms. He previously worked on underwater robots, rovers and UAVs. He is currently conducting research on navigation in GPS denied environments using ROS.

Part IV
Contributed ROS Packages

SROS1: Using and Developing Secure ROS1 Systems

**Ruffin White, Gianluca Caiazza, Henrik Christensen
and Agostino Cortesi**

Abstract SROS1 is a proposed addition to the ROS1 API and ecosystem to support modern cryptography and security measures. An overview of current progress will be presented, explaining each major advancement, including: over-the-wire cryptography for all data transport, namespaced access control enforcing graph policies/restrictions, and finally process profiles using Linux Security Modules to harden a node's resource access. This chapter not only seeks to raise community awareness of the vulnerabilities in ROS1, but to provide clear instruction along designed patterns of development for using proposed solutions provided by SROS1 to advance the state of security for open source robotics subsystems.

Keywords ROS · Secure communications · Access control · Robotics

1 Introduction

Cybersecurity is quickly becoming a pervasive issue in robotics, especially as robots become more ubiquitous in society. With the advent of industrial automation, autonomous vehicles, robot-assisted surgery, commercial surveillance platforms, home service robots, among many other robotics domains, security of these sub-

R. White (✉) · H. Christensen
Contextual Robotics Institute, University of California, San Diego 9500 Gilman Drive,
La Jolla, CA 92093, USA
e-mail: rwhitema@eng.ucsd.end

H. Christensen
e-mail: hichristensen@eng.ucsd.edu
URL: http://www.jacobsschool.ucsd.edu/contextualrobotics

G. Caiazza · A. Cortesi
Ca' Foscari University, Venice Dorsoduro 3246, 30123 Venezia, Italy
e-mail: 840009@stud.unive.it

A. Cortesi
e-mail: cortesi@unive.it
URL: http://www.unive.it

© Springer International Publishing AG, part of Springer Nature 2019
A. Koubaa (ed.), *Robot Operating System (ROS)*, Studies in Computational
Intelligence 778, https://doi.org/10.1007/978-3-319-91590-6_11

systems should be considered vital, as they all provide vectors for cyber threats to manifest into real-world risks. Even without the hazards associated with industrial-strength robot arms or high-speed driverless semi trucks, personal robots–with their promise to integrate with the Internet Of Things–could become targets for breaches in privacy and sources of identity theft, similar to smartphones and PCs [19].

The Robot Operating System (ROS), a standard robotic middleware now widely-adopted, due in part to its active community, provides a communication layer abstracted above a host operating system to construct a heterogeneous compute cluster for robots [24]. As an open source initiative, ROS was developed to support large-scale software integration efforts and to simplify code reuse among robots with wildly varying hardware.

However, the original open source development of ROS was targeted to those attributes most valued by robotics researchers, including: flexible computational graphs, modular publish and subscribe networks, and rapid software prototyping. Now that ROS and derived robot platforms are out-growing the realm of research and translating to commercial and industrial sectors support for network security, identity authorization, and scoping resource permissions have quickly risen to the forefront of newly requested features.

To address these issues, we propose a set of new security features to the core of ROS's codebase, Secure ROS (SROS).[1] First announced publicly at ROSCon 2016[2] by authors White and Quigley [28], SROS is a proposed addition to the ROS API and ecosystem to support modern cryptography and security measures in an effort to address existing vulnerabilities.

The reaming of this chapter will provide readers further background motivating this work, as well as instruction in using and developing with current features in SROS, including: native Transport Layer Security (TLS) support for all IP/socket level communication within ROS, signed X.509 public key certificates permeating distributed verification of node permissions, as well as convenient user-space tooling to auto generate node key pairs, audit ROS networks, and construct/train access control policies. In addition, AppArmor profile templates are also introduced, allowing users to harden or quarantine ROS based processes running on Linux.

Overview

- **Background**: Details how ROS1 is vulnerable, what attack surfaces exist, as well as additional literature and approaches in mitigating those risks.
- **Architecture**: Outlines the general structure and additions SROS1 introduces to secure the original ROS1 API and transport stack.
- **Using SROS1**: Provides a user orientated introduction in applying encryption, access control, and kernel modules to secure native ROS1 applications.
- **Developing SROS1**: Provides a developer orientated inspection of the underlying implementation as well as design rational for plugin architecture.

---

[1]http://wiki.ros.org/SROS.

[2]https://discourse.ros.org/t/announcing-sros-security-enhancements-for-ros.

- **Discussion and Future Work**: Includes possible improvements and expansive tooling that may carryover across to the next generation for a SROS2 ecosystem. Then finishing with closing remarks on the strength and weakness of SROS1, and summery of development and contributions.

2 Background

In recent years several research efforts have been carried out to address cyber-physical security threats for industrial connected systems and for industrial robotic applications [7]. Specifically, several challenges continue to arise for both ubiquitous consumer connected devices [9, 14] and for industrial robotic applications [5, 13, 15]. Although, these scenarios have been individually studied, we believe that they should be considered as two sides of the same coin. In fact, considering the spread of connected devices and infrastructures, the necessity of security across the whole scope is becoming an urgent issue. For example, we can see how the availability of connected features on the modern automobile can be exploited either from the cyber-physical or the logical point of view. Specifically, Koscher et al. [18] discuss the intrinsic vulnerabilities of modern cars, highlighting the carelessness of some implementation decisions, while Foster et al. [16] present a survey of telematic failures and threats posed both locally and remotely. Among the work on automotive exploitation, Valasek and Miller [6] present how serious these threats can be in a real-world scenario. That said, the possibility of compromising a larger infrastructure through a relatively small number of malicious agents–in this case a Intelligent Transport System (ITS) environment (i.e. smart-city) with cars–appears to be no longer a mere theoretical threat.

A considerable amount of research has already addressed the security of ROS and the publish-subscribe paradigm. In the literature we find two possible ways to tackle the publish-subscribe module: from the *communication channel* (i.e. the exterior transport methods) or from the *communication paradigm* (i.e. the interior application mechanisms).

Akerberg et al. [1] discuss possible limitations and necessary modifications to ensure security during the migration of industrial wired networks to wireless ones. Their framework ensures end-to-end integrity and authentication measures by utilizing the black channel concept. However, in the authors' opinion, basing security solely on the communication channel is not ideal–in part, because the immediate susceptibility of such an over reliant system surfaces whenever the security of the communication channel breaks down, as demonstrated repeatedly for wireless security protocols such as WEP in 2002 [25] and more recently in 2017 for WPA [26]. Alternatively, by leveraging on the publish-subscribe paradigm we are able to build more general security solutions. However, as shown by Wang et al. [27] the idea of working solely on the publish-subscribe system is not trivial; as they note, given the nature of publish-subscribe systems, keeping content confidential while routing information, or managing subscriptions without revealing subscription functions is extremely challenging.

An interesting overview of ROS's vulnerabilities, is set out by McClean et al. [4]. Vulnerabilities that arose by challenging visitors to DEFCON to hack into their 'honeypot' robot are detailed. Although their catalog fails to completely cover all the attack surfaces, their overview surely presents a good opportunity to begin an evaluation of the weak points of ROS across large number of vulnerabilities. As noted by the authors, the idea of open testing the robots helped us more effectively discover to what degree we need to secure both ROS and, more importantly, the host operating system.

Additionally, other researchers have focused their work on enforcing security measurements in ROS. Intuitively, the easiest way to ensure secrecy of a network is by using encryption; concerning this, Lera et al. [19] present an analysis of the use of ROS under encryption. In contrast to other discussions, in their paper they directly evaluate the performance of ROS from both the computing and the communications points of view. Specifically, they developed a testbed robot and tested its performance by encrypting small fixed messages, arbitrary messages, and video footage with 3DES. Still, the use of 3DES as a benchmark degrades CPU performance. One interesting observation that emerges from their paper is that all in all ROS performs easily under encryption.

An interesting project addressing safety and security is ROSRV [17]. The concept behind ROSRV is to place new middleware components, namely 'RVMaster' and 'monitor', between the original ROSMaster and the other nodes. The new entities act as a run-time monitor to filter out and log all the requests and operations sent in the graph. The novelty of this solution is that we need not modify ROS to enforce the new security enforcement; it's sufficient to define a specific access control policy profile for any given robot. Still, since this solution works independently of ROS, it fails to leverage any built-in features of ROS functionality; and this solution requires reliance solely on IP addresses. This strong limitation makes the solution vulnerable to exploits that are running on the same machine and make it extremely challenging to port the solution across different networks. Furthermore, because it's centralized–all the nodes live in the same multi-thread process–scalability of this solution is really not feasible.

Still, solutions mentioned so far fall short of addressing either Authorization or Authentication, both of which are critical. Doczi et al. [11], present a solution that tries to implement protection for both and to embed itself in ROS as one of the nodes. The idea is to deploy a new 'AA_node' in the graph that allows the other nodes to authenticate themselves via pre-defined *username* and *password*. All in all, their idea is to build an Application Level Gateway that authenticates a given node and verifies whether the data packages are transmittable or not. Specifically, once a node is authenticated, it will expose a token to the special wrapped ROS API in order to perform the requested operations. With this solution, the communication is always evaluated as to whether it comes from a trusted source or not. Still, the use of an extra node to regulate communication in the graph and a Database to store login information introduces a strong single point of failure (SPOF) with further privacy concerns.

Similar to our proposed solution, Dieber et al. [4, 10] propose a security architecture intended for use on top of ROS at the application level. By means of an external authorization server and the use of X.509 certificates for node authentication, they ensure that only valid registered nodes are authorized to execute some operations. Furthermore, by using cryptographic methods they ensure data confidentiality and integrity.

Specifically, each node has a X.509 certificate with a unique serial number. Leveraging these certificates, the nodes authenticate themselves and are verified if the requesting node is authorized to execute a given operation. However, this static architecture is based on the assumption that the user manually generates and distributes the certificates and registers a list of nodes in the authorization server; moreover, this solution delegates to the user the distribution of lists of certificate serial numbers of the nodes that are authorized to query each other. Our approach presents additional user orientated tools to help automate the security provisioning process, thereby reducing the chances of human error or overlooking exposed attack surfaces.

Finally, for a more complete comparative analysis of research mentioned here, including SROS1, the authors would like to point readers to a survey by Portugal et al. [23] that provides a valuable review of the pros and cons of each approach in securing robotic applications using ROS. This survey examines five separate initiatives for securing ROS, and provides quantitative performance and security comparisons. While this survey concludes that SROS was undoubtedly the most secure initiative tested, it also shows that SROS still has room for improvement in regards to performances, due its present lack of a C++ implementation.

One might argue that Virtual Private Networks (VPN) or Software-Defined Networks (SDN) could be used to mitigate ROS's existing vulnerabilities. However, resorting to such external infrastructure comes with considerable trade-offs. One advantage of VPNs and SDNs may be superior tunneling across distant networks. Since ROS1 natively allocates open ports during run-time, constructing a quarantined Demilitarized Zone (DMZ) via an ephemeral port firewall can be tedious and endless. Additionally, professional grade routers and switches can provide industry support in offloading any cryptographic overhead from dedicated robotic hardware. These existing infrastructures are quite mature in specification and implementation, and offer the additional benefits derived from thorough documentation and experienced communities.

However relying on such external security measures may continue to introduce additional vulnerabilities. These include session hijacking, since the ROS1 protocol and API remain exposed to malicious actors who may reside within or later infiltrate the private network layer. System complexity is also increased, requiring supplementary network configuration and maintenance, while at the same time creating an additional SPOF. Also, one final drawback is the lack of fine grain control in defining interconnection access in a data driven context centered around ROS1 related resources and namespaces. Again, using such black channel approaches are often all or nothing, providing manufacturers little flexibility in exposing only partial elements of the robotic software stack to the public, while simultaneously safeguarding critical surfaces of the same system from malicious actors. Given that these external security layers alone are incapable of providing the necessary application layer security, the ROS1 software stack requires direct enhancement to address this issue.

3 Architecture

In ROS1, nodes intercommunicate through a standard API via XML-RPC, a remote procedure call protocol using XML encoding, as well as message/service data exchanges using transport libraries such as ROSTCP or ROSUDP for serialization over IP sockets, as visualized in Fig. 1. A glaring deficiency in the traditional infrastructure is the fact that all network traffic is transmitted in clear text. Additionally, no integrity checking is preformed on received packets other than basic message type continuity and API syntax validity, i.e. there are no means to verify payload information was unaltered in transit, nor is identity a consideration, as the standard API is inherently anonymous. This makes ROS1 a prime target for packet sniffing and man-in-the-middle attacks, resulting in an absence of native network confidentiality and data integrity.

In SROS1, all network communication is encrypted using Secure Sockets Layer (SSL), or more specifically Transport Layer Security (TLS). This is done through the use of a Public Key Infrastructure (PKI), whereby each ROS1 node is provided an X.509 certificate, equivalently an asymmetric key pair, signed by a trusted certificate

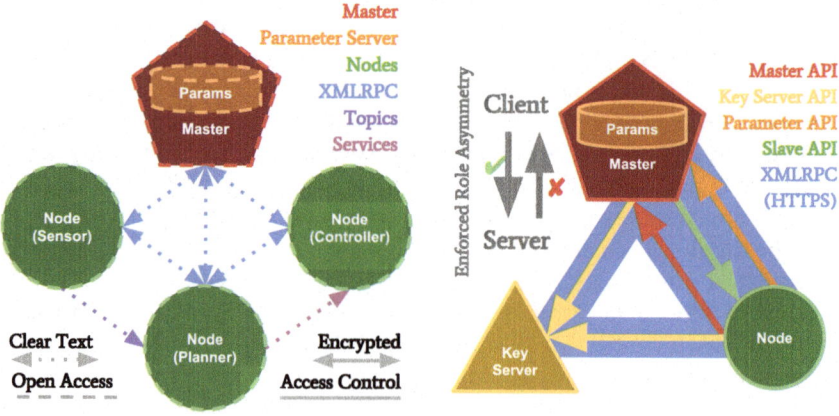

(a) Traditional ROS1 Graph Topology   (b) Secured ROS1 API via SROS

Fig. 1 Left, a high level visualization of a typical distributed computation graph in ROS1. Node communication is peer-to-peer over clear text. Master service provides pub/sub namespace gateway and hosts global parameter server. Right, both XMLRPC and message transport APIs are encrypted via SROS1, with strictly enforced asymmetric API access for specific roles, i.e. participants granted client access to a given API can only connect to peer provisioned as API server, and vice versa. Optional keyserver for PKI distribution is also shown

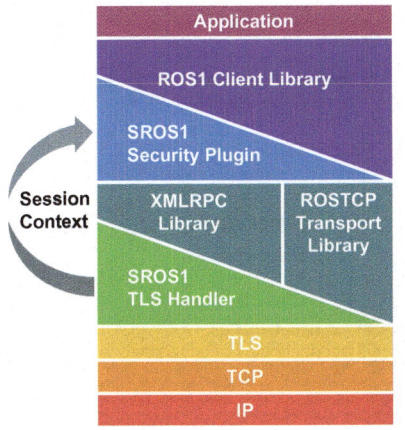
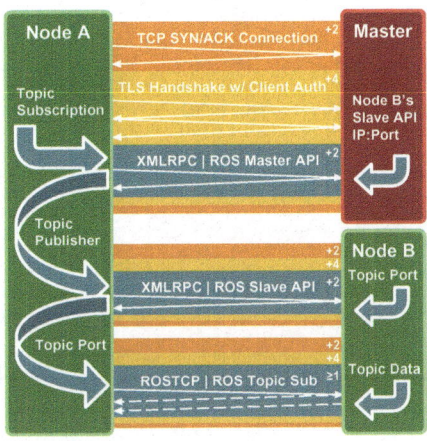

(a) SROS library shims for ROS1 stack (b) Topic subscription ladder diagram

Fig. 2 Left, a hierarchical representation of the network/software stack with ROS1. SROS1 shims remain minimally invasive within the native client library, thus maintaining the existing ROS1 API. As the SROS1 TLS handler establishes and preserves the session's context, i.e. adjoining participant's certificate, through the XMLRPC and ROS transport callbacks, API calls are evaluated by the security plugin before and after execution. Right, is a network event ladder diagram of a topic subscription; first negotiating with the master, then twice again with the publisher, all conducted within three sequential TLS sessions

authority (CA). These additions are shimmed between ROS1 libraries, enabling users with preexisting code and projects that leverage ROS1 abstraction layers to also seamlessly support SROS1. These alterations to ROS1 stack and wire protocol are visualized in Fig. 2. SROS1 essentially inserts itself between any socket connect or accept event, as well as before and after any API's request is sent or received, enabling a controlled and de-anonymized API. Further details on the stock ROS1 API are found in the wiki documentation.[3]

The additional security does come with its own complexities, and in the world of cybersecurity, the mantra remains that "security without usability is essentially insecure," whenever there are users who avoid adopting it. So, we have also developed tooling to support and simplify the tasks of using PKI in ROS. SROS1 provides a keyserver for certificate generation, as well as a low-level API to distribute ciphered certificates to nodes during setup, with the excerpted config file of which shown in Fig. 3. By customizing the keyserver config, users can control almost every aspect in the automation of PKI elements.

The SROS1 keyserver remains separate from the core ROS1 API, eliminating the need for sensitive private Certificate Authority (CA) keys from ever existing on the target/robot, and leaving signed certificates immutable. Additionally, the keyserver provides a customizable configuration, where users can curtail certificate and CA properties on a node by node or namespace basis, e.g. key algorithm, bit length, fingerprint, as well as CA info, extensions, restrictions, hierarchy, etc.

---

[3]http://wiki.ros.org/ROS/Technical%20Overview.

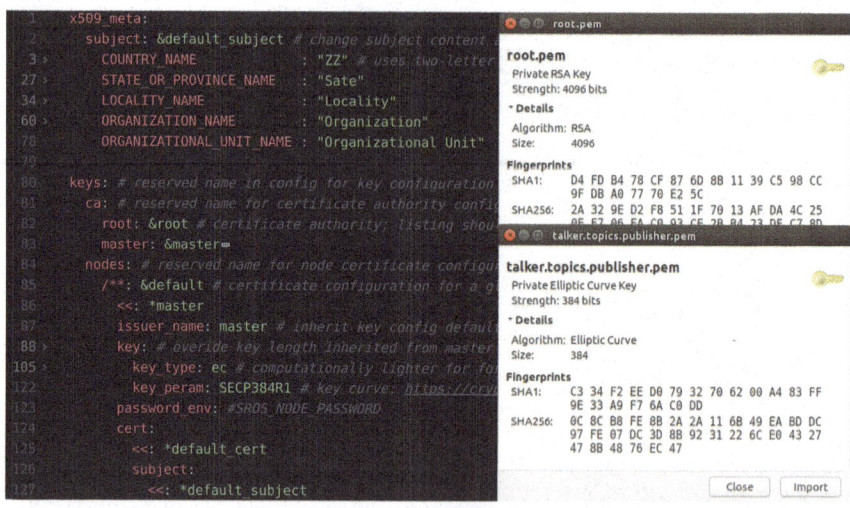

Fig. 3 Right, an example key-store configuration file used to describe and automate PKI setup. SROS1 provides a default config sufficient for basic use, but enables advanced users to customize every applicable certificate attribute: cryptographic and signature algorithm, signing CA hierarchy, X.509 extensions, etc. The annotated config is too large to show here in its entirety, but is provided in the user's .ros/sros directory. Right, an example of how one may customize key length/type between strong signatures for CAs versus lightweight payload for end use certificates

The keyserver is not a node itself, nor is it a service tied to any particular ROS1 node, unlike the parameter server and rosmaster. This allows for the keyserver to run separately from ROS1, be it from a different user process or a different machine entirely. In fact, the only time the keyserver should be up is during initial setup. Thereafter nodes should have their own keys locally in the respective key-store. In truth, using a keyserver is completely optional should advanced users have their own methods of generating and distributing PKI elements. However beginners may appreciate the features and integration that the SROS1 keyserver provides; it may be fully or partially utilized as users see fit.

As mentioned previously, SROS1 utilizes TLS, and TLS utilizes PKI, ergo SROS1 needs a way of generating and distributing PKI elements, including: asymmetric keys, CAs, and signed certificates with adequate permissions for embedded access control. In practice, this is not a simple task, and tackling it manually is unacceptably tedious. Generating and signing X.509 certificates correctly using tools like OpenSSL can be a challenge and a barrier for basic users, while simultaneously bookkeeping the policy permissions can be an even more formidable challenge. With this in mind, SROS1 was designed with a keyserver to better simplify the use and development of SROS1 enabled systems for end users. The keyserver in SROS1 provides integration with the rest of the SROS1 work flow, as well as conservative default configurations to smooth the PKI learning curve, while preventing users from shooting themselves in the foot.

4   Using SROS1

From a user perspective, we now proceed through two examples of securing a native ROS1 application from scratch using SROS. Before continuing with either, it is important to grasp first the rudimentary process or development cycle used to reach a steady state of policy that enforces a secure ROS graph, while leaving it functional for the target application. As shown in Fig. 4, the cycle is roughly divided into three consecutive stages: Amend, Test and Audit. It should be noted that constructing a policy for either the SROS1 graph or process follows this same cycle, but for brevity we now discuss this in terms of the SROS1 graph access policy.

During the Amend stage, we'll update a policy that will be used in re-generating the signed cryptographic artifacts and documents necessary for establishing secure and authorized ROS API exchanges. The policy basically encapsulates a concise definition of permissible API actions available for any ROS node. Next, we test the policy by running the ROS application stack as intended, i.e. as once released into production, while monitoring all interactions among the ROS nodes from an API perspective. Finally, we audit the observed exchanges during testing and determine the minimal set of accessed permissions required for expected operation. Note that the default SROS1 security plugin is governed by a Mandatory Access Control (MAC) approach, i.e. only API actions explicitly granted to participants by the security policy will be allowed during run time.

Within SROS1, the tooling provided–such as the auto-policy training and the keyserver–will help expedite this cycle of development, perhaps blurring the boundaries between each stage in the cycle thanks in part to ROS1's centralized gateway, i.e. logging from the rosmaster node. However, the authors feel it is important to distin-

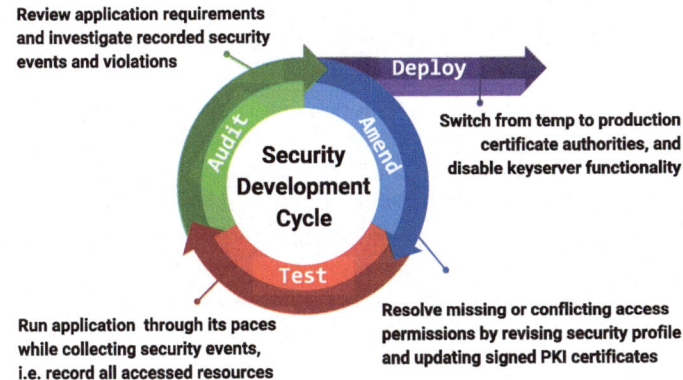

Fig. 4 The security development cycle as afforded within SROS. Users securing ROS based applications should iteratively audit and test security profiles as constructed to avoid inadvertently hindering intended deployment or provisioning greater access permissions than minimally required which may leave APIs vulnerable

guish each stage, as such differentiation will be advantageous for users transitioning to SROS2, since similar security infrastructures within a completely decentralized network may necessitate a more staggered development cycle.

In the first example, we explore how the keyserver interacts with the rest of SROS1 through a common bootstrapping example. When setting up an existing ROS1 application to use SROS1, we here assume the user has no preexisting PKI, so when nodes start under SROS1, they will have no PKI elements to draw from to connect to rosmaster, let alone to each other. In this case we would spawn a SROS1 keyserver beforehand, and express this onto bootstrapping ROS1 nodes by setting the environment variables.

- Upon start-up of the keyserver, it loads a designated configuration file, then checks the designated key-store for CAs and its own nodes-store. In this case the keyserver would find neither, but has been configured to make them if needed. And so the keyserver will first generate the required CAs, then generate and sign its own node-store using those CAs so that it may finally accept connections.
- Nodes that start under SROS1 will first check their local designated key-stores for their respective node-stores. If this does not exist, and the node has been enabled to contact the designated keyserver, it will attempt to connect and request for one using TLS's client mode, given the node has no PKI element to work with. It should be noted here that a variety of security options are available at this point of the exchange, e.g., a node may also be mandated that the accepting key-server's certificate be signed by an existing trusted local CA (server authentication), and/or the keyserver may be mandated to do the same for a connecting node that uses a intermediate client cert (client authentication).
- Once the keyserver and node are securely connected via TLS, the node may, if designated to do so at run-time, send an XML-RPC request for public certificates for its own key-store's trusted CA-path. Next the node will request personal PKI elements, including its public certificate and private key, to initialize its own node-store. This request is done by providing the node's own namespace that it has been mapped to by ROS.
- The keyserver then uses this given namespace and searches for the applicable key recipe specified by the user's keyserver configuration. This helps the keyserver decide on how or if it should respond, what algorithms should be for the keys, what policies should be embedded into the certificate, what CA should issue the signature, if or what cipher should be used to lock the private key, and much more.
- Once the node receives the PKI elements, it will save them to its own node-store, then attempt to load it; this may involve deciphering the private key using the local secret provided at the node's run-time and shared only with the keystore.
- Thereafter, nodes will attempt to connect to rosmaster, a node itself that goes through the same process abive, and then later nodes via TLS handshakes and scrutinizing peer certificates.

4.1 Securing ROS Graphs

As of this writing, SROS1 has not yet been mainlined into the official ROS release. However for ease of experimentation and development, SROS1 source code is made public, as well as are pre-built binaries available through distributed Linux Container images. More installation instructions can be found on the SROS wiki. For the remainder of the tutorial, we'll follow through using Docker to maintain the tutorial's reproducibility. To double check all is well before beginning, you may wish to test your setup with the following single command:

```
$ docker run --rm -it osrf/sros bash -c "\
    sroskeyserver & sleep 3 && \
    sroslaunch rospy_tutorials talker_listener.launch"
# or static image:tag for posterity: ruffsl/sros1:ros_book_v3
...
[INFO] [WallTime: 1515022662.068996] hello world 1515022662.07
[INFO] [WallTime: 1515022662.069891] /listener I heard
↪   hello world 1515022662.07
[INFO] [WallTime: 1515022662.168968] hello world 1515022662.17
```

The docker command above simply downloads and launches a SROS enabled container, then executes the bash sequence for starting the SROS keyserver and as well as the classic talker listener ROS example. The sleep command is used to give the keyserver a chance to bootstrap its keystore before servicing requests from the example node pair. From the command output you should observe the keyserver's initialization of the keystore, including the certificate generation for the master, roslaunch, rosout, and talker and listener nodes, debug output from the default security plugin for each API exchange, and lastly the message printouts from the talker listener example, securely messaging over TLS.

Now that we've confirmed the setup is operational, let us recreate this secure setup from scratch, step by step rather than reusing the default configuration shipped with the example. A terminal session recording of this example can also be viewed here[4] or with step by step commentary also presented in the accompanying ROSCon 2016 talk.[5]

We'll start by terminating our previous session and launching a new container, then removing the existing ROS configuration directory and starting the keyserver again.

---

[4]https://asciinema.org/a/88519.
[5]https://vimeo.com/187705073#t=494s.

```
$ docker run -it --net=host osrf/sros byobu
$ rm -rf /root/.ros
$ sroskeyserver
Starting an XML-RPC server to bootstrap SSL key distribution...
Certificate generated: root
Certificate generated: master
sleeping until keyserver has generated the initial keyring...
```

The first command above drops the shell into a byobu session. Byobu is simply a friendly variant of tmux or screens, allowing us to manage multiple bash sessions: with the <F2> key creating a new shell sessions, and <F3 / F4> keys cycling between them. The -it docker argument is used again to make the session interactive, while the −net=host argument is used to bind the container to host machine's network interfaces, allowing the reader to easily eavesdrop on the SROS1 network packets with monitoring tools, e.g. Wireshark as in Fig. 5.

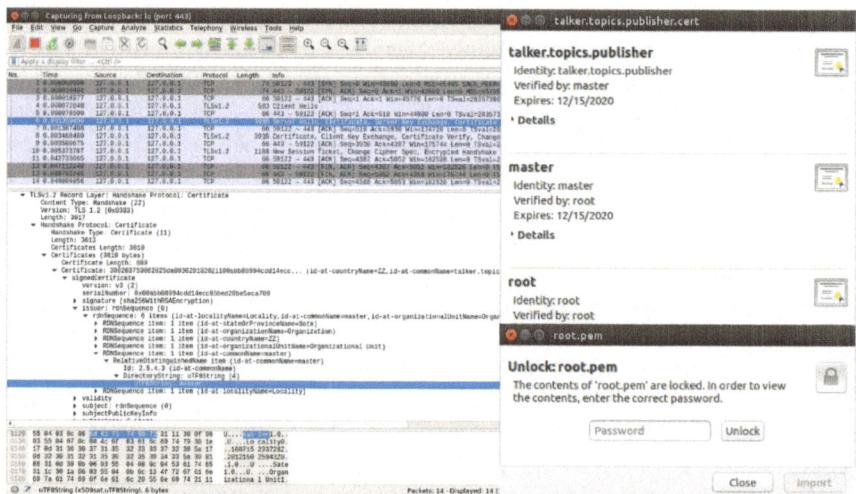

Fig. 5 Left, an example TLS handshake in SROS1 shown via Wireshark network packet capture. Both client and server side authentication is supported, with demonstrated PKI on the right used to form chains of trust, verifying the validity and authenticity via authorized certificate authorities. Ciphered private keys are natively supported but can be decrypted through run time environment secrets

Once the keystore is up and running, we can take a look at the SROS configuration directory that also stores the default keysore location. Within the keystore are CAs, the TLS capath for quick CA look-up needed for OpenSSL, as well as the credentials for keystore itself.

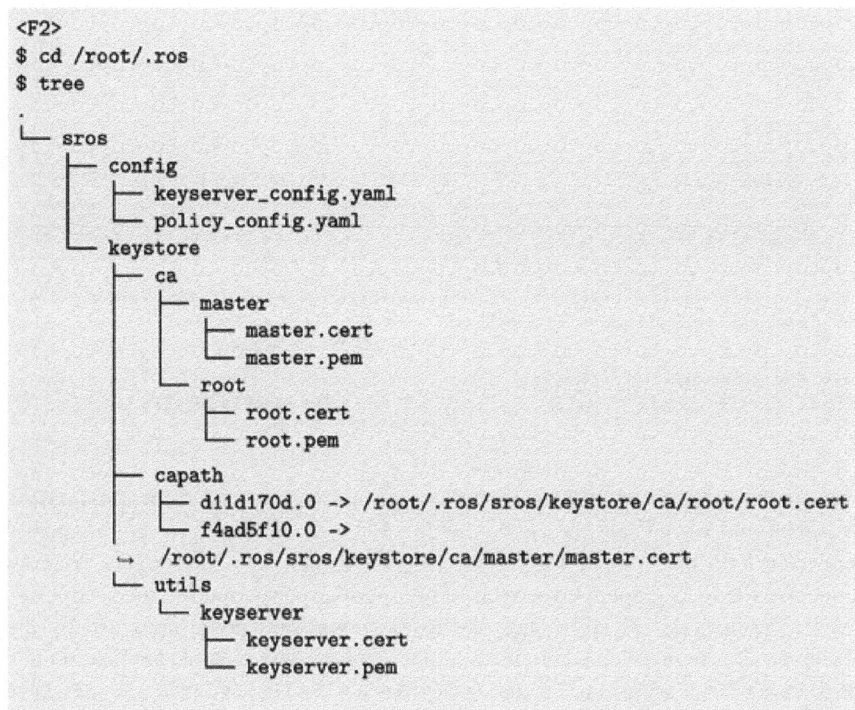

```
<F2>
$ cd /root/.ros
$ tree
.
└── sros
    ├── config
    │   ├── keyserver_config.yaml
    │   └── policy_config.yaml
    └── keystore
        ├── ca
        │   ├── master
        │   │   ├── master.cert
        │   │   └── master.pem
        │   └── root
        │       ├── root.cert
        │       └── root.pem
        ├── capath
        │   ├── d11d170d.0 -> /root/.ros/sros/keystore/ca/root/root.cert
        │   └── f4ad5f10.0 ->
        ↪   /root/.ros/sros/keystore/ca/master/master.cert
        └── utils
            └── keyserver
                ├── keyserver.cert
                └── keyserver.pem
```

Now we can start `sroscore` with the `policy_mode` set to `train` in order to learn the application's ROS graph typology from demonstration. Additional `policy_modes` are available within default SROS1 security plugin, such as `audit` or `enforce` (implicit default) that provide varying degrees of security for different run-time circumstances. For example, `train` mode is in fact a forgiving variant of `audit` mode, in that security event violations are reported but not raised or prevented, allowing partial policies to learn without prematurely impeding the application being trained from. However, when deployed, setting SROS1 to `enforce` mode rather than `audit` would be prudent in order to reduce logging overhead to only violation events and to strictly abide by the policies encountered.

```
$ sroscore --policy_mode train
initializing node's keystore: /root/.ros/sros/keystore/nodes/roslaunch
all startup certificates are present
...
started roslaunch server http://29e3a2f6f1b6:44307/
ros_comm version 1.12.0
SUMMARY
========
PARAMETERS
* /rosdistro: kinetic
* /rosversion: 1.12.0
NODES
auto-starting new master
process[master]: started with pid [726]
initializing node's keystore: /root/.ros/sros/keystore/nodes/master
all startup certificates are present
...
started core service [/rosoutpy]
initializing node's keystore: /root/.ros/sros/keystore/nodes/rosoutpy
```

From the abbreviated ROS1 printout for starting sroscore, we see that rosmaster, roslaunch and rosout, all core nodes within ROS ecosystem, have automatically connected with the keyserver and have initialized their own keystore to securely connect to the ROS graph. However most permissions provisioned to the certificates given to each node are still blank due to the initial null policy assumed by the keyserver. This is revealed when first attempting to run our example application in a third session, as it seems the listener node does not connect and print the messages broadcast by the talker node.

```
$ sroslaunch rospy_tutorials talker_listener.launch
...
NODES
  /
    listener (rospy_tutorials/listener.py)
    talker (rospy_tutorials/talker.py)
ROS_MASTER_URI=https://localhost:11311
core service [/rosoutpy] found
process[listener-1]: started with pid [923]
process[talker-2]: started with pid [924]
initializing node's keystore: /root/.ros/sros/keystore/nodes/listener
all startup certificates are present
initializing node's keystore: /root/.ros/sros/keystore/nodes/talker
all startup certificates are present
[INFO] [WallTime: 1475961387.938241] hello world 1475961387.94
```

This is due to SROS's secure default `policy_mode` behaviour, and because both nodes are independent of sroscore as it was set to train mode prior, thus we need to inform the default security plugin within each node that no access control should be enforced during training. This can be done by passing an environment variable appropriately:

```
SROS_POLICY=NONE sroslaunch rospy_tutorials talker_listener.launch
...
[INFO] [WallTime: 1514938858.938553] hello world 1514938858.94
[INFO] [WallTime: 1514938858.940335] /listenerI heard hello world
↪  1514938858.94
[INFO] [WallTime: 1514938859.038408] hello world 1514938859.04
```

Now that the example application has been sufficiently demonstrated to the training framework, i.e. having the application preform all API exchanges as intended when once deployed, we terminate all running process, including the application, sroscore, and the keyserver, clear out the previous iteration of keystore for the nodes involved. The resulting policy profile can be found within the `sros` configuration folder, and is shown in full in Fig. 6.

To reload the amended policy and redistribute updated credentials, we bring the keyserver back online as well all the relevant nodes involved. This time, any alterations from the default `policy_mode` can be omitted, thereby testing the amended policy in earnest with access control strictly enforced.

```
$ less sros/config/policy_config.yaml
$ sed -i '/\/roslaunch\/uris\/host_/c\        \/roslaunch*:'
↪  sros/config/policy_config.yaml
```

While still auditing, we could take the opportunity to make sure learned permissions generalize across anonymous instances of certain core nodes, as with roslaunch.

```
# In 3 separate sessions
$ sroskeyserver
$ sroscore
$ sroslaunch rospy_tutorials talker_listener
...
[INFO] [WallTime: 1514938859.038408] hello world 1514938859.04
```

Now that keystore has been repopulated, with the policy satisfactorily and stable, we can remove keyserver from the network and deploy the credentials.

```
 1  nodes:                          26        /rosversion:
 2    /listener:                    27          allow: w
 3      parameters:                 28        /run_id:
 4        /enable_statistics:       29          allow: rw
 5          allow: r                30    /rosoutpy:
 6        /tcp_keepalive:           31      parameters:
 7          allow: r                32        /enable_statistics:
 8        /use_sim_time:            33          allow: r
 9          allow: r                34        /tcp_keepalive:
10      services:                   35          allow: r
11        /listener/get_loggers:    36        /use_sim_time:
12          allow: x                37          allow: r
13        /listener/set_logger_level: 38    services:
14          allow: x                39        /rosoutpy/get_loggers:
15      topics:                     40          allow: x
16        /chatter:                 41        /rosoutpy/set_logger_level:
17          allow: s                42          allow: x
18        /rosout:                  43      topics:
19          allow: p                44        /rosout:
20    /roslaunch:                   45          allow: ps
21      parameters:                 46        /rosout_agg:
22        /rosdistro:               47          allow: p
23          allow: w                48    /talker:
24        /roslaunch/*:             49      parameters:
25          allow: w                50        /use_sim_time:
                                    51          allow: r
```

Fig. 6 A complete policy trained from the classic talker/listener example. Note the number of permission involved in even simple ROS graphs comprised of two node messaging over one /chatter topic. Many API permissions learned may be non-obvious to end users, but remain necessary for full ROS functionality. Note the single change of regular expression on line 24 added after manual audit allowing anonymous named roslaunch nodes to reuse the same permissions

```
$ tree ~/.ros/sros/nodes
nodes
├── listener
│   ├── listener.cert
│   └── listener.pem
├── master
│   ├── master.cert
│   └── master.pem
├── roslaunch
│   ├── roslaunch.cert
│   └── roslaunch.pem
├── rosoutpy
│   ├── rosoutpy.cert
│   └── rosoutpy.pem
└── talker
    ├── talker.cert
    └── talker.pem
```

For more complex graph typologies, users may expect to iterate further, gradually building a entire application policy through editing and testing sub-policies from smaller components, while stepping back periodically to audit and optimize any auto generated provisions.

4.2 Securing ROS Processes

For networked systems, application security is arguably the single-most important area of concern [3], and as ROS sits squarely in this domain serving as a networking middleware for robotic applications, all facets of its functions should be secured. However ROS packages come from varying sources; vetting each dependency in this high-level stack for unknown vulnerabilities can be impractical for many users. Enabling MAC for ROS node processes can help mitigate zero-day-exploits enabling malicious agents to gain shell access or invoke local privilege escalation, as well as quarantine unforeseen failures during run-time, such as accidentally overwriting another node's data archive.

AppArmor is one such implementation of MAC building from the security modules available in the Linux kernel, and is widely adopted in Ubuntu, ROS's primary release target. AppArmor is also well documented and user-friendly relative to alternatives such as SELinux, making it a suitable MAC of choice for ROS. SROS provides a profile library for AppArmor, composed of modular primitives to quickly build custom profiles for ROS nodes. These include the minimal permissions necessary for core ROS features, such as the inter-process signalling needed for roslaunch to manage other nodes, shared library access for nodes written in python or C++, network access for socket communication, etc. Using the SROS profile library, users can focus on defining MAC pertinent to the application, e.g. prescribing nodes' access to a serial bus, camera peripherals, or e-stop interfaces.

To demonstrate the application and deployment of the SROS profile library, we proceed with an example using the profile enclosed in Fig. 7. In addition to an AppArmor and ROS installation on our preferred Linux host, we also need to install the SROS profile library within AppArmor's configuration directory, as detailed here in the installation wiki.[6]

Once installed, we go ahead and create a new file within the AppArmor profile configuration directory (location of which may vary by Linux distribution) and paste in the example profile from Fig. 7. More examples are also provided at the SROS profile library repo.[7]

---

[6]http://wiki.ros.org/SROS/Tutorials/InstallingAppArmorProfilesForROS.

[7]https://github.com/ros-infrastructure/apparmor_profiles.

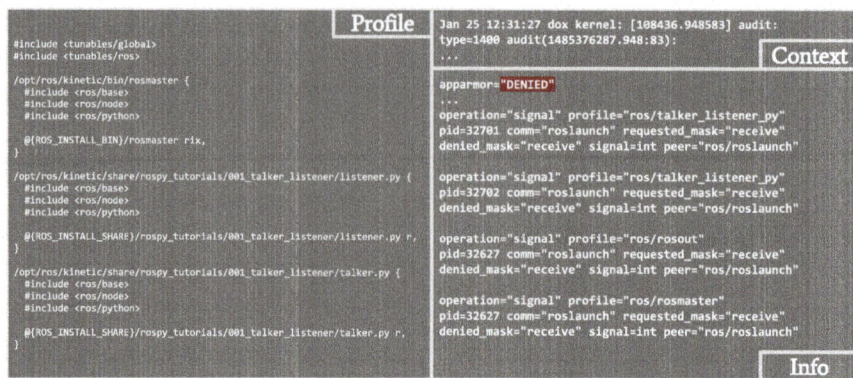

Fig. 7 Left, an example AppArmor file suited for minimal ROS1 talker and listener demo. SROS AppArmor profile library enables users to write custom profiles, providing default *includes* and *turntables* for ROS directories and run-times. Here only rosmaster and tutorial nodes are provisioned. But not roslaunch, preventing an interrupt signal to secured processes, as shown in syslog on right

```
$ sudo nano /etc/apparmor.d/rosmaster_talker_listener
$ tree /etc/apparmor.d/ros
/etc/apparmor.d/ros
├── base
├── node
├── nodes
│   └── roslaunch
└── python
```

Editing finished, we might like to investigate the definition of the profile just added. The example here starts by including two global tunables that mainly define common variables used by the profile library to specify ROS libraries and configuration directories. If we wish change the default location of such directories, or if we are using a differently installed release of ROS, those relatively imported files within the tunable directory are what we need to customize before continuing.

Next in the profile are three separate sub-profiles that each relate to the executable they govern, i.e. the rosmaster, talker, and listener nodes. More specifically the syntax follows the structure where an attachment specification (as a regular expression string) is used to bind the sub-profile to executables of matching paths. Within the scope of the profile, further imports, rules and additionally nested profiles may be defined. As the AppArmor profile language is extensive and evolving; the abbreviated AppArmor language documentation provides a quick introduction.[8]

In this example, each node is provisioned minimal access to necessary network and process signalling capabilities through the base import, as well as access to common

_____
[8]https://gitlab.com/apparmor/apparmor/wikis/QuickProfileLanguage.

ROS shared libraries via the `node` import. Special permissions for the `python` kernel are also provisioned to python executables. Lastly, read and/or execution access are provisioned to the relevant executable files themselves.

Now that the profile has been added, we need to ensure it is loaded into the security kernel module by either restarting the AppArmor system service or explicitly parsing the designated profile.

```
$ sudo service apparmor restart
# OR
$ sudo apparmor_parser -r /etc/apparmor.d/rosmaster_talker_listener
```

As the profile is now enforced by the kernel's security layer, we test this by running each process in the ROS talker and listener example separately.

```
# In 3 separate sessions
$ rosmaster
$ /opt/ros/kinetic/share/rospy_tutorials/001_talker_listener/listener.py
$ /opt/ros/kinetic/share/rospy_tutorials/001_talker_listener/talker.py
```

Note the use of `rosmaster`, as opposed to the more modern form of `roscore`. If we wish to additionally secure the `roscore` command, and its implicit `roslaunch` command counterpart, we must amend our profile to specifically include them. Without explicitly provisioning permissions in the MAC framework, ROS tools perhaps otherwise over constrained may no longer function as intended.

```
profile ros/roslaunch @{ROS_INSTALL_BIN}/roslaunch {
  #include <ros/nodes/roslaunch>
  @{ROS_INSTALL_BIN}/roslaunch rix,
}
profile ros/roscore @{ROS_INSTALL_BIN}/roscore {
  #include <ros/nodes/roslaunch>
  @{HOME}/.rnd r,
  @{ROS_INSTALL_BIN}/roslaunch rix,
  @{ROS_INSTALL_BIN}/roscore rix,
}
```

The example above now works as expected, with each ROS node given enough permission to join and communicate over the ROS graph. If we now alter the talker to preform a potentially malicious task, such as engaging sections of the host's filesystem outside its intended scope, as demonstrated within the talker node:

```
if __name__ == '__main__':
+    with open('/var/crash/evil.sh', 'w') as f:
+        f.write('echo evil laughter!\n'
+                'rm -rf /var/crash/* /\n')
    try:
        talker()
    except rospy.ROSInterruptException:
        pass
```

We can immediately restart the talker process and observe the python stack trace produced. This is due in part to the illegal operation intercepted by the kernel due to the insufficient policy permissions:

```
$ /opt/ros/kinetic/share/rospy_tutoris/001_talker_listener/talker.py
Traceback (most recent call last):
  File "/opt/ros/kinetic/share/rospy_tutorials/.../talker.py", line 53,
  ↪ in <module>
    with open('/var/crash/evil.sh', 'w') as f:
IOError: [Errno 13] Permission denied: '/var/crash/evil.sh'
```

This demonstrates how an enforced profile prevents a malicious node from touching log archives still accessible to the process owner, but denied to those outside the process's provisioned scope. Readers should note how control and the flexibility of using such security modules far exceeds the capabilities afforded by traditional Unix user level permissions. A prime example used within SROS1 itself would be to protect private PKI credentials for each node on disk, such that only the respective node process could read its own private key. A recorded terminal session of the example can be viewed here.[9]

5 Developing SROS1

As previously shown in the architecture section and Fig. 2, the main additions to fortifying the ROS1 framework pertain to the SROS1 software shims that are introduced at two core levels in the ROS1 client library stack–the first being at the socket layer before reaching the XMLRPC or data transport layers, and the second before reaching the API callbacks in the client library itself. This approach enables SROS1 to establish the TLS socket and preserve the accompanied session context to then be governed by a security plugin for additional access control, thus simultaneously supporting authentication, authorization, and encryption.

A primary motivation for this work was to provide security for current industrial robot systems. Consequently, the approach chosen in designing SROS1 was a plugin framework, anticipating that corporations may inevitably require custom level security implementations; i.e. interpretation of proprietary policy and governance definitions, or use of advanced CA-path provision and verification. Developers may provide their own plugin logic for governing the ROS1 API by extending the template classes within SROS1. This follows suit with similar pluggable security strategies used in other federated pub/sub frameworks such as the Data Distribution Services [22] (DDS) Security Specification[10] from the Object Management Group (OMG), as now used in SROS2.

However, for general users and to facilitate more immediate adoption, a default security plugin is provided with SROS1 to provide a standard example to build upon

---

[9]https://asciinema.org/a/88531.

[10]http://www.omg.org/spec/DDS-SECURITY/About-DDS-SECURITY/.

or use outright, affording the community a consistent MAC option. The remainder of this section delves deeper into the design of the default SROS1 plugin.

5.1 Secure Transport

For securing the network transport, TLS was chosen over other encrypted transport technologies, such as Internet Protocol security (IPsec), or implementing a custom authentication system for several important reasons. First, while the mathematics behind modern public key encryption and symmetric block ciphers are straightforward, the implementations of them are often not. Errors or weaknesses in implementation can lead to unforeseen vulnerabilities or side channel attacks [21], thus dissuading the authors from rolling our own authentication framework. Second, the authors wanted to choose a transport technology that has already garnered sufficient adoption and vetting by academia and industry so that any security vulnerabilities that might arise would attract notice and be addressed by the greater security community. Robotic developers would not be burdened with the sole responsibility for addressing these vulnerabilities. Third, the authors needed a transport approach sufficiently high level in the network stack that it could be monitored and controlled from within most ROS client libraries. Lastly, the authors chose TLS because it enjoys sufficient support for multiple operating systems and programming languages as well as continuing feature development, optimization and security updates [12] so as not to artificially restrict the future development of SROS1.

Perhaps one downside in selecting TLS is the lack of multi-broadcast transport or controllable Quality of Service (QOS), beneficial for both large subscriber count messaging or communication over lossy networks. Thus SROS1 client transport for message data is limited to ROSTCP, given that TLS requires a reliable transport to ensure received packet blocks can be deciphered. Missing packets would not only interfere with present message, but also with consecutive blocks as well due to cipher block chaining used. However, for users with more intensive use-cases needing realtime or multi-broadcast, the authors would like to recommend migrating to SROS2, as OMG's DDS Security spec is now supported in ROS2 which seeks to satisfy such requirements. By contrast, SROS1 seeks to resolve the more immediate security issue in current ROS1 applications, as well as to extend the longevity of such legacy robot code bases.

5.2 Access Control

ROS uses namespaces to define most locations and resources in the computational graph, including message topics, services, parameters, and path names of nodes themselves. Other than a requirement for unique node names, ROS1 provides no degree of access control between subjects and objects; e.g restricting what namespaces a node may register, what topics can be published/subscribed, parameters read/written to, services called/advertised, or what internal ROS APIs maybe invoked in the graph. Although this flexible aspect of ROS enables rapid prototyping and debugging, the absence of any safeguards leaves ROS graphs susceptible to rogue or compromised participants, spoofing message data and master/slave functions such as resource registration or process termination. This unsettles the deployment of ROS in industry settings, as there can be no guarantees in enforcing a specified graph topology once deployed.

SROS1 introduces access control through the use of PKI by embedding policy definitions in X.509 certificate extensions. Again, because these certificates are signed, any attempt to modify or elevate a node's own permissions will void the CA's signature, and thus fail immediately during the TLS handshake for incoming and outgoing p2p connections. A node's policy is defined by the presence of object identifier (OID) fields, where each OID maps to some specific allowed or denied action, and the value of the field uses a regex like syntax to define the namespace scope of the action in question. See Fig. 8. This policy profiling resembles that of AppArmor's mandatory access control (MAC) [3], where permissions may be aliased across a sub-tree path, but then also revoked for specific nested sub-trees. In fact, SROS1's default globbing style is the same as is used in AppArmor, since the regular expression syntax is more oriented towards scoping path names and is more legible for auditing purposes.

In the default plugin, MAC is simply achieved by first enumerating through the list of denied permissions (an expected OID element with in the peer's certificate) that are pertinent to the API involved. If a matching (i.e. by regular expression) deny rule is found, the event is rejected outright. If no matching deny rule is found, then

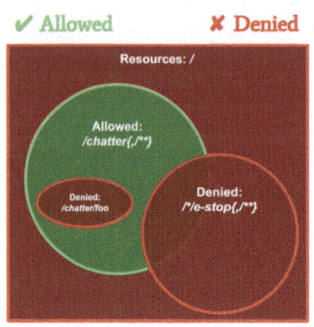

(a) Permitted Namespace Venn diagram

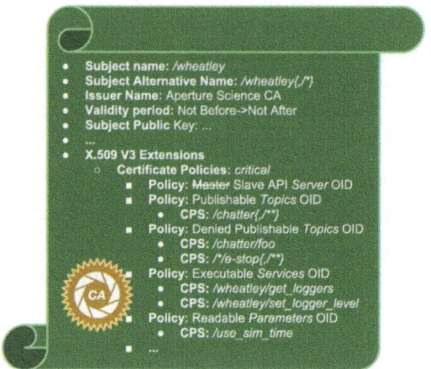

(b) Certificate Embodying Restrictions

Fig. 8 Using Mandatory Access Control (MAC) global security policy, deny by default, explicit permission to resource is required and adequate scope must be satisfied. Conflicts in allowed and denied scopes are resolved by denying intersection overlaps. Path globbing is used to formulate scope via wildcards or regular expressions. The X.509 certificates issued are embedded with object identified policy extentions that embody the access control restrictions over API roles: server/client and allowed/denied resources: parameters, services, and topics

similarly allow rules are searched as well. If no allow rules are matched as well, then the event is rejected by default. The worst case look-up time for this strategy could be exacerbated if the peer is provisioned a large number of individual permissions for that API.

The ability to re-prioritize the evaluation order of such polices was explored, as is achieved in the default DDS security plugin by collapsing allow and deny rules into a single ordered list. However, for simplicity, the more conservative use of two separate OID look-up lists was chosen to avoid deny rules being accidentally ignored due to user error in order mismatch. The authors believe the use of regular expressions will help curb the need for verbose enumerations of permissions.

By taking a closer look at the general ROS1 API, as summarized in Fig. 9, we find three main sub APIs: The Master API–used by nodes to register published/subscribed topics, advertise and look-up services, and to make general graph state inquiries; The Slave API–used by the master to control and relay relevant graph state changes to nodes; and finally, The Parameter API–enabling the storage and distribution of global space parameters, also hosted by the master.

To protect these APIs and the objects they expose, the subject initiating the API transaction must be checked for authorization. This is done by shimming around ROS1's original XML-RPC server and passing up the certificate context from the TLS session. From the connected socket, we can deduce that the TLS handshake was successful, as well as the validity and authenticity of the caller's certificate. The applicable X.509 extensions and policy payloads are determined by the standardized object identifier, then evaluated against the API request and potentially the sanitation of the API response.

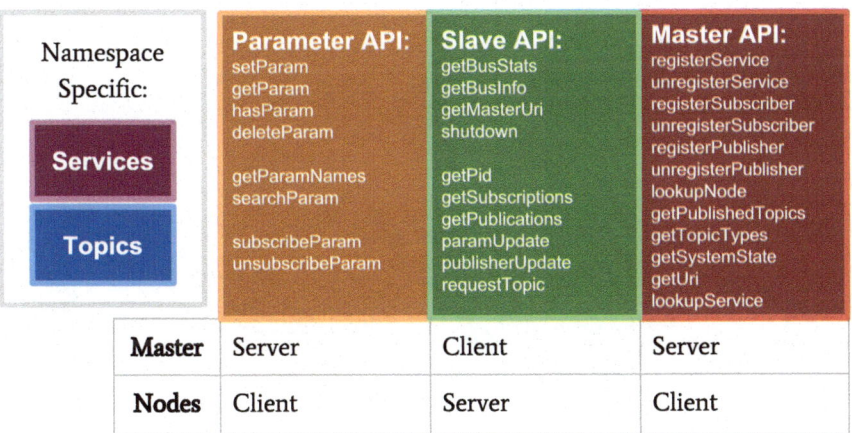

Fig. 9 Complete listing of the official ROS1 API and associated role for each type of ROS1 participant. The ROS1 Master acts as a server for both the Parameter and Master API, and with the roles reversed for Nodes and Slave API. Given this original asymmetric design, SROS additionally hardens the API by restricting client and server exchanges via enforced singular roles. For example, this prevents malicious ROS1 nodes from shutting down other nodes via the slave API

Figure 10 renders such exchanges. Similar mechanics are introduced not only for API servers, but also for the client side and in p2p transport as well. See Fig. 11. Both figures visualize the flow of the API (from left to right, in a clockwise direction) with the pre/post processing of authorization for the request barricaded around the fulfillment step of the native ROS1 library. As an example, Fig. 10a depicts a node sending the master a parameter API request. The node's credentials are extracted from the session context and used in checking the permission of the API accessed. If allowed, the API request is fulfilled by the native ROS1 client library. If a returned response value is appropriate, the response is additionally sanitized in accordance with the node's permissions before finally transmitting the reply.

To adequately encompass the operations for these resources and enable meaningful policy descriptions, we define a general set of permissions and roles specific to respective ROS1 domains and APIs. See Fig. 12. This is then mapped onto the set of API calls, defining the necessary checks and governance for a complete SROS1

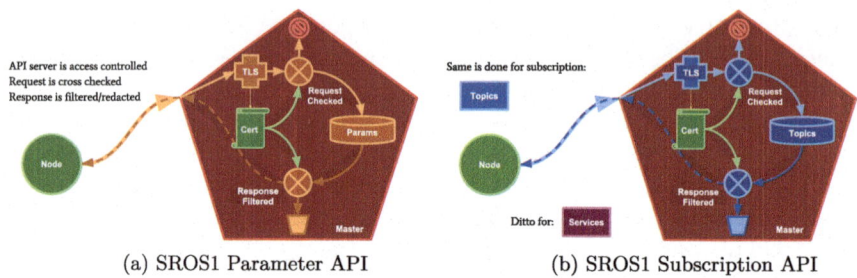

(a) SROS1 Parameter API   (b) SROS1 Subscription API

Fig. 10 Many core ROS1 features are reliant on generous APIs, e.g. recursive parameter retrieval. Thus revoking the entire API call would be non ideal. A data-centric approach, although invasive, helps balance both original API use case and newer access control paradigm. Here Master API calls are vetted prior and post execution by extracting the client certificate from the TLS context, then sanitizing or rejecting XML-RPC data before and after ROS1 library handles

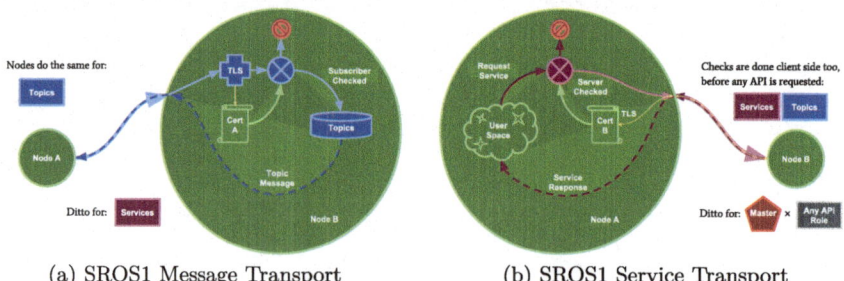

(a) SROS1 Message Transport   (b) SROS1 Service Transport

Fig. 11 For message and service transport, SROS1 supports both server and client side authentication. In ROS1, nodes first use XML-RPC to negotiate transport protocols and ports. Both participants scrutinize the other's credentials and access control before proceeding; client checking the server before sending request, and vice versa for server before acknowledging with response. This prevents malicious nodes from circumventing the check in SROS1 Master for subscriptions

Validation Terminology:

Permission Required	Description
master	Is a master or rosmaster node
slave	Is a slave or regular node
read	Read a scope of parameters
write	Write to a scope of parameters
publish	Publish to a scope of topics
subscribe	Subscribe to a scope of topics
call	Call or request a service
execute	Execute or advertise a service

Legend Key	Description
Permission Required	API called must respect Permission type granted
Caller ID Matched	Caller ID must must respect Subject Alternative Name scope
Resource Checked	Received arguments must respect Allowed & Denied, resource scope
Response Sanitized	Returning responses must respect Allowed & Denied resource scope

Fig. 12 To further harden the ROS1 API, more complete aliases overlapping permission domains of each API call are required. To better represent the boundaries of separation, we introduce some additional terminology for defining API access. Note: Caller ID in API call must match the scope of subject name in certificate

access control implementation. This matrix of object permissions and subject qualifiers is tabulated in Fig. 13.

This security feature also introduces its own complexities, since flawlessly defining a policy profile that encapsulates all of the expected exchanges in a ROS graph would be a daunting task to conduct manually. So again, in line with the authors' focus on security and usability, tooling for learning policy profile through demonstration has also been incorporated into SROS1. SROS1 provides a varying degree of run-time modes, including audit, enforce and complain, again borrowing feature designs from AppArmor. This provides developers a method to auto generate, or amend profiles through granular logging of access events and violation attempts.

Currently this assistive process is conducted within the master, as the graph topology and initiative exchanges are all routed through the master. Thus when configured, a complete encompassing profile can be learned through run time demonstration, amending the policy configuration file as needed. However, our future goals are to migrate such functions outside the master and into separate stages. This could be conducted by reusing ROS1 logging infrastructure to report security related exchanges for later auditing. This is expounded upon further in the future work section.

6 Discussion and Future Work

Despite the progress made, there remains much to be done by way of design and implementation. First, the current policy syntax, although easily machine readable and mutable, is not suitably expressive nor succinct. Policies profiles should natively permit the composition of hierarchical primitives, smaller components that can be

Fig. 13 API access is
contingent upon the call's
intrinsic purpose and if
permissible by scope.
Parameter API: one can only
mutate scope that is writable,
or see what is readable.
Some calls are common in
checking parameter exists
before writing and require
overlap. Slave API: only
permissible to ROS1 Master,
with exception of
requestTopic given
slave-to-slave subscription.
No restrictions are placed on
master's slave access as
ROS1 already forfeits its
critical arbitration. Master
API: requires more thorough
checking and sanitizing but
is quite compartmentalized.
Asterisks denote sections of
the API where balancing
security with user's use case
is particularly sensitive,
requiring further
investigation

Parameter API	Permission Required	Caller ID Matched	Resource Checked	Response Sanitized
setParam	write	✔	✔	
deleteParam	write	✔	✔	
getParam	read	✔	✔	
hasParam	read	✔	✔	
getParamNames	read ∪ write*	✔	✔	✔
searchParam	read ∪ write*	✔	✔	✔
subscribeParam	read	✔	✔	
unsubscribeParam	read	✔	✔	

Slave API	Permission Required	Caller ID Matched	Resource Checked	Response Sanitized
getBusStats	master			
getBusInfo	master			
getMasterUri	master			
shutdown	master			
getPid	master			
getSubscriptions	master			
getPublications	master			
paramUpdate	master			
publisherUpdate	master			
*requestTopic	master slave	✔	✔	✔

Master API	Permission Required	Caller ID Matched	Resource Checked	Response Sanitized
registerService	execute	✔	✔	
unregisterService	execute	✔	✔	
registerSubscriber	subscribe	✔	✔	
unregisterSubscriber	subscribe	✔	✔	
registerPublisher	publish	✔	✔	
unregisterPublisher	publish	✔	✔	
lookupNode	*pub ∪ sub	✔	✔	✔
getPublishedTopics	subscribe	✔	✔	✔
getTopicTypes	subscribe	✔	✔	✔
getSystemState	slave	✔	✔	**✔
getUri	slave	✔	✔	
lookupService	call	✔	✔	✔

unit verified then reused to compose larger and further complex expressions. To achieve this, we intend to adopt a similar syntactic language to that used in AppArmor which has proved quite flexible for experienced users, yet intuitive for beginners. See Fig. 14.

To further our efforts toward automating the constriction and validation of general access control policies, additional security event handling is necessary to adequately provide a completeness of context in which to apprise and tailor permissions. Currently, graph topology and inter communication requests are monitored directly from within the ROS1 Master. However such centralized handling fails to account for numerous p2p interactions that may accrue exterior to the Master. Additionally, such an approach does not translate well with the design paradigms for SROS2, where the absence of a centralized arbiter necessitates auto-discovery. To this end, we would like to standardize security event log messaging, such that each node may self report audited exchanges and attempted violations over the logging channels or local storage.

By building from existing design patterns for the ROS1 log format, see Fig. 15, we can reuse current logging exchange and recording mechanics, while avoiding regression for pre-existing ROS1 libraries and log introspection utilities. To further aid the intractability of SROS tools and implementation, a semantic as well as syntactic standard should also be established, see Fig. 16. This would enable ROS1 centric objects, actions, and subjects to be properly identified, and be agnostic of the specific implementation of the SROS1 governance, e.g. which node attempted what API on a given resource. See Fig. 16.

To further build upon the previous road-map, again, we would like to take an another cue from the AppArmor community by providing an equivalent profile generation command line experience for guiding users through the process of building

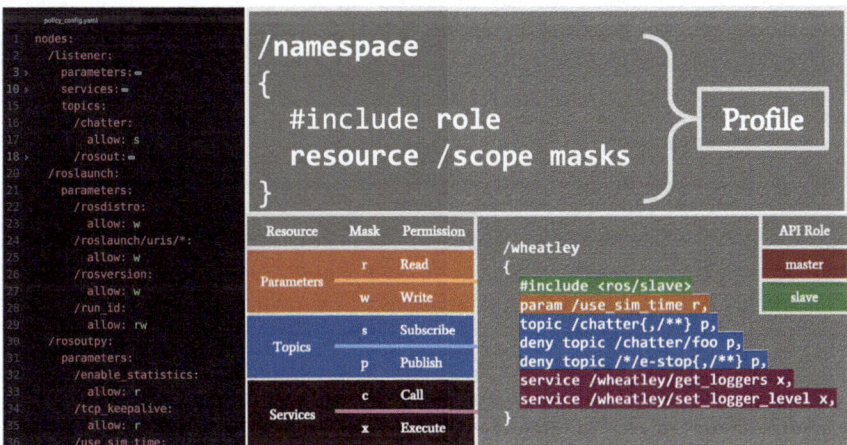

Fig. 14 Left, current YAML policy syntax. Right, new proposed language inspired from equivalent MAC approach in AppArmor. Profiles are titled with subject scope and encapsulates rules either *#included* or defined inline. Resource types make rules explicit to specific resource. Scope defines globbing namespace for permission. Permissions are specified via masks; masks are also resource explicit. Deny is used to revoke permissions, superseding any applicable allow

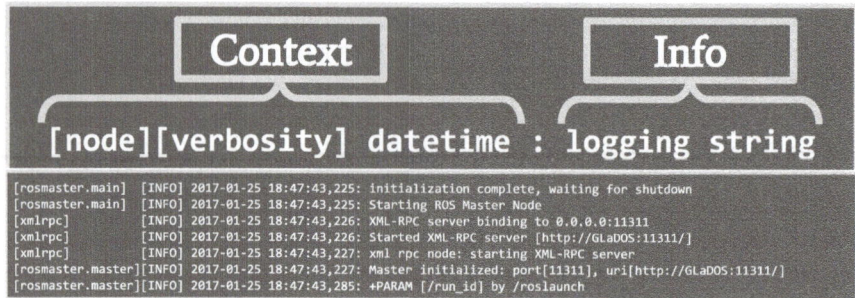

Fig. 15 Standard ROS1 logging format with contexts including node of origin, verbosity level, yyyy-MM-dd HH:mm:ss,fff date timestamp and log message info

Verbosity Level	Message Purpose	
		`[rosmaster.master][INFO]    2017-12-31 12:34:56,789: sros="STATUS" operation="runtime_mode" mode="audit"`
INFO	**Mode Status**	`[rosmaster.master][DEBUG] 2017-12-31 12:34:56,795: sros="AUDIT" operation="registerPublisher" node="/wheatley" resource="topic" path="/chatter"`
		`[rosmaster.master][DEBUG] 2017-12-31 12:34:56,850: sros="AUDIT" operation="registerService" node="/wheatley"  resource="service" path="/wheatley/get_loggers"`
DEBUG	**"AUDIT"**	`[rosmaster.master][DEBUG] 2017-12-31 12:34:56,880: sros="AUDIT" operation="registerService" node="/wheatley" resource="service" path="/wheatley/set_logger"`
WARN	**"COMPLAIN"**	`[rosmaster.master][WARN] 2017-12-31 12:38:57,789: sros="COMPLAIN" operation="getParam" node="/wheatley" resource="parameter" path="/use_sim_time"`
ERR	**"DENIED"**	`[rosmaster.master][ERR]    2017-12-31 12:34:57,839: sros="DENIED" operation="registerPublisher" node="/wheatley" resource="topic" path="/chatter/foo"`

Fig. 16 Proposed SROS logging format. Given set security mode at start-up, varying degrees of events could be logged for run-time debugging or post analysis audits. ERR denotes denied attempted violations in enforce mode. WARN: allowed but unauthorized events from to complain mode. DEBUG: resources accesses denoted by audit mode. INFO: report status of current security mode

custom profiles from scratch or debugging and modifying existing ones, see Fig. 17. Even with a clearer and more transparent profile language, locating and cross referencing conflicting rules and permissions for large scale networks can become too time consuming for manual audits. Thus a helper interface for resolving minimal yet satisfiable scope or suggesting alternate amendments by condensing and interpreting whole archives of security logs would be most valuable in security adoption and contribute to system validation.

To improve adoption and gather more contributors, we would like to streamline the distribution of SROS1 by mainlining with official distribution release of ROS1. Before this however, at a minimum we will need to port SROS1 security plugin framework into the roscpp client library to provide adequate support for current ROS1 users. Adding roscpp support will take significant investment, as the legacy C++ XMLRPC library underneath, minimally maintained for over a decade, does

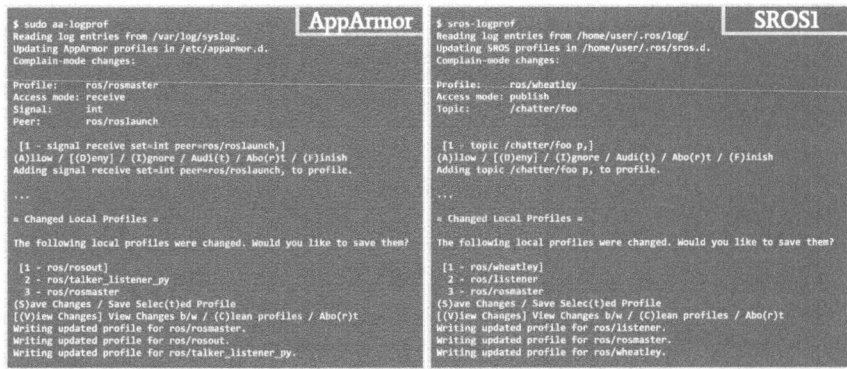

Fig. 17 Left, example interactive session with AppArmor's profiling CLI, used to audit/debug security event and edit profiles. Right, proposed SROS CLI tooling to support similar auto profile generation and log inspection. CLI tools can be used at run-time by monitoring log streams, or ex post factor via archived logs

not yet support TLS. Given its age and non-development, vulnerabilities from within XMLRPC spec like remote code injection or brute force exploits that have impacted PHP and Wordpress servers [20] would not be completely mitigated by SROS1 plugins. Given the old library is perhaps irrevocably intertwined with roscpp code base, replacing the old XMLRPC library with one that is actively maintained or back-porting TLS compatibility may take time. The authors would welcome any contributors who would like to join in our efforts.

Finally, our broader goal is a realistic framework for security features, such as auto-provisioning cryptographic assets and audit verification mechanisms, that is agnostic of middleware release or transport vendor. The ends are twofold: supporting both ROS1 and ROS2, benefiting both old and new robotic frameworks from a commonality of contributions, and maintaining independence from wire-level implementations. The first implies the second, but to stress this, much of ROS2's design has striven to decouple transport from the user-space libraries. To follow suit with SROS development would permit plugable support for future ROS2 communication layers, such as those using ZeroMQ, Protobuf, M2M, etc.

Currently there remain two main limitations to our approach. First, by using the same X.509 certificates for both transport security and access control, we inevitably expose potentially sensitive content during the public handshake procedure, e.g. policy meta-data, clues to participant resources, and partial system graph topology. Although this does afford efficient signature inspection, since only one chain of CA signatures must be recursively verified and checked for revocation, it nevertheless impedes complete system obscurification from external observers.

A simple solution might be to cipher in sensitive payloads to the certificate, with perhaps a common one-time-pad, then distribute the symmetric key when provisioning the rest of the keystore and run-time secrets. Perhaps a better approach would entail segregating Identity Certificates (IC) from Permissions Document (PD). This is being adopted for SROS2 via the default DDS security plugin; documents for establishing authentication and authorization are separate yet partially coupled so as to postpone divulging permission credentials until after a secure channel is established.

This implicates the second limitation as well; because of the monolithic nature of IDs and PDs bound into a single artifact, revocation of either necessitates revocation of both. The decoupling of the IDs and PDs is especially useful when attributes and identities are expected to have differentiating life cycles. This is another feature SROS2 seeks to support by utilizing DDS. Furthermore, numerous PDs could be granted to an ID such that participants need only exchange a minimal set of PDs to avoid divulging more information than required.

One reason this was not pursued here was due to the difficulty of retrofitting such a stateful exchange of PDs underneath the XML-RPC standard, itself a stateless protocol. Provisioning XML-RPC implementations with compatible TLS sockets was seen as a more feasible way to reproduce all the various ROS1 client libraries, given that TLS is well supported in most languages. However, further investigation into a combined approach using TLS with separate PDs could be valuable, and would be a welcome contribution.

In light of these remaining challenges, we would like to note that though our primary task in prototyping SROS1 was to enable legacy support for security in ROS1, an equally important objective of the authors was to explore and research security practices and access control patterns suitable for ROS and robotic related networks. We hope the lessons learned from our efforts will irrevocably benefit the next generation of ROS, ensuring security will come to reside as a first class citizen in ROS2.

Acknowledgements The authors would like to thank the Open Source Robotics Foundation for helping support the design and development of the SROS work presented. This work has been partially supported by CINI Cybersecurity National Laboratory within the project FilieraSicura: Securing the Supply Chain of Domestic Critical Infrastructures from Cyber Attacks.

References

1. J. Åkerberg, M. Gidlund, T. Lennvall, J. Neander, M. Björkman, *Efficient integration of secure and safety critical industrial wireless sensor networks* (EURASIP J. Wirel. Commun, Netw, 2011)
2. J. Åkerberg, M. Gidlund, T. Lennvall, J. Neander, M. Björkman, Efficient integration of secure and safety critical industrial wireless sensor networks. EURASIP J. Wirel. Commun. Netw. **2011**(1), 100 (2011)
3. M. Bauer, Paranoid penguin: an introduction to novell apparmor. Linux J. **2006**(148), 13 (2006)
4. B. Breiling, B. Dieber, P. Schartner, Secure communication for the robot operating system, in *2017 Annual IEEE International Systems Conference (SysCon)* (2017), pp. 1–6
5. M. Cheminod, L. Durante, A. Valenzano, Review of security issues in industrial networks, in *IEEE Transactions on Industrial Informatics*, vol. 9 (2013)
6. C.M. Chris Valasek, *Remote Exploitation of an Unaltered Passenger Vehicle*. Technical report, IOActive, (2015)

7. A. Cortesi, P. Ferrara, N. Chaki, Static analysis techniques for robotics software verification, in *Proceedings of the 44th International Symposium on Robotics, IEEE ISR 2013, Seoul, Korea (South), October 24–26, 2013* (2013), pp. 1–6

8. S.K. Datta, R.P.F. Da Costa, C. Bonnet, J. Härri, ONEM2M architecture based IOT framework for mobile crowd sensing in smart cities, in *2016 European Conference on Networks and Communications (EuCNC)* (IEEE, New York, 2016), pp. 168–173

9. T. Denning, C. Matuszek, K. Koscher, J.R. Smith, T. Kohno, A spotlight on security and privacy risks with future household robots: attacks and lessons, in *Proceedings of the 11th International Conference on Ubiquitous Computing, UbiComp'09* (2009), pp. 105–114

10. B. Dieber, S. Kacianka, S. Rass, P. Schartner, Application-level security for ROS-based applications, in *2016 IEEE/RSJ International Conference on Intelligent Robots and Systems (IROS)* (IEEE, 2016), pp. 4477–4482

11. R. Dóczi, F. Kis, B. St, V. Pser, G. Kronreif, E. Jsvai, M. Kozlovszky, Increasing ros 1.x communication security for medical surgery robot, in *2016 IEEE International Conference on Systems, Man, and Cybernetics (SMC)* (2016), pp. 4444–4449

12. B. Dowling, M. Fischlin, F. Günther, D. Stebila, A cryptographic analysis of the TLS 1.3 handshake protocol candidates, in *Proceedings of the 22nd ACM SIGSAC Conference on Computer and Communications Security, CCS '15* (ACM, New York, 2015), pp. 1197–1210. http://doi.acm.org/10.1145/2810103.2813653

13. D. Dzung, M. Naedele, T.P. von Hoff, M. Crevatin, Security for industrial communication systems. Proc. IEEE **93**, 1152–1177 (2005)

14. W.K. Edwards, R.E. Grinter, At home with ubiquitous computing: Seven challenges, in *Proceedings of the 3rd International Conference on Ubiquitous Computing, UbiComp '01* (2001), pp. 256–272

15. P.E. Eric Byres, D. Hoffman, The myths and facts behind cyber security risks for industrial control systems, in *VDE Kongress* (2004)

16. I. Foster, A. Prudhomme, K. Koscher, S. Savage, Fast and vulnerable: a story of telematic failures, in *Proceedings of the 9th USENIX Conference on Offensive Technologies. WOOT'15* (2015)

17. J. Huang, C. Erdogan, Y. Zhang, B. Moore, Q. Luo, A. Sundaresan, G. Rosu, Rosrv: Runtime verification for robots, in *Proceedings of the 14th International Conference on Runtime Verification*. LNCS, vol. 8734 (Springer International Publishing, Berlin, 2014), pp. 247–254

18. K. Koscher, A. Czeskis, F. Roesner, S. Patel, T. Kohno, S. Checkoway, D. McCoy, B. Kantor, D. Anderson, H. Shacham, S. Savage, Experimental security analysis of a modern automobile, in *Proceedings of the 2010 IEEE Symposium on Security and Privacy, SP '10* (2010), pp. 447–462

19. F.J.R. Lera, J. Balsa, F. Casado, C. Fernández, F.M. Rico, V. Matellán, Cybersecurity in autonomous systems: Evaluating the performance of hardening ROS. Málaga, Spain-June **2016**, 47 (2016)

20. I. Muscat, Web vulnerabilities: identifying patterns and remedies. Netw. Secur. **2016**(2), 5–10 (2016), http://www.sciencedirect.com/science/article/pii/S1353485816300162

21. M. Nemec, M. Sys, P. Svenda, D. Klinec, V. Matyas, The return of coppersmith's attack: practical factorization of widely used RSA moduli, in *24th ACM Conference on Computer and Communications Security (CCS'2017)* (ACM, New York, 2017), pp. 1631–1648

22. G. Pardo-Castellote, OMG data-distribution service: architectural overview, in *23rd International Conference on Distributed Computing Systems Workshops, 2003. Proceedings* (2003), pp. 200–206

23. D. Portugal, M.A. Santos, S. Pereira, M.S. Couceiro, *On the security of robotic applications using ROS, in Artificial Intelligence Safety and Security* (CRC Press, Boca Raton, 2017)

24. M. Quigley, K. Conley, B. Gerkey, J. Faust, T. Foote, J. Leibs, R. Wheeler, A.Y. Ng, *ROS: an open-source robot operating system, in ICRA Workshop on Open Source Software*, vol. 3 (Japan, Kobe, 2009), p. 5

25. A. Stubblefield, J. Ioannidis, A.D. Rubin et al., Using the fluhrer, mantin, and shamir attack to break wep, in *NDSS* (2002)

26. M. Vanhoef, F. Piessens, Key reinstallation attacks: Forcing nonce reuse in WPA2, in *Proceedings of the 2017 ACM SIGSAC Conference on Computer and Communications Security, CCS '17* (ACM, New York, 2017), pp. 1313–1328. http://doi.acm.org/10.1145/3133956.3134027

27. C. Wang, A. Carzaniga, D. Evans, A.L. Wolf, Security issues and requirements for internet-scale publish-subscribe systems, in *Proceedings of the 35th Annual Hawaii International Conference on System Sciences, 2002. HICSS* (IEEE, New York, 2002), pp. 3940–3947

28. R. White, M. Quigley, *SROS: Securing ROS Over the Wire, in the Graph, and Through the Kernel* (ROSCon, Seoul South Korea, 2016), https://vimeo.com/187705073

29. R. White, M. Quigley, H. Christensen, *SROS: Securing ROS over the wire, in the graph, and through the kernel, in Humanoids Workshop: Towards Humanoid Robots OS* (Cancun, Mexico, 2016)

30. W. Xu, S. Bhatkar, R. Sekar, Taint-enhanced policy enforcement: a practical approach to defeat a wide range of attacks, in *USENIX Security Symposium* (2006), pp. 121–136

Ruffin White is a Ph.D. student in the Contextual Robotics Institute at University of California San Diego, under the direction of Dr. Henrik Christensen. Having earned his Masters of Computer Science at the Institute for Robotics and Intelligent Machines, Georgia Institute of Technology, he remains an active contributor to ROS and a collaborator with the Open Source Robotics Foundation. His research interests include mobile robotics, with an focus on secure sub-systems design, as well as advancing repeatable and reproducible research in the field of robotics by improving development tools and standards for robotic software.

Gianluca Caiazza is a Ph.D. student in the Advances in Autonomous, Distributed and Pervasive systems (ACADIA) in security studies at Ca' Foscari University under the supervision of Professor Agostino Cortesi. His research interests include logical analysis of APIs, analysis of complex systems and reverse engineering, always along the line of cybersecurity. He is also passionate about connected and smart devices/infrastructure, specifically within the Consumer and Industrial IoT field.

Dr. Henrik I. Christensen is a Professor of Computer Science at the Department of Computer Science and Engineering University of California San Diego. He is also Director of the Institute for Contextual Robotics. Prior to his coming to the University of California San Diego he was the founding director of the Institute for Robotics and Intelligent machines (IRIM) at Georgia Institute of Technology (2006–2016). Dr. Christensen does research on systems integration, human-robot interaction, mapping and robot vision. He has published more than 300 contributions across AI, robotics and vision. His research has a strong emphasis on "real problems with real solutions." A problem needs a theoretical model, implementation, evaluation, and translation to the real world.

Professor Agostino Cortesi is a Full Professor at Ca' Foscari University of Venice. Recently, he served as Dean of the Computer Science program, and as Department Chair. He also served 8 years as Vice-Rector of Ca' Foscari University, taking care of quality assessment and institutional affairs. His main research interests concern programming languages theory and static analysis techniques, with particular emphas is on security applications. He is also interested in investigating the impact of ICT on different social and economic fields (from Tourism to E-Government to Social Sciences). He has published more than 100 papers in high level international journals and proceedings of international conferences. He served as member of several program committees for international conferences (e.g., SAS, VMCAI, CSF) and on editorial boards of scientific journals (Computer Languages, Systems and Structures, Journal of Universal Computer Science).

GPU and ROS the Use of General Parallel Processing Architecture for Robot Perception

**Nicolas Dalmedico, Marco Antônio Simões Teixeira,
Higor Santos Barbosa, André Schneider de Oliveira,
Lucia Valeria Ramos de Arruda and Flavio Neves Jr**

Abstract This chapter presents a full tutorial on how to get started on performing parallel processing with ROS. The chapter starts with a guide on how to install the complete version of ROS on the Nvidia development boards Tegra K1, Tegra X1 and Tegra X2. The tutorial includes a guide on how to update the development boards with the latest OS, and configuring CUDA, ROS and OpenCV4Tegra so that they are ready to perform the sample packages included in this chapter. The chapter follows with a description on how to install CUDA in a computer with Ubuntu operating system. After that, the integration between ROS and CUDA is covered, with many examples on how to create packages and perform parallel processing over several of the most used ROS message types. The codes and examples presented on this chapter are available in GitHub and can be found under the repository in https://github.com/ air-lasca/ros-cuda.

Keywords Parallel processing · CUDA · ROS · GPU

N. Dalmedico (✉) · M. A. Simões Teixeira · H. S. Barbosa
A. S. de Oliveira · L. V. Ramos de Arruda · F. Neves Jr
Federal University of Technology - Parana, Av. Sete de Setembro,
Curitiba 3165, Brazil
e-mail: ndalmedico@alunos.utfpr.edu.br

M. A. Simões Teixeira
e-mail: marcoteixeira@alunos.utfpr.edu.br

H. S. Barbosa
e-mail: higorsantos@alunos.utfpr.edu.br

A. S. de Oliveira
e-mail: andreoliveira@utfpr.edu.br

L. V. Ramos de Arruda
e-mail: lvrarruda@utfpr.edu.br

F. Neves Jr
e-mail: neves@utfpr.edu.br

© Springer International Publishing AG, part of Springer Nature 2019   407
A. Koubaa (ed.), *Robot Operating System (ROS)*, Studies in Computational
Intelligence 778, https://doi.org/10.1007/978-3-319-91590-6_12

1 Introduction

In the field of robotic perception it's common to face great amounts of data acquired through several different sensors. This kind of data requires a lot of processing to become useful for navigation or recognition. These operations may take a high time when processed using regular CPUs, mainly when performing tasks such as sensor fusion, object recognition or object tracking. For mobile robots, a small reaction time means being able to avoid an accident or to represent less risk to the people that work alongside them, meaning that smaller processing time is an attribute of robot building that must always be sought.

The most common procedure to ensure the processing requirements is to employ a graphics processing unit (GPU), where a large quantity of data is processed in parallel fashion in order to achieve low computation time. GPU applies the same functions on many inputs at the same time using multiple processing cores. One can also use a matrix as input for processing instead of an image. This kind of architecture is called GPGPU (General Processing GPU), taking advantage of the parallel processing power to apply a function to large amounts of data. GPGPU is very useful in robot perception as many modern sensors no longer capture planar data (single 2D images) but spatial data (as laser scans and point clouds that capture multiple 3D points). The interpretation of this kind of data usually requires high processing power.

The popularization of the GPGPU technique and the coming of new ways to represent spacial data (e.g., laser scan, point cloud, point cloud 2) creates new possibilities of applications when used together with systems like ROS, allowing the creation of robots with a bigger degree of autonomy. This tutorial chapter aims on tackling the ROS/GPGPU integration by presenting the following:

- For users of the Nvidia Development Boards (TK1, TX1 and TX2), the full installation of ROS on those kits will be described in the second section.
- Tutorials for users of other Nvidia GPU will also be included in the second section as well as an explanation on how CUDA GPU programming is done.
- The third section introduces CUDA programming and how to use it alongside ROS, this is done with a series of simple examples that allow the understanding for beginners of both frameworks.
- The fourth and fifth section present conclusions and acknowledgements, respectively.

To help the reader to follow all the tutorials described in this chapter, a Git repository was created under the username 'air-lasca' called 'ros-cuda'. The repository is open for everyone and its URL is as follows: https://github.com/air-lasca/ros-cuda. The repository contains the packages created in the Sect. 3, as well as other files that may be of use for the readers. It's also important to notice that any problem regarding the GitHub repository can be informed as feedback in the repository's issue tab at https://github.com/air-lasca/ros-cuda/issues.

1.1 Related Work

Since the 2000's, great interest was shown in the application of graphics processing units (GPU) to aid in general purpose computation, with the goal of splitting the computational load on central processing units (CPU). This technique of sharing GPU and CPU processing tasks was called GPGPU (General Purpose GPU) and it allowed new approaches when tackling heavy-cost processing in the fields of numerical calculus, artificial intelligence, computer vision and robot perception. One of the first studies published in the field of GPGPU can be found in [1], where the author develops a new programming environment for graphics processors and applies matrix multiplication to test the efficiency of the new architecture, comparing it to CPU processing.

With the introduction of the concept and demonstration of favorable results to this new approach, new studies started to be made by 2004, and new environments started to be developed, as for example, the BrookGPU [2], presenting not only a new specifically oriented coding language for GPGPU, but also a debugger that could simulate a virtual GPU on CPU. Also in 2004, the range of GPGPU applications was expanded when in [3] the architecture was used for database operations. In [4], a comparison between parallel processing and single core processors was made for a simulation of gas dispersion, where the authors claim to have reached a processing time five times smaller on GPU than on CPU.

With the growing use of the technique, environments and application programming interfaces (API) oriented to graphics computation started to include resources for general purpose parallel programming as well. One example is the OpenGL project, that started on 1992, a free access library focused on helping on the development of programs that make use of graphics processors, which started to received several new packs of functions and tools for general purpose applications [5]. OpenGL and other graphics program API's were superseded when Nvidia developed its own software platform called CUDA (Compute Unified Device Architecture), which integrated Direct3D and OpenGL and provided functions focused on parallel programming for C and C++ languages. The introduction of CUDA made using GPU resources simpler for software working on the supported devices. More information on CUDA can be found in [6].

From the new tools developed, many researchers did many processing time comparison between CPU and GPGPU architecture, most of them confirming the superiority of GPGPU for large amounts of data processed. A meta study of the evolution of comparative works between CPU and GPU can be found in [7], where the performance of GPU and CPU computation for a variety of processes is tested and compared with previous papers to make a more sensible estimation of the advantages, and the GPU was found to be able to execute these processes three times faster on average.

Other comparative papers that tested the benefits of GPGPU architecture are presented. In [8] the authors tested the process of finding vector derivatives in GPU and CPU, where GPU shows processing speed an order of magnitude higher than

CPU. In [9] Monte Carlo 2D and 3D simulations are made while evaluating the processing time, where parallel processing executes the simulations 30 times faster than on sequential processing. In [10] is carried out the operation of Nearest Neighbor Search (NNS) in both CPU and GPU for point clouds of various sizes and concluded that the processing on GPU architecture is 90 times faster for point clouds with more than 80000 points. Lastly, there is also the work [11] where GPU is used in optimization through evolutionary algorithms applied on point clouds for feature detection.

The relation between robot perception and general purpose GPU processing is already well explored. It's possible to find comparative studies assaying the performance of GPU for many functions used in perception and visual processing. In [12], GPU is used for accelerating processing of point clouds for autonomous vehicles. The authors details each step of the perception processing, presenting the processing time measured for each step in both CPU and GPU. It's valid to point out that some functions are executed over 100 times faster on GPU. The complete set of operations (which includes registering, detecting vehicles and lanes) takes 10 times less time on GPGPU architecture. In [13] the authors implement obstacle detection comparing both CPU and GPU and conclude that GPU shows a smaller computation time only when the number of operations is high. In [14] a surface normal vector extraction is made over a point cloud and the processing time over GPU and CPU is compared, showing computation a 100 times faster on GPU.

In 2011 the Point Cloud Library (PCL) was introduced, a library with a pack of functions and procedures created to ease the processing of point clouds. The PCL already has integration with GPU processing, which can be seen in [15]. The functions that came with the library allows the processes of filtering, mapping, surface detection, fitting, tracking, fusion, registering and shape reconstruction. The fusion of perception sources in point clouds is a common technique to reduce the number of redundant points and, therefore, speed up subsequent processing. One can also fuse point clouds with image data to generate a colored point cloud (usually referred to as RGB-D fusion). Registering a point cloud means aligning point clouds from different perception sources to match the environment. Several works use GPGPU for point cloud processing, for example in [16] the authors implemented a near real time fitting and tracking of primitive shapes for obstacle detection for a humanoid robot. In [17] RGB-D fusion is performed through GPU for mapping. In [18] the depth map from several perception sources is fused in real time using GPU for surface reconstruction. In [19] a GPGPU RGB-D fusion through point clouds is used in the tracking and reconstruction of objects shape and color. Lastly, in [20] NNS is applied over point cloud in GPU for k-d trees generation, achieving real time processing.

2 Installation and Setup

With the advance of deep learning and robot perception, the use of graphics processing units (GPU) on mobile robots becomes mandatory. Every robot with more than one visual perception source needs a graphics processor to be able to deal with the heavy computing requirements involved in performing tasks like sensor fusion or surface normal vector extraction. The first step to perform these operations is to integrate the robot's core operating system with the GPU.

This chapter is dedicated to guide the reader on how to set up an Nvidia Tegra development board (Nvidia TK1, TX1 and TX2) as well as a guide for a regular Nvidia GeForce GPU. The objective is to help the reader to learn how to make a clean install of the operating system, install any dependencies necessary and then install ROS of a distribution of her or his choice.

2.1 Nvidia Tegra Development Boards

The Nvidia Tegra is a family of processors that pack a CPU and GPU oriented for embedded application. Their small size allows them to be used in mobile devices while delivering computation power that can rival a desktop. To allow a better developing ambient, Nvidia distributes a development kit along with every version released of the Tegra family. As of now, the last three released version are the Tegra TX2, TX1 and TK1 (TK1 being the oldest). These development boards are perfect to code and test GPGPU applications since they are easy to setup GPU and CPU embedded boards compatible with CUDA, OpenGL and ROS, much similar to a multi-core CPU desktop with an Nvidia GeForce GPU. The table shown in Table 1 has all the hardware specifications for the Tegra kits.

The development boards are embedded-oriented and very small in comparison to their hardware, which makes them a good choice when developing GPGPU software for mobile robots. The following sections will teach how to configure your Tegra to work with ROS. If you're reading this chapter to learn about ROS and CUDA integration and will use an Nvidia GeForce instead, you may skip these sections and jump to the Sect. 2.5, which talks about the installation of ROS and CUDA on a regular Ubuntu PC with an Nvidia GeForce GPU.

2.2 Configuring Your Jetson Tegra

This tutorial will teach how to configure your Jetson Tegra development board to work with ROS, more specifically, the Jetson Tegra K1 (TK1), with OpenCV4Te-gra

Table 1 Table of hardware specifications for the Nvidia development kits

Nvidia kit	CPU	GPU technology	CUDA cores	RAM	Year
TK1	ARM Cortex-A15	Nvidia Kepler	192	2GB	2014
TX1	ARM Cortex-A57	Nvidia Maxwell	256	4GB	2015
TX2	2 Denver cores+ARM-A57	Nvidia Pascal	256	4GB	2017

and CUDA 6.5. Even though the TK1 was used as example, this tutorial should work with any of the Tegra development boards (TK1, TX1 and TX2). When first receiving your Jetson Tegra it will already be bootable with a Linux 4 Tegra (L4T) already installed, but the version it runs will probably be outdated. A full system reinstall is needed for ROS to work properly and to benefit from all features offered by the boards. For that, the JetPack tool is needed. The JetPack (current version is 3.0) is a tool that installs operating systems, developer softwares, libraries and APIs for the Jetson TK1, TX1 and TK2. One can find more information about the software on the web page http://docs.nvidia.com/jetpack-l4t/index.html. Another great source of information about the Jetson Tegra boards is the JetsonHacks blog available in [21], which also has tutorials regarding the Nvidia Tegra boards configuration. This tutorial is the same for all three boards.

You will need the following hardware to configure your Jetson:

- A personal computer (PC) with Ubuntu 14.04 × 64 and 10GB of free space on hard drive
- An Nvidia Jetson Tegra development board (it is advised that you start the tutorial with the Tegra turned on)
- The Nvidia Jetson programming cable
- A router connected to the internet and ethernet cables that can reach both the PC and the Tegra.

To start, go to https://developer.nvidia.com/embedded/jetpack and click Download to get the latest version of the JetPack. Move the .run file to your Home folder. As of the day this chapter is being written, the latest version of the JetPack is 3.0, and therefore, the name of the file is JetPack-L4T-3.0-linux-x64.run, if you have downloaded a file that has a different name, use your file name on the commands instead. Open a terminal and type:

```
$ chmod +x JetPack-L4T-3.0-linux-x64.run
```

This command gives executable permission to the file.

Next, run the JetPack executable by typing (Fig. 1a):

```
$ ./JetPack-L4T-3.0-linux-x64.run
```

An introducing window (Fig. 1b) should pop up with the JetPack. Click 'Next'.

It will inform you on the directory that will be used for installation. Click 'Next' again.

You can now select the device which you will be configuring (Fig. 2b), for this example it's the Jetson TK1. Select the board you are using and click 'Next'.

The JetPack shall prompt you for your password so it have permission to edit files. Type your PC password and click 'Next'.

The Component Manager window (Fig. 2b) will come up. By default, all tools, software and operating systems will be set to install. You can leave it like that, unless

(a)

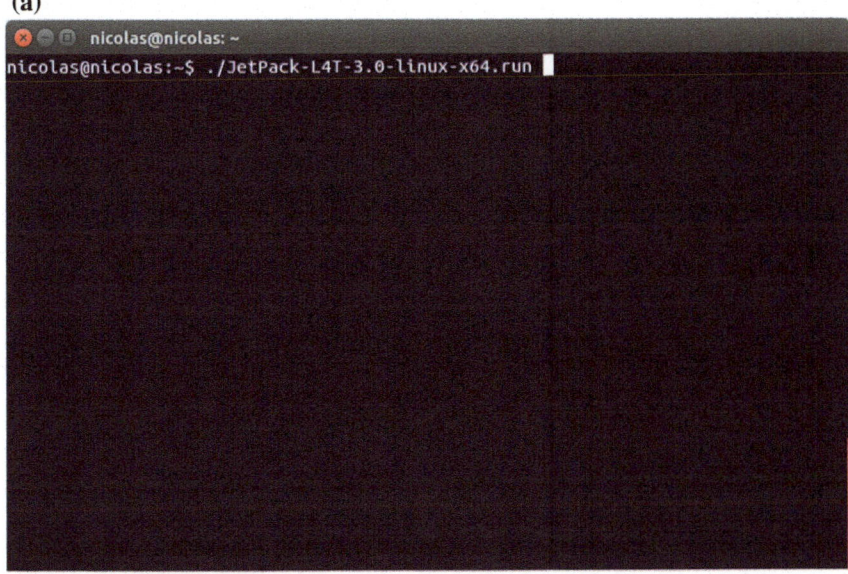

(b)

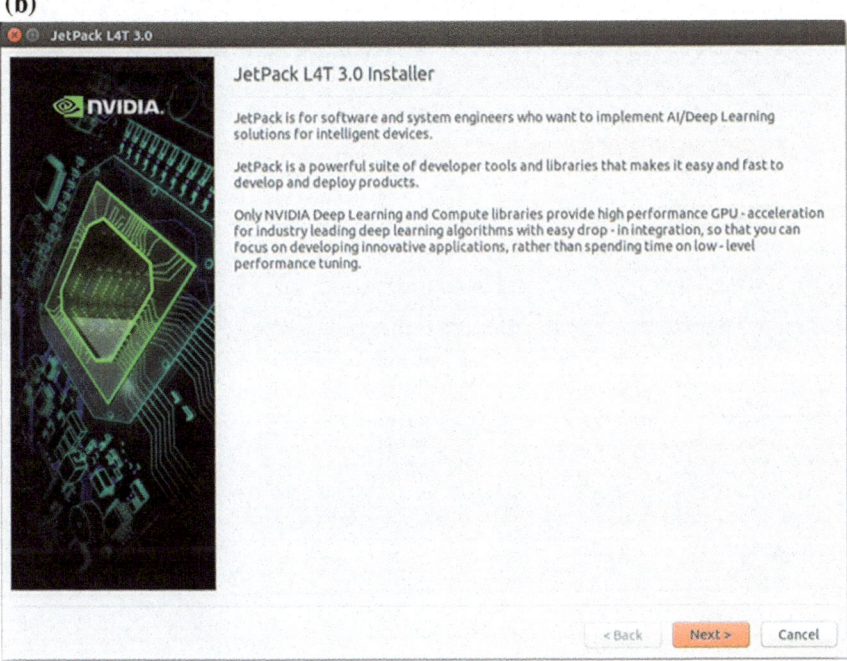

Fig. 1 a Command to start JetPack. b JetPack first window

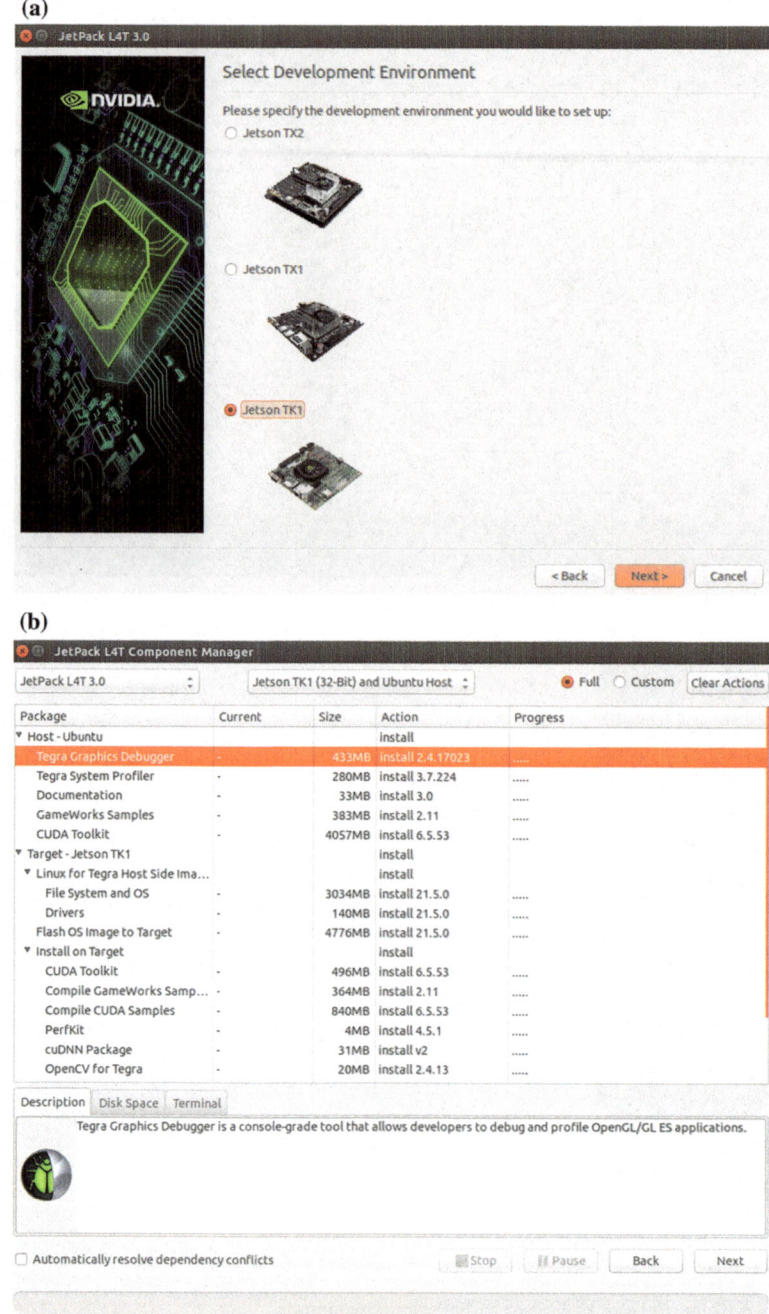

Fig. 2 **a** JetPack device selection window. **b** JetPack component manager window

you don't want a software from the list to be installed. Click 'Next' again and Nvidia will ask you to read the terms of service (Fig. 3a) and accept them before continuing, check all and click 'Accept'.

JetPack will start to download all the software marked to install. The duration of this step depends on your internet connection. Each downloaded package will be marked as 'Pending install'. When all files finished downloading, it will begin the installation (Fig. 3b). This process usually takes about 3 min. During this process, the 'Terminal' embedded in the JetPack window will show some information about the software being installed. Attention is required as some softwares might prompt the user for extra information. After finishing installing the softwares, JetPack will present a 'Completed Installation' screen, press 'Next' to continue.

Now JetPack will ask which network layout is preferred. If you connected both the Tegra and PC to the same network, leave the first setup selected and press 'Next'. If your computer has more than one network interface, you have the option to connect the Tegra directly to the PC through it, then select layout two and press 'Next'. For this tutorial we will use layout one. If Jetpack finds something wrong with the connection, it will prompt you the IP address, user name and password of the development board. The IP address can be found typing 'ifconfig' on a terminal in the Tegra. The user name and password are 'ubuntu/ubuntu' by default for all boards.

JetPack will ask which Network Interface has the connection with the Tegra (Fig. 4a). Usually it will be able to detect which one it is and you can just click 'Next'. If you have doubts upon which network interface you are connected, typing 'ifconfig' on a terminal will list them and present some information. It is valid to note that the 'lo' interface is the loopback interface used for testing only.

The JetPack will present a post-installation screen where it informs you of the following actions (Fig. 4b). Clicking 'Next' will bring up an X-Term terminal window with the following steps (Fig. 5a).

If your Tegra development board is on, turn it off and disconnect from power. It's important that the power supply plug is disconnected from the Tegra. After this, with the Tegra without power, connect the programming cable (USB A to micro USB 2.0 B cable) from the PC to the Tegra. Then, connect to the power supply. After this, press and release the 'Power' button to turn it on (Fig. 5b), press and keep pressed the 'Recovery' button, press and release the 'Reset' button, and then release the 'Recovery' button. This will put the development board in recovery state, where it will be powered on but nothing will appear on its screen.

To verify that the recovery state was initialized successfully, open a terminal window on the PC, type:

```
$ lsusb
```

Check if one of the connections listed have 'NVidia Corp' as its name (Fig. 6a). If 'NVidia Corp' is on the list of USB connections, return to the X-Term terminal windows brought up by JetPack and hit 'Enter'.

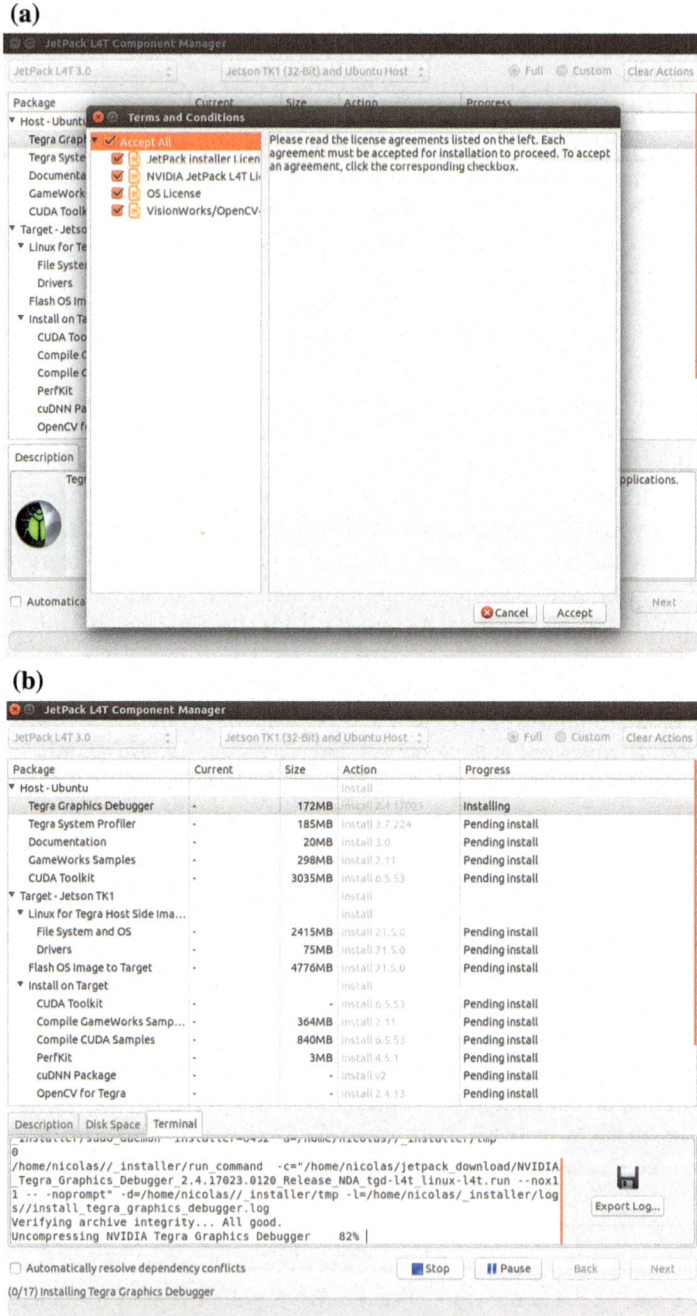

Fig. 3 **a** JetPack terms and conditions prompt window. **b** JetPack installing components

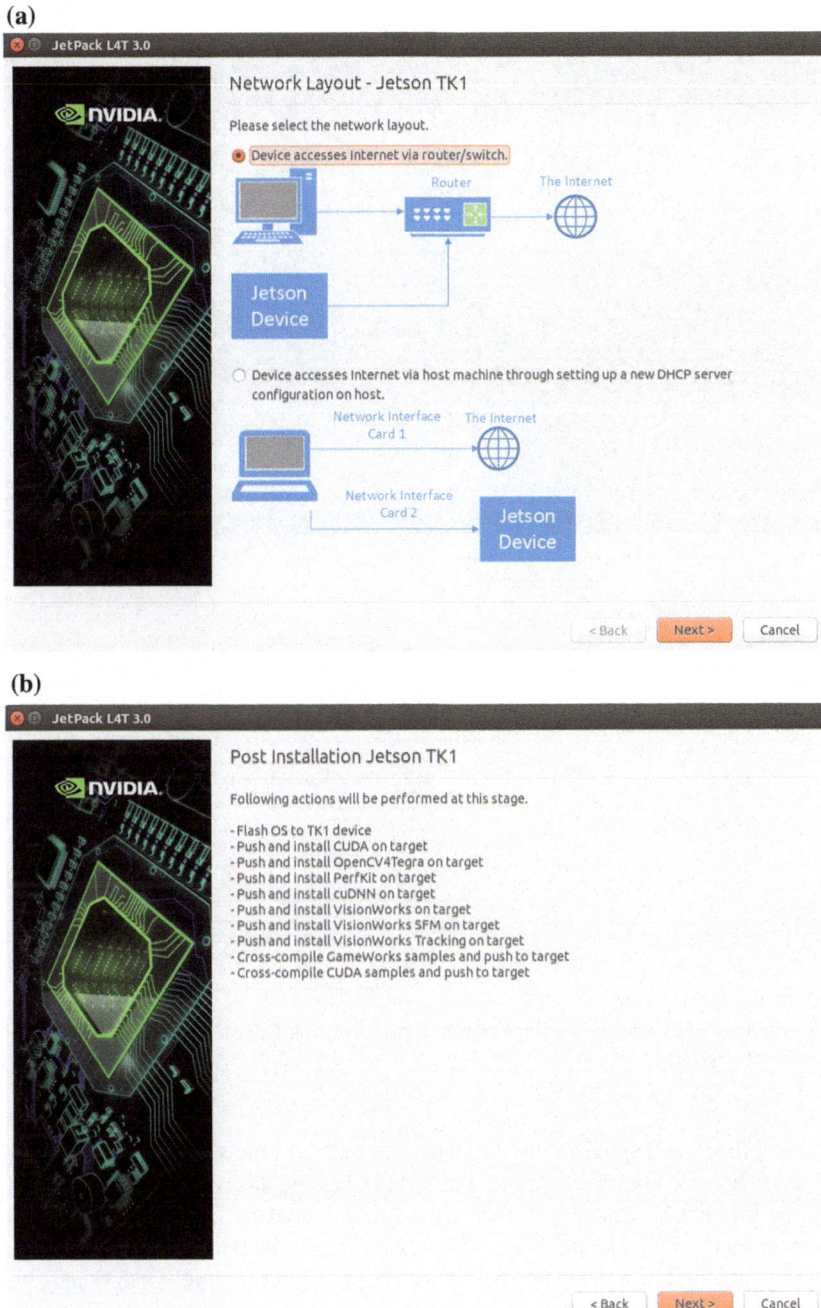

Fig. 4 **a** JetPack network layout selection window. **b** JetPack post-installation instructions

(a)

(b)

Fig. 5 **a** JetPack USB recovery mode prompt X-Term window. **b** Tegra Development Board recovery mode step-by-step

Now, JetPack will upload to the Tegra memory the programs that were set to install (Fig. 6b). This step takes between 15 and 30 min. During this process some windows might be brought up on your PC while the JetPack does its job, and Tegra will be turned on by the PC at some point. When the Tegra turns on, it does not mean the process is finished, it will be finished when the X-Term terminal window tells you so. Since the board will turn on and the process takes a while, the Tegra might enter sleep mode. If this happens move the mouse or press a key on the Tegra keyboard to turn it back on. While on sleep mode, the connection will be closed with the PC and the process will be paused. A counter of five will decrease every time the connection

(a)

```
nicolas@nicolas: ~
nicolas@nicolas: ~                    ×    nicolas@nicolas: ~              ×
nicolas@nicolas:~$ lsusb
Bus 002 Device 002: ID 8087:0024 Intel Corp. Integrated Rate Matching Hub
Bus 002 Device 001: ID 1d6b:0002 Linux Foundation 2.0 root hub
Bus 001 Device 003: ID 0955:7140 NVidia Corp.
Bus 001 Device 002: ID 8087:0024 Intel Corp. Integrated Rate Matching Hub
Bus 001 Device 001: ID 1d6b:0002 Linux Foundation 2.0 root hub
Bus 004 Device 001: ID 1d6b:0003 Linux Foundation 3.0 root hub
Bus 003 Device 003: ID 192f:0416 Avago Technologies, Pte. ADNS-5700 Optical Mous
e Controller (3-button)
Bus 003 Device 002: ID 045e:00dd Microsoft Corp. Comfort Curve Keyboard 2000 V1.
0
Bus 003 Device 001: ID 1d6b:0002 Linux Foundation 2.0 root hub
nicolas@nicolas:~$
```

(b)

```
Post Installation
Please put your device to Force USB Recovery Mode, when your are ready, press En
ter key
To place system in Force USB Recovery Mode:
1. Power down the device. If connected, remove the AC adapter from the device. T
he device MUST be powered OFF, not in a suspend or sleep state.
2. Connect the Micro-B plug on the USB cable to the Recovery (USB Micro-B) Port
on the device and the other end to an available USB port on the host PC.
3. Connect the power adapter to the device.
4. Press and release the POWER button to power on device. Press and hold the FOR
CE RECOVERY button: while pressing the FORCE RECOVERY button, press and release
the RESET button; wait two seconds and release the FORCE RECOVERY button.;
5. When device is in recovery mode, lsusb command on host will list a line of "N
Vidia Corp"

copying bctfile(/home/nicolas/TK1/Linux_for_Tegra_tk1/bootloader/ardbeg/BCT/PM37
5_Hynix_2GB_H5TC4G63AFR_H5TC4G63CFR_RDA_924MHz.cfg)... done.
copying bootloader(/home/nicolas/TK1/Linux_for_Tegra_tk1/bootloader/ardbeg/u-boo
t.bin)... done.
          populating kernel to rootfs... done.
          populating jetson-tk1_extlinux.conf.emmc to rootfs... done.
done.
Making system.img...
```

Fig. 6 **a** JetPack in recovery mode showing on terminal after 'lsusb' command. **b** JetPack starting to flash Tegra drive

drops (Fig. 7a). If the counter reaches zero, JetPack will assume a failure and stop the process. After all is done, the X-Term terminal window will ask you to close the window to continue.

After closing the X-Term window (Fig. 7b), it will bring the JetPack window up again with an 'Installation Complete' screen where you will be able to choose whether to keep the files or not. If you have plans of reinstalling your Tegra development board software again soon you might want to keep the files to skip the download steps in the future. Closing this window finishes the upgrade process.

(a)

(b)

Fig. 7 **a** JetPack recovering from a closed connection. **b** JetPack done flashing the board

2.3 ROS on Nvidia Tegra Development Boards

If you followed every step on the previous tutorial setting up your Nvidia Tegra development board, you should have the latest version of Linux4Tegra installed on the board. If, for some reason, you decided to install an older version (older than L4T 2.4.8), you will need the Grinch Kernel to install ROS on your board. This tutorial covers the installation process for all boards (TK1, TX1 and TX2). There is a script named 'ROS-Tegra.sh' that will install ROS in your Tegra board available in the GitHub repository, inside the /Files/Scrips folder, but it's highly recommended that you follow every step from the following tutorial instead of running the script if it's your first time performing the installation.

First, ROS requires that the system locale is defined. So, in the terminal, type:

```
$ sudo update-locale LANG=C LANGUAGE=C LC_ALL=C
   LC_MESSAGES=POSIX
```

Next, you must tell your computer to allow packages from packages.ros.org. For that, type:

```
$ sudo sh -c 'echo "deb http://packages.ros.org/
   ros/ubuntu trusty
main" > /etc/apt/sources.list.d/ros-latest.list'
```

Set up your keys:

```
$ sudo apt-key adv --keyserver hkp://ha.pool.sks-
   keyservers.net:80
--recv-key 421
   C365BD9FF1F717815A3895523BAEEB01FA116
```

Update the package lists:

```
$ sudo apt-get update
```

Now, the installation of ROS. The Nvidia Tegra boards, even though they are ARM based systems, have a very strong hardware and can easily handle more than the bare bones installation of ROS, even the TK1, which is not the case for most other ARM boards. Because of this, the desktop installation is advised, since it comes with RQT and RVIZ, which are very useful tools. Now is the time to decide which ROS distribution will be installed. The Kinetic Kame version is highly recommended since it is the stable release that will be supported longer (as of the time this chapter is being written). To install another distribution of ROS, change the next command lines to the desired distribution (e.g. indigo, lunar) instead of 'kinetic'. To continue with the ROS desktop installation, type:

```
$ sudo apt-get install ros-kinetic-desktop
```

If you feel you don't need the full installation and prefer just the bare bones of ROS, instead type:

```
$ sudo apt-get install ros-kinetic-ros-base
```

After the ROS installation ends, rosdep must be installed and updated. Type:

```
$ sudo apt-get install python-rosdep
$ sudo rosdep init
$ rosdep update
```

Setup ROS environment variables by typing:

```
$ echo "source /opt/ros/kinetic/setup.bash" >> ~
    /.bashrc
$ source ~/.bashrc
```

Install rosinstall, since it is used a lot but distributed separately:

```
$ sudo apt-get install python-rosinstall
```

To be able to use RVIZ, the GTK_IM_MODULE should be unset or RVIZ will crash every time it starts, so it's best to do that in your bashrc:

```
$ echo "unset GTK_IM_MODULE" >> ~/.bashrc
```

Now, ROS still needs its own workspace. Create one by typing:

```
$ mkdir -p ~/catkin_ws/src
$ cd ~/catkin_ws/
$ catkin_make
```

This will compile the workspace and create a /build and /devel folder. Now, all that is left to do is source your new devel/setup.bash. Instead of USERNAME insert your Linux user name (if you did the installation without changing the default user name, it should be 'ubuntu').

```
$ echo "source /home/USERNAME/catkin_ws/devel/
    setup.bash" >>
~/.bashrc
$ source ~/.bashrc
```

With this tutorial ROS should be almost fully working alongside CUDA in TK1, but there's still some gimmicks left to do so that OpenCV can work properly alongside ROS.

2.4 OpenCV4Tegra and ROS on Tegra Boards

The OpenCV that is installed on OpenCV4Tegra for TK1 and TX1 is based on version 2.4.12, but ROS searches for OpenCV 2.4.8. Because of that, a few changes on libraries must be made so that ROS can find the OpenCV4Tegra to compile OpenCV packages. The following tutorial was based on the one found in [22]. Attention is required for the next steps to not corrupt library files. It's strongly advised that a backup of the files being edited is done so that it is possible to revert any changes and start over if something goes wrong. To ease the process, the Git repository will contain the changed files under 'ros-cuda/Files' in the URL https://github.com/air-lasca/ros-cuda, but be aware that, since Nvidia constantly updates their software, it may happen that the files become obsolete, and even this tutorial may not be necessary in the future. When this tutorial was created, OpenCV4Tegra's last version was 2.4.13, and it was still necessary. If your OpenCV4Tegra have a newer version (you can check during installation, on JetPack Component Manager window), you will have to search for the release notes of the new version to verify if the bug was fixed.

The files that need to be edited are:

- /opt/ros/kinetic/lib/pkgconfig/cv_bridge.pc
- /opt/ros/kinetic/lib/pkgconfig/image_geometry.pc
- /opt/ros/kinetic/share/cv_bridge/cmake/cv_bridgeConfig.cmake
- /opt/ros/kinetic/share/image_geometry/cmake/image_geometryConfig. cmake

To create a backup file of those files, you can use the command 'cp' to create a copy with a different name, for example:

```
$ sudo cp /opt/ros/kinetic/lib/pkgconfig/
   cv_bridge.pc/opt/ros/kinetic/lib/pkgconfig/
   cv_bridge.pc-bak
```

This will create a file named cv_bridge.pc-bak in the same location as the original file. If there's ever a need to restore the original file, one can delete the old corrupted cv_bridge.pc and rename cv_bridge.pc-bak to cv_bridge.pc.

Use your favorite text editor to open and modify the files. Gedit is a good example:

```
$ sudo gedit /opt/ros/kinetic/lib/pkgconfig/
   image_geometry.pc
```

Now, repeat the next steps on each of the four files listed:

- Erase appearances of '/usr/lib/arm-linux-gnueabihf/libopencv_ocl.so.2.4.8'. You can use the Ctrl+F shortcut to find the string in the files. Be careful to keep the code structure the same (for example, keeping the symbols ':' or ';' that divide multiple parameters). This is to remove references of OpenCL since Nvidia, as of the day this chapter is being written, does not provide it for the development boards.
- Replace occurrences of '/usr/lib/arm-linux-gnueabihf/' for '/usr/lib'. Nvidia has a different path for OpenCV4Tegra.
- Replace occurrences of '2.4.8' for '2.4.12' because of the different versions.

If those steps were followed correctly, ROS packages that use OpenCV should now find and use the optimized OpenCV4Tegra instead, allowing the developer to embed the board on a robot, or use it to test OpenCV ROS packages that will later work in other versions of OpenCV.

2.5 ROS and CUDA on a Regular Ubuntu PC

The following tutorial teaches how to install ROS and CUDA on an Ubuntu PC with an Nvidia GeForce GPU compatible with CUDA. The tutorials will guide you towards installing a complete version of each software on your machine. If you have a computer with different hardware and OS than the one used for this tutorial (Ubuntu 16.04 64-bit and a GeForce CUDA-enabled GPU) you can check http://wiki.ros.org/ROS/Installation to see the differences for each OS, distribution and hardware. Installing CUDA first is advised as it is harder to fix a corrupted installation of CUDA than it is of ROS.

To install CUDA first check if your computer is identifying the graphics card and that it is a supported model. For that type:

```
$ lspci | grep -i nvidia
```

Check which version is your Nvidia graphics card and then go to https://developer. nvidia.com/cuda-gpus and see if the model is on the list. If you find your graphics card model on the list, it is CUDA-capable and you can proceed with the installation.

Go to the website https://developer.nvidia.com/cuda-downloads and inform your operating system (for this tutorial, we are using Linux), the architecture (x86_64), the distribution (Ubuntu), the version (16.04) and how you want to install CUDA. For this tutorial we will proceed with the network deb installation (deb (network)). Download the installer. Open a terminal and navigate to your Download directory (or where you saved the file) and then type:

```
$ sudo dpkg -i cuda-repo-ubuntu1604_8.0.61-1
  _amd64.deb
```

If the file you downloaded have other name, use it instead. After this, update your sources with:

```
$ sudo apt-get update
```

And then proceed to install CUDA by typing:

```
$ sudo apt-get install cuda
```

It may take a while since the total size of the files is about 2GB, but it will mostly depend on your network speed. After it finished downloading and installing CUDA, we need to edit the PATH variable to include the path to CUDA libraries. Type:

```
$ export PATH=/usr/local/cuda-8.0/bin${PATH:+:${
  PATH}}
```

We also need to change the LD_LIBRARY_PATH environment variable, so that executables can find shared libraries without referencing the path in its header. The path to be added change whether your computer's architecture is x86 or x64 bits. For x64 computers type:

```
$ export LD_LIBRARY_PATH=/usr/local/cuda-8.0/
  lib64${LD_LIBRARY_PATH:+:${LD_LIBRARY_PATH}}
```

For x86 computers, type:

```
$ export LD_LIBRARY_PATH=/usr/local/cuda-8.0/lib$
  {LD_LIBRARY_PATH:+:${LD_LIBRARY_PATH}}
```

Restart your computer. After that, CUDA should have been successfully installed in your computer. To check if this is true, type:

```
$ nvcc --version
```

Now we need to install ROS, to do that one can follow this tutorial or run the script named 'ROS-Ubuntu.sh' found inside the GitHub repository inside the /Files/Scripts folder.

First tell Ubuntu to accept software from packages.ros.org by typing:

```
$ sudo sh -c 'echo "deb http://packages.ros.org/
  ros/ubuntu
```

```
$(lsb_release -sc) main" > /etc/apt/sources.list.
    d/ros-latest.list'
```

Then, set up your keys:

```
$ sudo apt-key adv --keyserver hkp://ha.pool.sks-
    keyservers.net:80
--recv-key 421
    C365BD9FF1F717815A3895523BAEEB01FA116
```

Update the package index:

```
$ sudo apt-get update
```

We will install ROS full since some packages and tools (e.g. RVIZ and RQT) that come with it are very important for future examples:

```
$ sudo apt-get install ros-kinetic-desktop-full
```

Next, start rosdep and update it:

```
$ sudo rosdep init $ rosdep update
```

Then, set your environment:

```
$ echo "source /opt/ros/kinetic/setup.bash" >> ~
    /.bashrc $ source
~/.bashrc
```

Create a workspace by typing:

```
$ mkdir -p ~/catkin_ws/src
$ cd ~/catkin_ws
$ catkin_make
```

Source your new setup file. Instead of USERNAME insert your Linux user name.

```
$ echo "source /home/USERNAME/catkin_ws/devel/
    setup.bash" >> ~/.bashrc $ source ~/.bashrc
```

With this, ROS should have been successfully installed on your computer. To check, type:

```
$ echo $ROS_PACKAGE_PATH
```

If the response include '/home/USERNAME/catkin_ws/src/opt/ros/kinetic/ share', ROS has its workspace properly set. You can start ROS by typing 'roscore'.

These tutorials explained how to get both CUDA and ROS on your system. In the next section we will see how we can use the two softwares in the same application to take advantage of GPU processing in robot computation.

3 CUDA and ROS

Now we will see how we can create ROS packages with CUDA libraries included in order to use the GPU to perform heavy processing. This next tutorial will be explained in a very simple manner as to make it easy for beginners in both ROS and CUDA. All the codes from every example package are available in the GitHub repository at https://github.com/air-lasca/ros-cuda, but, if you are learning the basics with this tutorial, it is highly advised that you build the packages yourself without the help from the repository.

First, let's create a ROS package. Navigate to your package folder by typing:

```
$ cd ~/catkin_ws/src
```

Let's call this new package 'roscuda_template' so that it can be used as a blank project package later. To create it, type:

```
$ catkin_create_pkg roscuda_template std_msgs
    roscpp rospy
```

This command will create a folder called 'roscuda_template', with two other folders inside called 'src' and 'include'. There will also be two new files inside called 'CMakeLists.txt' and 'package.xml'. Now let's tell ROS that the package will use CUDA libraries so that when we compile the package we have everything needed. Open the 'CMakeLists.txt' file with your favorite text editor and make the following changes.

Add the next line in the header, after the line 'project(roscuda_template)':

```
find_package(CUDA REQUIRED)
```

This line will tell the CMake build system that the CUDA package is required for build. Then, find out what is the compute capability of the Nvidia GPU you are using. To do that, visit the website https://developer.nvidia.com/cuda-gpus and find your GPU and the compute capability number. Add the next lines in the Build section. Instead of compute_XX, use the number of the compute capability of your GPU (e.g. if it's 6.1, use compute_61):

```
set(CUDA_NVCC_FLAGS "-arch=compute_XX" CACHE
   STRING "nvcc flags"
FORCE)
set(CUDA_VERBOSE_BUILD ON CACHE BOOL "nvcc
   verbose" FORCE)
set(LIB_TYPE STATIC) cuda_add_library(SampLib ${
   LIB_TYPE}
src/template.cu)
```

The first three lines set variables that affect the CUDA_ADD_LIBRARY call. The CUDA_NVCC_FLAGS variable define flags for the NVCC compiler. In this case it is used to define the virtual architecture for which the CUDA files are compiled. Other NVCC flags can be set here by changing the string argument (different flags are separated by ';', e.g. "-arch=compute_30;-code=sm_30"). The CUDA_VERBOSE_BUILD tells NVCC to use verbose, meaning it will show more information when compiling. The third line sets the library type to static since we would not want external dependencies a template. Do not worry about the CUDA file we are adding to the library, we will create it later. The result of these lines when the package is built will be a LibSampLib.a in your catkin_ws/devel/lib folder. Now we need to add the library as a dependency and link it to the build. On the Build section of the file, insert the following line:

```
add_executable(roscuda_template_node src/main.cpp
   )
add_dependencies(roscuda_template_node SampLib)
```

Again, overlook the C++ file we are adding as executable, we will create it later. Lastly, add the following lines to link the package, the catkin libraries and the new CUDA library. It should look like this:

```
target_link_libraries(roscuda_template_node
   ${catkin_LIBRARIES}
   SampLib
)
```

These lines will do exactly what they tell. Create the package node, add the library as a dependency and link it to the package being compiled. With this, we are ready to use CUDA code alongside our package, but we still have no code to compile.

Use your favorite text or code editor to create two files file with the names we used in the CMakeLists.txt (template.cu and main.cpp). Do that inside the package in your ROS workspace, in the 'roscuda_template/src' folder. Modify the blank template.cu file you created and add the code from Code 1.

```
1 #include <stdio.h>
2 #include <stdlib.h>
3 #include <cuda.h>
```

```
4  #include <cuda_runtime.h>
5
6  __global__ void kernel()
7  {
8
9  }
10
11 void cudamain()
12 {
13        kernel<<<1,1>>>();
14        return;
15 }
```

Code 1 Basic CUDA code (template.cu)

Inside the blank main.cpp file, add the code from Code 2.

```
1  #include "ros/ros.h"
2
3  void cudamain();
4
5  int main(int argc, char **argv)
6  {
7        ros::init(argc, argv, "roscuda_basic");
8        ros::NodeHandle n;
9        ros::Rate loop_rate(1);
10
11       while(ros::ok())
12       {
13             cudamain();
14             ros::spinOnce();
15             loop_rate.sleep();
16       }
17       return 0;
18 }
```

Code 2 Basic C++ coda (main.cpp)

If you examine the code you will realize it is doing nothing but initializing a ROS node and calling an empty CUDA function. It is supposed to be like this, since we only want to test if everything is working properly. Let's also create a launch file to start the node. Having a launch file in the template is important since once you pick the template to start a new project, a launch file will allow you to set up parameters easier than running packages with 'rosrun'. Create another file named 'roscuda_template.launch' inside the package folder ('/roscuda_template'). Inside it insert the following lines:

```
<launch>
  <node pkg="roscuda_template" type="
      roscuda_template_node" name="
      roscuda_template" respawn="true" output="
      screen">
  </node>
</launch>
```

Save the files and, on a terminal, build your workspace. Type:

```
$ cd ~/catkin_ws
$ catkin_make
```

If the steps were followed correctly the package should be compiled without any errors. You can start the node by typing:

```
$ roslaunch roscuda_template roscuda_template.
    launch
```

Now, let's take a look at the code to understand what is happening.

In the CUDA file, there is a function called kernel() that is defined and called with a different notation than other C and C++ functions. That is because it uses a notation introduced with the CUDA library, and it means it is executed on the device. A function that is declared preceded by the keyword '__global__' indicates a function that is executed on the 'device', which means the GPU. Code outside of these functions are executed in linear fashion like regular C code and are called 'host' code (e.g. cudamain() and main()). These functions have to be called from the host to be executed, and this is why it also have a different notation, the triple angle brackets '<<< 1, 1 >>>'. The meaning of the numbers in the brackets will be explained in the next example.

3.1 Example 1: ROS/CUDA Vector Addition

We saw how CUDA and ROS can be integrated and compiled together in a single package, but the package itself was "empty", and not actually doing anything. Now let's give the kernel function a use so that we can understand how memory allocation works on the device.

The following code was based on Nvidia's CUDA C/C++ basics tutorial [23]. Let's change the two .cu and .cpp files. If you want to organize the examples better, you can create a new package with a different name (using the last tutorial). Change your template.cu (or create a new file vectoradd.cu) to the content of Code 3.

```
1  #include <stdio.h>
2  #include <stdlib.h>
3  #include <cuda.h>
4  #include <cuda_runtime.h>
5
6  #define THREADS_PER_BLOCK 512
7
8  // Global function means it will be executed on
       the device (GPU)
9  __global__ void add(int *in1, int *in2, int *out)
10 {
11        int index = threadIdx.x + blockIdx.x *
              blockDim.x;
12        out[index] = in1[index] + in2[index];
13 }
14
15 void random_ints(int *i, int size)
16 {
17        for(int k=0; k<size; k++)
18        {
19              i[k]=rand()%50;
20        }
21 }
22
23 int *testmain(int num)
24 {
25        int *in1, *in2, *out; // host copies of
              inputs and output
26        int *d_in1, *d_in2, *d_out; // device copies
              of inputs and output
27        int size = num * sizeof(int);
28
29        // Alloc space for device copies of three
              vectors
30        cudaMalloc((void **)&d_in1, size);
31        cudaMalloc((void **)&d_in2, size);
32        cudaMalloc((void **)&d_out, size);
33
34        // Alloc space for host copies of the three
              vectors and setup input values
35        in1 = (int *)malloc(size); random_ints(in1,
              num);
36        in2 = (int *)malloc(size); random_ints(in2,
              num);
37        out = (int *)malloc(size);
38
39        // Copy inputs to device
40        cudaMemcpy(d_in1, in1, size,
              cudaMemcpyHostToDevice);
41        cudaMemcpy(d_in2, in2, size,
              cudaMemcpyHostToDevice);
42
43        // Launch add() kernel on GPU
```

```
44    add<<<num/THREADS_PER_BLOCK,
          THREADS_PER_BLOCK>>>(d_in1, d_in2, d_out)
          ;
45    // Wait for the GPU to finish
46    cudaDeviceSynchronize();
47    // Copy result back to host
48    cudaMemcpy(out, d_out, size,
          cudaMemcpyDeviceToHost);
```

```
49
50    // Cleanup
51    free(in1); free(in2); free(out);
52    cudaFree(d_in1); cudaFree(d_in2); cudaFree(
          d_out);
53    return out;
54 }
```

Code 3 Vector addition CUDA code (vectoradd.cu).

This code will apply random vector addition. Usually it's bad to allocate and define a vector inside the function that is also modifying it, but this is only to be used as an example. Insert the content of Code 4 in your main.cpp code.

```
1  #include "ros/ros.h"
2  #include <time.h>
3  #include "std_msgs/Int32.h"
4  #include "std_msgs/Float32.h"
5
6  #define N (1024*1024)
7
8  int *testmain(int num, int threads);
9  int size=10;
10 std_msgs::Float32 msg_time;
11
12 void size_Callback(const std_msgs::Int32& msg)
13 {
14     size=msg.data;
15     if(size<1){size=1;}
16     if(size>100){size=100;}
17 }
18
19 int main(int argc, char **argv)
20 {
21     ros::init(argc, argv, "roscuda_vectoradd");
22     ros::NodeHandle n;
23     ros::Publisher time_pub = n.advertise<
           std_msgs::Float32>("/time", 1);
24     ros::Subscriber size_sub = n.subscribe("/
           size", 100, size_Callback);
25
26     clock_t start, end;
```

```
27    double cpu_time_used;
28    int *p;
29
30    while(ros::ok())
31    {
32        start = clock();
33        p = testmain(size*N,THREADS_PER_BLOCK);
34        end = clock();
35
36        cpu_time_used = ((double) (end - start)
              ) / CLOCKS_PER_SEC;
```

```
37        cpu_time_used = ((double) (end - start)
              ) / CLOCKS_PER_SEC;
38        msg_time.data=cpu_time_used;
39        time_pub.publish(msg_time);
40
41        ros::spinOnce();
42    }
43    return 0;
44 }
```

Code 4 Processing time evaluation for vector addition code (main.cpp).

Let's take a look at what the code is doing. In the new main.cpp file, we start two topics, one subscriber and one publisher. The subscriber sets the length of a vector (multiplied by 1 million) and have '/size' as topic name. It is started with an initial value of 10 (meaning a vector with 10 million int values). The '/size' topic can only assume values between 1 and 100. The publisher is a Float32 topic called '/time' that informs the time taken for processing the vector addition operation between two vectors with random values with the length informed by '/size'. Inside the main loop the package starts a clock, calls the .cu code, calculates the time taken for processing and updates the ROS node.

In the template.cu, we now initialize three pointers (in1, in2 and out) to use for storing the three vectors on the host (CPU), and other three pointers (d_in1, d_in2 and d_out) to store the vectors in the device (GPU). The function cudaMalloc() is used to allocate GPU memory for the device vectors, and works almost exactly like malloc(), with the difference that the pointer is passed as argument instead of as return. The memory of the host pointers is allocated and the two vectors are filled with random integers (through the function random_ints()). To copy the content of the vectors from the host to the device, the function cudaMemcpy() is used. This function have as arguments the device pointer, the host pointer, the size in bytes and the kind of transfer (*cudaMemcpyHostToDevice* or *cudaMemcpyDeviceToHost*) and is needed before and after our function since we need to input the data to the device and transfer the result back to the host after processing. Similarly, every memory allocated on the device must be freed in the end using the cudaFree() function, which works exactly like free().

After the allocation and definition of the vectors, the function add() is called. Observe that now instead of '$<<< 1, 1 >>>$' we have '$<<< N, M >>>$'. These two numbers set the parallelism of the CUDA function being called. The N number represents the number of blocks used, and M represents the number of threads per each block.

The number of threads per block is easy to define. Usually it can be constant between all applications, between 64 and 512, and a multiple of 32. Kernels (GPU functions) issue instructions in warps (warps are 32 threads). If you pick a block size of 100, the GPU will issue commands to 128 threads, but 28 will be wasted. Good first choices of threads per block are 128, 256 and 512. It's better to leave a constant THREADS_PER_BLOCK defined as it is on the template.cu file. The N number shall be calculated based on your chosen threads per block. N multiplied by M should give the amount of iterations needed if using a 'for' loop instead, with N rounded up, meaning that N should be the total amount of data to be processed divided by the number of threads per block.

To run the package, compile and run it by typing:

```
$ cd ~/catkin_ws $ catkin_make $ roslaunch
    roscuda_vectoradd
roscuda_vectoradd.launch
```

Open another terminal so that you can echo the value of the '/time' topic with:

```
$ rostopic echo /time
```

This will show the time taken to perform the vector addition using CUDA. You can change the size of the vector to 30 or other number of your choice (between 1 and 100) by publishing on the '/size' topic typing:

```
$ rostopic pub /size std_msgs/Int32 "data: 30"
```

You will notice the processing time will increase and the '/time' topic will report with a lower frequency. In the next example we will see better ways to guarantee the correct definition of the number of threads and blocks. We will also tweak the add() kernel function to apply an average and use it to filter a noisy laserscan input.

3.2   Example 2: ROS/CUDA Laserscan Filter

The next code presented will show how to apply an average over the three last frames of a laserscan topic. The laserscan topic was captured in a bag file and it is available in the Git repository https://github.com/air-lasca/ros-cuda, inside the folder named 'roscuda_laserscan_filter'. You will need to download the 'laserscan_noise.bag' file so that you can test the package this tutorial will cover. The bag contains data collected

(a)

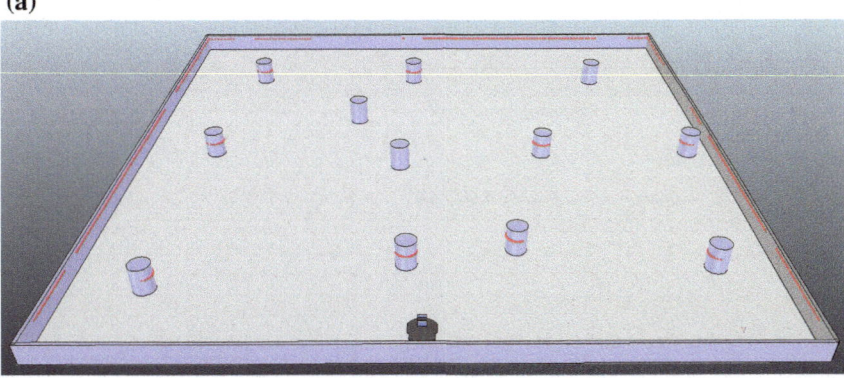

(b)

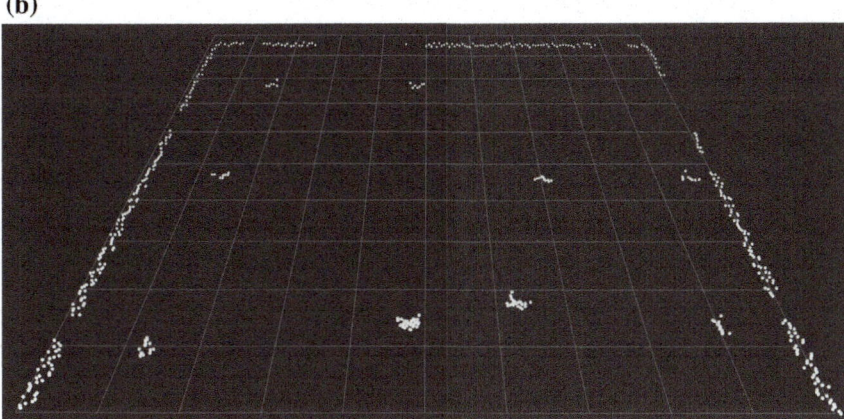

Fig. 8 **a** Simulated arena with P3AT robot and laserscan created on the software V-Rep. **b** Laserscan topic with added noise plotted on software Rviz

from a simulated Sick LMS-200 [24] sensor with artificially added white noise (7 dBm). The message contains the distance data of 511 points acquired inside a simulated arena (Fig. 8a). These messages were collected through the software V-Rep [25]. Upon the clean laserscan message from V-Rep a white noise is added over its ranges vector. The laserscan with noise can be visualized with the software Rviz (Fig. 8b).

What we will do in this new package is create a CUDA function that performs an average filter over the last three frames of the laserscan with noise, with the goal of reducing the magnitude of the noise. Copy the content of codes 5 and 6 into two new files inside a new package called 'roscuda_laserscan_filter'.

```
1  #include <stdio.h>
2  #include <stdlib.h>
3  #include <cuda.h>
```

```
 4  #include <cuda_runtime.h>
 5
 6  #define THREADS_PER_BLOCK 128
 7
 8  __global__  void av3(int n, float *in1, float *in2
        , float *in3, float *out)
 9  {
10      int index = threadIdx.x + blockIdx.x *
            blockDim.x;
11      // Guarantees that index does not go beyond
            vector size and applies average
12      if (index<n)
13      {
14              out[index] = (in1[index] + in2[
                    index] + in3[index])/3;
15      }
16  }
17
18  float *average3(int num, float *in1, float *in2,
        float *in3, float *out)
19  {
20      // Device copies of three inputs and output,
            size of allocated memory, num of threads
            and blocks
21      float *d_in1, *d_in2, *d_in3, *d_out;
22      int size = num * sizeof(float);
23      int thr, blk;
24      // Alloc memory for device copies of inputs
            and outputs
25      cudaMalloc((void **)&d_in1, size);
26      cudaMalloc((void **)&d_in2, size);
27      cudaMalloc((void **)&d_in3, size);
28      cudaMalloc((void **)&d_out, size);
29      // Copy inputs to device
30      cudaMemcpy(d_in1, in1, size,
            cudaMemcpyHostToDevice);
31      cudaMemcpy(d_in2, in2, size,
            cudaMemcpyHostToDevice);
32      cudaMemcpy(d_in3, in3, size,
            cudaMemcpyHostToDevice);
33      // Calculates blocks and threads and launch
            average3 kernel on GPU
34      blk=floor(num/THREADS_PER_BLOCK)+1;
```

```
35      blk=floor(num/THREADS_PER_BLOCK)+1;
36      thr=THREADS_PER_BLOCK;
37      av3<<<blk,thr>>>(num, d_in1, d_in2, d_in3,
            d_out);
38      // Wait for the GPU to finish
39      cudaDeviceSynchronize();
40      // Copy result back to host and cleanup
```

```
41        cudaMemcpy(out, d_out, size,
              cudaMemcpyDeviceToHost);
42        cudaFree(d_in1); cudaFree(d_in2); cudaFree(
              d_in3); cudaFree(d_out);
43        return out;
44 }
```

Code 5 Laserscan filtering package CUDA code (laserscan.cu).

```
1  #include "ros/ros.h"
2  #include "sensor_msgs/LaserScan.h"
3  #include "std_msgs/Float32.h"
4
5  //Since we already know the laserscan used has
      fixed size, we will not bother using dynamic
      size
6  #define SIZE 511
7
8  float *average3(int num, float *in1, float *in2,
      float *in3, float *out);
9  float in1[SIZE], in2[SIZE], in3[SIZE], out[SIZE];
10 sensor_msgs::LaserScan msg_laser;
11
12 void laserscan_Callback(const sensor_msgs::
      LaserScan& msg)
13 {
14     msg_laser=msg;
15     // Cycle vectors (in3=in2, in2=in1, in1=new)
16     for(int i=0;i<SIZE;i++)
17     {
18         in3[i]=in2[i];
19         in2[i]=in1[i];
20         in1[i]=msg.ranges[i];
21     }
22 }
23
24 int main(int argc, char **argv)
25 {
26     ros::init(argc, argv, "
          roscuda_laserscan_filter");
27     ros::NodeHandle n;
28     ros::Publisher laser_pub = n.advertise<
          sensor_msgs::LaserScan>("/laserscan/
          filtered", 1);
29     ros::Subscriber laser_sub = n.subscribe("/
          laserscan/raw", 100, laserscan_Callback);
30
31     // Initializes the vectors with zeros
32     for (int i = 0; i < SIZE; ++i)
33     {
34         in1[i]=in2[i]=in3[i]=out[i]=0;
```

```
35          in1[i]=in2[i]=in3[i]=out[i]=0;
36      }
37      while(ros::ok())
38      {   // Get new message and perform average
39          ros::spinOnce();
40          average3(SIZE, in1, in2, in3, out);
41          // Assign frame_id and ranges size to
                 be able to publish and visualize
                 topic
42          msg_laser.header.frame_id="
                 LaserScanner_2D";
43          msg_laser.ranges.resize(511);
44          // Assign values
45          for(int i=0;i<SIZE;i++)
46          {
47              msg_laser.ranges[i]=out[i];
48          }
49          laser_pub.publish(msg_laser);
50      }
51      return 0;
52  }
```

Code 6 Laserscan filtering package C code (main.cpp).

Now let's review the code. Starting on the main.cpp file (Code 6), we define the SIZE of the laserscan ranges vector since it's constant. This is because the bag file was captured with simulated data on a small arena (smaller than the maximum range of the sensor), we have fixed size and no invalid range scanning. Then we initialize the variables (three vectors to store the inputs and one for the output), the laserscan message that will provide the filtered scan and the call of our CUDA 'average3' function.

On the main function (Code 6, line 24 onward), we initialize the ROS node, the publisher and the subscriber. Then we initialize all the vectors with zeros and go into the main loop (line 36). Inside the main loop we get incoming ROS messages, which will call the laserscan callback function. In the callback we cycle through the vectors, the 'in3' vector is discarded and receives the values of the 'in2' vector, the 'in2' vector receives data from the 'in1' vector, and the 'in1' vector receives the new reading. Finishing the cycle will bring the execution back to the main loop, which will then call the 'average3' function located inside the .cu file.

Inside the 'laserscan.cu' file, we have the 'average3' function (Code 5, line 18) which takes the size of the laserscan range vector as parameter, as well as the four vectors involved in the average, three vectors as input and one as output (in our case, 'in1', 'in2', 'in3' as input and 'out' as output). The function starts initializing the pointers that will store the vectors on the device (GPU) and allocating memory for them on the device with cudaMalloc(). The three inputs are then copied to the device vectors using cudaMemcpy(). Now, to guarantee no errors on the thread and block numbers, they are calculated outside the kernel function call using two integer variables ('blk' and 'thr'). After this the kernel function 'av3' is called. The function 'av3' (line 8) initializes the index and uses an 'if' to guarantee no number beyond the index will be computed, for that it is necessary to take the number of computations 'num' as parameter. Inside the 'av3' function a simple average of the last three frames of the laserscan is applied and stored in 'd_out'. The program waits for the CUDA computation to finish (line 38) using cudaDeviceSynchronize(), copy the result back to 'out' and free the allocated device memory with cudaFree().

Back to the main function (Code 6, line 41), we assign a frame and the size of the ranges vector for our message, so that we can plot the result in Rviz and insert the new range values. A 'for' loop copies the content of the 'out' vector inside our message and after it is finished, the message is published and the loop starts again.

If you have downloaded the bag file, move it inside your catkin workspace (~/catkin_ws) so that we have easy access to it. Let's compile the package, run the bag in loop and launch the package to see if everything is working. To do that, on a terminal type:

```
$ cd ~/catkin_ws
$ catkin_make
$ rosbag play -l ~/catkin_ws/laserscan_noise.bag
```

This will compile the package and play the bag in loop. The bag contains a topic called '/laserscan/raw' of type LaserScan. You can see the laserscan message with noise using Rviz, to do that type 'rviz' in a new terminal, set the fixed frame to 'LaserScanner_2D', click in add, sort by topic and find '/laserscan/raw' to visualize it.

We should then launch the package to visualize it. Open a new terminal and type:

```
$ roslaunch roscuda_laserscan_filter
    roscuda_laserscan_filter.launch
```

The package will start and create a topic called '/laserscan/filtered'. Add it on the Rviz the same way the raw topic was added earlier. You will be able to see that, despite still having visible amount of noise, the magnitude of the noise decreased. Opening the messages from each topic on Matlab allow us to measure the SNR (signal to noise ratio), interpreting the range vectors as a signal. The average value of the error was calculated comparing the original range vector without noise to the one with added noise and to the filtered range vector. The range vector with added noise had an SNR 39.4 dB, and after the filtering the SNR raised to 43.1 dB, resulting in a gain of 3.7 dB over the noise. The filter successfully reduces the uncertainty from the signal, allowing the scan to be more precise.

In the next example we will see how to apply a filter to robot vision, making the task of extracting characteristics directly from the robot's camera easier and faster.

3.3 Example 3: ROS/CUDA Sobel Image Filter

For the next example, let's perform a Sobel filtering over a Microsoft Kinect RGB camera output using CUDA. To do that, we will need to use the OpenCV library through the 'vision_opencv' stack for ROS. If you installed the ROS desktop full version, the stack should already be included in your ROS libraries and no additional installation is required. If you are following the examples on a Nvidia Tegra development board, you will need to follow the OpenCV4Tegra and ROS integration tutorial on Sect. 2.4 before proceeding.

To test if the filter is working you will need an input video feed. Again, a bag file is available if you do not have any other source of ROS image topic, though it is advised that you first try the package with the bag file since it is sure to work that way, before trying to apply it over other video feed. The bag file called 'image_raw.bag' can be found in the Git repository https://github.com/air-lasca/ros-cuda, inside the 'roscuda_img_filter' folder.

The bag file was recorded using V-Rep. A Microsoft Kinect was pointed towards a simulated person walking (Bill, from V-Rep standard model database). The topics collected by the bag file were all the topics related to the Kinect's RGB camera. No spacial points were collected.

A Sobel image filter is used primarily for edge detection. The filter consists of two masks applied over the image, their values are then combined to create the resultant filtered image. These two masks simulate a gradient operation applied vertically and horizontally over the image. The two matrix that are convoluted over the image are as follow:

$$
V = \begin{bmatrix} -1 & 0 & 1 \\ -2 & 0 & 2 \\ -1 & 0 & 1 \end{bmatrix} \qquad W = \begin{bmatrix} 1 & 2 & 1 \\ 0 & 0 & 0 \\ -1 & -2 & -1 \end{bmatrix}
$$

This means that each pixel of the filtered image is the result of the original eight surrounding pixels multiplied by the respective coefficient of the two matrix and added together. This results in a problem to calculate the value for pixels on the edge of the image since they have less than eight surrounding pixels. There are many workarounds for this, like repeating the image borders or ignoring just the absent pixels, but to simplify our example we will just ignore the border pixels.

To create the package we will have to add dependencies we have not used before. They are the 'cv_bridge' and 'image_transport' packages. So, when using 'catkin_create_pkg' the resulting command line will be:

```
$ catkin_create_pkg roscuda_img_filter
    sensor_msgs cv_bridge roscpp std_msgs
    image_transport
```

We still have to edit the CMakeLists file to include the CUDA library, but it is the same process as the other packages. Create the two files containing the CUDA and C++ code. This example was based on the cv_bridge C++ tutorial available on http://wiki.ros.org/cv_bridge/Tutorials. The CUDA file should have the content of Code 7.

```
1  #include <stdio.h>
2  #include <stdlib.h>
3  #include <cuda.h>
4  #include <cuda_runtime.h>
5  #include <cv_bridge/cv_bridge.h>
6
7  #define THREADS_PER_BLOCK 128
8
9  __global__ void mask(double num, uchar *in, uchar
       *out, int n)
10 {
11     int index = threadIdx.x + blockIdx.x *
           blockDim.x;
12     int aux1,aux2;
13     // Guarantees that index does not go beyond
           vector size and applies filter ignoring
           image edge
14     if (index>(n+1) && index<(num-(n+1)))
15     {
16
17                 aux1 = (in[index-n-1] + in[index-
                       n]*2 - in[index+n-1]
18                     + in[index-n+1] - in[index+n
                           ]*2 - in[index+n+1]);
19                 aux2 = (in[index-n-1] + in[index
                       -1]*2 +  in[index+n-1]
20                     - in[index-n+1] - in[index
                           +1]*2 - in[index+n+1]);
21                 out[index] = abs(aux2-aux1);
22     }else{if(index<num){out[index]=0;}}
23 }
24
25 int sobelfilter(cv::Mat src, cv::Mat dst)
26 {
27     int m = src.rows;
28     int n = src.cols;
29     double num=n*m;
30     // Define the vectors that store the image
           data (pre and post filter)
31     std::vector<uchar> v,w;
32     v = src.reshape(1,num);
33     w.reserve(num);
34     // Device copies of v, w, size of allocated
           memory, threads and blocks
35     uchar *d_in, *d_out;
36     double size = num * sizeof(uchar);
37     int thr, blk;
38     // Alloc memory for device copies of v, w
```

```
39    cudaMalloc((void **)&d_in, size);
40    cudaMalloc((void **)&d_out, size);
41    // Copy inputs to device
```

```
42    cudaMemcpy(d_in, v.data(), size,
         cudaMemcpyHostToDevice);
43    // Calculates blocks and threads and launch
         mask kernel on GPU
44    blk=floor(num/THREADS_PER_BLOCK)+1;
45    thr=THREADS_PER_BLOCK;
46    mask<<<blk,thr>>>(num, d_in, d_out, n);
47    // Wait for the GPU to finish
48    cudaDeviceSynchronize();
49    // Copy result back to host, then from
         vector to image, and cleanup
50    cudaMemcpy(w.data(), d_out, size,
         cudaMemcpyDeviceToHost);
51    memcpy(dst.data, w.data(), size);
52    cudaFree(d_in); cudaFree(d_out);
53    return 1;
54 }
```

Code 7 Sobel image filtering package CUDA code (sobel.cu)

The C++ file should look like Code 8.

```
1 #include <ros/ros.h>
2 #include <image_transport/image_transport.h>
3 #include <cv_bridge/cv_bridge.h>
4 #include <opencv2/highgui/highgui.hpp>
5
6 static const std::string OPENCV_WINDOW_1 = "Image
      window";
7 static const std::string OPENCV_WINDOW_2 = "Gray
      image window";
8 static const std::string OPENCV_WINDOW_3 = "
      Filtered image window";
9 int sobelfilter(cv::Mat src, cv::Mat dst);
10 //FRAMES ARE 640 x 480 PIXELS
11
12 class ImageConverter
13 {
14   ros::NodeHandle nh_;
15   image_transport::ImageTransport it_;
16   image_transport::Subscriber image_sub_;
17   image_transport::Publisher gimage_pub_;
18   image_transport::Publisher image_pub_;
19
20 public:
21   ImageConverter()
22     : it_(nh_)
```

```
23    {
24      // Subscribe to input video feed and publish
          output videos
25      image_sub_ = it_.subscribe("/image/cam", 1,
26        &ImageConverter::imageCb, this);
27      gimage_pub_ = it_.advertise("/image_converter
          /gray_video", 1);
```

```
28      image_pub_ = it_.advertise("/image_converter/
          output_video", 1);
29
30      cv::namedWindow(OPENCV_WINDOW_1);
31      cv::namedWindow(OPENCV_WINDOW_2);
32      cv::namedWindow(OPENCV_WINDOW_3);
33    }
34
35    ~ImageConverter()
36    {
37      cv::destroyWindow(OPENCV_WINDOW_1);
38      cv::destroyWindow(OPENCV_WINDOW_2);
39      cv::destroyWindow(OPENCV_WINDOW_3);
40    }
41
42    void imageCb(const sensor_msgs::ImageConstPtr&
        msg)
43    {
44      cv_bridge::CvImagePtr cv_ptr;
45      try
46      {
47        cv_ptr = cv_bridge::toCvCopy(msg,
            sensor_msgs::image_encodings::BGR8);
48      }
49      catch (cv_bridge::Exception& e)
50      {
51        ROS_ERROR("cv_bridge exception: %s", e.what
            ());
52        return;
53      }
54
55      cv::Mat destiny, gray;
56
57      // Shows input stream
58      cv::imshow(OPENCV_WINDOW_1, cv_ptr->image);
59      cv::waitKey(3);
60
61      // Converts to grayscale and updates
          publisher
62      cv::cvtColor(cv_ptr->image, gray, CV_BGR2GRAY);
63      cv_ptr->image = gray;
64      sensor_msgs::ImagePtr gray_msg = cv_bridge::
          CvImage(std_msgs::Header(), "mono8", gray)
          .toImageMsg();
```

```
65    // Update Grayscale GUI Window
66    cv::imshow(OPENCV_WINDOW_2, cv_ptr->image);
67    cv::waitKey(3);
68    // Assign destiny with the same size as image
         , defines encoding, fill it with zeros
69    destiny = cv::Mat::zeros( cv_ptr->image.size
         (), CV_8UC1);
70
71    // Applies the filter (function from the .cu
         file)
72    sobelfilter(cv_ptr->image,destiny);
73
74    // Output modified video stream and publish
         messages
75    sensor_msgs::ImagePtr dst_msg = cv_bridge::
         CvImage(std_msgs::Header(), "mono8",
         destiny).toImageMsg();
```

```
76       cv::imshow(OPENCV_WINDOW_3, destiny);
77       cv::waitKey(3);
78       image_pub_.publish(dst_msg);
79       gimage_pub_.publish(gray_msg);
80    }
81  };
82
83  int main(int argc, char** argv)
84  {
85    ros::init(argc, argv, "roscuda_img_filter");
86    ImageConverter ic;
87    ros::spin();
88    return 0;
89  }
```

Code 8 Image filtering package C++ code (main.cpp)

Now let's go over the code to understand what it is doing. Starting on main.cpp (Code 8, lines 1-4), we include all the libraries we will use. The 'image_transport' library eases ROS image topics manipulation, the 'cv_bridge' library allows us to use OpenCV with ROS, and the 'HighGUI' (High-level Graphical User Interface) library allows us to create and manipulate UI windows. Then, we define the three windows we use to show the process of image filtering, which will be the input RGB image, the grayscale image and the output filtered image. We also define the call for the C++ function inside the .cu file.

Then, after line 12, we create a class named ImageConverter. This class is responsible to initialize the node handle, the publishers and subscribers. The constructor function defines the topics for the publishers and the subscriber and creates the UI windows. The destructor is responsible for destroying the UI windows. With the definition of the subscriber, we define a callback function for it called imageCb, which is also inside the ImageConverter class (line 41). The callback creates an image pointer that will receive the ROS message and creates two new Mat objects (OpenCV matrix used to manipulate images) to store the output and the gray image. The GUI window to show the input is updated and then the input image is converted to grayscale (line 62). After that, we initialize the output image with zeros the same size as the input, define its encoding and call the function on the .cu file that will apply the filter.

In the .cu file (Code 7, line 25) we get the dimensions of the image and then define two vectors (for input and output) with type unsigned char. The reason why we use unsigned char is because the gray image is a matrix which each pixel has values ranging from 0 to 255 (255 being black), and 'uchar' behaves the same way. We reshape the image matrix to a vector using the reshape() property, and then allocate the output vector to have the same number of pixels. After that we define the two pointers that will store the data in the device and allocated device memory for them with cudaMalloc(). The content of the input vector is then copied to the device using cudaMemcpy(). The kernel function is called using the number of points, the device input and output vectors and the number of columns of the image.

Inside the kernel function (Code 7, line 9), we define the index and two auxiliary integers. We use an 'if' to not apply the filter on the first and last rows of the image to avoid dealing with special cases (since the mask would be applied with less than eight surrounding pixels). We calculate the resultant from each mask and add them to the resulting pixel for the output image (lines 17–21) and insert a zero if there are not eight surrounding pixels. We wait for the GPU to finish (line 48), copy back the result, free the device memory and return.

Back to the callback function on main.cpp (Code 8, line 78), we update the output message and its GUI window, and publish the messages, starting everything over again when a new frame arrives on the subscriber.

Now, let's test the code to see the filter in action. Compile the package with 'catkin_make' like you did with previous examples. Also, to visualize the filter being applied, we need to start the bag file. Navigate to the folder where the bag file is located and type:

```
$ rosbag play -l image_raw.bag
```

This will start the bag file, and create the topics of the video feed (480x640p) from a simulated Microsoft Kinect that will be the input for our package. Now we can just start the package using 'roslaunch' or 'rosrun'.

```
$ roslaunch roscuda_img_filter roscuda_img_filter
  .launch
```

If the bag file was already running, three windows will pop up showing Bill walking from left to right. The first window will be the colored image view which is the input of the package, the second window will show the video feed in grayscale, and the third will show the Sobel border detection filter applied (Fig. 9).

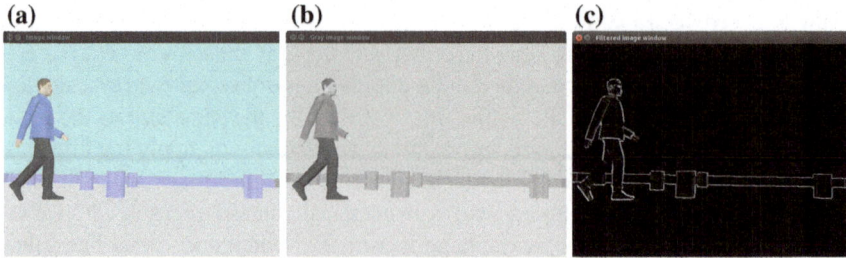

Fig. 9 **a** Video feed of simulated arena with Bill walking through it. **b** Video feed after being converted to grayscale. **c** Video feed after Sobel edge detection filter applied

Fig. 10 **a** Video feed of a real Microsoft Kinect in a high contrast environment. **b** Video feed of real Microsoft Kinect after being converted to grayscale. **c**: Video feed of real Microsoft Kinect after Sobel filter applied

Any video stream input that uses ROS 'sensor_msgs/Image' type will work as input for the filter. Here is another example of the filter at work over a real Kinect RGB stream from a real robot (Fig 10).

It is also very important to notice that, since the example works for cameras of different resolutions and that all the image processing is done inside the CUDA kernel, one can edit the package to apply many different kinds of image filters that uses convolution masks without having to worry much about encoding (since OpenCV can handle that very easily).

4 Conclusion

The chapter aims to introduce a newcomer to GPU processing all he or she needs to know to get started, and how to do that alongside ROS. Parallel processing is a very powerful tool to use when dealing with large amounts of data, and since robots and their sensors are getting more complex with time, using GPU to tackle the heavy computation required for extracting characteristics becomes more and more necessary. The chapter shows how the user can configure an Nvidia Tegra development board and use it embedded in mobile robots, and also how one can create CUDA ROS packages to take full advantage of the parallel processing of the boards.

Acknowledgements The projects of this chapter were partially funded by National Counsel of Technological and Scientific Development of Brazil (CNPq), by Coordination for the Improvement of Higher Level People (CAPES) and by National Agency of Petroleum, Natural Gas and Biofuels (ANP) together with the Financier of Studies and Projects (FINEP) and Brazilian Ministry of Science and Technology (MCT) through the ANP Human Resources Program for the Petroleum and Gas Sector - PRH-ANP/MCT PRH10-UTFPR. We gratefully acknowledge the support of NVIDIA Corporation with the donation of the Tegra X1 and Tegra K1 development boards used for this chapter.

References

1. C.J. Thompson, S. Hahn, M. Oskin, Using modern graphics architectures for general-purpose computing: a framework and analysis, in *Proceedings 35th Annual IEEE/ACM International Symposium on Micro Architecture, (MICRO-35)* (IEEE, New York, 2002), pp. 306–317
2. I. Buck, T. Foley, D. Horn, J. Sugerman, K. Fatahalian, M. Houston, P. Hanrahan, Brook for GPUs: stream computing on graphics hardware. ACM Trans. Graph. (TOG) **23**(3), 777–786 (2004). ACM
3. N.K. Govindaraju, B. Lloyd, W. Wang, M. Lin, D. Manocha, Fast computation of database operations using graphics processors, in *Proceedings of the 2004 ACM SIGMOD International Conference on Management of Data* (ACM, 2004), pp. 215–226
4. Z. Fan, F. Qiu, A. Kaufman, S. Yoakum-Stover, Gpu cluster for high performance computing, in *Proceedings of the ACM/IEEE SC2004 Conference on Supercomputing* (IEEE, New York, 2004), pp. 47–47
5. A. Barak, T. Ben-Nun, E. Levy, A. Shiloh, A package for opencl based heterogeneous computing on clusters with many gpu devices, in *2010 IEEE International Conference on Cluster Computing Workshops and Posters (CLUSTER WORKSHOPS)* (IEEE, New York 2010), pp. 1–7
6. Nvidia, Compute unified device architecture programming guide, 2007
7. V.W. Lee, C. Kim, J. Chhugani, M. Deisher, D. Kim, A.D. Nguyen, N. Satish, M. Smelyanskiy, S. Chennupaty, P. Hammarlund et al., Debunking the 100X GPU vs. CPU myth: an evaluation of throughput computing on CPU and GPU. ACM SIGARCH Comput. Archit. News **38**(3), 451–460 (2010)
8. P. Micikevicius, 3d finite difference computation on GPUS using CUDA, in *Proceedings of 2nd Workshop on General Purpose Processing on Graphics Processing Units* (ACM, 2009), pp. 79–84
9. T. Preis, P. Virnau, W. Paul, J.J. Schneider, Gpu accelerated monte carlo simulation of the 2d and 3d ising model. J. Comput. Phys. **228**(12), 4468–4477 (2009)
10. D. Qiu, S. May, A. Nüchter, GPU-accelerated nearest neighbor search for 3d registration, in *International Conference on Computer Vision Systems* (Springer, Berlin, 2009), pp. 194–203
11. R. Ugolotti, G. Micconi, J. Aleotti, S.Cagnoni, GPU-based point cloud recognition using evolutionary algorithms, in *European Conference on the Applications of Evolutionary Computation* (Springer, Berlin, 2014), pp. 489–500
12. L.M.F. Christino, Aceleração por gpu de serviços em sistemas robóticos focado no processamento de tempo real de nuvem de pontos 3d, Ph.D. dissertation, Universidade de São Paulo
13. K.B. Kaldestad, G. Hovland, D.A. Anisi, 3d sensor-based obstacle detection comparing octrees and point clouds using CUDA. Model. Identif. Control **33**(4), 123 (2012)

14. M. Liu, F. Pomerleau, F. Colas, R. Siegwart, Normal estimation for pointcloud using GPU based sparse tensor voting, in *2012 IEEE International Conference on Robotics and Biomimetics (ROBIO)* (IEEE, New York, 2012), pp. 91–96

15. R.B. Rusu, S. Cousins, 3D is here: Point cloud library (PCL), in *2011 IEEE International Conference on Robotics and Automation (ICRA)* (IEEE, New York, 2011), pp. 1–4

16. P. Michel, J. Chestnutt, S. Kagami, K. Nishiwaki, J. Kuffner, T. Kanade, GPU-accelerated real-time 3D tracking for humanoid locomotion and stair climbing, in *IEEE/RSJ International Conference on Intelligent Robots and Systems, IROS 2007* (IEEE, New York, 2007), pp. 463–469

17. P. Henry, M. Krainin, E. Herbst, X. Ren, D. Fox, RGB-D mapping: using kinect-style depth cameras for dense 3D modeling of indoor environments. Int. J. Robot. Res. **31**(5), 647–663 (2012)

18. P. Merrell, A. Akbarzadeh, L. Wang, P. Mordohai, J.-M. Frahm, R. Yang, D. Nistér, M. Pollefeys, Real-time visibility-based fusion of depth maps, in *IEEE 11th International Conference on Computer Vision, ICCV 2007* (IEEE, New York, 2007), pp. 1–8

19. C. Choi, H.I. Christensen, RGB-D object tracking: A particle filter approach on GPU, in *2013 IEEE/RSJ International Conference on Intelligent Robots and Systems (IROS)* (IEEE, New York, 2013), pp. 1084–1091

20. P.J.S. Leite, J.M.X.N. Teixeira, T.S.M.C. de Farias, V. Teichrieb, J. Kelner, Massively parallel nearest neighbor queries for dynamic point clouds on the GPU, in *21st International Symposium on Computer Architecture and High Performance Computing, SBAC-PAD'09* (IEEE, New York, 2009), pp. 19–25

21. JetsonHacks, Jetsonhacks - developing for Nvidia jetson, [Online] (2017), http://www.jetsonhacks.com/

22. W. Lucetti, Ros hacking for opencv on Nvidia jetson tx1 & jetson tk1, [Online], (2016), http://myzharbot.robot-home.it/blog/software/ros-nvidia-jetson-tx1-jetson-tk1-opencv-ultimate-guide/

23. C. Zeller, Cuda c/c++ basics, Nvidia Corporation, Supercomputing Tutorial, 2011, pp. 9–11

24. SICK, Lms200 technical description, [online][retrieved sep. 11, 2014], 2003

25. E. Rohmer, S.P. Singh, M. Freese, V-rep: a versatile and scalable robot simulation framework, in *2013 IEEE/RSJ International Conference on Intelligent Robots and Systems (IROS)* (IEEE, New York, 2013), pp. 1321–1326

Connecting ROS and FIWARE: Concepts and Tutorial

Raffaele Limosani, Alessandro Manzi, Laura Fiorini, Paolo Dario
and Filippo Cavallo

Abstract Nowadays, the Cloud technology permeates our daily life, spread in various services and applications used by modern instruments, such as smartphones, computer, and IoT devices. Besides, the robotic field represents one of the future emerging markets. Nevertheless, these two distinct worlds seem to be very far from each other, due to the lack of common strategies and standards. The aim of this tutorial chapter is to provide a walkthrough to build a basic Cloud Robotics application using ROS and the FIWARE Cloud framework. At the beginning, the chapter offers step-by-step instructions to create and manage an Orion Context Broker running on a virtual machine. Then, the `firos` package is used to integrate the ROS topic communication using publishers and subscribers, providing a clear example. Finally, a more concrete use case is detailed, developing a Cloud Robotics application to control a ROS-based robot through the FIWARE framework. The code of the present tutorial is available at https://github.com/Raffa87/ROS_FIWARE_Tutorial, tested using ROS Indigo.

Keywords Tutorial ROS FIWARE firos

1 Introduction

Nowadays, the ROS framework [1] is considered the "de facto" robotic standard in academic research and it is widely adopted all over the world [2]. Moreover, in the last years, the ROS-Industrial project is trying to extend the advanced capabilities of ROS also to manufacturing automation. Therefore, it is reasonable to forecast a much

R. Limosani (✉) · A. Manzi · L. Fiorini · P. Dario · F. Cavallo
The BioRobotics Institute, Scuola Superiore Sant'Anna, Viale Rinaldo Piaggio, 34,
56026 Pontedera, PI, Italy
e-mail: r.limosani@santannapisa.it
URL: https://www.santannapisa.it/en/institute/biorobotics/biorobotics-institute

© Springer International Publishing AG, part of Springer Nature 2019
A. Koubaa (ed.), *Robot Operating System (ROS)*, Studies in Computational
Intelligence 778, https://doi.org/10.1007/978-3-319-91590-6_13

449

intense use of ROS in the near future for developing complex services involving several interacting robots. However, the ROS framework is conceived to develop systems that run on machines connected to the same LAN network. To overcome this limitation, different attempts have been made to enhance the ROS capabilities through the use of Cloud resources, leveraging the so-called "Cloud Robotics". One of the definitions of this concept is "any robot or automation system that relies on either data or code from a network to support its operation, i.e., where not all sensing, computation, and memory are integrated into a single standalone system" [3].

Several research groups have focused their efforts on cloud robotic challenges: for example, in [4] authors extended the computation and information sharing capabilities of networked robotics by proposing a cloud robotic architecture, while in [5] the key research issues in cloud network robotics are discussed analyzing a case study in a shopping mall. One of the first application of the Cloud Robotics paradigm is the DAvinCi [6] project developed in 2010, which focus more on the computational side rather than in communication between robot and server. Another example is represented by Rapyuta [7], an open-source Cloud Robotics platform, developed during the RoboEarth project [8]. Nevertheless, the overall performances do not differ much from other solutions [9]. Furthermore, cloud infrastructures have been developed to teach ROS,[1] focusing on the release of easy-to-use development instruments.

Despite the aforementioned researches, robotics and cloud technologies still appear two distinct worlds in terms of standards, communities, instruments, and infrastructures. Although the robotic field represents an emerging future market, the Cloud is currently an active business. In fact, most of the big Hi-Tech companies offer services over the cloud, in the form of Software-as-a-Service, or Platform-as-a-Service, including the Google Cloud Platform,[2] the Amazon Web Services[3] and the Microsoft Azure.[4]

The importance of the Cloud technology is also underlined by the European Union, which founded the universAAL project[5] [10], and, more recently, the FIWARE[6] [11] program, which provides a middleware platform for the development of applications for Future Internet.

FIWARE provides an enhanced OpenStack-based cloud environment plus a rich set of open standard APIs that make it easier to connect to the Internet of Things, process and analyze Big data and real-time media or incorporate advanced features for user interaction.

Even if a link between ROS and FIWARE has been already implemented[7] [12], its use is still difficult due to a lack of tutorials and documentation for users with

---

[1]http://www.theconstructsim.com/, visited on December 2017.

[2]https://cloud.google.com/, visited on December 2017.

[3]https://aws.amazon.com/, visited on December 2017.

[4]https://azure.microsoft.com, visited on December 2017.

[5]http://universaal.aaloa.org/, visited on December 2017.

[6]https://www.fiware.org, visited on December 2017

[7]http://wiki.ros.org/firos, visited on December 2017.

different backgrounds (i.e. experts in ROS without previous experience in FIWARE or vice-versa).

The aim of this tutorial chapter is to provide a step-by-step walkthrough about the integration of ROS with FIWARE, starting from the creation of a basic virtual machine over the Cloud. It shows how to handle the communication using subscribers and publishers introducing the `firos` package. At the end, a concrete Cloud Robotics example is provided to control a simulated robot using the FIWARE framework.

In details, this chapter covers the following aspects:

- basic concepts about ROS and FIWARE;
- introduction to FIWARE and the `firos` package;
- creation and management of a FIWARE Context Broker;
- how to use the FIWARE Context Broker;
- how to connect ROS and FIWARE;
- how to develop a robot control application through FIWARE.

At the end of the tutorial, despite the initial background, the reader will be able to manage communication among ROS and FIWARE, allowing to enhance robotic projects with Cloud capabilities. The final aim is to provide instruments and documentation that will improve the collaborations between the robotic and the Cloud worlds.

The remainder of this tutorial chapter is organized as follows. The basic concepts of the ROS middleware and the FIWARE framework are provided in Sect. 2 and in Sect. 3 respectively. A walkthrough about the creation and the use of the FIWARE Orion Context Broker and *firos* are detailed in Sects. 4 and 5, showing a first "*Hello world*" example. A more complex and concrete use case is described in Sect. 6, explaining how to develop a basic Cloud Robotics application using FIWARE to send a goal and receive data from a simulated robot. Finally, the Sect. 7 concludes the chapter and discusses further developments.

2 ROS Concepts

ROS[8] is an open-source, meta-operating system for robot control and development. The term "meta-operating" is used to indicate its similarity with an operating system, including hardware abstraction, low-level device control, implementation of commonly-used functionality and message-passing between processes. On the other hand, ROS is not an operating system in a strict sense; it currently runs on Unix-based platforms.

As the main feature, ROS provides a robust infrastructure that simplifies communications among processes. The ROS runtime "graph" is a peer-to-peer network of processes (potentially distributed across machines) that are loosely coupled using

---

[8]More information about ROS can be found at http://wiki.ros.org/ROS, visited on December 2017.

a common communication infrastructure. ROS implements different styles of communication, including synchronous RPC-style communication over services, asynchronous streaming of data over topics, and data storage through the Parameter Server.

2.1 ROS Graph

The Computation Graph[9] is the peer-to-peer network of ROS processes that are processing data together. The basic Computation Graph concepts of ROS are nodes, Master, Parameter Server, messages, services, topics, and bags, all of which provide data to the Graph in different ways.

For the purpose of this tutorial, only details about nodes, Master, messages, and topics are reported.

Nodes are processes that perform computation. ROS is designed to be modular at a fine-grained scale; a robot control system usually comprises many nodes. For example, one node controls a laser range-finder, one node controls the wheel motors, one node performs localization, one node performs path planning, one node provides a graphical view of the system, and so on.

Master provides name registration and lookup to the rest of the Computation Graph. Without the Master, nodes would not be able to find each other or exchange messages.

Messages are used by the nodes to communicate with each other. A message is simply a data structure, comprising typed fields. Standard primitive types (integer, floating point, boolean, etc.) are supported, as are arrays of primitive types. Messages can include arbitrarily nested structures and arrays.

Topics Messages are routed via a transport system with publish/subscribe semantics. A node sends out a message by publishing it to a given topic. The topic is a name that is used to identify the content of the message. A node that is interested in a certain kind of data will subscribe to the appropriate topic. There may be multiple concurrent publishers and subscribers for a single topic, and a single node may publish and/or subscribe to multiple topics. In general, publishers and subscribers are not aware of each others' existence. The idea is to decouple the production of information from its consumption. Logically, one can think of a topic as a strongly typed message bus. Each bus has a name, and anyone can connect to the bus to send or receive messages as long as they are the right type.

---

[9]Deeper information on ROS Graph can be found at http://wiki.ros.org/ROS/Concepts, visited on December 2017.

The Master acts as a nameservice in the ROS Computation Graph. Nodes connect to other nodes directly; the Master only provides lookup information, much like a DNS server, allowing nodes to dynamically create connections as new nodes are run. Refer to the official ROS website[10] for details about the framework installation.

3 FIWARE Concepts

The *FIWARE platform*[11] provides a rather simple yet powerful set of APIs that ease the development of smart applications in multiple vertical sectors. The specifications of these APIs are public and royalty-free. Besides, an open source reference implementation of each of the FIWARE components is publicly available so that multiple FIWARE providers can emerge faster in the market with a low-cost proposition.

For the purpose of this tutorial, reader will be leaded to be a user of FIWARE Lab, which is a non-commercial sandbox environment where innovation and experimentation based on FIWARE technologies take place. Entrepreneurs and individuals can test the technology as well as their applications, exploiting Open Data published by cities and other organizations. FIWARE Lab is deployed over a geographically distributed network of federated nodes leveraging on a wide range of experimental infrastructures.

In details, the FIWARE framework contains a rich library of components, called Generic Enablers (GE), that allow developers to put into effect functionalities such as the connection to the Internet of Things or Big Data analysis. By combining them, it is possible to develop modular and complex applications.

The GEs offer a number of general-purpose functions, including:

Data/Context Management   to easy access, gather, process, publish, and analyze context information (e.g. Comet, Cygnus, Kurento, Orion).
IoT Services   to use, search, and access IoT devices and sensors, handling various protocols.
Web-based User Interface   to give tools for 2D/3D graphics, and Geographical Information System (GIS) Data Provider.
Security   to implement security and privacy requirements (e.g. PEP Proxy, Key-Rock, AuthZForce).
Cloud Hosting   to provide computation, storage, and network resources (e.g. Docker, Murano, Bosun, Pegasus).

The deployment of a GE is referred as an *instance*. FIWARE allows to fully customize new instances or use a set of pre-configured instances available in the framework. For the purpose of this tutorial walkthrough, the remainder will focus on the Orion Context Broker, which is one of the key component of a FIWARE

---

[10]http://wiki.ros.org/ROS/Installation, visited on December 2017.
[11]https://www.fiware.org, visited on December 2017.

application. In addition, this section also introduces `firos`, which allows to connect a ROS node with Orion.

3.1 The Orion Context Broker

The Orion Context Broker[12] is a C++ implementation of the NGSIv2 REST API binding developed as a part of the FIWARE platform. The NGSI specification is an information model that uses the concept of entities to virtually represent physical objects in the real world. Any information about physical entities is expressed in the form of attributes of virtual entities.[13]

The Orion Context Broker allows to manage all the whole lifecycle of context information including updates, queries, registrations and subscriptions. It implements an NGSIv2 server to manage context information and its availability. Using the Orion Context Broker, the user is able to create context elements and manage them through updates and queries. In addition, it is possible to subscribe to context information and automatically receive a notification when some condition occurs (e.g. a context element has changed).

3.2 Firos

`firos` [12] represents the link between ROS and FIWARE. In particular, it is a ROS node that communicates with the Orion Context Broker to publish and listen robot data. In other words, `firos` works as a translator between the robotics field and the cloud world, transforming ROS messages into NGSI to publish them in the cloud, and vice-versa.

4 Walkthrough: How to Use the Orion Context Broker

As introduced, after a short description of ROS and FIWARE concepts, the aim of the paper is to provide working tutorial where the reader can understand new concepts and tools through practical example. Three walkthroughs are presented, in a sort of growing difficulty:

1. in the first one, the use of Orion Context Broker as "container" of data is explained, after a detailed step-by-step guide on the creation in FIWARE platform;
2. in the second one, an "Hello World" example is depicted showing the mechanism of communication-based on `firos` ROS node;

---

[12]http://fiware-orion.readthedocs.io/en/master/index.html.

[13]More information can be found at http://aeronbroker.github.io/Aeron.

3. in the third one, a more concrete example is presented showing how FIWARE
   and ROS can be used to create a cloud-robotic application where commands
   (e.g. navigation goals) can be remotely sent by a third party and feedbacks (e.g.
   odometry) can be remotely received.

4.1   First Step: Create a FIWARE Lab Account

The first step to begin the walkthrough is to become a FIWARE Lab user, therefore
create an account at https://account.lab.fiware.org. Mark the option "*I want to be
a trial user*". This procedure will create a trial account lasting 14 days. Check the
provided e-mail and confirm the account. To proceed with the tutorial, we need to
upgrade the trial account to be able to instantiate the required GE. Hence, log in, go
to the account settings and click on *Account Status*. Here it is possible to request a
Community Account upgrade.

For the purpose of this tutorial, the reader can select the *default quota* on the
relative upgrade form. It is worth to mention that the account upgrade process is not
immediate and can require up to two days.

4.2   Create a Orion Instance

This section covers the creation of a Orion Context Broker instance. The FIWARE
catalog contains a set of pre-configured instances to easy deploy specific Virtual
Machine (VM) on the Cloud. The use of pre-configured instances is not mandatory
and the user can also manually create its own personalized instance. The present tutorial will use a pre-configured Orion instance, which automatically creates a Virtual
Machine (VM) equipped with the Context Broker running on a centOS machine.

Create the instance To create the new Orion instance, move to the *Cloud* tab on the
FIWARE GUI and click on the *Launch New Instance*.

Instances

Choose an Orion Context Broker image from the list (*orion-psb-image-R5.4*) and launch it.

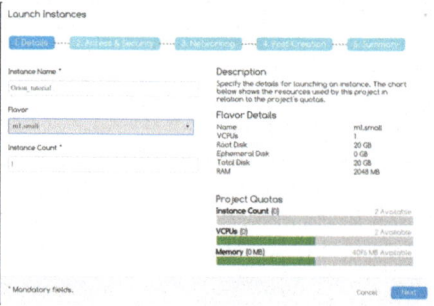

After that, a *Launch Instances* window related to its settings will appear. First of all, provide the instance *details*, choosing a name and selecting its `flavor` (i.e. resources of the VM). For the aim of this tutorial, it is enough to choose the `m1.small` option, which allocates a VM with 1 CPU, 2 Gb RAM, and 20 Gb of disk space:

After that, you can switch to the *Access and Security* tab to generate the key-pairs needed for the secure SSH remote access to the instance. By clicking on `Create Keypair` the VM will be configured for the remote connection and the relative key-pair will be download from the host computer:

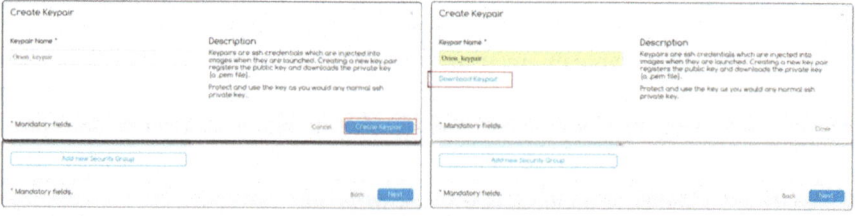

Use the new created keypair and check the `default` option in the *Security Group* field:

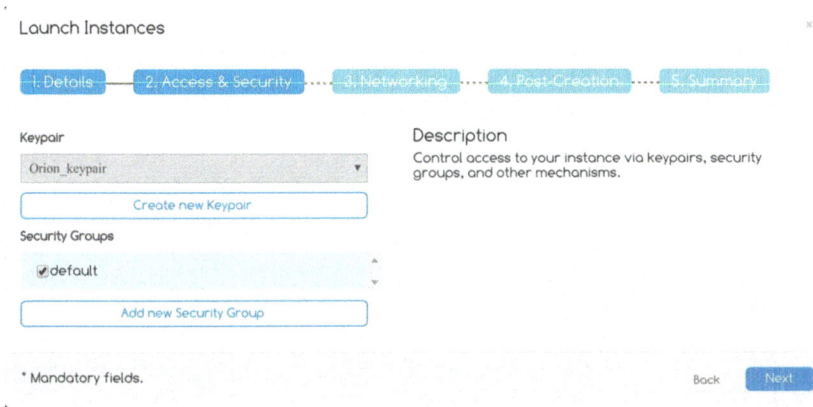

Afterwards, use the *Networking* tab to assign a network by moving the created instance from `Available Networks` to `Selected Networks`:

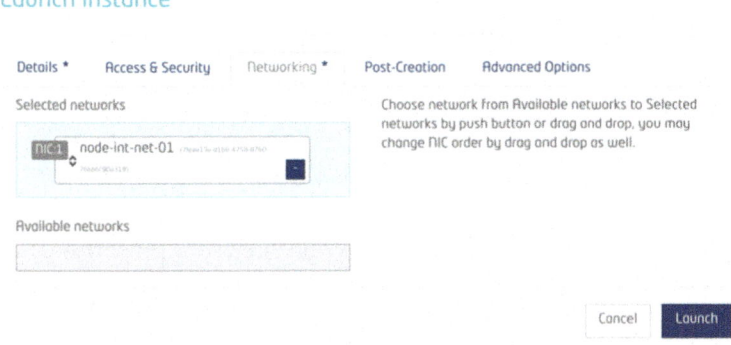

Eventually, continue through the final steps *Post-Creation* and *Summary* to launch the new instance. For the purpose of the tutorial, in these steps is not needed any additional information.

Connecting to the instance To remotely connect to the newly created instance, first a public IP must be assigned. Therefore, move to the `Security` tab and click on `Allocate IP to Project` button:

Now, use the default `public-ext-net-01` option for `Pool` and click on `Allocate IP`:

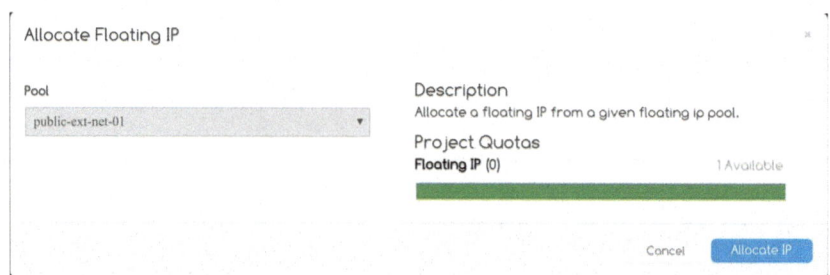

Finally, you can associate the IP with the instance selecting it and clicking on
`Actions: Associate IP`

After the confirmation, the created VM is accessible from the Internet. In the remainder, we provide the detailed instruction to connect to the instance using an SSH client for Linux-based operating system.[14]

1. Open the Terminal
2. Locate the key-pair associated to the instance when launching it
   (e.g. `Orion_keypair.pem`)
3. Modify the key-pair permissions to make it not publicly viewable:

```
$ chmod 400 Orion_keypair.pem
```

4. Connect to the instance using its public IP:

```
$ ssh -i Orion_keypair.pem username@public_ip
```

Now, you have the remote access to the instance, and in the next, the tutorial will cover the necessary steps on how to use the Orion Context Broker.

4.3 Orion Context Broker Example Case

Before discussing how to integrate FIWARE with robotics, let's introduce an Orion example case[15] to fully understand its capabilities.

This section will show how to manage the context information of a building with several rooms using the Orion Context Broker. The rooms are `Room1`, and `Room2` equipped with two sensors: temperature and (atmospheric) pressure.

---

[14]Windows user can use the PuTTY client or similar, please refer to the relative documentation.

[15]Taken from http://fiware-orion.readthedocs.io/en/master/user/walkthrough_apiv2/index.html.

The Orion Context Broker interacts with a *context producer* applications (which provide sensor information) and a *context consumer* application (which processes that information, e.g. to show it in a graphical user interface). We will show how to act both as producer and consumer.

Starting the broker Start the broker (as root or using the `sudo` command) on the FIWARE instance:

```
$ sudo /etc/init.d/contextBroker start
```

On normal behavior the output is:

```
Starting...
contextBroker (pid 1372) is running...
```

In case you need to restart the broker, execute the following:

```
$ sudo /etc/init.d/contextBroker restart
```

Issuing commands to the broker To issue requests to the broker, this tutorial uses the `curl` command line tool, because it is often present in any GNU/Linux system and simplifies the explanation. Of course, it is not mandatory, and the reader can use any REST client tool instead (e.g. RESTClient). Indeed, in a real case, a developer will probably interact with the Orion Context Broker using a programming language library implementing the REST client.

The basic patterns for all the `curl` examples in this document are the following:

POST

```
curl localhost:1026/<operation_url> -s -S [headers]' -d @-
 <<EOF
[payload]
EOF
```

PUT

```
curl localhost:1026/<operation_url> -s -S [headers] -X PUT -d
@- <<EOF
[payload]
EOF
```

PATCH

```
curl localhost:1026/<operation_url> -s -S [headers] -X PATCH -d
@- <<EOF
[payload]
EOF
```

GET

```
curl localhost:1026/<operation_url> -s -S [headers]
```

DELETE

```
curl localhost:1026/<operation_url> -s -S [headers] -X DELETE
```

Regarding the `headers`, it is possible to include the following ones:

Accept header, to specify which payload format you want to receive in the response. User should explicitly specify JSON:

```
curl ... --header 'Accept: application/json' ...
```

Content-Type header, when a payload is needed for requests (i.e. POST, PUT or PATCH). Also in this case, user should explicitly specify JSON:

```
curl ... --header 'Content-Type: application/json' ...
```

Regarding the aforementioned commands, please take into account the following remarks:

- most of the time we are using multi-line shell commands, using EOF to mark the beginning and the end of the multi-line block;
- in some cases (GET and DELETE), we omit `-d @-` as they don't use payload;
- we assume that the broker is listening on port 1026. Adjust this in the `curl` command line in case of different setting;
- to pretty-print JSON in responses, we use the `json.tool` from the json Python module:

```
(curl ... | python -m json.tool) <<EOF
...
EOF
```

Context Management At this step, we are ready to create both producer and consumer applications, using the Orion Context Broker. At the beginning, the broker starts in an empty state. Therefore, we need to make it aware of the existence of certain entities. In particular, we are going to create the entities for Room1 and Room2, each one with two attributes (temperature and pressure). We create the entities using the POST /v2/entities operation. First of all, we declare the Room1 entity with 23 °C and 720 mmHg of temperature and pressure respectively:

```
curl localhost:1026/v2/entities -s -S --header 'Content-Type: \
application/json' -d @- <<EOF
{
  "id": "Room1",
  "type": "Room",
  "temperature": {
    "value": 23,
    "type": "Float"
  },
  "pressure": {
    "value": 720,
```

```
    "type": "Integer"
  }
}
EOF
```

The entity is defined with an `id` and a `type`, and contains 2 attributes with a `value` and a `type` field. It is important to know that the Orion Context Broker does not perform any type check. In other words, it will accept a temperature value either if it is formatted as a float like 25.5 or as a string like *hot*. This means that the developer has to take care of the correctness of the data types during the implementation. Upon receipt of this request, the broker will create the entity in its internal database, it will set the values for its attributes and it will respond with a 201 Created HTTP code.

In the same way, we create the `Room2` entity, setting temperature and pressure to 21 °C and 711 mmHg respectively:

```
curl localhost:1026/v2/entities -s -S --header 'Content-Type: \
application/json' -d @- <<EOF
{
  "id": "Room2",
  "type": "Room",
  "temperature": {
    "value": 21,
    "type": "Float"
  },
  "pressure": {
    "value": 711,
    "type": "Integer"
  }
}
EOF
```

Now, we can act as a consumer application, wanting to access the context information stored by the Orion Context Broker to do something interesting with it (e.g. show a graph with the room temperature in a graphical user interface). In this case, we use the `GET /v2/entities/{id}` request. To retrieve the context information of the `Room1` use the following command:

```
curl localhost:1026/v2/entities/Room1?type=Room -s -S \
    --header 'Accept: application/json' | python -m json.tool
```

Actually, you don't need to specify the type, as in this case there is no ambiguity using just the ID, so you can also do:

```
curl localhost:1026/v2/entities/Room1 -s -S \
    --header 'Accept: application/json' | python -m json.tool
```

In both cases, the response includes all the attributes belonging to `Room1`:

```
{
  "id": "Room1",
  "pressure": {
```

```
     "metadata": {},
     "type": "Integer",
     "value": 720
  },
  "temperature": {
     "metadata": {},
     "type": "Float",
     "value": 23
     },
     "type": "Room"
}
```

It is also possible to use the `keyValues` option in the request:

```
curl localhost:1026/v2/entities/Room1?options=keyValues -s -S \
   --header 'Accept: application/json' | python -m json.tool
```

which produces a compact response, including just the attribute values:

```
{
   "id": "Room1",
   "pressure": 720,
   "temperature": 23,
   "type": "Room"
}
```

A third way consists in requesting specific attributes using the `values` option plus a list of attribute name. To do so, use the `attrs` URL parameter to specify the order. For instance, using this command (temperature first, pressure second):

```
curl 'localhost:1026/v2/entities/Room1?options=values&attrs=
temperature,
   pressure' -s -S \
   --header 'Accept: application/json' | python -m json.tool
```

produce the following response:

```
[
   23,
   720
]
```

Finally, note that requesting a non-existing entity will produce the following error:

```
{
   "description": "The requested entity has not been found. \
      Check type and id",
   "error": "NotFound"
}
```

The same happens for non-existing attribute:

```
{
    "description": "The entity does not have such an attribute",
    "error": "NotFound"
}
```

Another useful way to get all the context information is to list all the entities omitting the id in the GET command:

```
curl localhost:1026/v2/entities -s -S \
    --header 'Accept: application/json' | python -m json.tool
```

which, in our case, return both Room1 and Room2:

```
[
    {
        "id": "Room1",
        "pressure": {
            "metadata": {},
            "type": "Integer",
            "value": 720
        },
        "temperature": {
            "metadata": {},
            "type": "Float",
            "value": 23
        },
        "type": "Room"
    },
    {
        "id": "Room2",
        "pressure": {
            "metadata": {},
            "type": "Integer",
            "value": 711
        },
        "temperature": {
            "metadata": {},
            "type": "Float",
            "value": 21
        },
        "type": "Room"
    }
]
```

At this step, we are able to produce and consume information using the Orion Context Broker. In the next, this tutorial will focus on the integration between ROS and FIWARE.

5   Walkthrough: Connecting ROS and FIWARE

The link between ROS and FIWARE is represented by the `firos` package [12]. For the aim of this tutorial, we refer to our fork version of the original package available at https://github.com/Raffa87/ROS_FIWARE_Tutorial.[16] The forked package contains a simplified version of the code with utilities already set for the present tutorial. At the end, the reader will be able to send and receive data using ROS topics and the FIWARE Orion Context Broker.

5.1   Install Firos

First of all, clone the repository into your ROS workspace

```
$ git clone https://github.com/Raffa87/ROS_FIWARE_Tutorial
```

build the package

```
$ catkin_make
```

The `firos` package is now installed and needs to be configured according to the specific application.

5.2   Configuring FIROS

The configuration files are located at the `src/firos/config` folder. Open the `config.json` file and fill the following parameters:

address: the public IP of the FIWARE instance (e.g. 217.123.12.123)
port: the port used by the Orion Context Broker (default: 1026)
interface: the network interface used by the PC (e.g. `wlan0`, `eth0`)

Open the file *robots.json* to declare the ROS topics that will be used by `firos`. For instance, using the following configuration file:

```
{
   "end_end_test":{
        "topics": {
           "p1": {
               "msg": "std_msgs.msg.String",
               "type": "publisher"
           },
           "s1": {
               "msg": "std_msgs.msg.String",
```

---

[16]The original `firos` package is available at https://github.com/Ikergune/firos.

```
            "type": "subscriber"
        }
    }
}
```

will configure firos to transmit the data received on the s1 topic to the specified Orion Context Broker and to republish on the corresponding topic the change of information on p1 of end_end_test entity.

5.3 Run FIROS

To execute the firos node:

```
$ rosrun firos core.py
```

The output of a running node with the previous configurations will be like the following:

```
Initializing ROS node: firos
Initialized
Starting Firos setup...
---------------------------------
Generating Message Description Files
Successfully generated
Starting Firos...
---------------------------------
Getting configuration data
Generating topic handlers:
  -end_end_test
    -p1
    -s1
Subscribing on context broker to ROBOT end_end_test and topics:[]
Connected to Context Broker with id 59ba8a5bb05d744325f22525
Subscribed to end_end_test's topics

Press Ctrl+C to Exit

Serving HTTP on 0.0.0.0 port 10100 ...
```

5.4 From ROS Topic to Orion Context Broker

At this step, we can check the connection between the firos node and the Orion Context Broker by publishing on the specified s1 topic:

```
$ rostopic pub /s1 std_msgs/String "hello world" __ns:=end_end
_test
```

After that, the `firos` node modifies the attribute of the entity `end_end_test` in the Orion Context Broker. We can verify it by querying Orion in the following way:

```
curl contextbroker_ip:1026/v2/entities/end_end_test -s -S \
    --header 'Accept: application/json' | python -mjson.tool
```

The response will look like as:

```
{
    "COMMAND": {
        "metadata": {},
        "type": "COMMAND",
        "value": ""
    },
    "id": "end_end_test",
    "s1": {
        "metadata": {},
        "type": "std_msgs.msg.String",
        "value": "{%27firosstamp%27: 1505398290.893342,
            %27data%27: %27hello world%27}"
    },
    "type": "ROBOT"
}
```

As we can see from the response, the communication between ROS and FIWARE was successful. In particular, the use and configuration of `firos` automatically create a new entity of type `ROBOT` that has an attribute named as the subscribed topic whose value is automatically set and updated according to the data flowing on the topic.

5.5 From Orion Context Broker to ROS Topic

In a similar way, we can use `firos` to gather data from the context broker to ROS topics. The *robots.json* configuration file already specifies a subscriber for the p1 topic in the `end_end_test` namespace. However, although the `firos` node automatically creates subscriber attributes for the specified entity, it does not do the same for publishers. Therefore, we have to add a publisher attribute to our entity in the Context Broker:

```
curl contextbroker_ip:1026/v2/entities/end_end_test/attrs -s -S \
    --header 'Content-Type: application/json' -d @- <<EOF
{
    "p1": {
        "metadata": {},
        "type": "std_msgs.msg.String",
        "value": "test"
    }
```

```
}
EOF
```

Now, to verify the communication, open another terminal on the machine where `firos` is running and echo the topic:

```
rostopic echo /end_end_test/p1
```

After that, modify the value of `p1` attribute on the Orion Context Broker using the `PATCH` command:

```
curl contextbroker_ip:1026/v2/entities/end_end_test/attrs -s -S \
    --header 'Content-Type: application/json' -X PATCH -d @- <<EOF
{
    "p1": {
            "metadata": {},
            "type": "std_msgs.msg.String",
            "value": "{%27data%27: %27`echo $RANDOM`%27}"
    },

     "COMMAND": {
            "type": "COMMAND",
            "value": ["p1"]
        }
}
EOF
```

The new value will be printed in the `rostopic echo` terminal. At this point, we have demonstrated how to connect ROS with FIWARE, both on subscription and publication side. However, concerning the subscription case, it is worth to note that the Orion Context Broker has to be able to reach the IP address of the machine running the `firos` node. In other words, the robot has to be on the same network of the cloud resource. This, of course, represents a huge limitation, and the remainder will face this visibility issue.

6   Walkthrough: How to Control a Robot Through FIWARE

So far we have seen how to practically use the FIWARE Context Broker (Sect. 4) and how to use the `firos` package to simply integrate FIWARE with ROS using publishers and subscribers (Sect. 5). In this section, we go a step forward providing a more concrete Cloud Robotics example. We will see how to send a goal command to a robot, and look at the odometry data passing through the FIWARE Orion instance. To be close to a real Cloud Robotics application, we propose a system in which the control station is not in the same network of the FIWARE cloud resource (i.e. we do not setup a dedicated VPN). To solve the visibility issue between different machines over the Internet, we implement a specific module on the robot that periodically asks for commands to execute (polling-mode).

In details, the following step will be implemented and explained:

- create a new entity in the Orion Context Broker to store new commands for a mobile platform through the PUT and PATCH commands;
- use Gazebo to simulate a Clearpath[17] Husky mobile platform equipped with a navigation stack to move in the environment;
- use firos to update odometry information in the Orion Context Broker;
- implement a python script to periodically poll new command from the Orion Context Broker through the GET command;
- retrieve updated odometry data from the Orion Context Broker through the GET command.

In the following, we provide a detailed walkthrough that details the proposed system.

6.1 Create a New Entity to Store Robot Commands

First of all, let's create a new bare entity on the Orion Context Broker where to store the commands for the robot:

```
curl contextbroker_ip:1026/v2/entities -s -S \
    --header 'Content-Type: application/json' -d @- <<EOF
{
  "id": "Husky_fiware_command",
  "type": "Robot_command",
  "command": {
    "value": "test",
    "type": "String"
  }
}
EOF
```

We have created a Husky_fiware_command entity having a command attribute which is defined through a value and type field. We will see later, how the value field will contain the information for a goal command, formatted as a ROS message, while the latter will contain the ROS message type.

---

[17]https://www.clearpathrobotics.com, visited on December 2017.

6.2 Simulate a Husky Mobile Platform

In our example, we will use a simulated Husky platform using the Gazebo software. We decided to use it because it is well documented[18] and easy to reproduce. Hence, for simulating the robot we use the `husky_gazebo` and `husky_navigation` packages. In this walkthrough, we use the `husky_fiware` namespace, so our tutorial repository, already contains the launch files for those packages that refer the ROS topics to our specific namespace. Therefore, start the pre-configured Husky simulation environment where `ns` is already specified, also as argument of the `gazebo_ros/spawn` node to keep consistency between simulated robot and running ROS nodes.

```
$ roslaunch fiware_polling_command husky_playpen_fiware_demo.
launch
```

and the `husky_navigation` stack in the `husky_fiware` namespace:

```
$ roslaunch husky_navigation move_base_mapless_demo.launch __ns:=husky_fiware
```

The Gazebo simulator will open showing a Husky robot ready to execute commands.

The rationale beyond the use of namespace is related to the `firos` implementation: `firos` node could be used to simultaneously manage several robots: for each robot, a different entity and topics are published and subscribed using the specified namespace.

6.3 Update the Odometry on Orion Context Broker

To update the odometry information on the Orion Context Broker, we have to change the `firos` config file, as we did in Sect. 5.2. Hence, modify the `robots.json` file to specify that the `odometry/filtered` topic has to be forwarded to the Orion Context Broker. The modified file looks like:

```
{
    "end_end_test":{
        "topics": {
            "p1": {
                "msg": "std_msgs.msg.String",
                "type": "publisher"
            },
            "s1": {
                "msg": "std_msgs.msg.String",
                "type": "subscriber"
            }
        }
    }
```

---

[18]http://wiki.ros.org/husky_navigation/Tutorials, visited on December 2017.

```
},
"husky_fiware":{
    "topics": {
        "odometry/filtered": {
            "msg": "nav_msgs.msg.Odometry",
            "type": "subscriber"
        }
    }
  }
}
```

We declare a subscriber on the `odometry/filtered` topic expecting a ROS `nav_msgs.msg.Odometry` message. Now, you can run the `firos` node:

```
$ rosrun firos core.py
```

To complete the robot side programs, run the `fiware_polling_command.py` script that periodically request (through a GET) the value of the `command` attribute of the `Husky_fiware_command` entity. The script also parses the command, publishing it on the relative navigation stack topic. The code of the script is the following:

```python
#!/usr/bin/env python
2
    import time
4   import requests
    import json
6   import rospy
    from move_base_msgs.msg import MoveBaseGoal
8   from geometry_msgs.msg import PoseStamped

10  def fiware_polling():
        context_broker_ip = IP_ADDRESS
12      pub = rospy.Publisher('/husky_fiware/move_base_simple/goal',
            PoseStamped, queue_size=10)
14      rospy.init_node('fiware_polling_goal', anonymous=True)
        rate = rospy.Rate(1) #1hz
16      while not rospy.is_shutdown():
            headers = {
18              'Accept': 'application/json',
            }
20          r = requests.get('http://' + context_broker_ip +
                ':1026/v2/entities/Husky_fiware_command', headers=headers)
22          j = json.loads(r.text)
            if ("PoseStamped" in j['command']['type']):
24          try:
            msg = PoseStamped()
26          msg.header.frame_id = j['command']['value']['header']['frame_id']
            msg.pose.position.x = j['command']['value']['pose']['position']['x']
28          msg.pose.position.y = j['command']['value']['pose']['position']['y']
            msg.pose.position.z = j['command']['value']['pose']['position']['z']
30          msg.pose.orientation.x = j['command']['value']['pose']['orientation']['x']
            msg.pose.orientation.y = j['command']['value']['pose']['orientation']['y']
32          msg.pose.orientation.z = j['command']['value']['pose']['orientation']['z']
            msg.pose.orientation.w = j['command']['value']['pose']['orientation']['w']
34          pub.publish(msg)
                rate.sleep()
36          except TypeError:
```

```
        continue
38
        headers = {
40          'Content-Type': 'application/json',
        }
42      data = '{"command": {"value": "none","type": "String"}}'
        r = requests.patch('http://' + context_broker_ip + ':1026/v2/entities/
44          Husky_fiware_command/attrs', headers=headers, data=data)

46 if __name__ == '__main__':
        try:
48          fiware_polling()
        except rospy.ROSInterruptException:
50          pass
```

In details, it implements a ROS node, which instantiate a publisher for the navigation stack of the robot (line 12). Then, it continuously asks the Context Broker about the `command` attribute through a GET request (line 20). Then, it checks if the type field is equal to the expected `PoseStamped` (line 23). After that, it parses the retrieved value and fills a new `geometry_msgs/PoseStamped` message that is published on the `/husky_fiware/move_base_simple/goal` topic (lines 25–34). At the end, it reset the `command` attribute on the Orion Context Broker using `PATCH` (line 43).

6.4 Store Robot Commands on the Context Broker

New commands can be set by a third-party application (e.g. user interface) through a PATCH message, modifying the value of the `command` attribute of the `Husky_fiware_command` entity. A possible message looks like the following:

```
curl contextbroker_ip:1026/v2/entities/Husky_fiware_command/attrs
  -s -S \
    --header 'Content-Type: application/json' \
      -X PATCH -d @- <<EOF
{
    "command": {
        "value": {
            "header" : {
                "seq" : 1,
                "time" : 0.0,
                "frame_id" : "base_link"
            },
            "pose" : {
                "position" :
                {
                    "x" : 0.0,
                    "y" : -1.0,
                    "z" : 0.0
                },
                "orientation" :
                {
```

```
                           "x" : 0.0,
                           "y" : 0.0,
                           "z" : 0.0,
                           "w" : 1.0
                     }
               }
        },
        "type": "PoseStamped"
  }
}
EOF
```

Here, we define a goal point which is at 1 m on the right from the current robot pose (i.e. referred to the `base_link` frame). In particular, the `value` field is formatted like the `PoseStamped` ROS message.

6.5 Demo Execution

At this point, everything is ready for the execution of the demo. The reader can send the `PATCH` command explained before. Looking at the Gazebo window, the robot will start to move to the given point. Naturally, we can also check the odometry value that is continuously updated by the `firos` node. This information can be easily retrieved by querying the Orion Context Broker, as shown in Sect. 5.4. Nevertheless, the value of the `odometry_filtered` attribute is quite complex compared to the previous example. In fact, it contains several data, such as position, twist, and relative covariance matrix. Hence, it is not so easy to read by humans. For this purpose, our repository contains a utility python program that parses this odometry message and outputs only the data relevant to this example case (i.e. x and y position). So, launch the program using:

```
$ rosrun fiware_polling_command get_odometry_x_y.py
```

In this way, the reader can easily check the updated odometry values.[19] We have demonstrated a basic Cloud Robotic infrastructure.

To control a ROS-based robot using the FIWARE framework. In our example, we used command line tools to control the robot to provide a deep insight of the adopted technologies. Obviously, a real Cloud Robotic application which aims to control a robotic system will have a graphical user interface that can be easily implemented using web technologies (e.g. HTTP-based) or with pre-configured FIWARE instances (see Sect. 3).

---

[19]A video demonstration of this tutorial is available at https://youtu.be/czhD-krCRZc, visited on December 2017.

7 Conclusion and Future Works

Nowadays, the importance of the Cloud technologies is attested by the wide adoption of this technology in our everyday life. All the main Hi-Tech companies provide a cloud-based solution in the form of Software-as-a-Service, Platform-as-a-Service, and Infrastructure-as-a-Service. On the other hands, robotics is one of the future emerging markets. Despite this, a clear solution that links these two distinct worlds is not clearly affirmed yet.

The present chapter proposes the use of the FIWARE framework for the development of Cloud Robotics applications using ROS-based robots. In particular, it plainly shows how to integrate this Cloud technology through detailed walkthrough tutorials. The chapter presents the steps to create and use a FIWARE Orion Context Broker running on a VM in FIWARE platform, and its integration with the ROS framework using the `firos` package. The reader has the possibility to follow the walkthrough developing a simple *Hello World* example. Finally, a more concrete Cloud Robotics application to control a simulated robot sending a goal and reading odometry data is presented.

The use of the FIWARE framework allows to easily build modular Cloud application, enhancing the proposed system with additional features, the so-called FIWARE Generic Enablers. Among this, a developer can extend the system with a powerful web-based user interface, employ tools for analyzing context information, and modules designed for addressing security and privacy requirements. In addition, FIWARE can represent a common infrastructure between IoT and robotics.

Furthermore, another aspect that emerges from this chapter is the well-know visibility issue between machines connected on different networks across the Internet. This problem can be addressed in several ways as, for instance, setting a dedicated VPN. However, it can be painful and needs to be configured for every machine acting on the system. The solution detailed in this tutorial proposes the development of a *polling* node, which periodically queries the Cloud resource asking for commands. Clearly, this method can be suitable for most of the IoT devices that typically perform simple actions. More realistically, Cloud Robotics applications need a different and optimized mechanism. A viable option is represented by the modern web-based communication technologies such as WebSockets that implements a low-latency, bidirectional communication layer between clients (web browsers) and servers. Concerning ROS, the Robot Web Tools (RWT) [13] project already offers a ROS interface for the WebSocket transport layer through the `rosbridge_suite` package. A practical use case example of the RWT adoption in a Cloud Robotics teleoperation system can be found at [14].

An extension of the FIWARE framework with the development of a module that integrates the `rosbridge_suite` will concern our future works. It will enhance a FIWARE robotic application with real-time capabilities, allowing to optimize the streaming of a huge amount of data.

Acknowledgements This work was supported by the ACCRA Project, founded by the European Commission—Horizon 2020 Founding Programme (H2020-SCI-PM14-2016) and National

Institute of Information and Communications Technology (NICT) of Japan under grant agreement No. 738251.

References

1. M. Quigley, K. Conley, B. Gerkey, J. Faust, T. Foote, J. Leibs, R. Wheeler, A.Y. Ng, Ros: an open-source robot operating system. in *ICRA Workshop on Open Source Software, Kobe*, vol. 3 (2009), p. 5
2. T. Foote, Community metrics report, http://download.ros.org/downloads/metrics/metrics-report-2017-07.pdf. Accessed on Dec 2017
3. B. Kehoe, S. Patil, P. Abbeel, K. Goldberg, A survey of research on cloud robotics and automation. IEEE Trans. Autom. Sci. Eng. **12**(2), 398–409 (2015)
4. G. Hu, W.P. Tay, Y. Wen, Cloud robotics: architecture, challenges and applications. IEEE Netw. **26**(3), 21–28 (2012)
5. K. Kamei, S. Nishio, N. Hagita, M. Sato, Cloud networked robotics. IEEE Netw. **26**(3), 28–34 (2012)
6. R. Arumugam, V.R. Enti, L. Bingbing, W. Xiaojun, K. Baskaran, F.F. Kong, A.S. Kumar, K.D. Meng, G.W. Kit, Davinci: a cloud computing framework for service robots. in *2010 IEEE International Conference on Robotics and Automation (ICRA)* (IEEE, 2010), pp. 3084–3089
7. G. Mohanarajah, D. Hunziker, R. D'Andrea, M. Waibel, Rapyuta: a cloud robotics platform. IEEE Trans. Autom. Sci. Eng. **12**(2), 481–493 (2015). April
8. M. Waibel, M. Beetz, J. Civera, R. d'Andrea, J. Elfring, D. Galvez-Lopez, K. Häussermann, R. Janssen, J.M. Montiel, A. Perzylo et al., Roboearth. IEEE Robot. Autom. Magaz. **18**(2), 69–82 (2011)
9. D. Hunziker, M. Gajamohan, M. Waibel, R. D'Andrea, Rapyuta: the roboearth cloud engine. in *2013 IEEE International Conference on Robotics and Automation (ICRA)* (IEEE, 2013), pp. 438–444
10. S. Hanke, C. Mayer, O. Hoeftberger, H. Boos, R. Wichert, M.-R. Tazari, P. Wolf, F. Furfari, Universaal–an open and consolidated aal platform, *Ambient Assisted Living* (Springer, Berlin, 2011), pp. 127–140
11. T. Zahariadis, A. Papadakis, F. Alvarez, J. Gonzalez, F. Lopez, F. Facca, Y. Al-Hazmi. Fiware lab: managing resources and services in a cloud federation supporting future internet applications. in *2014 IEEE/ACM 7th International Conference on Utility and Cloud Computing* (2014), pp. 792–799
12. F. Herranz, J. Jaime, I. González, Á. Hernández, Cloud robotics in fiware: a proof of concept. in *International Conference on Hybrid Artificial Intelligence Systems* (Springer, Berlin, 2015), pp. 580–591
13. R. Toris, J. Kammerl, D.V. Lu, J. Lee, O.C. Jenkins, S. Osentoski, M. Wills, S. Chernova. Robot web tools: efficient messaging for cloud robotics. in *2015 IEEE/RSJ International Conference on Intelligent Robots and Systems (IROS)* (IEEE, 2015), pp. 4530–4537
14. A. Manzi, L. Fiorini, R. Limosani, P. Sincak, P. Dario, F. Cavallo. Use case evaluation of a cloud robotics teleoperation system (short paper). in *2016 5th IEEE International Conference on Cloud Networking (Cloudnet)* (IEEE, 2016), pp. 208–211

Raffaele Limosani received the Master Degree in Biomedical Engineering at University of Pisa on July 2011 and received the Ph.D. in Biorobotics (cum laude) from the Scuola Superiore Sant'Anna in November 2015. During his Ph.D., he was a visiting researcher at the ATR (Advanced Telecommunications Research Institute International) Laboratory, Japan. Currently he is a post-doc at the BioRobotics Institute of the Scuola Superiore Sant'Anna. His research fields are

Robotic Navigation, Human Robot Interaction and Mobile Manipulation, especially in unstructured environments.

Alessandro Manzi received the MSc in Computer Science from the University of Pisa, Italy in 2007, and received the Ph.D. in BioRobotics (cum laude) from the Scuola Superiore Sant'Anna in 2017. Currently, he is a post-doc at the BioRobotics Institute of the Scuola Superiore Sant'Anna. His research interests include machine learning, computer vision, data processing from depth cameras, robotic navigation, and perception.

Laura Fiorini received the Master Degree (with honours) in BioMedical Engineering at University of Pisa on April 2012 and the Ph.D. in Biorobotics (cum laude) from the Scuola Superiore Sant'Anna in February 2016. Currently she is a post-doc at the BioRobotics Institute of the Scuola Superiore Sant'Anna. She was a visiting researcher at the Bristol Robotics Laboratory, UK. Her research interests include Ambient Assisted Living, Cloud Service Robotics, ICT system for cognitive activation, pattern recognition, signal processing and experimental protocol.

Paolo Dario received his Dr. Eng. Degree in Mechanical Engineering from the University of Pisa, Italy, in 1977. He is currently a Professor of Biomedical Robotics at Scuola Superiore Sant'Anna in Pisa. He has been Visiting Professor at prestigious universities in Italy and abroad, like Brown University, Ecole Polytechnique Federale de Lausanne, Waseda University, University of Tokyo, College de France, Zhejiang University. He was the founder and the Coordinator of the BioRobotics Institute of Scuola Superiore Sant'Anna, where he supervises a team of about 120 researchers and Ph.D. students. He is the Director of Polo Sant'Anna Valdera. His main research interests are in the fields of BioRobotics, medical robotics, micro/nanoengineering. He is the coordinator of many national and European projects, the editor of special issues and books on the subject of BioRobotics, and the author of more than 200 scientific papers.

Filippo Cavallo MScEE, Ph.D. in Bioengineering, is Assistant Professor at BioRobotics Institute, Scuola Superiore Sant'Anna (Pisa, Italy), focusing on cloud and social robotics, ambient assisted living, biomedical processing, wireless and wearable sensor systems. He participated in various National and European projects, being project manager of Robot-Era, AALIANCE2 and Parkinson Project. He was visiting researcher at the the EndoCAS Center of Excellence, Pisa; at the Takanishi Lab, Waseda University, Tokyo; at Tecnalia Research Center, Spain. He was granted from the International Symposium of Robotics Research Committee as Fellowship Winner for best Ph.D. thesis in Robotics; from the Regional POR FSE 2007–2013 for a 3-years Research position at The BioRobotics Institute; from the ACCESS-IT 2009 for the Good Practice Label in Alzheimer Project; from the Well-Tech Award for Quality of Life with the Robot-Era Project. He is author of various papers on conferences and ISI journals.

Enabling Real-Time Processing for ROS2 Embedded Systems

Lucas da Silva Medeiros, Ricardo Emerson Julio, Rodrigo Maximiano Antunes de Almeida and Guilherme Sousa Bastos

Abstract Our research aims to integrate FreeRTPS, a portable and minimalist RTPS (Real-Time Publisher-Subscriber), an implementation that provides an option for embedded ROS2 (Robot Operating System), applications where RAM(Random Access Memory)/ROM(Read-Only Memory) size is a critical factor, with FreeRTOS, a free real-time operating system for microcontrollers and small microprocessors. As a result, we have a portable system that enables sensing and the possibilities of real-time processing, while communicating with ROS2 nodes in small and low-cost devices. Even having tools to implement internal real-time processing the system not ensure that the communication with other nodes will have real time constraints, once that we look for processing and not communication real-time. Real-time processing is an important component especially in Robotics where many applications require some data processing as it comes in, what means that they need processing with time requirements. The chapter shows some concepts, how the system was developed, how to implement it on the STM32F4 microcontroller and some tests to show its capabilities. The main system was developed under Ubuntu 16.04, with a STM32F4 microcontroller, and the portability test was made under Microsoft Windows 10, with a Texas Instruments LM3S Stellaris board. All presented components are published on the wiki ROS link: http://wiki.ros.org/FreeRTPS%2BFreeRTOS.

L. da S. Medeiros (✉) · R. E. Julio · R. M. A. de Almeida · G. S. Bastos
System Engineering and Information Technology Institute, Federal University of Itajuba, Av.
BPS, 1303. UNIFEI - IESTI - Pinheirinho, 37500-903 Itajuba, Minas Gerais, Brazil
e-mail: lukas.eco@gmail.com
URL: http://www.unifei.edu.br

R. E. Julio
e-mail: ricardoej@unifei.edu.br

R. M. A. de Almeida
e-mail: rodrigomax@unifei.edu.br

G. S. Bastos
e-mail: sousa@unifei.edu.br

© Springer International Publishing AG, part of Springer Nature 2019
A. Koubaa (ed.), *Robot Operating System (ROS)*, Studies in Computational
Intelligence 778, https://doi.org/10.1007/978-3-319-91590-6_14

Keywords Real-time · Embedded · ROS2 · FreeRTPS · FreeRTOS

1 Introduction

Nowadays, we have embedded systems everywhere: cell phones, clocks, smart TVs, robots and so on. Despite the amount of resources and the ease in developing embedded applications, most of these applications can just be implemented with sequential coding. However, some applications require a multitask processing where the system performs multiple tasks over a certain period of time by executing them concurrently, actually one task each time in short slices of time. Some systems need real-time resources, in a system with real-time capabilities it is crucial that a result be presented in the "right time".

Robotic systems need to be responsive. In mission-critical applications a delay of less than a millisecond on the system can cause a catastrophic failure [1]. Real-time software ensures correct computing at the correct time, in other words a system with fast processing is not necessarily a real-time system [2]. That is, it is necessary that the system has determinism, so that the data is delivered in the correct time [1].

Thus, we have real-time operating systems (RTOS). An RTOS is an operating system (OS) intended to serve real-time applications. It offers resources to ensure real time processing, but still requires effort from the programmer for that. An RTOS just schedules the tasks in such way that if the priorities are assigned in the right way, then it is possible to prove that all tasks will comply with their real-time requirements. Some methods such as rate-monotonic, deadline-monotonic, early-deadline-first and others assign a different priority to each task to offer the time constraint requirements. We must keep in mind that for any non-trivial real-time systems, there are usually more than a single task with real-time requirements. The real time test applied to the proposed system shows that the integration made here do not lose the FreeRTOS scheduler capabilities. FreeRTOS is a simple and small RTOS for embedded systems, it was designed to try to provide a deterministic execution of a task, i.e., the operating system gives preference to the task that has been configured with maximum priority by the user.

Embedded and real-time systems are within the interests of both the robotics and ROS community, as can be seen in the presentation at ROSCon 2013, "Bridging ROS to Embedded Systems: A Survey" by Morgan Quigley [3], and the papers presented at ROSCon 2015, "ROS 2 on 'small' embedded systems" by Morgan Quigley [4] and "Real-time Performance in ROS 2" by Jackie Kay and Adolfo Tsouroukdissian [5].

In embedded systems, one of the main applications using real-time systems is in digital control system. A control system is a method widely used in many systems such as position control, velocity, flow and others. A control system provides the output that the operator wishes to obtain from a given process [6].

However, in order to obtain a stable control in the digital world it is necessary that the system has determinism in computing the output value of the process. One of the ways to guarantee the time restrictions imposed by these controllers is through the use of real-time methods and operating systems. Real-time operating systems provide scheduling policies that produce a certain behavior of the application [7] but, as explained before, one method must be used to determine the system tasks priorities to ensure real-time restrictions.

One way to obtain real-time systems on embedded world is using real-time operating systems. They provide real-time kernels that offers system determinism when well implemented, with less memory consumption and low latency of context switching. However, some methods must be used together with the RTOS to enable real-time processing, like rate-monotonic, deadline-monotonic, early-deadline-first and others with a complete analysis of CPU usage and resources used by the real-time task [8]. You can find applications where an RTOS should run not just a real-time task set (RTTs), but also a non-real-time task set (NRTT) [7]. These types of systems that deal with a heterogeneous set of tasks, involving RTTs and NRTTs, are known as Mixed Critical Systems.

With the interest of integrating the embedded systems in ROS applications, several messaging methodologies were developed. To show The **rosserial** and **rosbridge** communicate with other nodes through a bridge node, which converts the message from a specific protocol to the ROS protocol [9, 10]. **UROSnode** and **rosc** implement ROSTCP in their code, so they can communicate directly with the system master and other nodes [11, 12]. With the change of the communication middleware in ROS2 to DDS, the development of FreeRTPS, a compact RTPS implementation for embedded devices that have few RAM and ROM was started [13].

FreeRTPS is a free, portable and minimalist RTPS implementation. It aims to be a "small" RTPS implementation capable of fitting in a variety of micro controllers using only on-chip memory and flash and provide embedded ROS2 applications in low-cost devices with hardware restrictions.

Although FreeRTPS is a minimalist implementation, depending on the complexity of the tasks assigned to the micro controller, the system may not be able to guarantee real-time constraints. Seeking to meet the real-time requirements and complement some features in this minimalist version of an RTPS, the integration project of FreeRTPS with the real-time operating system FreeRTOS was started.

This chapter will show this integration between these two embedded systems. As a result it is possible integrate micro controllers and small microprocessors running real-time applications, at processor level, but not in communication level, with ROS2, given that ROS is a tool used by a huge quantity of people and companies around the world. It will also be possible to integrate the ROS in an area that has been growing too much in recent times, the Internet of Things (IoT), which has many applications developed in embedded systems.

With real time resources it will also be possible to distribute actuator control processes in small, low consumption and low cost nodes. Since applications involving large systems generally use a central computer with high processing power to handle a number of critical processes.

Therefore, this chapter will present a discussion about real-time applications using ROS2 and embedded systems, how integrate FreeRTOS and FreeRTPS and some validation tests on an embedded low-cost device.

A key contribution of this chapter is the integration of FreeRTOS and FreeRTPS where some concepts about real-time applications using ROS2 will be presented. The problem of using embedded real-time ROS applications in low-cost devices will be addressed and a feasible solution for this problem is discussed.

The complete system and the tutorials present in this chapter can be downloaded at https://github.com/Expertinos/FreeRTPS-FreeRTOS. The instructions to install and run the system and examples can be found at http://wiki.ros.org/FreeRTPS %2BFreeRTOS.

After a brief discussion about the motivation of this chapter, we will introduce the following topics:

- A simple example on embedded real-time applications;
- A summary of ROS2 and its main changes;
- A summary of FreeRTOS and FreeRTPS;
- A review about how build embedded applications using ROS2, FreeRTOS and FreeRTPS;
- A case study on how to use the package in an embedded real-time application using ROS2;
- And, a discussion about the results.

2   Background

"Embedded systems are microprocessor systems designed for a specific purpose or application, with generally limited memory and limited processing resources. In most cases, they are systems designed for applications that do not require human intervention." [14, p.4].

Portability, size and energy consumption are desirable items in micro controllable products. The low power consumption involves not only economical but also greater autonomy, important items in applications in the field of robotics [15].

2.1   Operating system

In general, operating systems have three main responsibilities: managing available memory and coordinating process access to it, managing and coordinating process execution through some criteria and mediating communication between hardware peripherals and processes [16]:

The kernel is the software that contains the core components of an operating system, such as: process scheduler, memory management, input and output managementa and file system management [17].

Operating systems use the multitasking mechanism to pretend that processes are running simultaneously [18]. This method performs a fast switching process, executing a portion of each process at a time.

When using an operating system, applications are implemented as tasks, so the applications are managed by the kernel through its process scheduler component, which defines the execution sequence of the processes according to the defined schedule algorithm. Each scheduler has its process selection policy and performs better in certain types of environments.

To use operating systems with embedded systems many implementations were developed, one of them is FreeRTOS. The resources offered by FreeRTOS make it good to use in the development of embedded systems applications, it has items like free license, certificates, low size (ranging from 6 KBytes to 12 KBytes) and has support to a lot of architectures. FreeRTOS works on a modified GPL, General Public License [15]. The modified license clause allows the proprietary code to be closed if the application just link FreeRTOS original code and do not change any line of the FreeRTOS code [19]. In addition, FreeRTOS offers features such as SafeRTOS with RTOS certificate for critical systems (paid version) [20].

FreeRTOS has a lot of additional implementations to add more functionalities, the FreeRTOS+UDP is one of them. FreeRTOS+UDP is a small and efficient implementation of the UDP(User Datagram Protocol)/IP(Internet Protocol) stack IPv4, which was developed for use with FreeRTOS in embedded systems [21]. This implementation adds to the FreeRTOS the capability of communicating with other devices over an Ethernet link.

Real-time systems have the characteristic that a correct result also depends on meeting temporal requirements, differing from non-real-time systems (conventional systems) that a processing do not need to meet temporal requirements [22].

With the increasing use of the term real-time to designate systems that are not, strictly speaking, real-time systems, a classification was created to differentiate them [1, 15, 22], we have:

- hard real time: real-time systems in the strict sense, that is, response correction is associated with the timing specification that needs to be met throughout the operation. A single violation of time constraint means failure of the system;
- firm real time :systems in which breach of time specifications, while tolerated, implies a degradation of system performance;
- soft real time: systems in which sporadic violations of time specifications may occur, however, most often the specifications are met. In a way, it is as if the temporal specification were viewed as a medium to be maintained, as long as the media is maintained there is no degradation in the performance of the system.

A digital controller is a example of system that needs hard real time. A control system can be defined as a set of components that changes the behavior of a process, so that it produces a that desired response. It is defined as a plant or process the system we wish to control [6]. Control systems that do not use feedback to adjust the actuator value are known as open loop control. From the moment that feedback is used to adjust the output of the actuator, the system is called the closed loop control [23].

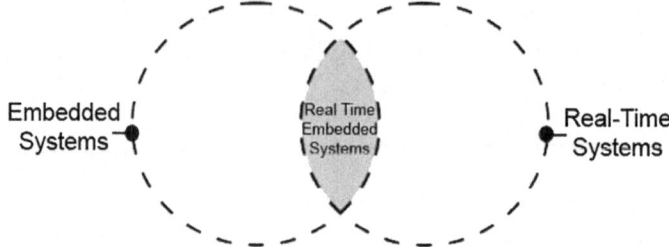

Fig. 1 Embedded real-time systems [24]

It is observed that the closed-loop control has a better behavior than the open-loop, mostly in systems susceptible to disturbance, because knowing the current value of the output of the system, the controller can compensate and act in a better way.

As an example of a real time soft system, we have the reproduction of a video or audio. In this case, the sporadic violation of response time is often imperceptible to the human senses.

The relationship between embedded systems and real-time systems can be represented as follows: not every embedded system has real-time behavior and not every real-time system can be embedded. However, there is a zone of intersection between the two systems, in this region are the systems known as real-time embedded systems, Fig. 1, [24].

Digital PID Controller

As the digital PID controller has time constraints to ensure a stable output, one of the tests made in this chapter uses a PID to show that the FreeRTOS+FreeRTPS system did not lose the FreeRTOS scheduler capabilities. To do this all care was taken to ensure that the system used in the test just operate between the operation range, causing the system to work as a linear system. The PID controller, Proportional Integral Derivative, uses the difference between the desired and current value of some system variable to adjust the process output and achieve the desired value. It combines proportional (P), integral (I) and derivative (D) control modes [25]. Setting the I or D gains to zero, the control can turn into PI or PD control system, each one having your benefits for different behaviors.

The Laplace transfer function of the PID controller is represented by the Eq. 1:

$$G_c(s) = K_p + K_d s + \frac{K_i}{s} \tag{1}$$

Where K_p, K_d e K_i represent the constants of the proportional, derivative and integral gains, respectively. These values are adjusted, depending on the controlled system, to obtain low values of overshoot, error and response time.

To use Eq. 1 in discrete systems, the Z-transform is used. The bi-linear transformation is one of the mathematical methods that allows to turn equations, that represent analogous dynamic systems, to the digital world, so that the system starts to be analyzed in constant intervals of time T (sample time) [26]. At this point it is necessary to use a real-time system, since it is necessary to respect the time T to perform the correct processing.

Applying the bi-linear transformation in the transfer function of a PID controller in 's', Fig. 1, and then applying the inverse Z transform to the discrete time domain [25], the following equation is obtained Eq. 2:

$$
\begin{aligned}
U(N) = {}& U(N-2) + k_p[e(N) - e(N-2)] + \\
& k_d \frac{2}{T}[e(N) - 2e(N-1) + e(N-2)] + \\
& k_i \frac{T}{2}[e(N) + 2e(N-1) + e(N-2)]
\end{aligned}
\tag{2}
$$

In this work, the equation obtained above will be used to carry out the validation of the proposed system.

2.2 Computer Network

The system developed in this work communicates with other ROS2 applications distributed over the Ethernet network and using the UDP/IP protocol for exchange of messages.

Data Distributed System (DDS)

DDS provides a subscriber publisher-type transport, similar to ROS transport. It uses an Object Management Group (OMG) defined Interactive Data Language (IDL) for message, definition and serialization. The DDS discovery model is a distributed discovery system, this allows the DDS to communicate with other nodes without the need for a Master, which makes the system more flexible and fault tolerant [27].

DDS is implemented by default in UDP and does not depend on a transport medium or reliable hardware for communication. In this way, DDS ends up needing to develop methods to increase reliability, but in return it has portability and control over its behavior. Through reliability parameters, QoS, it offers flexibility in controlling communication behavior.

Since the beginning of ROS1 development several new middleware technologies have emerged relevant to ROS [28]. Middleware is software that links applications, often incompatible. This way it is possible to enable communication between different architectures [17]. With the help of these technologies it is now possible to build a ROS-like middleware system using open source libraries. Through this approach, it can then benefit in a number of ways, such as:

- Less code to maintain, especially non-specific robotic codes;
- Take advantage of library resources; and
- Improvements made to these libraries.

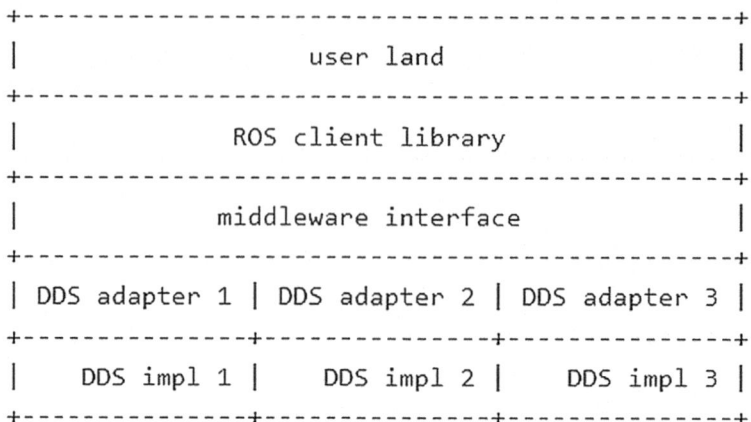

```
+-------------------------------------------------+
|                  user land                      |
+-------------------------------------------------+
|               ROS client library               |
+-------------------------------------------------+
|              middleware interface               |
+-------------------------------------------------+
| DDS adapter 1 | DDS adapter 2 | DDS adapter 3 |
+---------------+---------------+---------------+
|   DDS impl 1  |   DDS impl 2  |   DDS impl 3  |
+---------------+---------------+---------------+
```

Fig. 2 The middleware interface [29]

There are several DDS implementations and each of these implementations have their pros and cons. In this way, ROS2 has the objective of supporting multiple DDS implementations, thus allowing the user to define the best implementation for their application. For this, a middleware abstraction interface was inserted between the ROS Client Library (RCL) and the DDS implementations, Fig. 2.

The RCL don't show to the user any DDS implementation, to hide the complexity of DDS and its APIs [30].

Among the supported ROS Middleware (RMW) implementations to date, we have:

- eProsima Fast RTPS;
- RTI Connext;
- PrismTech Opensplice;
- OSRF FreeRTPS.

RTPS

RTPS is a pub-sub protocol with reliable and best effort communication over unreliable data transport media, such as UDP. RTPS is standardized by OMG as an interoperability protocol for DDS implementations, a widely used standard in aerospace and defense for real-time applications [31].

Among the advantages of the protocol we can mention:

- performance and quality of service, enabling reliable and best-effort communications for real-time applications using standard IP networks;
- extensibility, enabling backwards compatibility and interoperability with other DDS protocols;
- plug and play connectivity, allowing automatic discovery of other network participants and being able to log in and out of the network at any time;

- modularity, allowing to balance the requirements of reliability and punctuality in the delivery of data;
- scalability, allowing the system to grow into a broad network of publishers and subscribers;
- strongly typed, preventing programming errors from compromising the remote nodes in the network.

FreeRTPS

FreeRTPS aims to be a free, portable and minimalist RTPS implementation. Developers are looking to provide an option for ROS2 applications in embedded systems, where ROM and RAM memory sizes are critical [13].

The system was developed with the purpose of portability, so there is no implementation dependency. Compiling and running FreeRTPS on a micro-controller requires a cross-compiler. At present, FreeRTPS only runs on STM32 micro-controllers of ST company. All requirements and steps to make use of the system are in the github repository of the ROS2 [32].

The FreeRTPS project offers the chance for ROS to enter the world of embedded systems with a direct communication, without the need of using an intermediate process to "translate" the messages from a different protocol to ROS protocol. Despite the system's processing power in the STMDISCOVERY development board, the only microcontroller supported so far, the system may not be able to provide time requirements depending on the complexity of the embedded software.

2.3 Related Work

The rosserial protocol is aimed at point-to-point ROS communications over a serial transmission line, such as serial port or socket [9]. This tool provides support for the Arduino platform, embedded Linux systems, Microsoft Windows operating system applications, mbed platforms, and some micro controllers from the Texas Instruments Tiva family. In order for the embedded device to send and receive messages, a host node must exist on the host computer. This interface node is responsible for bridging the device using the rosserial library with the rest of the ROS network. This interface has implementations in python (rosserial_python), C++ (rosserial_server) and Java (rosserial_java).

Rosbridge provides an API in JavaScript Object Notation (JSON), a text pattern based on a subset of the JavaScript language for data exchange [33], for ROS functionality in non-ROS systems [10]. This system has the rosbridge_library library, which is responsible for receiving JSON messages to convert them to ROS and vice versa, the rosapi API, which provides functions in JSON for non-ROS systems, and finally rosbridge_server, which provides a Websocket connection for systems to send their requests in JSON.

UROSnode is a compact ROS client written in C that can run on modern microcontrollers such as ARM Cortex Mx. It provides the key features necessary for a ROS node to be run [11]. UROSnode was programmed with an RTOS operating system in mind, so it relies on some primitives of an operating system, such as processes and mutexes. Currently it is ported to ChibioOS, an operating system for embedded devices, and POSIX.

Rosc is a ROS client implementation in C and without any dependency, which aims to support small embedded systems as well as any operating system. It aims to become an efficient and highly portable implementation of ROS middleware [12].

In the stm32 repository is a system that includes: an embedded development board STM32F4Discovery, the embedded operating system FreeRTOS, an embedded ROS middleware and an embedded ROS client library. The embedded client library has the ability to create nodes, publishers, subscribers, and define ROS messages [34].

In the ROS 2.0 NuttX repository is the prototype of an embedded system using: ROS2, Tinq (DDS implementation) and the NuttX operating system in a microcontroller of the ST family, STM32F4 [35].

3 System Development

As seen before in related work topic, the embedded system implementations presented, which offer real-time resources and direct communication with ROS nodes, have low publication rate or larger size. The other implementations that use a 'bridge' to send/receive data to/from ROS nodes offer a bit more latency of communication due to the intermediate processing.

An attempt was made to develop an integration between FreeRTPS and FreeRTOS thus bringing together the main qualities of the two systems. It is expected to obtain with the final system a tool with good performance, real-time resources and ease of use.

3.1 Proposed Model

We want to obtain in this work a system that offers: ease of use, ease of portability and real-time processing. For this, we propose the following model for development, Fig. 3. It has two levels at vertical: one level is composed by infrastructure zone and the other one shows application zone. At horizontal we have 3 levels: one representing the process, one to represent the queue and the last one representing the libraries.

In this architecture we have the FreeRTOS operating system managing all the processes, including: users processes and FreeRTOS+FreeRTPS processes. With an operating system we can take better advantage of the processing resources offered by the micro controller. Once that, every time that one process is in paused state we

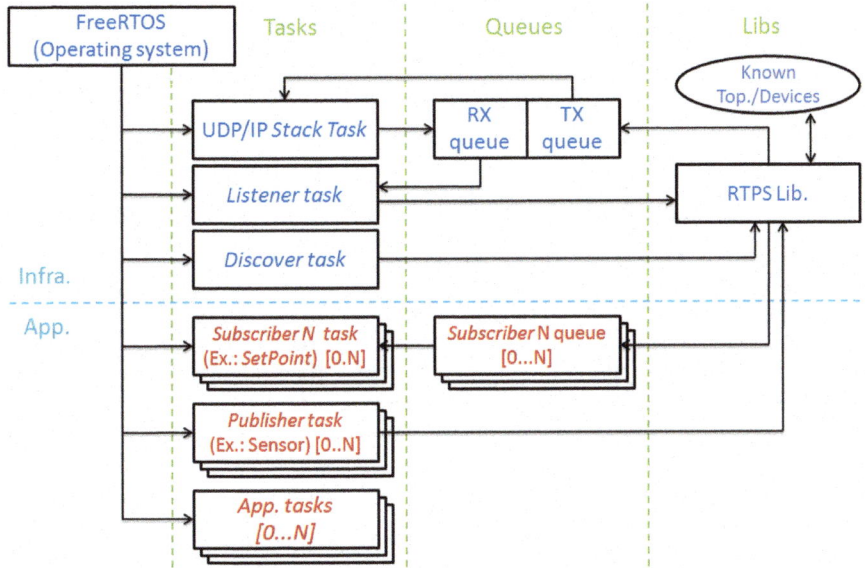

Fig. 3 Simple model of the proposed architecture

can use the microcontroller processing time to execute other processes or we can put the microcontroller on a low power state, if there are no tasks running.

At infrastructure vs process level, we have:

- UDP/IP stack task running the code that manages incoming and outgoing messages, this task gets all the packets stored in the TX queue and put them to the network and stores all the messages received at the network in the RX queue for further processing.
- The listening task gets all incoming messages in the RTPS port socket and decodes them, to identify other nodes or receive data from them. When receiving a discover message the RTPS library stores the new discovered node. When receives a data message the RTPS library store the data at the respective subscriber queue topic.
- Discover task calls routines from RTPS library to send RTPS discover messages at the network, the RTPS discover routine encode a RTPS message and puts it in the TX queue.

At application versus tasks level, we have:

- Subscriber task 1…N that represents all subscriber handle tasks. Each subscriber's task has its own queue where the task looks for new arrived data.
- Publisher task 1…N represents all tasks that calls publish routines from RTPS library. When a publish routine is called the RTPS verifies if it knows a node subscribing that topic, creates a formatted message and send it to the UDP/IP TX queue.

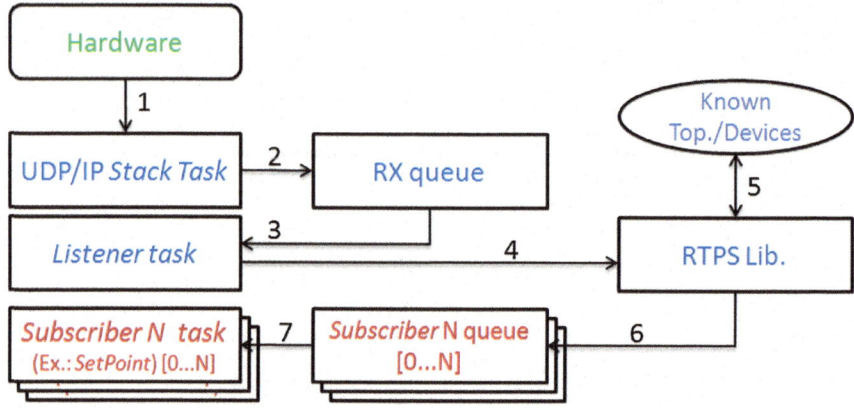

Fig. 4 Path of incoming messages

- Application tasks represent tasks running general processing, like: display printing, button status and others. Application real-time tasks must be well implemented to ensure the real time requirements for all of them.

The path of an incoming message is presented in Fig. 4. First the hardware receives a message and filters it to verify if it is a valid message to put it in the UDP/IP stack. After that, the UDP/IP task decodes the Ethernet frame and stores it in its UDP/IP RX queue. When the Listener task is active and running, it gets all the messages in UDP/IP RX queue addressed to RTPS socket port. Next the listener calls a RTPS routine to decode the RTPS message and store a new known device, if it is a discover message, or put the data in a subscriber queue, if the message is a data type. Finally, the subscriber going to an active state and, when start to execute, it reads the arrived data and execute the programmed code to handle the data, after this the subscriber handle task go into an inactive state again until a new data arrives.

The path of an outgoing data is presented in Fig. 5. In discover case, the discover task calls a RTPS routine to encode a RTPS discover message and puts it in the UDP/IP TX queue. In the publisher case the system calls a RTPS routine that verifies if the device has any nodes subscribing the published topic, encodes a data message and puts it in the UDP/IP TX queue. When the UDP/IP stack task runs and the UDP/IP TX queue has any messages to send, the UDP/IP stack encodes the message in an Ethernet frame and dispatches it to the hardware. Finally, the hardware puts the message on the network.

If we have processes that must execute in real time we need to use methods like rate-monotonic, deadline-monotonic, early-deadline-first and others to determine the task priorities, keeping in mind that we must coding the program right and analyze the CPU usage to ensure time necessities.

Another advantage of using this system is that the user does not need to worry about the middleware operation, the user just uses the publisher e subscriber functions as it

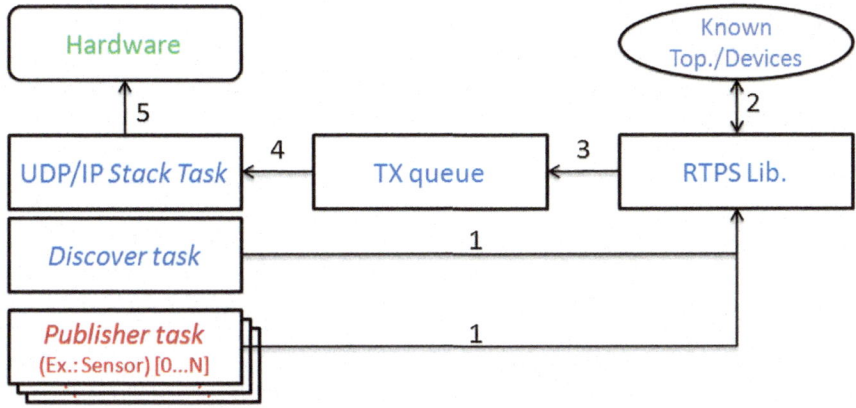

Fig. 5 Path of outgoing messages

needs, and leave the rest for the system background tasks, that are running discovery and decoding RTPS messages.

On our model we created a process for discovery operations and other for listening operations. The processes are situated outside of the user code, so they start and stay running in the background. Every time that user publish a message, the message will be formatted in RTPS protocol and sent to the UDP stack process. The UDP stack will be responsible to format put the packet over Ethernet network.

In the other hand, every time that a RTPS message arrives at the device, the UDP process puts the data into a queue for further processing. When the listen process starts to run it will verify if there are new RTPS messages on his port. If there are data available the listen process decodes the RTPS message following the RTPS protocol rules.

In this model we have two main RTPS functions at user level, one of them is the publisher. To use this function the user must create a pub with its type and topic. After that, just call the publisher function with the message and the related pub, as shown in the following code.

```
1 //Create a publisher that publish at the topic "
    ↪ topic" and has the String type.
2 frudp_pub_t *pub;
3 pub = freertps_create_pub( "topic",
    ↪ std_msgs__string__type.rtps_typename );
4
5 //freertps_publish publish a message "message"
    ↪    (String type), with size of
    ↪ message_length   (integer type) at the
    ↪ publisher "pub"    (frudp_pub_t type)
6 freertps_publish( pub, message, message_length );
```

Listing 1 Example of how to publish a message

To use the subscriber the user must define a static void routine, that will work as a task of the systems to handle the arrived messages. To register a subscriber the user must use the routine "freertps_create_sub" and specify the topic name, the subscriber data type, the static void routine that was created to handle the data and the size of the queue used to store the arrived data.

The routine used to handle the subscribed data must have the following format:

```
1   static void subTaskFormat( void *parameters ){
2       //Queue responsible for handle receiver
            ↪ information on "topic"
3       QueueHandle_t sQueue = ( QueueHandle_t )
            ↪ parameters;
4       //Struct to handle the data
5       Message xMessage;
6
7       while( true ){
8           //portMAX_DELAY tells that xQueueReceive
                ↪ will be blocked until data arrive in
                ↪ sQueue
9           if( xQueueReceive( sQueue, &xMessage,
                ↪ portMAX_DELAY ) == pdPASS ){
10              //Handle the arrived message
11          }
12      }
13  }
```

Listing 2 Subscriber handle task format

And the following code presents how to use the function "freertps_create_sub" to register a subscriber.

```
1   freertps_create_sub( "topic",
        ↪ std_msgs__string__type.rtps_typename,
2   subTaskFormat, queue_size );
```

Listing 3 Function to create a subscriber

The subscriber handle task puts itself into inactive state when no data is present on its queue, it only keeps active when data is present in the subscriber queue. Thus, the subscriber handle don't consume CPU, Central Processing Unit, processing while no data arrives. The task responsible for putting the data into the subscriber queue is the listen process. When a data message is sent by another node and the main application of the current node has subscribed that message topic, the message arrives at the current device and it is decoded by the listen process, that put the decoded data in the respective subscriber queue.

The original structure of the FreeRTPS can be seen in the Fig. 6.

The structure of the system after the modification is present in the following image, at Fig. 7.

Fig. 6 FreeRTPS original model

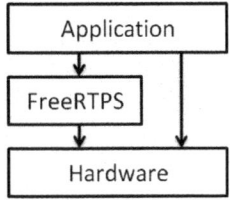

Fig. 7 FreeRTPS+Free-RTOS modified model

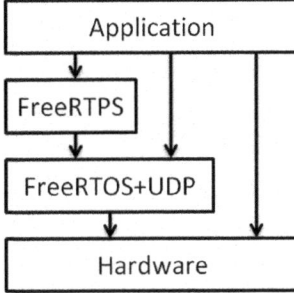

As we can see, in the proposed model the user has access to the hardware, to the UDP stack and to the FreeRTPS. But the FreeRTPS just have access to the UDP stack.

More information available on https://github.com/Expertinos/FreeRTPS-Free RTOS.

3.2 Implementation

Since FreeRTPS has a linear code implementation, with the exception of interrupts generated by the timers and Ethernet peripherals, the first step in integrating the real-time operating system is to break the code into smaller, independent processes, so they can be managed by the operating system.

The RTOSListenTask process was created for the listening and interpreting routine. This process calls the frudp_listen function to check for incoming new messages on the FreeRTPS ports, and closes the process cycle with a delay of 1ms. Already for the discovery routine, the RTOSDiscotickTask routine was created. The discovery routine consists of calling the FreeRTPS frudp_disco_tick function, which sends messages with participant (device) information and its endpoints (topics), and ends the process task with a period of 1s. The period of listen task is set by default with 1ms and the discovery task to 1 second value and they can be changed as the programmer needs, but keeping in mind that that reducing the period time will increase the processing consumption leaving less processing time for the rest of the system. The minimum period value to be used is 1ms.

```
1   static void RTOSDiscotickTask ( void *
        ↪ parameters ) {
2     while ( true ) {
3       vTaskDelay ( 1000 / portTICK_PERIOD_MS );
4       frudp_disco_tick ();
5     }
6   }
7   static void RTOSListenTask ( void * parameters )
        ↪ {
8     while ( true ) {
9       vTaskDelay ( 1 / portTICK_PERIOD_MS );
10      frudp_listen ();
11    }
12  }
```

Listing 4 Created task to listen messages and send discover messages

Finally, to integrate the UDP/IP stack efficiently with the FreeRTPS+FreeRTOS system, the entire FreeRTPS UDP/IP implementation was removed and replaced with FreeRTOS+UDP. This replacement will offer a better performance for the proposed system, since this implementation of the UDP stack complements the FreeRTOS operating system.

The first step is to change the code of the function responsible for adding the FreeRTPS communication ports, the frudp_add_ucast_rx routine, in udp.c file. To do this, we added the FreeRTOS_setsockopt function of FreeRTOS+UDP, which enables the desired ports for sending and receiving messages in the UDP/IP stack. With this, the system becomes able to send and receive messages through the RTPS protocol.

The interface between FreeRTPS and the UDP/IP stack happens in the frudp_listen and frudp_tx functions, both in the udp.c file. The frudp_listen function reads the RTPS messages received by the UDP protocol and then calls the frudp_rx function (udp.c file), so that this message is unpacked and processed according to the RTPS protocol header information. This routine receives as a parameter the time interval, in microseconds, provided by the user to process the messages that arrived and were stored in the circular buffer of the implemented UDP stack. The processing is done through the enet_process_rx_ring (enet.c) function. During this time the main program does not run.

The modification of the frudp_listen routine just get the messages received by the UDP/IP stack, from FreeRTOS+UDP, and then send them to the frudp_rx function, which has the rule set to decode RTPS protocol messages. The new frudp_listen routine has the following form:

```
1   void frudp_listen ( void ) {
2     struct freertos_sockaddr xSourceAddress;
3     int32_t iReturned = 0;
```

```
4    for( int i = 0; i < g_enet_allowed_udp_ports_wpos;
        ↪   i++ ){
5      do{
6        iReturned = FreeRTOS_recvfrom( xSocket[ i ],
             ↪  ucBuffer, 1600, 0, &xSourceAddress, (
             ↪  uint32_t *)(sizeof(xSourceAddress)) );
7        if( iReturned > 0 ){
8          if( frudp_rx( ucBuffer, iReturned ) ){}
9        }
10       }while( iReturned > 0 );
11     }
12   }
```

Listing 5 Modified FreeRTPS routine to listen messages

The original frudp_tx routine receives a formatted RTPS message as a parameter, depending on the type of information that would be sent through the protocol, along with the IP address and destination port. Then the message is passed to the enet_send_udp_ucast function (file enet.c), which is responsible for inserting it into a UDP/IP packet and ending in an Ethernet packet. After packaged in an Ethernet frame, the message is submitted over the network.

To insert the FreeRTOS+UDP/IP stack into the frudp_tx routine we need to indicate the UDP port that is sending the message. The original version has defined the source port of the message as the same as the destination. To improve the compatibility of this function in the further development of FreeRTPS, we added the 'srcPort' field in the parameters. The modified routine is shown next:

```
1    bool frudp_tx( const uint32_t dstAddr, const uint16_t
        ↪   dstPort, const uint16_t srcPort, const uint8_t
        ↪   *data, const uint16_t length ){
2      struct freertos_sockaddr xDestinationAddress;
3      int32_t iReturned;
4      xDestinationAddress.sin_addr = FreeRTOS_htonl(
          ↪  dstAddr );
5      xDestinationAddress.sin_port = FreeRTOS_htons(
          ↪  dstPort );
6      for( int i = 0; i < g_enet_allowed_udp_ports_wpos;
          ↪   i++ ){
7        if( g_enet_allowed_udp_ports[ i ] == srcPort ){
8          iReturned = FreeRTOS_sendto( xSocket[ i ], data
              ↪  , length, 0, &xDestinationAddress, sizeof
              ↪  ( xDestinationAddress ) );
9          if( iReturned == length ){
10           return true;
11         }
12       }
13     }
14     return false;
15   }
```

Listing 6 Modified FreeRTPS routine to send messages

As we can see, a loop was inserted to check which output port will be used to send the message. Once identified, the socket responsible for that port is selected and the FreeRTOS_sendto function is called from the FreeRTOS+UDP/IP. This function is responsible for packaging the message in the UDP protocol and dispatch the message on the network.

Next code modification consists of editing the routine that adds the ports used by the FreeRTPS protocol, originally in the file enet.c.

In the modified version the activation of the ports is still done by the routine frudp_add_ucast_rx, but now is in the file udp.c. In the routine, the FreeRTOS_bind function binds a socket to a local port, and thus enables receiving and sending messages by combining the local IP with the selected port on the network. Modified routine is shown below:

```
1   bool frudp_add_ucast_rx( const uint16_t port ){
2     TickType_t xReceiveTimeout_ms = 0;
3   for( int i = 0; i < g_enet_allowed_udp_ports_wpos; i++
        ↪    ){
4   if( g_enet_allowed_udp_ports[i] == port )
5   return true;
6   }
7     if( g_enet_allowed_udp_ports_wpos >=
          ↪ ENET_MAX_ALLOWED_UDP_PORTS )
8       return false;
9     g_enet_allowed_udp_ports[
          ↪ g_enet_allowed_udp_ports_wpos] = port;
10    xSocket[g_enet_allowed_udp_ports_wpos] =
          ↪ FreeRTOS_socket( FREERTOS_AF_INET,
          ↪ FREERTOS_SOCK_DGRAM, FREERTOS_IPPROTO_UDP );
11    FreeRTOS_setsockopt( xSocket[
          ↪ g_enet_allowed_udp_ports_wpos], 0,
          ↪ FREERTOS_SO_RCVTIMEO, &xReceiveTimeout_ms, 0 )
          ↪ ;
12    if( xSocket[g_enet_allowed_udp_ports_wpos] !=
          ↪ FREERTOS_INVALID_SOCKET ){
13      xBindAddress[g_enet_allowed_udp_ports_wpos].
            ↪ sin_port = FreeRTOS_htons( port );
14      if( FreeRTOS_bind( xSocket[
            ↪ g_enet_allowed_udp_ports_wpos], &
            ↪ xBindAddress[g_enet_allowed_udp_ports_wpos],
            ↪ sizeof( &xBindAddress[
            ↪ g_enet_allowed_udp_ports_wpos]) ) != 0 )
15        return false;
16    }else
17      return false;
18    g_enet_allowed_udp_ports_wpos++;
19    return true;
20  }
```

Listing 7 Modified FreeRTPS routine to open ports to listen RTPS messages

The enet.c file, which was responsible for performing UDP/IP actions in the original FreeRTPS was removed. The file arm_trap.c was also removed, leaving the unregistered interrupts on account of the specific micro controller file. The bswap.c file was also excluded, leaving the byte order functions now to the FreeRTOS equivalents functions. And finally, the routines of the system.c file were transferred to the udp.c file.

The original implementation of the udp.c file is divided into two levels, at one level are the general processing functions of the FreeRTPS protocol and at another level are the functions of receiving and sending the RTPS messages. The second level is divided into two udp.c files, one responsible for the POSIX interface and one for bare metal systems.

The POSIX implementation was removed and the routines were transferred from the udp.c file of the metal_common folder to the general file udp.c. The modifications were made with the intention of making the concept test development simpler, giving more time to develop and testing.

3.3 Implementation of Multicast Messages in FreeRTOS+UDP

FreeRTPS sends discovery messages on the network using multicast IP address, but was observed that version 9.0.0 of FreeRTOS+UDP does not implement multicast in its code. To enable this task we added a few lines of code in the FreeRTOS_UDP_IP.c file. Modifications were performed on the functions: prvGetARPCacheEntry (line 792), eConsiderFrameForProcessing (line 1305), and prvProcessIPPacket (line 1478).

The prvGetARPCacheEntry (FreeRTOS_UDP_IP.c - line 792) routine was modified as follows: a else conditional was added to the "if(*pulIPAddress == ipBROADCAST_IP_ADDRESS)":

```
1   else if( *pulIPAddress ==
        ↪  ipMULTICAST_IP_ADDRESS ){
2       memcpy( (void *)pxMACAddress,
            ↪  &xBroadcastMACAddress, sizeof(
            ↪  xMACAddress_t) );
3       eReturn = eARPCacheHit;
4   }
```

Listing 8 Modified FreeRTOS code to enable multicast

Doing this, the multicast sent messages going to have their destination MAC address filled correctly.

An additional else conditional was added to the eConsiderFrameForProcessing (FreeRTOS_UDP_IP.c - line 1305) routine:

```
1  else if ( pxEthernetHeader ->xDestinationAddress .
   ↪  ucBytes[ 2 ] == 0x5E ){
2     eReturn = eProcessBuffer;
3  }
```

Listing 9 Modified FreeRTOS code to enable multicast

This extra conditional tells to the UDP/IP stack filter that the arrived multicast packets can be placed in the UDP/IP list to be processed.

The prvProcessIPPacket (FreeRTOS_UDP_IP.c - line 1478) routine also received one more comparison in its "OR" conditional statement, the added comparison verifies if the received packet is a multicast message. The final conditional is presented below:

```
1  //(pxIPHeader ->ulDestinationIPAddress ==
   ↪  ipMULTICAST_IP_ADDRESS) was added to the
   ↪  conditional
2  if ( (pxIPHeader ->ulDestinationIPAddress == *
   ↪  ipLOCAL_IP_ADDRESS_POINTER) || (pxIPHeader
   ↪  ->ulDestinationIPAddress ==
   ↪  ipBROADCAST_IP_ADDRESS )||( *
   ↪  ipLOCAL_IP_ADDRESS_POINTER == 0) || (
   ↪  pxIPHeader ->ulDestinationIPAddress ==
   ↪  ipMULTICAST_IP_ADDRESS) )
```

Listing 10 Modified FreeRTOS code to enable multicast

This way, multicast messages will now be processed by the UDP/IP stack.

Finally, the definition of the variable ipMULTICAST_IP_ADDRESS was added to the FreeRTOS+UDP file FreeRTOS_IP_Private.h.

```
1  #define ipMULTICAST_IP_ADDRESS 0x0100FFEFUL
```

Listing 11 Modified FreeRTOS code to enable multicast

All modifications described above can be found in the code github https://github.com/Expertinos/FreeRTPS-FreeRTOS.

4   Using RTPS+RTOS on STMDISCOVERY

It is necessary to install some tools that are requirements to compile and write the code to the the STMDISCOVERY board. The steps are the same as on page http://github.com/ros2/freertps/wiki/Prerequisites (last access 08/2017), since we based our project on FreeRTPS.

However, since it is possible that the page changes over time, the installation of the necessary items will be described below and will also be available on https://github.com/Expertinos/FreeRTPS-FreeRTOS. The operating system used in this walk-through is Ubuntu 16.04.

Initially, we need to install the build-essential, cmake, and git packages through the following command on the terminal.

```
1   sudo apt-get install build-essential cmake git
```

Now you need to install the cross-compiling toolchain for ARM Cortex-M, it is responsible for compiling the source code of the application and generating the binary for the ARM architecture.

```
1   sudo apt-get install software-properties-common
      ↪ sudo
2   add-apt-repository ppa:team-gcc-arm-embedded/ppa
      ↪   sudo apt-get update
3   sudo apt-get install gcc-arm-embedded
```

The next step is to install OpenOCD, a tool that writes the program to the development board. In the next commands, a folder will be created in the root directory (where the source code of the application will be downloaded), the parameters of the tool will be configured and finally the installation will be done.

```
1   sudo apt-get install libtool autoconf automake
      ↪ pkg-config
2   libusb-1.0-0-dev libhidapi-dev
3    cd ~
4   git clone http://repo.or.cz/openocd.git cd
      ↪ openocd
5   ./bootstrap
6   ./configure --enable-stlink --enable-ftdi --
      ↪ enable-cmsis-dap
7   --prefix=/usr/local make -j4
8   sudo make install
```

The next step is to download the FreeRTPS+FreeRTOS system. For this you need to access the site https://github.com/Expertinos/FreeRTPS-FreeRTOS and download the FreeRTPS+FreeRTOS package. Once the download has finished, the content must be extracted into a folder at the user's desire.

Since the system still only supports ROS2 string messages, you only need to navigate to the system folder through the terminal and enter the 'make' command. When executing the 'make' command in the terminal, the applications 'talker', 'listener' and 'PID', contained in the directory 'freertps_freertos/apps', are compiled and their binary files are generated.

Finally, the application examples are programmed using the 'make program-listener-stm32f4_disco-metal', 'make program-talker-stm32f4_disco-metal' or program-pid-stm32f4_disco-metal' commands.

Currently, the ROS2 for desktop has 2 examples that can communicate with the applications described above. The examples of ROS2 are the 'listener__opensplice_ cpp' and 'talker_rmw_opensplice_cpp', the example listener subscribes the topic 'chatter', while the example talker publishes strings in the topic 'chatter'. To run these examples, you need to install the ROS2 in the machine through the tutorial described in the page https://github.com/ros2/ros2/wiki/Installation.

After ROS2 has been installed, the examples of the listener program can be executed through the following commands:

```
1  cd ~/ros2_ws
2  source install/setup.bash
       ↪ listener__rmw_opensplice_cpp
```

And to execute the talker, the following commands:

```
1  cd ~/ros2_ws
2  source install/setup.bash
       ↪ talker__rmw_opensplice_cpp
```

5 Tests

To evaluate the functionality and performance of the system some tests were developed for validation of real-time and system performance. For the validation of the real time a voltage control test of an RC circuit was made using a PID controller. To evaluate the performance of the system we verified the quantity of messages that the system is able to send and the quantity of resources needed, while executing the PID control.

5.1 Hardware

The original FreeRTPS system was developed on the STM32F4DISCOVERY development board. A board that has all the necessary circuits to operate the micro controller STM32F407VGT6, a circuit included to program the micro controller and some features (buttons, leds, accelerometer, audio output and connectors) [36]. The flash memory of this chip is 1 Mbyte with 192 KBytes of RAM. Its maximum main clock frequency can reach values of 168 MHz [37].

As the FreeRTPS communicates through UDP messages, it will be necessary to make use of the Ethernet peripheral of the micro controller. To make use of extra features on the development board, ST provides another ready-to-use board designed to be integrated with STM32F4DISCOVERY. The extra board that will be used in this work is the discovery base board, it has: a MicroSD memory card slot, an RJ45 10/100 Ethernet connector, connectors for the communication protocols and connectors for a camera and an LCD expansion cards of the kit DISCOVERY [38]

To verify the portability of the proposed system, the system was also implemented on a Texas Instruments development board, the Stellaris LM3S6965 Evaluation Board. This board has an ARM Cortex M3 micro controller and some peripheral devices such as: an LED display, buttons, LEDs, memory card, buzzer and an RJ45 Ethernet connector [39].

The microcontroller present in the Stellaris board can operate at the maximum frequency of 50 MHz, has 256 KBytes of flash and 64 KBytes of RAM. It can be programmed using the StellaresWare library provided by Texas Instruments, which provides routines for configuration and use of all peripherals in a more intuitive way [40].

5.2 Portability Test

In order to verify the performance and portability of the proposed system, this test will be made on another board, giving us the opportunity to verify if the system could be implemented on a different micro controller (with another architecture), provided by a different manufacturer and uses an IDE (Integrated Development Environment) to compile and write the program, differently from the original FreeRTPS system which uses the GCC and Ubuntu terminal for this job.

For the portability test, we used the Stellaris LM3S6965 Evaluation Board, which has a LM3S6965 microcontroller from Texas Instruments, and the IDE IAR Embedded Workbench for programming. The operating system used was Microsoft Windows 7 64 bits.

Unlike the test performed in Ubuntu 16.04, it was not necessary to define the micro controller setup files, that is, linker, loader and stack files. The setups of these files are already supplied by the IDE itself when a new project is created, so only the source and header files were needed. IDE IAR Embedded Workbench was chosen because it supports drivers from several companies and supports a lot of devices. The FreeRTOS is compatible with IAR and provides the necessary files to run its kernel.

Initially, we need to download and install the IDE IAR Embedded Workbench for ARM, the StellarisWare library (Stellaris family peripherals) and FreeRTOS+Free RTOS. The following are the links from where these files can be found:

- IAR Embedded Workbench 8.11: www.iar.com/iar-embedded-workbench/
- StellarisWare: www.ti.com/tool/SW-LM3S.
- FreeRTPS+FreeRTOS: http://github.com/Expertinos/FreeRTPS-FreeRTOS

StellarisWare files are installed by default in the root of the Windows system, but can be installed in another location. FreeRTOS+FreeRTPS files can be extracted into any folder that user has access. After installation, run the IAR Embedded Workbench application.

At this point, we create the project that will have the application's code and FreeRTPS+FreeRTOS system. Then select the item 'Project' in the menu bar and then 'Create New Project'. In the next window, Fig. 8, select the 'main' template in C language, and define the name and directory of the project.

The microcontroller startup code is then added, this code provides the interrupt table and standard device interrupt handlers. The StellarisWare library provides the startup files for the IDEs: IAR, CCS, CodeRed, and the GCC compiler. Then navigate to the directory of the microcontroller used, "StellarisWare\boards\ek-lm3s6965\

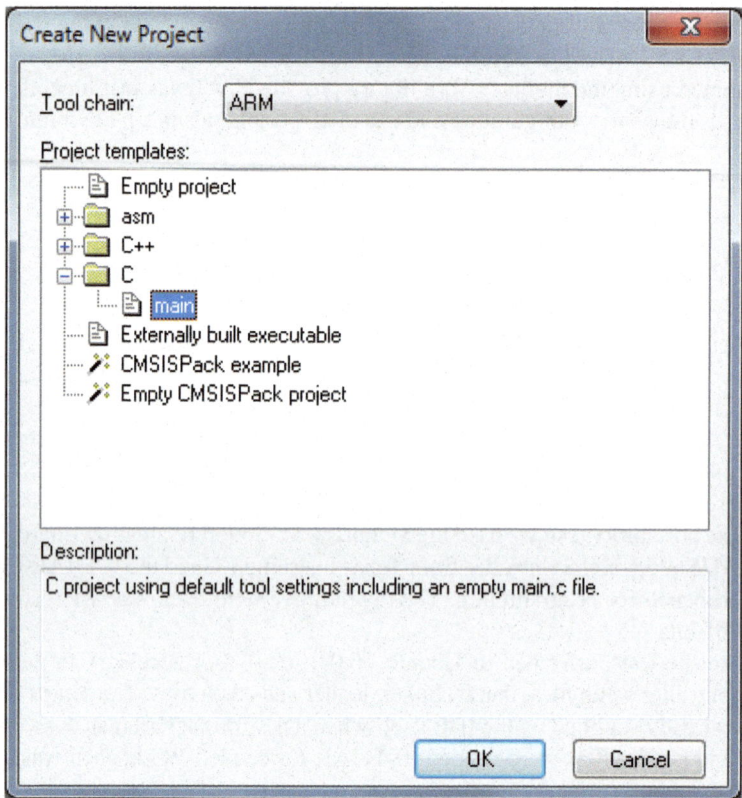

Fig. 8 Creating new project

blinky", and copy the startup_ewarm.c file to the previously created project directory. After copying the file to the directory you must add the copied files directories to the project, so that it enters in the list of files to be processed by the compiler. To do this, right click on the project and select the 'Add Files' item in the 'Add' tab.

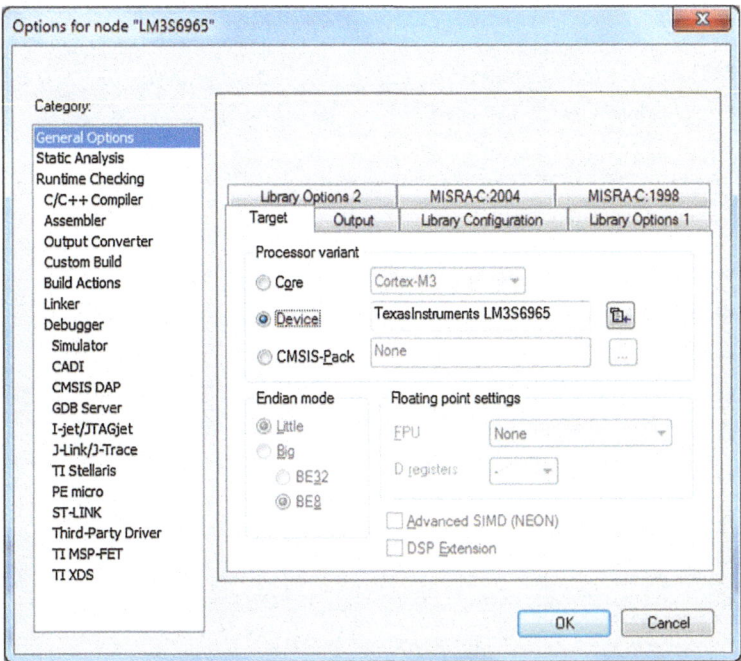

Fig. 9 Selection of microcontroller used

Now we select the target device of the project, so that the IDE manages the files needed to compile and write the program to a specific device. To do this, right-click on the project and select the 'Options' item. In the following window, Fig. 9, we open the category 'General Options' and then select the 'TexasInstruments LM3S6965' device in the 'Processor Variant' box on the tab 'Target' (right of the window).

At this point we insert the dependency files for the StellarisWare library into the project. Initially add the driverlib.a file to the project, the same process used to insert the startup file. The driverlib.a file is located in the "StellarisWare\driverlib\ewarm \Exe".

The source and header files of the StellarisWare and FreeRTOS+FreeRTPS library are now added. In this step, click on the 'Options' button and select the 'C/C++ Compiler' category. Next, select the 'Preprocessor' tab, which is located to the right of the window, and add the following header directories, as in Fig. 10.

- C:/StellarisWare;
- "FreeRTOS\FreeRTOS\include";
- "FreeRTOS\FreeRTOS\portable\IAR\ARM_CM3";
- "FreeRTOS\FreeRTOS_Plus_UDP\include";
- "FreeRTOS\FreeRTOS_Plus_UDP\portable\Compiler\GCC";
- "include/freertps";
- "$PROJ_DIR$".

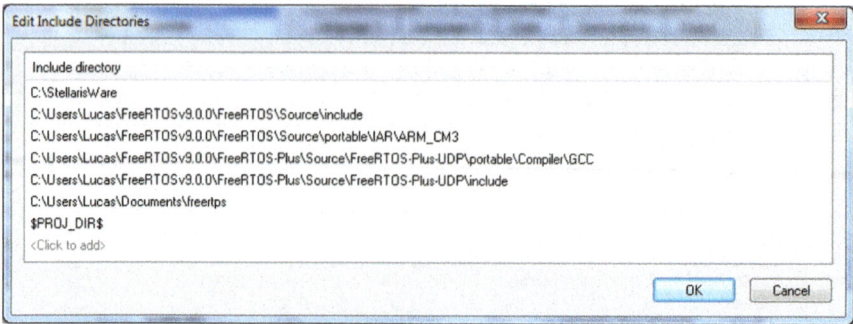

Fig. 10 C Compiler Preprocessor - Include directories

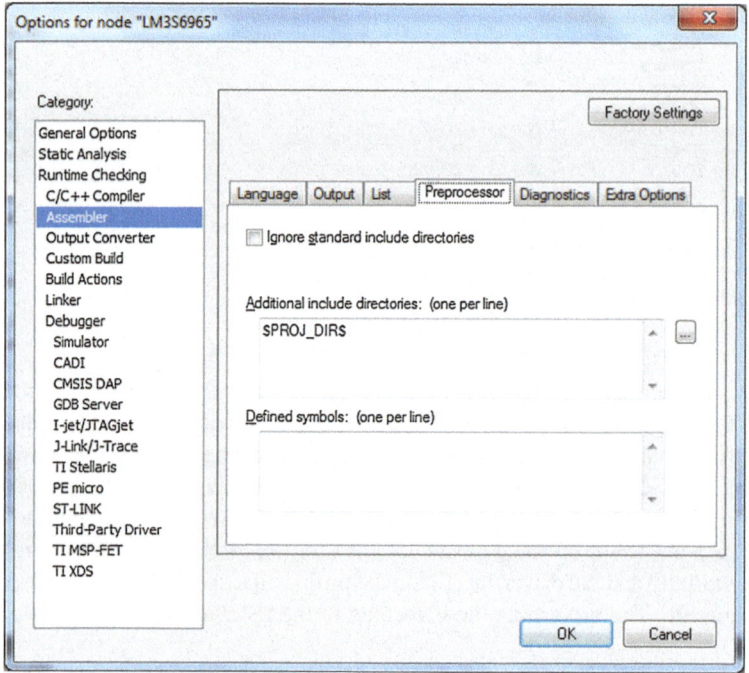

Fig. 11 Assembler Preprocessor - Include directories

It is also necessary to add the include directory at the tab 'Preprocessor' of the 'Assembler' category, Fig. 11.

To finish the options settings some parameters of the 'Debugger' category are edited. In the 'Setup' tab select 'Driver TI Stellaris', and in the 'Download' tab check the options 'Verify download' and 'Use flash loader', Fig. 12.

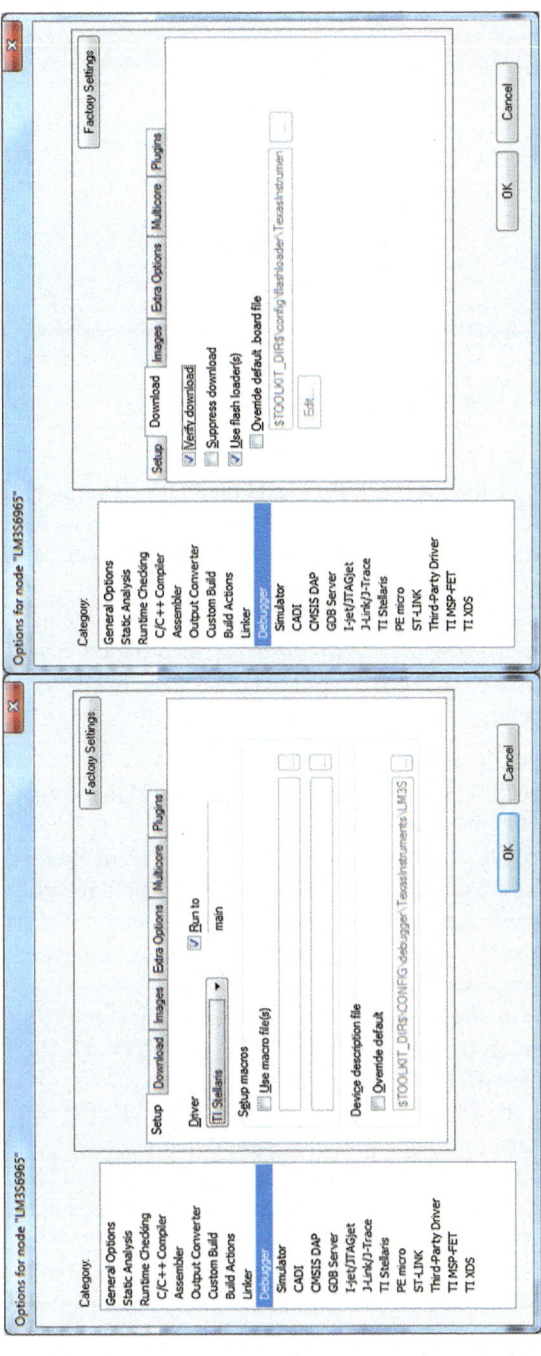

Fig. 12 Debugger Parameters

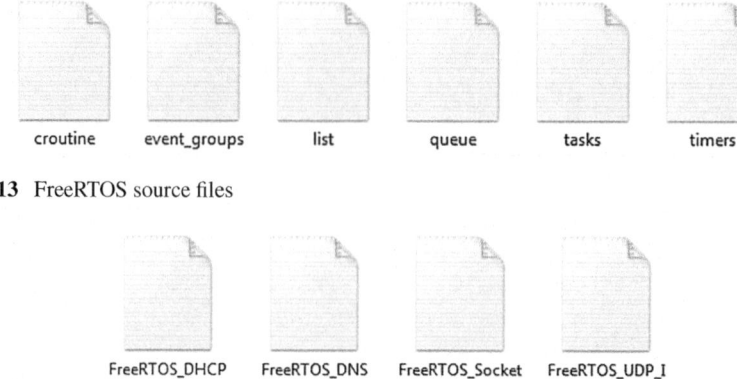

| croutine | event_groups | list | queue | tasks | timers |

Fig. 13 FreeRTOS source files

| FreeRTOS_DHCP | FreeRTOS_DNS | FreeRTOS_Socket s | FreeRTOS_UDP_I p |

Fig. 14 FreeRTOS+UDP source files

Finally, the source files of FreeRTOS and FreeRTPS are imported into the project. Again, right-click the project and use the 'add files' option. For FreeRTOS, the following files located in the directory "FreeRTOS\FreeRTOS", Fig. 13.

The files 'port.c' and 'portasm.s', from directory "FreeRTOS\FreeRTOS\portable\IAR\ARM_CM3", and the file heap4.c, from "FreeRTOS\FreeRTOS\portable\MemMang", also should be add.

For the FreeRTOS+UDP, add the following files located in the directory "FreeRTOS\FreeRTOS_Plus_UDP", Fig. 14:

The file 'BufferAllocation_2', found at directory "FreeRTOS\FreeRTOS_Plus_UDP\portable\BufferManagement", responsible for managing FreeRTOS memory allocation is also added.

Next, add the file 'LM3s6965NetworkInterface.c', available for download on the FreeRTPS+FreeRTOS GitHub repository. It is responsible for making the interface of the UDP stack with the hardware. On folder "FreeRTOS\FreeRTOS_Plus_UDP" there are some examples of network interface files for others micro controllers.

Finally, add the main application file, main.c, and the FreeRTPS source files. The main.c is available for download at FreeRTPS+FreeRTOS GitHub. The FreeRTPS files are found in FreeRTPS+FreeRTOS directory, Fig. 15

By clicking on the Download and Debug button, at the top of IAR window, the device is programmed through the USB port. When the device is connected to an Ethernet network it searches for a valid IP address, if there is a DHCP server on the network. In case there is no DHCP server, the static IP defined in the file 'main.c' is assigned to it.

With this test, it is possible to carry the new system for architecture micro controllers, family and different manufacturers, as long as compatible with FreeRTOS.

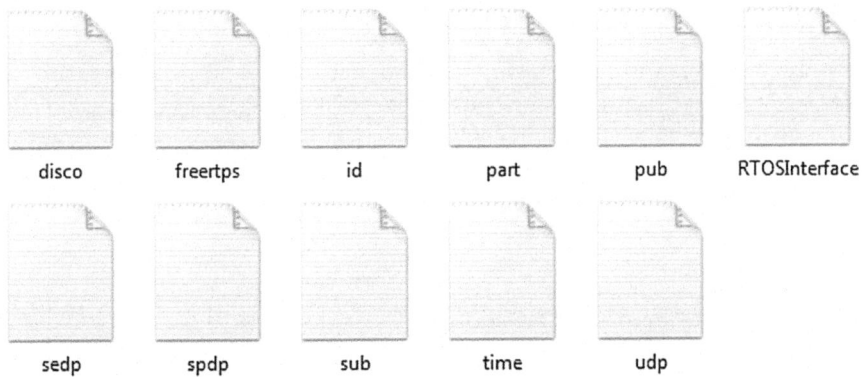

<table>
<tr><td>disco</td><td>freertps</td><td>id</td><td>part</td><td>pub</td><td>RTOSInterface</td></tr>
<tr><td>sedp</td><td>spdp</td><td>sub</td><td>time</td><td>udp</td><td></td></tr>
</table>

Fig. 15 FreeRTPS source files

5.3 Real Time Test

A system having tasks with different priorities is almost a requirement for proof that all of them can be scheduled and meet their time constraints. The system presented in this topic is a simple example just to proof that the FreeRTOS did not lose its capacity to handle tasks based on their priorities. Showing that the FreeRTOS still has all your capabilities, we can say that the FreeRTOS+FreeRTPS system has the necessary but not sufficient resources needed to implement a more complex real-time system. To implement a more complex real-time systems other care needs to be taken, like methods to determine their priorities, shared resources, CPU usage and others.

To perform system validation, a closed control loop using a digital PID controller (Eq. 2) was implemented with a sampling rate of 1000 Hz. The validation is done by comparing the output of the physical system with a simulation of a discrete PID controller also using the bilinear approximation method (Tustin) with the same parameters used in the physical test.

We used a voltage control of a simple RC circuit, we chose this system because it is simple and fast to assemble. The system values used in the test were assigned in a way that the system behave like a linear process, i.e. the output never reach the lower and higher voltage limits.

The chosen values were a resistor of 10 KΩ and a capacitor of 1uF, the connection of the components has the following structure, Fig. 16.

As a simple test to analyze the operating system scheduler capabilities, the control is made by one task that has the highest priority of the system, expecting to obtain the process scheduler preference, with no shared variables and other tasks with a low priority. The control task is an infinite loop, often 1000 Hz, which reads the current value of the system output voltage and then calculates the output value of the PWM (Pulse Width Modulation) so that the desired voltage value is reached. The frequency

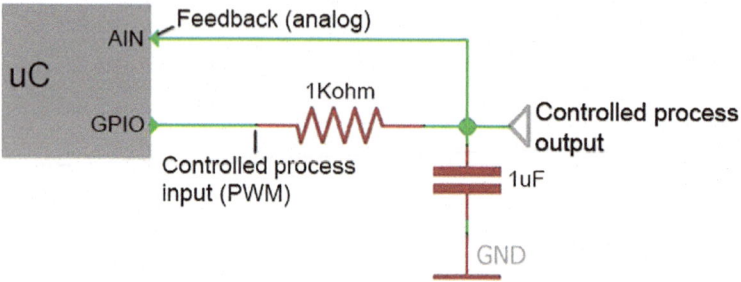

Fig. 16 Circuit used to perform PID controller test

used in the PWM in this test was 20000 Hz. Using a high frequency offer basically a DC output once that the RC time constant used is 10 ms.

The system has an output voltage between 0 V and 3 V and the values of desired voltage and current voltage are published on the network using 'uInt32' type. The voltage values are published in the 12 bits digital representation, this is because our publisher and subscriber are integer types and the voltage is read this way by the microcontroller, that has a ADC with 12 bits resolution and 3 V reference.

Two more tasks were created, one to publish messages, with the current voltage value of the controlled system, and another with a subscriber, which receives setpoints from the other ROS2 node. The setpoints are published by another board connected to the same network.

In this way, the subscriber and PID tasks from the program executing the PID control has the following form:

```
1  #include <stdio.h>
2  #include "freertps/freertps.h"
3  #include
4  "std_msgs/uint32.h"
5  #include "freertps/RTOSInterface.h"
6
7  //If use PID is set to 1 the program use closed
       ↪    loop control, if not the program use
       ↪    open loop control
8  #define usePID 1
9  //PID variables
10 #define KP    10.0
11 #define KI    500.0
12 #define KD    0.0
13 #define PIDsampleTime 0.001
14 //Max and Min voltage of the PWM wave
15 #define PWM_MAX_VOLTAGE 3.0
16 #define PWM_MIN_VOLTAGE 0.0
17 //Number of bits from PWM
18 #define PWM_BITS 16
19 //Number of bits from ADC
```

```
20  #define ADC_BITS 12
21  //PID control task
22  static void ctrlTask( void *parameters );
23  //Voltage publisher task
24  static void pubCurVoltTask( void *parameters );
25  //Topic setPoint, uint32, subscriber task
26  static void desiredVoltSubTask( void *
        ↪ parameters );
27  //PWM init function
28  void initPWM();
29  //ADC init function
30  void initADC();
31  //Current ADC value function
32  uint16_t getADCValue();
33  //Voltage publisher variable
34  frudp_pub_t *pub;
35  //PID voltage setpoint variable. Start with
        ↪ 2048 what means 1.5 volts
36  double setPoint = 0.0;
37  //Current analogic to digital value
38  double currentADCValue = 0.0;
39  //MAC address, must be different for each
        ↪ device
40  uint8_t ucMACAddress[ 6 ] = { 0x2C, 0x4D, 0x59,
        ↪  0x01, 0x23, 0x51 };
41  //Desired IP parameter if DHCP do not work
42  static const uint8_t ucIPAddress
        ↪ [4]={192,168,2,151};
43  static const uint8_t ucNetMask
        ↪ [4]={255,255,255,0};
44  static const uint8_t ucGatewayAddress[4]={
        ↪ 192,168,2,0 };
45  static const uint8_t ucDNSServerAddress
        ↪ [4]={208,67,222,222 };
45
46  //Main function
47  int main( void ){
48    //Clock configuration before FreeRTOS_IPInit
49    //Start FreeRTOS+UDP stack
50    FreeRTOS_IPInit( ucIPAddress, ucNetMask,
        ↪ ucGatewayAddress, ucDNSServerAddress,
        ↪ ucMACAddress );
51    //Start operating system scheduler
52    vTaskStartScheduler();
53    //Infinite loop, the program must not reach
        ↪ this loop
54    for(;;){}
55    return 0;
56  }
```

```
57
58   //Function to setup pubs, subs and others tasks
59   void setup( void ){
60     //Init ADC and PWM peripheral
61     initADC();
62     initPWM();
63     //Pubs here. Example pub =
           ↪ freertps_create_pub( topicName,
           ↪ typeName );
64     pub = freertps_create_pub( "currentVolt",
           ↪ std_msgs__uint32__type.rtps_typename );
65     //Subs here. Example freertps_create_sub(
           ↪ topicName, typeName, handlerTask,
           ↪ dataQueueSize );
66     freertps_create_sub( "desiredVolt",
           ↪ std_msgs__uint32__type.rtps_typename,
           ↪ desiredVoltSubTask, 10 );
67     //General tasks here. ctrlTask with greater
           ↪ priority
68     xTaskCreate( pubCurVoltTask, "pubCurVoltTask"
           ↪ , configMINIMAL_STACK_SIZE, NULL,
           ↪ tskIDLE_PRIORITY + 1, NULL );
69     xTaskCreate( ctrlTask, "ctrlTask",
           ↪ configMINIMAL_STACK_SIZE, NULL,
           ↪ tskIDLE_PRIORITY + 2, NULL );
70   }
71
72   //PID control task
73   static void ctrlTask( void *parameters ){
74     //Variable that holds the time at which the
           ↪ task was last unblocked
75     TickType_t xLastWakeTime = xTaskGetTickCount
           ↪ ();
76     //PID variables
77     double y0, y1, y2, e0, e1, e2;
78     const double kp = KP;
79     const double kd = KD * 2.0 / PIDsampleTime;
80     const double ki = KI * PIDsampleTime / 2.0;
81     //Init the variables
82     e0 = e1 = e2 = y0 = y1 = y2 = 0.0;
83     //PID loop
84     while( true ){
85   #if usePID == 1
86       //Getting the current ADC value
87       currentADCValue = getADCValue();
88       //Doing the calculations
89       y2 = y1;
90       y1 = y0;
91       e2 = e1;
```

```
92      e1 = e0;
93      e0 = setPoint - ( currentADCValue / ( ( 1
            ↪ << ADC_BITS ) - 1 ) );
94      //PID digital equation using bilinear (
            ↪ tustin) aproximation
95      y0 = y2 + kp * ( e0 - e2 ) + ki * ( e0 +
            ↪ 2.0 * e1 + e2 ) + kd * ( e0 - 2.0 *
            ↪ e1 + e2 );
96  #else
97      //open loop test, output just receives the
            ↪ desired value
98      y0 = setPoint;
99  #endif
100     //Saturating at maximum value in terms of
            ↪ volts output
101     if( y0 > PWM_MAX_VOLTAGE )
102       y0 = PWM_MAX_VOLTAGE;
103     //Saturating at minimum value in terms of
            ↪ volts output
104     if( y0 < PWM_MIN_VOLTAGE )
105       y0 = PWM_MIN_VOLTAGE;
106     //Setting the new PWM value to the
            ↪ controled system
107     //converting voltage to digial in PWM bits
108     TIM4->CCR4 = ( uint16_t )( y0 * ( ( 1 <<
            ↪ PWM_BITS ) - 1 ) / PWM_MAX_VOLTAGE );
109     //Delay sampleTime milliseconds to get the
            ↪ correct PID time control
110     vTaskDelayUntil( &xLastWakeTime, (
            ↪ PIDsampleTime * 1000.0 ) /
            ↪ portTICK_PERIOD_MS );
111  }}
112
113  //Function to receive messages from
         ↪ desiredVoltage topic
114  static void desiredVoltSubTask( void *
         ↪ parameters ){
115    //Queue responsible for handle receiver
           ↪ information on "setPoint" topic
116    QueueHandle_t sQueue = ( QueueHandle_t )
           ↪ parameters;
117    //Variables to handle the data
118    uint8_t  msg[ 128 ] = { 0 };
119    while( true ){
120      //portMAX_DELAY tells that xQueueReceive
             ↪ will be blocked until data arrive in
             ↪ sQueue
121      if( xQueueReceive( sQueue, msg,
             ↪ portMAX_DELAY ) == pdPASS ){
```

```
122        //Led toggle to see that data has been
               ↪ arrived
123        led_toggle();
124        //Getting the received setpointd as 12
               ↪ bit digital value
125        setPoint = *( ( uint32_t* )msg );
126        //Convert digital value to voltage
127        setPoint = PWM_MAX_VOLTAGE * setPoint / (
               ↪ ( 1 << ADC_BITS ) - 1 );
128 }}}
129
130 //Voltage publisher task
131 static void pubCurVoltTask( void *parameters ){
132    //FreeRTPS publisher variables
133    uint8_t cdr[ 20 ] = { 0 };
134    int cdr_len;
135    struct std_msgs__uint32 voltage;
136    while( true ){
137        //Getting the current voltage in digital
               ↪ value
138        voltage.data = currentADCValue;
139        //serialize the data
140        cdr_len = serialize_std_msgs__uint32( &
               ↪ voltage, cdr, sizeof( cdr ) );
141        //Publish the data
142        freertps_publish( pub, cdr, cdr_len );
143        //Task period
144        vTaskDelay( 10 / portTICK_PERIOD_MS );
145 }}
146
147 //Init ADC1 - Channel 11 - Pin C1
148 void initADC(){
149    //Enable PORT C clock
150    RCC->AHB1ENR |= RCC_AHB1ENR_GPIOCEN;
151    //Enable clocl for AD1
152    RCC->APB2ENR |= RCC_APB2ENR_ADC1EN;
153    //Enale port C as analog
154    GPIOC->MODER |= GPIO_MODER_MODER1_1 +
               ↪ GPIO_MODER_MODER1_0;
155    //Set ADC1 with channel 11
156    ADC1->SQR3 |= 0x0000000B;
157    //Enable AD1
158    ADC1->CR2 |= ADC_CR2_ADON;
159 }
160
161 //Init PWM - TIMER 4 - Channel 4 - PIN D15
162 void initPWM(){
163    //Enable PORT D clock
164    RCC->AHB1ENR |= RCC_AHB1ENR_GPIODEN;
```

```
165    //Enale port D as alternate function (PWM)
166    GPIOD->MODER |= GPIO_MODER_MODER15_1;
167    //Alternate function for Pin D15 (TIMER 4 PWM
         ↪ Channel 4)
168    GPIOD->AFR[ 1 ] |= 0x20000000;
169    //Enable TIMER 4, used to PWM
170    RCC->APB1ENR |= RCC_APB1ENR_TIM4EN;
171    //Timer 4 Auto-reload value
172    TIM4->ARR = 0x00000FFF;
173    //Set Timer 4 as PWM
174    TIM4->CCMR2 |= TIM_CCMR2_OC4M_2 |
         ↪ TIM_CCMR2_OC4M_1;
175    //TIMER 4 compare enable
176    TIM4->CCER |= TIM_CCER_CC4E;
177    //Enable TIMER 4
178    TIM4->CR1 |= TIM_CR1_CEN;
179  }
180
181  //Routine to get current ADC1 value
182  uint16_t getADCValue(){
183    //Start AD1 conversion
184    ADC1->CR2 |= ADC_CR2_SWSTART;
185    //Wait for the end of AD1 conversion
186    while( !( ADC1->SR & ADC_SR_EOC ) );
187    //Get the conversion result
188    return ADC1->DR;
189  }
```

Listing 12 Simple PID code example

The second device connected to the network, which publishes the desired voltage values of the system, has a periodic task that alternates the desired voltage between 1.2 V and 1.8 V, with a period of 200 ms. As previously described, the analog voltage values are converted into digital values with a 3 V reference and 12 bit representation. Therefore, the values that are actually published are: 1638 for the value of 1.2 V and 2457 for the value of 1.8 V. The code for this second device was implemented as follows:

```
1   #include <stdio.h>
2   #include "freertps/freertps.h"
3   #include "std_msgs/uint32.h"
4   #include "freertps/RTOSInterface.h"
5   //Number of bits from ADC
6   #define ADC_12_BITS 12
7   //ADC reference voltage
8   #define ADC_REF_VOLTAGE 3.0
9   //Voltage setpoint publisher task
10  static void pubTask( void *parameters );
11  //Voltage setpoint publisher variable
12  frudp_pub_t *pub;
```

```
13  //Device ethernet MAC address, must be
        ↪ different for each device
14  uint8_t ucMACAddress [6]={0x2C,0x4D,0x59,0x01,0
        ↪ x23,0x52};
15  //Desired IP parameter if DHCP do not work
16  static const uint8_t ucIPAddress
        ↪ [4]={192,168,2,152};
17  static const uint8_t ucNetMask
        ↪ [4]={255,255,255,0};
18  static const uint8_t ucGatewayAddress
        ↪ [4]={192,168,2,0};
19  static const uint8_t ucDNSServerAddress
        ↪ [4]={208,67,222,222};
20  //Main function
21  int main( void ){
22      //Do necessary clock configuration before
            ↪ FreeRTOS_IPInit
23      //Start FreeRTOS+UDP stack
24      FreeRTOS_IPInit( ucIPAddress, ucNetMask,
            ↪ ucGatewayAddress, ucDNSServerAddress,
            ↪ ucMACAddress );
25      //Start operating system scheduler
26      vTaskStartScheduler();
27      //Infinite loop, the program must not reach
            ↪ this loop
28      for(;;){}
29      return 0;
30  }
31  //Function to setup pubs, subs and others tasks
32  void setup( void ){
33      //Pubs here. Example pub =
            ↪ freertps_create_pub( topicName,
            ↪ typeName );
34      //Publisher for thetopic desiredVolt
35      pub = freertps_create_pub( "desiredVolt",
            ↪ std_msgs__uint32__type.rtps_typename );
36      //General tasks here. ctrlTask with greater
            ↪ priority
37      xTaskCreate( pubTask, "pubTask",
            ↪ configMINIMAL_STACK_SIZE, NULL,
            ↪ tskIDLE_PRIORITY + 1, NULL );
38  }
39
40  //Uint32 publisher task
41  static void pubTask( void *parameters ){
42      //FreeRTPS publisher variables
43      uint8_t cdr[ 20 ] = { 0 };
44      int cdr_len;
45      struct std_msgs__uint32 digital12bitsVoltage;
```

```
46    double desiredVolt;
47    while( true ){
48      //Desired voltage to reach
49      desiredVolt = 1.2;
50      //Converting 0V-3V volts to digital value
              ↪ in 12 bits
51      digital12bitsVoltage.data = ( uint32_t )(
              ↪ desiredVolt * ( ( 1 << ADC_12_BITS )
              ↪ - 1 ) / ADC_REF_VOLTAGE );
52      //Blink led each time that data is
              ↪ published
53      led_toggle();
54      //serialize the data
55      cdr_len = serialize_std_msgs__uint32( &
              ↪ digital12bitsVoltage, cdr, sizeof(
              ↪ cdr ) );
56      //Publish the data
57      freertps_publish( pub, cdr, cdr_len );
58      //Delay for 100 ms
59      vTaskDelayUntil( &xLastWakeTime, 100 /
              ↪ portTICK_PERIOD_MS );
60
61      //Desired voltage to reach
62      desiredVolt = 1.8;
63      //Converting 0V-3V volts to digital value
              ↪ in 12 bits
64      digital12bitsVoltage.data = ( uint32_t )(
              ↪ desiredVolt * ( ( 1 << ADC_12_BITS )
              ↪ - 1 ) / ADC_REF_VOLTAGE );
65      //Blink led each time that data is
              ↪ published
66      led_toggle();
67      //serialize the data
68      cdr_len = serialize_std_msgs__uint32( &
              ↪ digital12bitsVoltage, cdr, sizeof(
              ↪ cdr ) );
69      //Publish the data
70      freertps_publish( pub, cdr, cdr_len );
71      //Delay for 100 ms
72      vTaskDelayUntil( &xLastWakeTime, 100 /
              ↪ portTICK_PERIOD_MS );
73    }}
```

Listing 13 Publisher PID setpoints example code

The first test consists of obtaining the signals using an open-loop control, i.e. the system simply assigns the PWM the desired average voltage value. Since the STMDISCOVERY digital pins have a low logic level at 0V and high logic level at 3 V, the average output voltage for 0% duty cycle of the PWM is 0V and for 100%

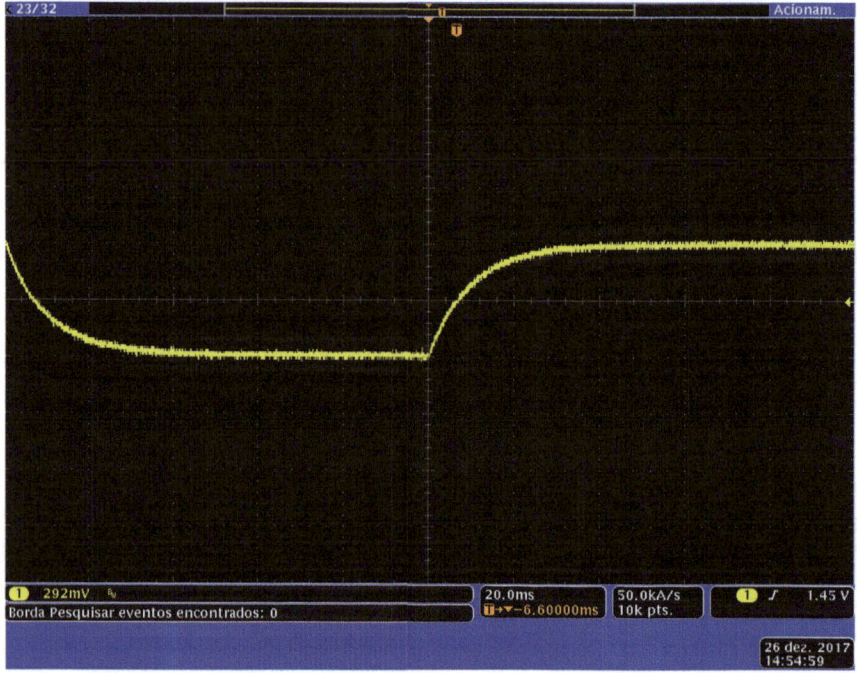

Fig. 17 Open-loop voltage control of as RC circuit

of duty cycle of the PWM the average voltage is 3 V. Thus, for the values of 1.2 V and 1.8 V, 4% and 60% respectively were applied. Because this development board has a 16-bit PWM, the applied digital values were 26214 (40%) and 39321 (60%).

With the assistance of an oscilloscope to check the voltage level of the capacitor during the operation of the system, the code execution of the boards with the open-loop control is started and the waveform shown in Fig. 17 is obtained.

The typical charge curve of a capacitor delineated by an exponential, with a time constant dependent on the RC circuit constant, is observed. In the case of the experiment, for the resistance value of $10\,K\Omega$ and capacitance of 10 uF, the time constant of 10 ms is obtained. It is noticed that the value accommodates between 30 ms and 40 ms after the change of the new output value of the PWM, consistent with the control theory, once that a RC circuit going stables between 4 5 times the RC constant

As the focus of the project was not to obtain a well-adjusted PID controller but to validate the real time. Proportional, integrative and derivative gains were assigned with values to obtain a good output response for a comparison between the closed-loop system and the simulation of a non-discrete equivalent controller. Programming the control board with the PID code gives the waveform of the Fig. 18:

It is observed that while the open loop control has a smoother rise and fall curve limited by the circuit time constant, the PID control offers a faster response waveform.

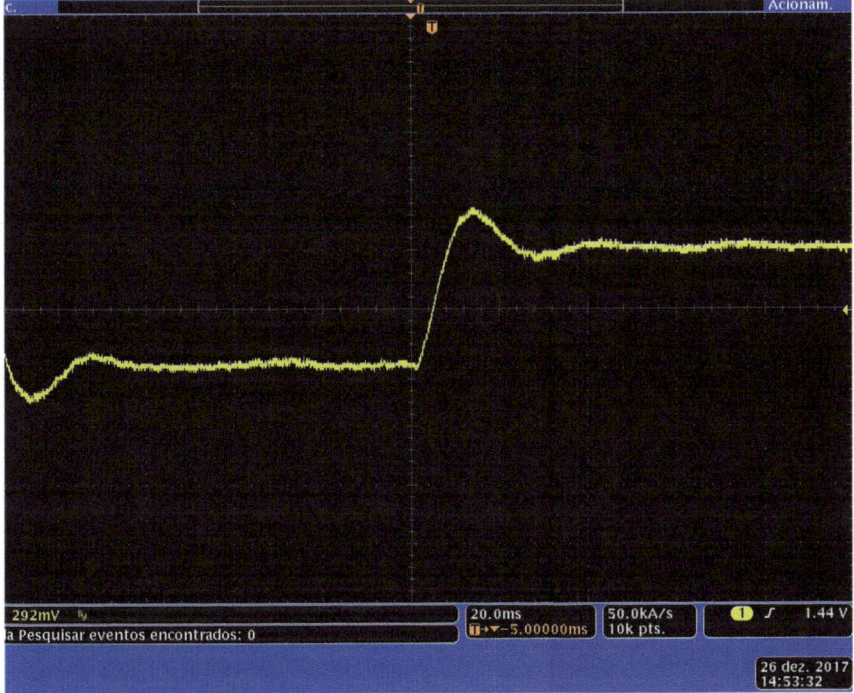

Fig. 18 Closed-loop voltage control of as RC circuit using a PID controller

This is due to the PID control apply a 100% duty cycle to the PWM when the desired voltage changes and then continues adjusting this value to keep the output at the desired level of 1.8 V, different from the open loop control that provided 60% duty cycle of the PWM all the time, percentage referring to the average voltage of 1.8 V for the board used.

The same happens for the lowest voltage level, the PID controller applies 0% duty cycle of the PWM at the beginning, forcing the output to have a faster voltage lowering, and then adjust the value of duty cycle in the time to maintain the output at 1.2 V, while the open loop control maintained 40% duty cycle of the PWM equivalent to the average voltage of 1.2 V, for as long as the desired voltage output was 1.2 V.

Finally, we performed the simulation of a PID controller acting on a system that mathematically represents the RC circuit tested. The program used to perform the simulation was the SciLab, where all the limitations of the physical circuit were applied to it: lower and upper saturation value, the parameters used in the controller, sampling frequency, PWM resolution, ADC and gains the controller. The model elaborated for simulation can be observed in the Fig. 19.

In the simulation it is observed that for the same desired voltage values and parameters used in the physical test, the method used in the discrete model presents a curve very close to the continuous model, Fig. 20.

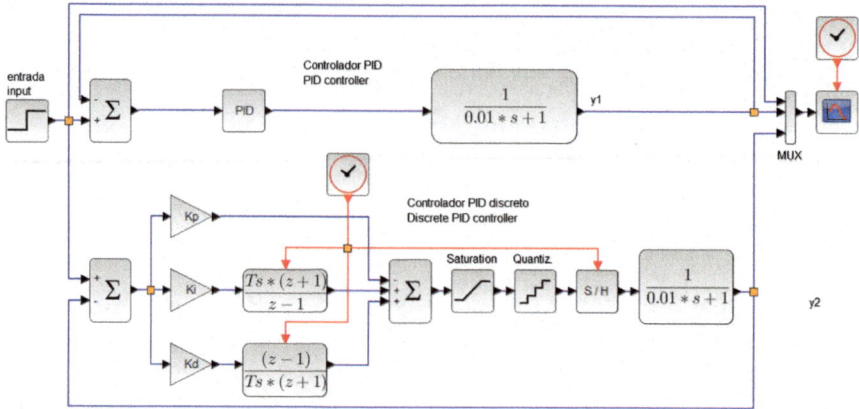

Fig. 19 Block model of a PID controller equivalent to the tested system

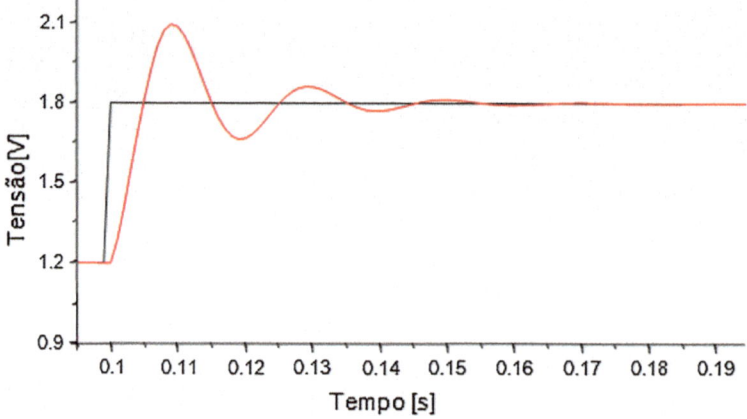

Fig. 20 Result obtained in the simulation using the parameters of the real test

Now, overlapping the curve obtained in the physical test along the curve obtained in the simulation, Fig. 21, it is possible to see that the digital control system provided a stable response close to its simulated continuous model, following the trends of simulated signal, which shows that the controller's time requirements are being met.

To finish the test and see if the operating system is maintain the control task with the correct time constraints, a second real time test was made where three digital pins were used as output to indicate when the tasks start and finish their loops. This way, using an oscilloscope, will be possible to see the period of all tasks. The code presented before responsible for the control system was modified as follows: the subscriber was removed, the general purpose pin D14 is set every time that control task start to run and clear when it finishes the task loop and the publisher task was modified as follow:

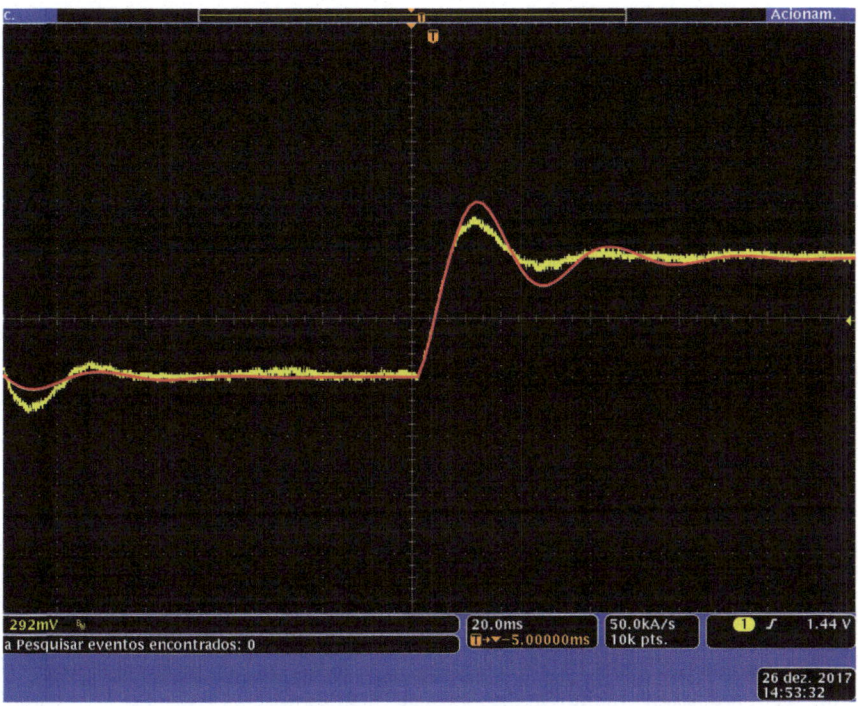

Fig. 21 Comparison of the simulated signal to the signal obtained in the real test

```
1    static void pubTask( void *parameters ){
2      TickType_t xLastWakeTime = xTaskGetTickCount
           ↪ ();
3      //FreeRTPS publisher variables
4      uint8_t cdr[ 20 ] = { 0 };
5      int cdr_len;
6      struct std_msgs__uint32 digital12bitsVoltage;
7      int i;
8      float aux;
9
10     while( true ){
11       //-----MODIFIED-----
12       //Set GPIO D12
13       setPin( 12 );
14       //-----MODIFIED-----
15
16       //Converting 0V to 3V volts to digital
             ↪ value in 12 bits
17       digital12bitsVoltage.data = ( uint32_t )(
             ↪ currentADCValue * ( ( 1 << ADC_BITS )
             ↪ - 1 ) / PWM_MAX_VOLTAGE );
```

```
18        //serialize the data
19        cdr_len = serialize_std_msgs__uint32( &
            ↪ digital12bitsVoltage, cdr, sizeof(
            ↪ cdr ) );
20        //Publish the data
21        freertps_publish( pub2, cdr, cdr_len );
22
23        //-----MODIFIED-----
24        //Add some math processing
25        for( i = 0; i < N_INTERATIONS; i++ ){
26          aux = aux * 1.1;
27        }
28        //Clear GPIO D12
29        clearPin( 12 );
30        //1ms period to has the same period as the
            ↪ main application task
31        vTaskDelay( 1 / portTICK_PERIOD_MS );
32        //-----MODIFIED-----
33
34  }}
```

Listing 14 Second real time test modified publisher

As can be seen, the task received setPin(12) at its start, clearPin(12) at its end and some math processing in the middle to consume more CPU.

Other task having the same code as the task 'pubTask' was created. The new task just uses a different GPIO pin to represent its state, this task uses the pin D13.

```
1   static void pubTask2( void *parameters ){
2     TickType_t xLastWakeTime = xTaskGetTickCount
          ↪ ();
3     //FreeRTPS publisher variables
4     uint8_t cdr[ 20 ] = { 0 };
5     int cdr_len;
6     struct std_msgs__uint32 digital12bitsVoltage;
7     int i;
8     float aux;
9
10    while( true ){
11      //-----MODIFIED-----
12      //Set GPIO D13
13      setPin( 13 );
14      //-----MODIFIED-----
15
16      //Converting 0V to 3V volts to digital
          ↪ value in 12 bits
17      digital12bitsVoltage.data = ( uint32_t )(
          ↪ currentADCValue * ( ( 1 << ADC_BITS )
          ↪ - 1 ) / PWM_MAX_VOLTAGE );
18      //serialize the data
```

```
19        cdr_len = serialize_std_msgs__uint32 ( &
            ↪ digital12bitsVoltage, cdr, sizeof(
            ↪ cdr ) );
20        //Publish the data
21        freertps_publish( pub2, cdr, cdr_len );
22
23        //Add some math processing
24        for( i = 0; i < N_INTERATIONS; i++ ){
25          aux = aux * 1.1;
26        }
27
28        //-----MODIFIED-----
29        //Clear GPIO D13
30        clearPin( 13 );
31        //-----MODIFIED-----
32        //1ms period to has the same period as the
            ↪ main application task
33        vTaskDelay( ( 1 ) / portTICK_PERIOD_MS );
34  }}
```

Listing 15 Second real time test modified publisher

The first run of the second real-time test consists in a system with the tasks 'ctrl-Task', 'pubTask' and 'pubTask2' controlling the pins D12, D13 and D14 respectively, to show the behave of each task. The task 'ctrlTask' received the priority 2 (higher), the other two tasks received the priority 1. The the variable N_INTERATIONS present in 'pubTask' and 'pubTask2' tasks received the value 15000. With these values the Fig. 22 was obtained, the yellow line represents the control task, the blue line represents one of the pub task and the purple one represents the other pub task.

When the application starts to run, it can be seen that the control task is running with 1 ms period and deviation of 19.34 nanoseconds, while the other two similar tasks take 274.4 microseconds to run each one. It is possible to see that the processor is in idle state when all measured pins is in the low logic state. This analysis showed that the control task, with a light load on the system, is running inside the period limit.

Now increasing the variable N_INTERATIONS to 30000, this way increasing the total processing need by the two pubTask, we obtained the Fig. 23. This figure shows that now the system needs more than 100% of CPU, since at no time all the three signals are in low logic level. While pubTasks (blue and purple) can't keep their time constraints the control task (yellow) still keep the 1 ms period.

Looking at the blue line is possible to see that every time that one of the pubTask1 (blue) start to run, the control task (yellow) takes the CPU after a while, to maintain its task period, leaving the pubTask1 (blue) to finish its processing just when the CPU is free again. So, even increasing the system load the control task still keep its period inside the period limit, showing that the scheduler organizes the tasks in the right way, even after the integration with FreeRTPS.

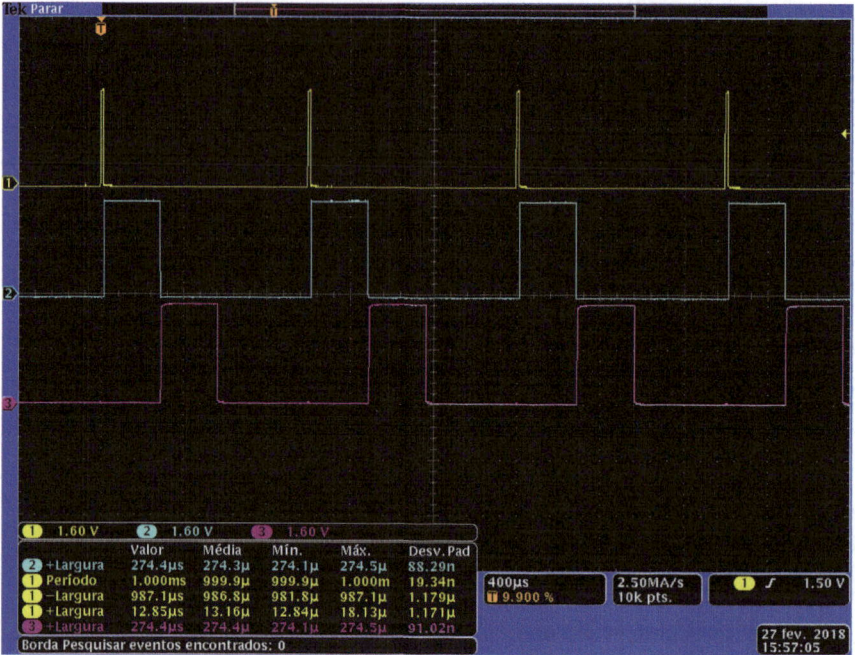

Fig. 22 Tasks processing with total CPU less than 100%

5.4 Performance Test

To verify the performance test of the system, a network structure was developed with a Dell Vostro 14-5480 computer, a STMDISCOVERY development board and a Stellaris LM3S6965 Evaluation Board. All connected by Ethernet cables to a TL-WR741ND TP-LINK switch through its 10/100Mbps LAN ports.

This test focuses on observing the publication rate of the boards while running the operating system, the FreeRTPS processes, and the voltage PID control of a RC circuit.

The frequency of the PID controller task was defined with a rate of 100 Hz. With the PID controller task, a task with a string publisher was implemented publishing 5-byte size messages with a non-delay between successive publications.

The STMDISCOVERY board used the topic 'stmtopic' to publish the messages, while Stellaris Eval Board used the topic 'titopic'. Then, two applications were started on the computer, each one subscribing one topic.

The Wireshark program was used to view and count the packets arriving at the computer's Ethernet port. This program was used because it offers a lot of information about the packets that go through the Ethernet port.

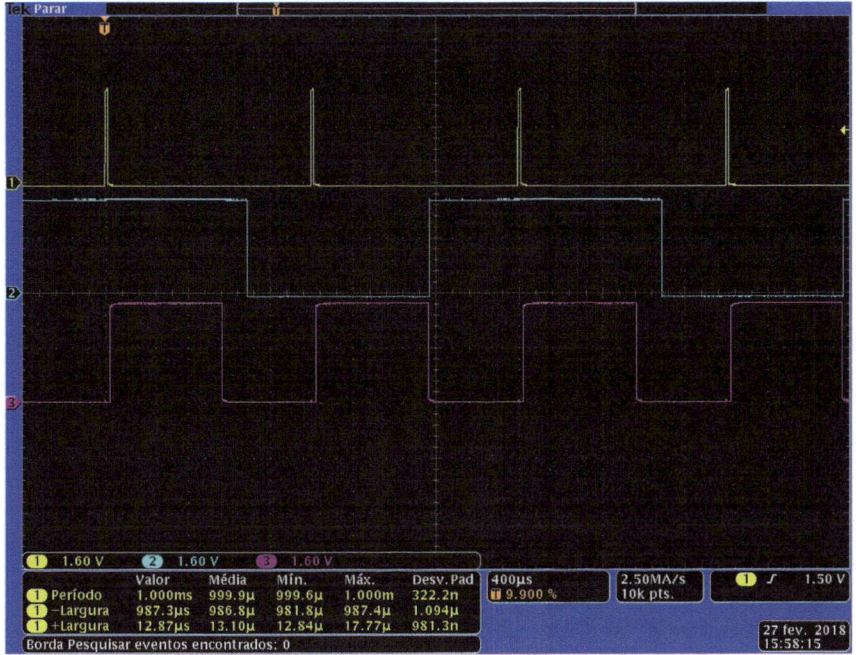

Fig. 23 Tasks processing with total CPU more than 100%

Table 1 Number of messages received from STMDISCOVERY

Time interval	Quantity
Between 6 e 7 s	25489 received messages
Between 7 e 8 s	25493 received messages
Between 8 e 9 s	25516 received messages
Between 9 e 10 s	25519 received messages
Between 10 e 11 s	25502 received messages
Average	25504 received messages

· Initially, only the messages that were received by STMDISCOVERY are filtered, and then the number of messages that were received in the interval of 1 second, Table 1

Doing the same process with the Stellaris Eval Board, the following result was obtained, Table 2.

As can be seen, the STMDISCOVERY board had an average capacity of 25504 messages per second while the Stellaris Eval Board reached the mark of 2739 messages per second. This difference was obtained by the disparity of the main clock of the micro controllers, while the STMDISCOVERY was at 168 MHz the Stellaris Eval Board was at 25 MHz. The dimension of the values obtained from the test applied to

Table 2 Number of messages received from Stellaris Eval Board

Time interval	Quantity
Between 6 e 7 s	2729 received messages
Between 7 e 8 s	2742 received messages
Between 8 e 9 s	2729 received messages
Between 9 e 10 s	2745 received messages
Between 10 e 11 s	2746 received messages

the two plates is approximately 9.3 times, observing that the ratio of the clock value from one board to another reaches 6.7 and the STM microcontroller has DMA on the ethernet peripheral, the number of messages comparing one test to the other is appropriate.

5.5 Resource Consumption

In the resource consumption test, one of the output ports of the micro controller is used in digital mode, that is, the pin can assume a high or low logic level state. Putting the function 'pin_low' as the first operation of all user processes and the function 'pin_high' command in the idle state process, it is obtained that the chosen digital pin will be at the low logic level whenever that the operating system is executing some user process, and in high logic level if it is not executing any.

Using an oscilloscope we observed the digital pin for a certain period of time. Analyzing the total time that the pin remained at the low logic level, we can see the percentage of time that the micro controller is effectively processing application code.

For this test, a network structure was assembled with a Dell Vostro 14-5480 computer, a STMDISCOVERY development board and a Stellaris LM3S6965 Evaluation Board development board, again connected by Ethernet cables to a TL-WR741ND TP-LINK switch .

The first test was elaborated with the following model: a process executing a PID control at a frequency of 1000 Hz and another process publishing messages of type String at a rate of 1000 Hz, with 5 bytes size, Fig. 24.

The obtained values are shown in the Table 3.

For the second test, only one process executing a PID control at a frequency of 1000 Hz was evaluated, Fig. 25.

Analyzing the data obtained from the second test, the following values are obtained, Table 4.

The last test consists of a process with 5 publishers, all of type String, at a rate of 1000 Hz and another process executing a PID control at a frequency of 1000 Hz, Fig. 26.

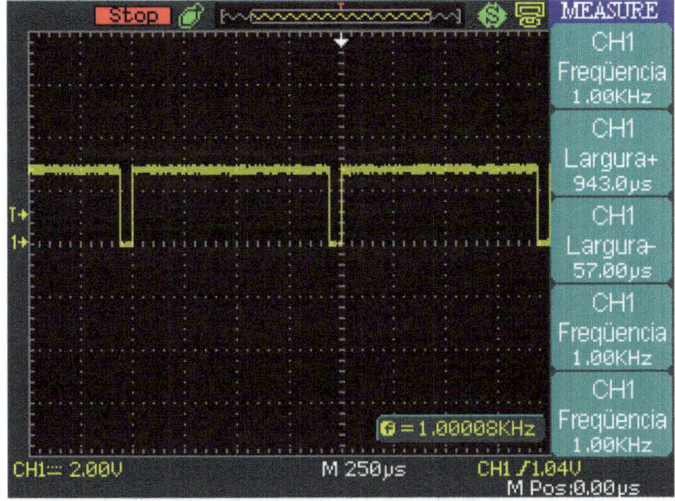

Fig. 24 Resource consumption with a PID process at 1000 Hz and 1 string publisher at 1000 Hz

Table 3 Resource consumption with a PID process at 1000 Hz and 1 string

Low state time [s]	High state time [s]	Low state percentage [%]
57 us	943 us	5

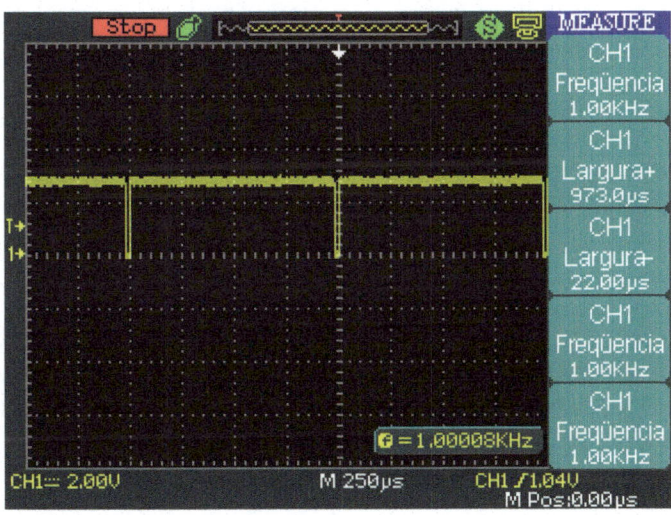

Fig. 25 Resource consumption with a PID process at 1000 Hz

Table 4 Resource consumption with a PID process at 1000 Hz

Low state time [s]	High state time [s]	Low state percentage [%]
22 us	973 us	2.2

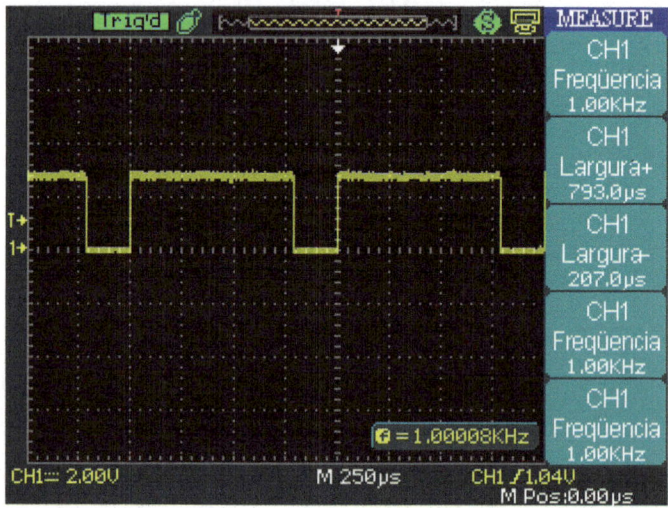

Fig. 26 Resource consumption with a PID process at 1000 Hz and 5 publishers publishing at 1000 Hz rate

Table 5 Resource consumption with a PID process at 1000 Hz and 5 publishers

Low state time [s]	High state time [s]	Low state percentage [%]
793 us	207 us	5

We obtain the following values, Table 5.

The first test shows that 50 us is needed to run one complete cycle of the PID and publisher process.

The second test tells us how much time is necessary to execute just one cycle of the PID process. Thus, it is possible to conclude that during 1 s, only 22 ms are necessary to execute the PID process 1000 times.

The third test presents how much time of processing each additional publisher uses. Comparing the results with the two previous tests, is possible to find a relationship between the number of publishers and the necessary processing time. Knowing that just the PID process uses 22 us, the remaining 35 us were used by the publisher to publish a message.

Using the necessary time for one publisher and multiplying it by 5, we can find a theoretical time for 5 publishers and thus compare it with the practical value found on the third test. The Eq. 3 shows that 5 publishers will supposed to take 197 us to execute.

$$(5\_publishers)x(35\,us) + (22us(PIDtask)) = 197\,us \qquad (3)$$

Comparing the value found in Eq. 3 with the value presented on the third test, can be seen that we have just a 5% of error.

6 Results

Analyzing the results obtained in the tests of performance and consumption of resources, it is concluded that the processing consumption inherent to FreeRTOS is not significant in the system as a whole, under the conditions established in this work, while the tools offered by the operating system add many resources to FreeRTPS, the most important of them the features for real-time implementation. It should be noted that the system does not necessarily guarantee real time, but rather tools to implement one. The capability of a application using this system offer real-time is under the responsibility of the developer, that must use the necessary methods and analyzes with the resources provided by the FreeRTPS+FreeRTOS system.

The portability test shows that it is possible to port the system between different devices without many code modifications. Reaching the goal of offering a portable system. It is also understood that the architecture developed for system programming proved to be easy to understand and implement new applications.

The data obtained from the performance test shows that the final system has a high capacity of publication, even while maintaining a real-time process at 1000 Hz. The STMDiscovery board running at 168 MHz offered an average of 25000 messages published per second, while the Stellaris Eval Board provided about 2700 publications per second, with its main clock at 25 MHz. By comparing the results obtained with the frequency of clock operation, it is possible to obtain an estimate of the minimum requirements of the device which will be used according to the characteristics of the application, because, as it was observed, the performance of the system is directly related to the processing capacity of the microcontroller. This demonstrates that the choice of the microcontroller must be taken into account, looking the quantity, speed and complexity of the tasks inserted in the system.

Comparing the results of the performance tests, it is also observed that when connected directly to the computer, the development board running the original FreeRTPS code offers a high publication rate. However, when the same system assigns to a router the system has a reduction of 30x. It was hypothesized that this variation is inherent in the way the UDP / IP packets of the original FreeRTPS system were implemented, causing the router to have to do more operations to distribute its messages.

During the portability test the IDE provided the compiled file size from the FreeRTOS+FreeRTPS system with an application running one subscriber task and one PID controller task at 1000 Hz. This application resulted a compiled bin file with 32 KBytes. This value presents us the possibility of porting the system to a variety of microcontrollers with less processing and memory resources. The compiled binary file generated by GCC in linux environment for STMDiscovery has reached the size of 74KBytes. Comparing the final size and results of the performance test, it is verified that the proposed system offers an excellent alternative to the projects of systems embarked for ROS.

Despite the limitations offered by the FreeRTOS+UDP license, it is noted that the features added by this UDP stack are considerable. Because it is integrated with the FreeRTOS process system, the whole stream of ethernet packet processing ends up being optimized to work on that system. In addition, this UDP stack provides several interesting features, such as: DHCP, ping, DNS, and message fragmentation.

The way that we implemented the subscriber handle task in the system, the handle routine does not stop the FreeRTPS listen execution to run itself. Original FreeRTPS does a callback to run the subscriber handle routine, and if it is too long the listen will wait for it to end and stop receiving new messages for a while. In the proposed system the listen just put the data at subscriber queue and continues with decoding data. When the scheduler puts subscriber handle to execute, it gets the data from its queue making the listen not dependent from subscriber handle routines.

Finally, although the focus of the work is not high performance, but rather validation of the proposed architecture, it turns out that the architecture offered satisfactory and functional results. Even the operating system adding processing consumption to the system, the message publishing rate remained at a very satisfactory level. Comparing with the related work [35], which offers a maximum publication of 50 messages per second, it is observed that the system response proposed in this work was relatively high, although it offers less resources than the implementation that was developed in NuttX.

References

1. J. Kay, *Introduction to Real-time Systems* (2017), http://design.ros2.org/articles/realtime_background.html
2. J.A. Stankovic, Misconceptions about real-time computing: a serious problem for next-generation systems. Computer **21**(10), 10–19 (1988). https://doi.org/10.1109/2.7053. ISSN: 0018-9162
3. M. Quigley, *Bridging ROS to Embedded Systems: A Survey* (2017), https://roscon.ros.org/2013/wp-content/uploads/2013/06/ros_and_embedded_systems.pdf
4. M. Quigley, *ROS 2 on 'small' Embedded Systems* (2017), https://roscon.ros.org/2015/presentations/ros2_on_small_embedded_systems.pdf
5. J. Kay, A. Tsouroukdissian, *Real-Time Performance in ROS2* (2017), https://roscon.ros.org/2015/presentations/RealtimeROS2.pdf
6. T. Nagle., C. Philips, *Digital Control Systems - Analysis and Design* (Pearson Prentice Hall, 1995)
7. F.E. Páez et al., FreeRTOS user mode scheduler for mixed critical systems, in *2015 Sixth Argentine Symposium and Conference on Embedded Systems (CASE)* (2015)
8. *Escalonamento em Sistemas de Tempo Real* (2018), http://www.ece.ufrgs.br/~fetter/ele213/sched.pdf
9. *rosserial* (2017), http://wiki.ros.org/rosserial
10. *rosbridge suite* (2017), http://wiki.ros.org/rosbridge_suite

11. *uROSnode: a middleware targeted to embedded systems* (2017), https://github.com/openrobots-dev/uROSnode
12. *rosc* (2017), http://wiki.ros.org/rosc
13. *FreeRTPS* (2017), https://github.com/ros2/freertps/wiki
14. R.M.A. de Almeida, Troca de contexto segura em sistemas operacionais utilizando técnicas dedetecção e correção de erros. Ph.D thesis. Universidade Federal de Itajubá, 2013
15. César Augusto Marcelino dos Santos. Sistema Dinâmico de Economia de Energia em RTOS. MA thesis. Universidade Federal de Itajubá (2017)
16. A. Silberschatz, P.B. Galvin, G. Gagne, *Operating System Concepts*, 9th edn. (Wiley, New Jersey, 2009)
17. H.M. Deitel, P.J. Deitel, D.R. Choffnes, *Sistemas Operacionais* (Pearson Education, 2005). ISBN: 85-7605-011-0
18. Multitasking (2017), http://www.freertos.org/implementation/a00004.html
19. *License Details* (2017), http://www.freertos.org/a00114.html
20. *FreeRTOS FAQ - What is This All About?* (2017), http://www.freertos.org/FAQWhat.html#WhyUseRTOS
21. *FreeRTOS+UDP* (2017), http://www.freertos.org/FreeRTOS-Plus/FreeRTOS_Plus_UDP/FreeRTOS_Plus_UDP.shtml
22. W.F. Lages. Sistemas de Tempo Real. Editora UFRGS (2014). ISBN: 978-85-386-0234-7
23. K. Ogata, *Engenharia de controle moderno* (Pearson Prentice Hall, 2011). ISBN: 9788576058106
24. Q. Li, C. Yao, *Real-Time Concepts for Embedded Systems* (CMP Books, Taylor & Francis, 2003). ISBN: 9781578201242
25. E. de Souza Leal, Projeto de uma controladora Single Loop com arquitetura ARM. MA thesis. Universidade Federal de Itajubá (2012)
26. M.S. Fadali., A. Visioli, *Digital Control Engineering: Analysis and Design*, Analysis and Design Series. (Elsevier Science, 2009). ISBN: 9780080922867
27. W. Woodall, *ROS on DDS* (2017), http://design.ros2.org/articles/ros_on_dds.html
28. *Why ROS2* (2017), http://design.ros2.org/articles/why_ros2.html
29. *ROS 2 middleware interface* (2017) http://design.ros2.org/articles/ros_middleware_interface.html
30. *Proposal for Implementation of Real-time Systems in ROS 2* (2017), http://design.ros2.org/articles/realtime_proposal.html
31. *RTPS Introduction* (2017), http://www.eprosima.com/index.php/resources-all/rtps
32. *Prerequisites* (2017), https://github.com/ros2/freertps/wiki/Prerequisites
33. *Introdução ao JSON* (2017), http://www.json.org/json-pt.html
34. *Program Real-Time ROS Nodes on STM32* (2017), https://github.com/bosch-ros-pkg/stm32
35. *ROS 2.0 NuttX prototype* (2017), https://github.com/bosch-ros-pkg/ros2_embedded_nuttx
36. *STM32F4DISCOVERY* (2017), http://www.st.com/en/evaluation-tools/stm32f4discovery.html
37. *STM32F407VG, High-performance foundation line, ARM Cortex-M4 core with DSP and FPU, 1 Mbyte Flash, 168 MHz CPU, ART Accelerator, Ethernet, FSMC* (2017), http://www.st.com/en/microcontrollers/stm32f407vg.html
38. *STMicroelectronics STM32F4DIS-BB* (2017), https://www.digikey.com/product-detail/en/stmicroelectronics/STM32F4DIS-BB/497-13545-ND/3878236
39. *Stellaris R LM3S6965 Evaluation Board, User's Manual* (2017), http://www.ti.com/lit/ug/spmu029a/spmu029a.pdf
40. *Stellaris LM3S Microcontroller* (2017), http://www.ti.com/product/LM3S6965

Lucas da Silva Medeiros studied Computer Engineer at Federal University of Itajuba (Unifei). He did his M.Sc. in Science and Computing Technology where worked in the System Engineering and Information Technology Institute (IESTI), Federal University of Itajuba (Unifei).

Ricardo Emerson Julio studied Computer Science at Federal University of Lavras (UFLA). He did his M.Sc. in Science and Computing Technology working in the System Engineering and Information Technology Institute (IESTI), Federal University of Itajuba (Unifei). Nowadays, he is a Ph.D. Student in Electrical Engineering at Unifei. His research focuses on multi-agent systems, robotics, communication and ROS. He is in expert on software development and programming with 8 years of industrial experience working in data mining area.

Rodrigo Maximiano Antunes de Almeida studied Automation and Control Engineering at Federal University of Itajub (Unifei), M.Sc. in Electrical Engineering at Unifei and a Ph.D. in Electrical Engineering at Unifei. Nowadays, he is an assistant professor at the Unifei and coordinator of Electronic Engineer. He has experience in Computer Science, with emphasis on Embedded Systems, working mainly in the following topics: embedded systems, real-time operating system, microcontrollers, automation and digital control.

Guilherme Sousa Bastos studied Electrical Engineering at Federal University of Itajuba (Unifei), M.Sc. in Electrical Engineering at Unifei, and Ph.D. in Electronic and Computation Engineering at Aeronautics Institute of Technology (ITA), with part of doctorate done at Australian Centre for Field Robotics (ACFR). Nowadays, he is associate professor at Unifei and coordinator of Computer Science and Technology. He has experience in Electrical Engineering and Automation of Electrical and Industrial Processes, acting on the following subjects: electrical hydro plants, mining automation, optimization, system integration and modeling, decision making, autonomous robotics, machine learning, discrete event systems, and thermography.

Part V
Interfaces for Interaction with Robots

`bum_ros`: Distributed User Modelling for Social Robots Using ROS

Gonçalo S. Martins, Luís Santos and Jorge Dias

Abstract In this chapter we present the ROS implementation of our Bayesian User Model, BUM. BUM is a distributed user modelling technique that can be easily implemented in several system topologies. It is able to infer the characteristics of multiple users from heterogeneous data gathered by multiple devices, such as social robots, ambient sensors and surveillance cameras. This chapter presents the BUM process and its implementation, emphasizing the essential and advanced ROS concepts used and extended to achieve the modularity and flexibility needed. Instructions on how to achieve our experimental set-ups are also presented, including a discussion on the role of ROS in the experimental success of the system, and illustrations of the results that can be achieved with our technique. This chapter serves as a thorough description and tutorial for the usage of our package, which can now be useful to the scientific community in user modelling and user-adaptive systems.

1 Introduction

Strong efforts are being made towards the inclusion of social robotic systems in unstructured domestic environments, where they are required to interact with the human users of the environment. These users can be extremely changeable,

---

[1]The `bum_ros` package can be found at https://github.com/gondsm/bum.

This work was developed in the context of the GrowMeUp project, funded by the European Union's Horizon 2020 Research and Innovation Programme - Societal Challenge 1 (DG CONNECT/H) under grant agreement N0 643647.

G. S. Martins (✉) · L. Santos · J. Dias
Institute for Systems and Robotics, University of Coimbra, 3030-790 Coimbra, Portugal
e-mail: gmartins@isr.uc.pt

L. Santos
e-mail: luis@isr.uc.pt

J. Dias
e-mail: jorge@deec.uc.pt
URL: http://www.ap.isr.uc.pt/

© Springer International Publishing AG, part of Springer Nature 2019
A. Koubaa (ed.), *Robot Operating System (ROS)*, Studies in Computational Intelligence 778, https://doi.org/10.1007/978-3-319-91590-6_15

531

unpredictable and demanding, constantly inventing and evolving new forms of inter-action, thus presenting a new and complex challenge for roboticists. Solving this problem represents a departure from the extremely strict, controlled conditions, within which robots have thus far operated: in the past, the user accommodated the robot; now, the robot *must* accommodate the user.

Humans are social beings, and being *known* by those we interact with is an impor-tant part of our social wellness. We expect that other social beings that interact repeat-edly with us, be they people, animals or robots, learn our individual characteristics and develop a *rapport* with us. This gathering of person-specific information for later usage in interaction is one of the main principles of User-Adaptive systems [1, 2].

In User-Adaptive systems, information about the user is integrated into the decision-making process, as illustrated in Fig. 1, so that the actions taken by the virtual agent are tailored to the user they are interacting with [3]. As such, they have been shown to excel in many different scenarios such as organizing shelves [4], service tasks such as delivering items [5] or guiding users to a certain location [6], showing promise for long-term application in social environments. Furthermore, their adaptive characteristics endow these systems with the ability to accommodate a wider variety of users, instead of solely the target audience they were designed for [1].

As mentioned above, adaptation to the user requires that the agent have have access to a *user model* [7] that it can use as a guide. The user model acts as a container for any and all information the system possesses on the user, and may also include the facilities for further collection or inference of information on the user. User models can vary widely in complexity, from single attributes of the user, such as their current intention [8], to representations of their personality [9].

In this work we focus on the "Learning" and "User Model" blocks of Fig. 1, presenting an implementation[1] of a user modelling technique that is able to fuse information from heterogeneous sources, learning the user's characteristics online. Our overarching goal is to achieve a system similar to that represented in Fig. 2, wherein a distributed network of interactive devices are able to gather heterogeneous information about the users of the collective system, which is fused by a centralized mechanism that ensures that each device has a unified representation of all users. A system of this nature allows for the seamless integration of information gathered by today's distributed devices, such as smartphones, smart homes and social robots, into a unified representation of the user that can be used by all devices for adaptivity.

Our technique was previously presented and scientifically validated in [10]. As such, the present work focuses on the implementation of this model, on the design decisions that were taken and, most importantly, on how ROS is an essential tool during all phases of the experimental demonstration and validation of this technique. We aim to present a detailed account of the system's architecture and functionality, serving as an introductory tool for the future usage of the bum_ros package by the ROS community.

This chapter is structured as follows:

- The remainder of this section presents the fundamental theoretical concepts, related work and the ROS concepts necessary for the comprehension of the text;

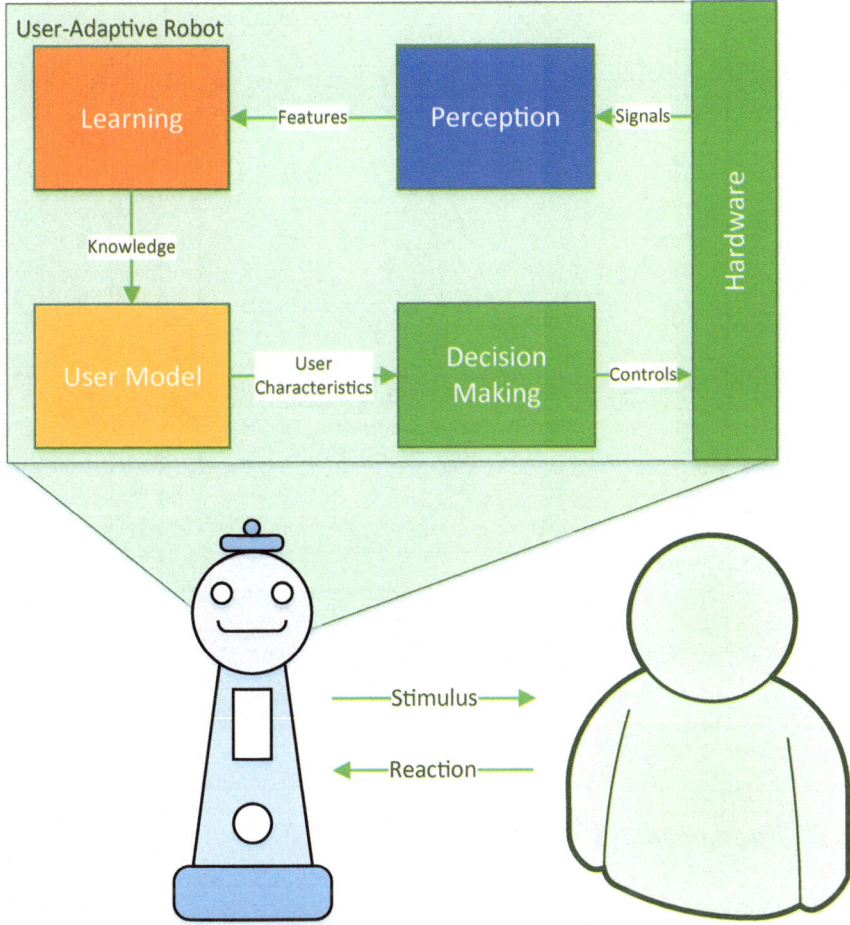

Fig. 1 An overview of a User-Adaptive Robot. By stimulating the user, the system gains access to their reactions, such as responses to questions. From this information, the robot perceives a set of features that are used for learning the characteristics of the user, stored in a user model. These are then used by a decision-making technique to synthesize adapted stimuli

- Section 2 presents an extended mathematical formulation of the system presented in [10], as well as the general architecture of the system;
- Section 3 presents the general ROS architecture and functionality of the system, including each of the ROS messages, topics and nodes that constitute the `bum_ros` package;
- Section 4 presents a guide on the usage of the package, including the various operation modes and system topologies that can be achieved;
- Section 5 presents the concrete set-up we have developed based on the `bum_ros` system, as well as some illustrative results;
- Section 6 presents a few concluding remarks, including our thoughts on possible future work.

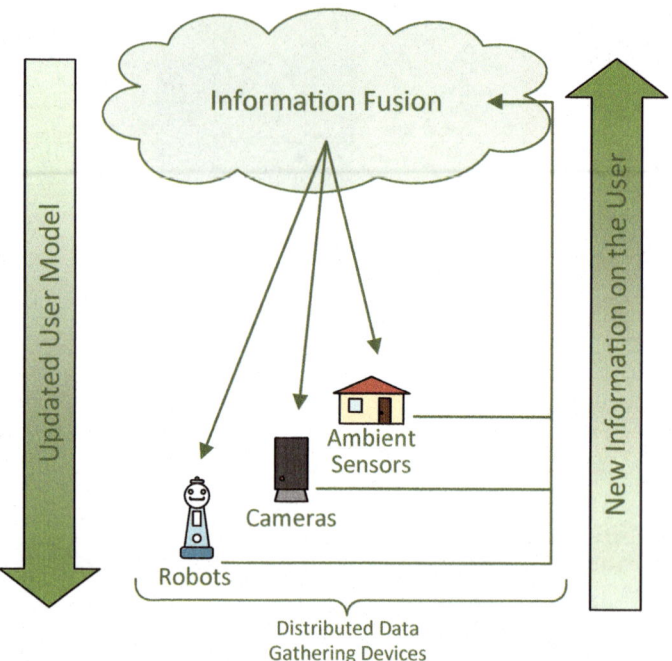

Fig. 2 An overview of the bum_ros system's basic functionality. Distributed devices gather heterogeneous data, which is processed and sent to a centralized fusion system, enabling the it to learn the characteristics of the user

1.1 Fundamental Concepts

For the purposes of this work, we define a *user model* as a container for information on the user. Concretely, we define user modelling as the combination of the user model with the surrounding techniques that are used to populate and maintain the information it contains. As mentioned before, user models can be extremely simple and still be effective. For instance, a single real value encoding the proficiency of a user in using a certain system can be extremely useful in the adaptation of said system to the user.

In this work, we assume that the model of the user is composed of atomic units called *characteristics*. Each characteristic encodes a single attribute of the user, a single dimension that represents part of what makes the user unique. In the simplistic example above, a single-characteristic user model could be employed, where the only characteristic in the model would be "user's proficiency with the system", which would be used by the system to adapt. If C_1 is a variable such that:

$$C_1 \in [0, 1] \tag{1}$$

we encode the user's proficiency level as a variable that exists in the range $[0, 1]$, where 0 would mean that the user is unfamiliar with the system, and 1 meaning that the user is very proficient in using the system. In this case the vector **C**, that contains the characteristics of the user, would be composed only of the C_1 variable:

$$\mathbf{C} = \{C_1\}. \tag{2}$$

In this context, **C** can be seen as our *user model*. The *user* is thus represented by the combination of all of their characteristics, which can also be conveniently combined into a graphical representation, as will be explored in Sect. 2.

Once the structure of the model has been defined, it is necessary to find ways to populate it with information on the user. The information about the user that is contained in user models can be gathered in two distinct ways: *explicitly* and *implicitly* [11]. In the explicit case, the user is directly asked about the information that the system wants to know. For instance, in the simple example above, the user could be asked how many times they had interacted with the system in the past, which would constitute *evidence*:

$$E_1 \in \mathbb{N} \tag{3}$$

representing how many times the user interacted with the system. The goal of the user modelling system is then, as will be detailed in Sect. 2, to determine **C** as a function of **E**. In this example, the adaptive behavior of the system interacting with the user could be, for instance, to lower the difficulty of the interaction for inexperienced users, as in video games, or to offer additional information that could be relevant to new users.

On the other hand, a system can opaquely infer the information it needs about the user, without their cooperation, knowledge or, sometimes, consent. In this case, the user model is populated with information that is inferred from sources such as their behavior, their remaining characteristics or their relationships with other users. An example would be the inference of a user's preferences towards a particular brand of products, performed with basis on information on the user's purchasing history. In this case, the user characteristics could be structured as follows:

$$\mathbf{C} = \{C_1, C_2, \ldots, C_n\}. \tag{4}$$

where C_i is a metric of the user's preference towards brand i. Evidence gathered could be represented as follows:

$$\mathbf{E} = \{E_1, E_2, \ldots, E_n\}. \tag{5}$$

where E_i is the number of products bought from brand i in a certain time period. The goal of the user modelling system would then be, as before, to determine **C** as a function of **E** (Sect. 2). In this example, the adaptive behavior of the system would then be to recommend new products for the user to purchase with basis on the estimated brand preferences.[2]

---

[2]Systems of this nature are known as Recommender Systems [12].

We aim to support both implicit and explicit data collection, i.e. collect information both by directly asking the user and by inferring their characteristics via observation. As such, our system makes no assumptions as to the gathering technique, and simply defines that characteristics are inferred from *evidence*, as in the **E** vectors in the examples above.

1.2 Related Work on User Modelling

User Modelling systems are a mature field in Human-Computer Interaction (HCI), as evidenced by early surveys such as [1, 2]. These systems are not as well developed and widespread in the Human-Robot Interaction (HRI) community, where there is a lack of specialized solutions. However, the field of User Modelling for Human-Robot Interaction is rapidly evolving, with several new approaches having been proposed in recent years.

The usage of generalized user profiles, also a popular choice in HCI, has been shown to be possible. In these works, users are split into particular, at least partially pre-made profiles, which are used by the system to quickly adapt to new users without the load of inferring their characteristics. These profiles can be fully constructed a priori, as in [13, 14] or can be parametrically constructed to allow further personalization, as in [15]. These systems offer, in terms of user modelling, the ability to quickly adapt to any user, suffering from the natural inaccuracy that results from each user not having their own fully-personalized profile. Furthermore, these systems are typically unable to deal with users as they evolve in time, and are unsuitable for long-term interactions.

Dynamic user models are able to deal with changing users by evolving as their users do. Their ability to change as the interaction progresses makes them suitable for use in dynamic situations, such as Human-Robot Collaboration [16]. By using the history of previous interactions, as in [17–20], artificial systems can progressively learn the characteristics of their users over longer timespans, thus potentiating their adaptive abilities and, thus, long-term possibilities. In this manner, adaptive systems gain additional *autonomy in interaction*, for instance becoming able to act in a satisfactory manner without any commands from the user.

User models can also contain deeply personal information, allowing for the exploration of effects such as affect or rapport [21]. Personality traits have been used in works such as [9, 22, 23] to achieve higher levels of compatibility with their users, with robots sometimes synthesizing their own personality to match with the user's. These works stem from research in the field of Personality Computing [24], a mature field in HCI, wherein systems aim to determine the personality traits of their users, or synthesize their own.

Our technique is also based on the concepts of interoperability, shell systems and user model servers. Model interoperability allows several models to operate in unison, allowing for an extensible holistic representation of the user [25]. Interoperability is an interesting concept when dealing with the newest interfaces available, such as robots, since it should allow, for instance, for older models to be integrated into newly-developed systems. Shell systems [7] consist of "empty" user modelling systems, essentially frameworks that allow for the definition of custom user models, essentially serving as vessels for domain knowledge to be integrated later on. User modelling servers [25, 26] consist of services operating on the client-server model, which maintain a user model remotely. This model can then be queried by client applications, both for information insertion and retrieval, thus avoiding the problem of interoperability via centralization of the user modelling effort.

1.3 Related ROS Concepts and Packages

ROS stands for Robot Operating System, a *de facto* misnomer in the sense that ROS is not a true operating system. In fact, ROS is essentially a set of libraries, packages and guidelines that allow for the streamlined development of software targeting robots. Its standards allow for the decoupling of packages to the point where software can be developed regardless of the robotic hardware it will run on, thus allowing for great flexibility and code reuse.

ROS *nodes* are processes that run on a computer, and which are linked against (or import) the ROS libraries. Thus, ROS nodes have access to the ROS API, allowing them to benefit from the ROS framework. In every other respect, ROS nodes are normal processes. The execution of nodes can be configured using *parameters*, which are stored globally and can be accessed by any node.

ROS nodes communicate with each other using *messages*. These messages are constituted by *fields*, much like a regular `struct` as found in many programming languages. Message delivery is *asynchronous*, meaning that nodes do not (usually) explicitly wait for messages; messages are delivered through callback functions that execute in parallel with the main thread. Messages carry all of the information that nodes need to communicate to each other, be it laser scans, metric maps or, in our case, information about the status of the user of a robot.

Custom messages, tailored to the application of a specific package, can be created using the same format and basic types as the messages in the standard ROS library. Furthermore, messages can contain other messages, allowing for the composition of generic messages into rich, highly specialized information containers.

Messages travel in *topics* with names such as `/scan` or `bum_ros/tuple`. All of the nodes that subscribe to a certain topic get all the messages that are published in that topic and, conversely, any node can publish on any topic. This carries a caveat: all nodes must agree on the message type that flows in each particular topic, since each topic can only support one message type.

Launch files allow for the simplified start-up of a ROS system via a condensed file-based interface. Launch files allow for the definition of a full ROS system, including nodes, topic remapping, parameters, *etc*, in a single file. This definition can then be used to launch the full system into operation via a single `roslaunch` command. Launch files can also be used for *namespacing* nodes, "pushing" their topic names and parameters to their own namespaces, allowing for the simple reutilization of full ROS systems.

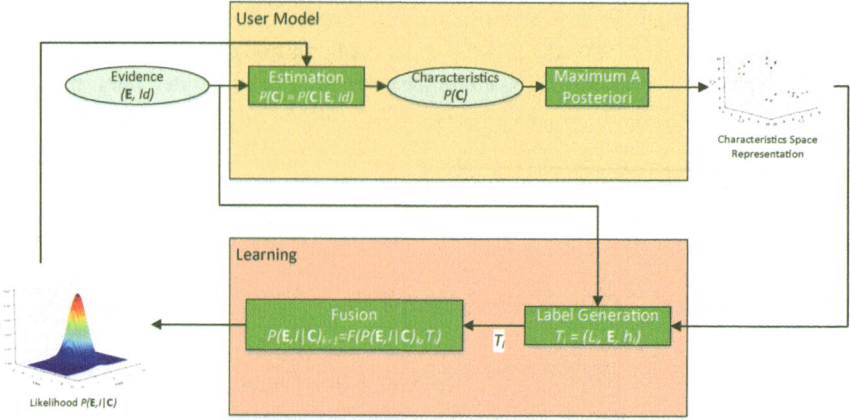

Fig. 3 An overview of the `bum_ros` system's mathematical foundations. Evidence gathered by the distributed system is input into an estimation step, which outputs the characteristics of the user. These characteristics, coupled with the input data, are then fed to the fusion mechanism, thus integrating this information and producing a new likelihood for the model

YAML [27] is a human-readable data serialization language, allowing for the definition of data formats that are easily parsed by computers and humans alike. YAML is used throughout ROS for many different functions, such as storing user preferences or configuration information.

This work is also built upon other state-of-the art software packages. The ROS library is accessed mainly through its Python [28] interface, `rospy` [29]. Matplotlib [30] is used for data visualization. Lastly, the ProBT package, developed by ProbaYes,[3] is used for modelling and computing the Bayesian inference techniques used in our model and system.

All of these subjects will be further illustrated as the system is described in the following sections. For further information about any particular subjects, the authors refer the interested reader to the ROS Wiki [31].

2 The BUM Process

2.1 Inference of User Characteristics

The aim of a user modelling system is to infer and maintain any relevant characteristics of the user. In our case, the BUM process outputs a vector of the user's characteristics, $\mathbf{C} \in \mathbb{R}^n$, where n is the number of user characteristics under study, as in the Estimation process of Fig. 3. As input, the process takes a vector of m evidence variables $\mathbf{E} \in \mathbb{R}^m$, as well as user's identity $Id \in \mathbb{N}$.

---

[3]ProbaYes can be contacted through http://www.probayes.com/en/.

The main output of the system is the distribution

$$P(\mathbf{C}|\mathbf{E}, Id) \propto P(\mathbf{C})P(\mathbf{E}, Id|\mathbf{C}), \tag{6}$$

which encodes the characteristics of user Id as revealed by the evidence. By using the user's identity as an additional evidence variable, the model is able to represent a population of users while still allowing for individual personalization.

The evidence processed by the system at this stage can be obtained from any underlying classification or estimation techniques. By applying techniques such as Dynamic Time Warping [32], the system can also be extended to operate on raw signals from sensors.

Assuming that each of the user's characteristics are statistically independent from one another, the estimation process is split into modules. Each module is responsible for estimating one of the components of the $\mathbf{C}$ vector, using isolated Bayesian Programs [33] that solve the non-vectorial form of Eq. 7 for each characteristic.

Additionally, we assume that evidence variables are independent from each other, allowing us to apply the Naïve Bayes paradigm when solving Eq. 7:

$$P(\mathbf{C}|\mathbf{E}, Id) \propto P(\mathbf{C}) \prod_{E_i \in \mathbf{E}} P(E_i, Id|\mathbf{C}). \tag{7}$$

The distribution $P(\mathbf{C}|\mathbf{E}, Id)$ provides an estimate of the characteristics of all of the users known to the system. By performing *maximum a priori* estimation on this distribution over user identities, we can extract the most likely characteristics of each user. Thus, we can define a characteristics space with points defined as:

$$\mathbf{U}_u = [C_1, C_2, \ldots, C_n], \tag{8}$$

where n is the number of characteristics being inferred, and u the identity of the user, and $\mathbf{U}$ is the vector containing all users. This space constitutes a *characteristics-space representation* of the user population. This representation can then be used by decision-making techniques to synthesize adapted stimulus to their users.

2.2 Learning of the User Model

The system operates in *steps*, with each step consisting of the processing of a batch of evidence received. Each of these steps produces a tuple of the form:

$$T_i = (L_i, \mathbf{E}, h_i), \tag{9}$$

where $L_i \in \mathbb{N}$ is the label obtained for characteristic C_i, via Maximum a Posteriori estimation:

$$L_i = \underset{x}{\operatorname{argmax}} \ P(C_i|\mathbf{E}, Id). \tag{10}$$

h_i is the entropy of the distribution $P(\mathbf{C}_i)$), which we approximate by the posterior obtained from the estimation step, $P(\mathbf{C}_i|\mathbf{E}, Id)$:

$$h_i = H(P(\mathbf{C}_i)) \approx H(P(\mathbf{C}_i|\mathbf{E}, Id)). \tag{11}$$

H is the entropy function, as defined in [34]:

$$H(P(C)) = -\int_C P(C) \log P(C)\, dx \tag{12}$$

These tuples represent, essentially, the results of the system's estimation, with the level of certainty inherent to that estimation.

The systems knowledge on the user population is stored on the likelihood $P(\mathbf{E}, Id|\mathbf{C})$, which is used by all Bayesian Programs to estimate the user's characteristics from evidence. This distribution is iteratively constructed from Gaussian kernels by performing

$$P(\mathbf{E}, Id|C_i = L_i)_{k+1} = \frac{1}{\psi}(P(\mathbf{E}, Id|C_i = L_i)_k + D) \tag{13}$$

where D is a Gaussian distribution, function of T_i, defined according to the label received. ψ is a normalization factor, ensuring that the resulting probability distribution is valid.

This constitutes our system's learning mechanism. In essence, by allowing each module (and, by extension, each device) to feed others with their own estimation results, we achieve a system that is similar in conceptual terms to classifier ensembles, but not explicitly modelled as such. This mechanism allows for two kinds of tuples: *hard* or *soft*, where soft evidence is that generated by the system itself, and hard evidence is produced by an external agent, such as a different module injecting labelled training data into BUM. Thus, our system can also learn in a supervised manner, with the possibility to inject ground-truth data asynchronously into the system.

Tuples containing soft evidence are fused by defining the distribution D as

$$D = P(\mathbf{E}, Id|C_i = c)_{observed} = \mathcal{N}(\mu, \Sigma) \tag{14}$$

where μ is defined according to the evidence received:

$$\mu = \mathbf{E} \tag{15}$$

and Σ is a covariance matrix where each diagonal element is defined by entropy:

$$\Sigma_{i,i} = F(h_i), \tag{16}$$

which is calculated in Eq. 11. Thus, these tuples are fused according to their level of entropy. Tuples that contain high uncertainty produce little impact on the likelihood, while tuples with high levels of certainty are able to significantly re-shape the system's likelihood.

Tuples containing hard evidence are fused by defining D as

$$D = \mathcal{N}(\mu, \alpha I) \tag{17}$$

where I is the identity matrix and α is a learning factor, which we set to 0.001 in our trials, and μ is defined as in Eq. 15.

2.3 Obtaining User Profiles

As seen in Sect. 1, user profiles can be an important tool in the quick adaptation to new users. In our case, generalized user profiles can be obtained by performing a clustering step on the characteristics-space representation of the users. The Expectation-Maximization [33] algorithm is used for clustering the users, producing a Gaussian mixture on characteristics space.

This subprocess results in a number of n-dimensional Gaussian distributions:

$$(\mu, \Sigma) = EM(\mathbf{U}), \tag{18}$$

where μ contains the means of the clusters, and Σ the respective covariance matrices. EM denotes the Expectation-Maximization algorithm. Each of these clusters can be regarded as a user profile, which represents a common "type" of user found by the system.

When the system is introduced to a completely new user, or if one of its modules fail making it impossible to estimate some of the characteristics, the system can match the user to one of the existing clusters as a first estimation of their characteristics. This matching process is achieved via a distance metric. Assuming a user given by a vector with incomplete characteristics:

$$\mathbf{U}_{inc} = [C_0, \ldots, C_{k-1}, C_{k+1}, \ldots, C_n], \tag{19}$$

i.e. as defined in Eq. 8 except for the missing kth component, C_k. We can define a distance measurement, based on the classical Euclidean distance, to each of the clusters already existing in the system:

$$d_i(\mathbf{C}, \Sigma) = \sqrt{\sum_{C_j \in \mathbf{U}_{inc}} (C_j - \Sigma_j)^2}, \tag{20}$$

where Σ_j is the element of cluster i corresponding to the characteristic C_j of the user. d_i corresponds to the Euclidean distance over the projections of clusters in $n - 1$-dimensional space, i.e. excluding the dimension that could not be estimated.

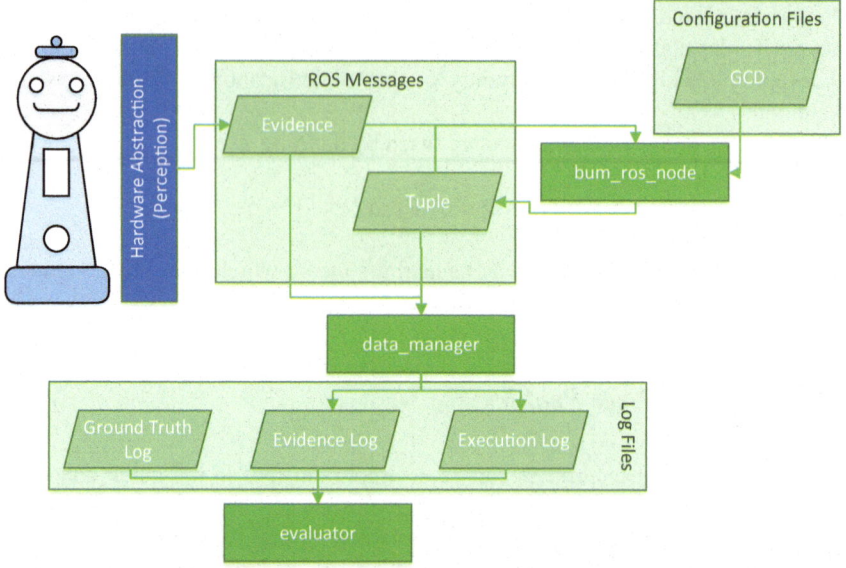

Fig. 4 An overview of the ROS implementation of BUM. Evidence and Tuples are propagated as custom ROS messages, and are processed by bum_ros nodes, which use them to estimate and learn the users' characteristics. This node is configured by a Global Characteristic Definition (GCD) file. All data processed and produced by the system is recorded by the data_manager, and can later be evaluated by the evaluator

The closest cluster to the user is selected by minimizing the distance:

$$\Sigma^* = \underset{\Sigma}{\operatorname{argmin}} \ d_i(\mathbf{C}, \Sigma) \tag{21}$$

and, thus, obtaining a matching cluster. Each of the missing characteristics can then be obtained by applying

$$C_j = \Sigma_j^* \tag{22}$$

to each missing characteristic.

3 The bum_ros System

The bum_ros system, depicted in Fig. 4, is composed of three main ROS nodes:

1. The bum_ros_node node, responsible for all of the main functions described in Sect. 2, including estimation and fusion;
2. The data_manager node, responsible for the management of all data produced and received by the system, including generating and playing back log files;
3. The evaluator node, mainly used for visualization and system evaluation purposes.

As seen in Sect. 2, the bum_ros system communicates through two main types of information: tuples containing estimation results, and evidence collected from the user. This information is contained in two custom ROS messages:

1. The Tuple message, containing the results of estimation as well as hard evidence received by the system. By default, these flow in the bum_ros/tuple topic;
2. The Evidence message, containing regular evidence received by the system, used for estimation. By default, these flow in the bum_ros/evidence topic.

The system maintains logs of its knowledge and of the information received, easing the repetition of experiments and evaluation. These are of three kinds, according to the data they store:

1. The Ground Truth Log (gt_log), which stores ground truth data created by a human annotator to be used for evaluation purposes;
2. The Evidence Log (ev_log), which stores all of the evidence received by the system, including any hard evidence received from an annotator in Tuple messages;
3. The Execution Log (exec_log), which stores all of the estimation results produced by the system.

A main configuration file, the Global Characteristic Definition (GCD), contains the definition of the structure of the problem, such as the number of user characteristics that will be estimated, as well as what specific variables they depend on. Additionally, it configures each individual bum_ros system as to its main mode of operation (further discussed in Sect. 4), e.g. whether it will perform data fusion or estimation only. As such, each individual instance of the bum_ros system expects to find the path to a GCD in its parameters, and execution will fail if it does not exist.

The system is supported by a hardware abstraction layer (HAL), as depicted in Fig. 4, which must be custom-built for each data-gathering device. The only requirement that bum_ros imposes on this layer is that it conform to the GCD when producing Evidence messages. No other synchronization methods are employed, and the system is able to operate in a completely asynchronous manner. In fact, the Evidence messages produced do not even need to correspond to the evidence needed by any particular characteristic, as we will see below. The HAL can also produce Tuple messages containing hard evidence, if the interaction they implement allows for explicit gathering of information.

In the simplest operation mode, with a single device, execution flows as follows:

1. The HAL produces an Evidence message;
2. This message is simultaneously logged to the log file and processed by the bum_ros_node, potentially producing new estimations (if enough evidence variables are received);
3. If estimation occurs, this produces a Tuple message, which is simultaneously logged to the Execution Log, and re-used by the bum_ros_node for fusion.

Other operation modes and possible topologies will be explored in Sect. 4.

3.1   The Global Characteristic Definition

The distributed devices must have a manner of knowing which evidence variables they
are processing, which characteristics to estimate, and the correspondence between
evidence and characteristic variables. Furthermore, each device would need to know
which characteristics, out of all of those estimated by the system, it would be itself
estimating, and also if it was expected to perform data fusion.

To tackle this issue, we developed the Global Characteristic Definition (GCD).
The GCD is composed of a single YAML file, as illustrated in Listing 1.1, which
contains dictionary entries with all the information the system needs, namely:

```yaml
 1  # Characteristics
 2  C:
 3    C1:
 4      input: [E1, E2]
 5      nclasses: 5
 6      description: "A user characteristic."
 7    C2:
 8      input: [E2, E3, E4]
 9      nclasses: 5
10      description: "Another user characteristic."
11  # Evidence
12  E:
13    E1:
14      nclasses: 2
15    E2:
16      nclasses: 5
17    E3:
18      nclasses: 3
19    E4:
20      nclasses: 3
21  # Specify the number of users
22  nusers: 10
23  # Configuration of the node
24  Config:
25    # Specify characteristics this node will work
    on
26    Active:
27      - C1
28      - C2
29    # Will this node fuse tuples?
30    Fusion: True
31    # Will it only fuse hard evidence?
32    Only_fuse_hard: False
33    # Will tuples be published on prediction?
34    Publish_on_predict: True
```

Listing 1.1 An example of a Global Characteristic Definition (GCD) file. This file is propagated
across all nodes running in the network, and provides the bare minimum information needed for
them to operate cooperatively. The file also allows for the configuration of the node itself, defining
whether it should fuse tuples, fuse only hard evidence or if tuples should be published on prediction.

- The C key, containing a number of subkeys, each implicitly defining the name of the characteristic, and explicitly defining its input variables, the number of classes it is classified into (if applicable) and a string containing a description. By explicitly indicating which characteristics depend on which evidence variables, we are able to determine, for each Evidence message, if the evidences received are enough to estimate any of the user characteristics;
- The E key, with the same structure of the C key, defining all available evidence variables with one sub-key per evidence variable available;
- The nusers key, specifying the number of users the system is expected to encounter, which is used for initial set-up. It can be set to a large number if a large number of users is expected to use the system;
- The Config key, which contains configuration information for the node, such as if it is expected to fuse information, to fuse only hard evidence or to publish its predictions.

The GCD is used numerous times during execution, for instance for the bum_ros_node node to determine if incoming evidence can be used to estimate any characteristics, or by the *data_manager* to determine the names of the various variables involved. The GCD must be transmitted to all devices a priori, during deployment, as no GCD synchronization mechanism was devised.

3.2 The *Evidence* and *Tuple* Messages

```
1    # A vector of the evidence values
2    int32[] values
3
4    # A vector of corresponding evidence IDs, in accordance
     with the current
5    # global characteristic description
6    string[] evidence_ids
7
8    # The ID of the active user
9    uint32 user_id
```

Listing 1.2 Definition of the Evidence.msg message. This message type is used to convey information gathered by the distributed devices to the main BUM node, allowing it to estimate user characteristics.

The Evidence message, illustrated in Listing 1.2, contains the main output of the HAL. It conveys a vector of evidence values, each corresponding to an evidence ID as defined in the GCD, as well as to the ID of the user to which the evidence corresponds. This information is matched to the GCD by the bum_ros_node node to determine which, if any, characteristics can be estimated from this message.

```
1    # The ID of the characteristic at hand
2    string char_id
3
4    # The classification value
5    int32 characteristic
6
7    # The evidence used to determine the characteristic,
     corresponding to this
8    # datapoint. It must be in the order specified in the GDC!
     (listeners
9    # assume this).
10   int32[] evidence
11
12   # The ID of the active user
13   uint32 user_id
14
15   # The entropy or certainty level of the classification
16   float32 h
17
18   # Signals whether this tuple is hard evidence or not
19   bool hard
```

Listing 1.3 Definition of the `Tuple.msg` message. This message is used to convey the estimation results produced by the main BUM node, as well as to convey hard evidence produced by an annotator or by other systems.

The `Tuple` message, illustrated in Listing 1.3, serves as a container for both the output of the estimation process, which is used as input by other `bum_ros` nodes, and for any hard evidence received by the system. It contains the ID of the estimated characteristic, to be matched against the GCD, as well as the corresponding value and the entropy generated during estimation. It also carries a boolean variable that indicates whether the tuple represents hard evidence, which is fused as containing much higher entropy so as to act in a supervised manner, as detailed in Sect. 2.

3.3 Log Files

The system generates log files for all received evidence, to record execution and ground truth. These logs are stored in YAML files as lists of dictionaries, with each list element corresponding to a single entity, be it an `Evidence` or `Tuple` message, or the individual values of a characteristic for all users.

ROS already provides us with a very competent logging tool, `rosbag`. However, we have found that this tool was lacking in several aspects:

1. Human Readability and edit-ability: `rosbag` logs are not easily edited directly, which was a necessary feature for our logs. By editing the logs, we are able to, for instance, interleave several evidence logs to simulate operation using a single device for several users, and vice-versa, without needing to repeat experiments;

```
 1    # Hard evidence contains both a key for each
 2    # encoded characteristic and the respective
 3    # evidence
 4    - C1: 4
 5      Evidence:
 6        E1: 1
 7        E2: 2
 8        E3: 3
 9      Identity: 1
10    # Soft evidence entries contain only the
11    # evidence and identity keys used for inference
12    - Evidence:
13        E1: 1
14        E2: 2
15        E3: 3
16      Identity: 1
17    #  ...
```

Listing 1.4 An example of the evidence log, usually stored as `ev_log.yaml`. This log maintains all of the evidence received by the system while executing. Each entry is kept as an item in a YAML list, allowing for easy reproduction and interleaving with logs obtained from different devices.

```
 1    # Each entry in the execution log contains:
 2    - C:                  # A dictionary of inferred
       characteristics
 3        C2: 0
 4      Entropy: 1.609437 # The entropy generated by the
       estimation
 5      Evidence: []       # The evidence values used for
       estimation
 6      Identity: 5        # The identity of the user in question
 7    - C:
 8        C1: 0
 9      Entropy: 1.12437
10      Evidence: []
11      Identity: 5
12    #  ...
```

Listing 1.5 An example of the execution log, usually stored as `exec_log.yaml`. This log maintains all of the estimations (tuples) produced by the system, as well as the evidence used and entropy generated. This log functions as a record of the execution of the node, and can be used for later evaluation of the node, or as input for a decision-making technique.

```
1    # One dictionary key for each possible characteristic
2    # with one entry per user
3    C1:
4        1: 2 # User 1 has C1 = 2
5        2: 2 # User 2 has C1 = 2
6        3: 2 # And so on
7        # ...
8    C2:
9        1: 0
10       2: 1
11       3: 4
12       # ...
```

Listing 1.6 An example of the ground truth log, usually stored as `gt_log.yaml`. This file maintains the "ground truth" of each characteristic with respect to the user they belong to. This allows for the evaluation of the system, by comparison with the `exec_log`, which is performed by the `evaluator`.

2. Easy parsing and conversion to other formats: the YAML format allows for the easy parsing of information programmatically, which is also partially achieved by the `rosbag` API. However, this application required higher flexibility and the potential conversion to other formats, for instance for dataset publication, which called for a slimmer logging technique;
3. Efficiency of storage: the `rosbag` tool excels at recording whole experiments using a particular robot. However, we were more interested in obtaining isolated information from the system itself, not including ROS facilities such as timestamps and custom data packing.

For these reasons, we opted for designing a lightweight, YAML-based logging system based on three log files: the Evidence log, the Execution log and the Ground Truth log.

The Evidence Log, illustrated in Listing 1.4, contains all of the evidence received by the system. Each YAML list element (denoted by a –) contains a dictionary which, for soft evidence, contains only an `evidence` key with each of the evidence variables involved. For hard evidence, a C key, corresponding to one of the characteristics on the GCD, is also stored, establishing the connection between input and output variables needed for training the model.

The Execution Log, illustrated in Listing 1.5, contains all of the classifications produced by the `bum_ros_node` node. Each list element contains a C key with a characteristic ID and the corresponding value, as well as the entropy generated during classification and the evidence values used, thus emulating the essential data present in the `Tuple` message. These allow the `evaluator` node to build a representation of the user population and update it as new estimations are received. This log can thus be used to evaluate the performance of the system after or during execution, using the `evaluator` node.

The Ground Truth Log, illustrated in Listing 1.6, contains the ground truth information that is used by the `evaluator` node to evaluate the performance of the system. This information is stored as a dictionary, with an entry for each characteristic. Within each entry, a dictionary maps between the user ID and the corresponding value for the current characteristic, thus representing the whole user population, according to the known characteristics, in a compact manner.

3.4 The *bum_ros_node* Node

The bum_ros_node node, illustrated in Fig. 5 contains all of the main functionality described in Sect. 2. Namely, this node is responsible for both of the estimation and fusion sub-processes. Using the GCD, various instances of this node can be run simultaneously on the same or in different computers, allowing for a multitude of operational topologies, which will be detailed in Sect. 4. In practical terms, through the GCD, the node can be configured to act as an estimator for certain characteristics, as merely a fusion node, or as both.

The execution of the node is controlled via callback functions, one for incoming Tuple messages, another for Evidence messages. For each incoming Tuple message, the node checks whether it is able to perform fusion. If it is, then the tuple is fused into the corresponding likelihood, according to its evidence dependencies specified in the GCD. This includes tuples created by the node itself; in fact, Tuple messages are not related to the system that generated them in any way, providing the independence necessary for the intended operation of the system. Tuples containing hard evidence are fused by the same method as regular tuples, with the difference of being attributed a high entropy value, as described in Sect. 2.

For each incoming Evidence message, the node compares the evidence list to all of the characteristics' dependencies on the GCD, and determines what, if any, characteristics can be estimated from the received evidence. If any characteristics can be estimated, estimation is performed for each individually, and a Tuple message is published for each individual estimation. This process is illustrated in the algorithm of Fig. 5.

3.5 The *data_manager* Node

The data_manager node is responsible for managing the logs described in the previous sections, as illustrated in Fig. 6. Its operation is split into three main modes:

- The playback mode, which is used for for playing back evidence in a provided evidence log file;
- The listen mode, used for listening for evidence and tuples, which are written to the evidence and execution log files, respectively;
- The dual mode, for playing back evidence from the log file while listening for results writing results to the execution log.

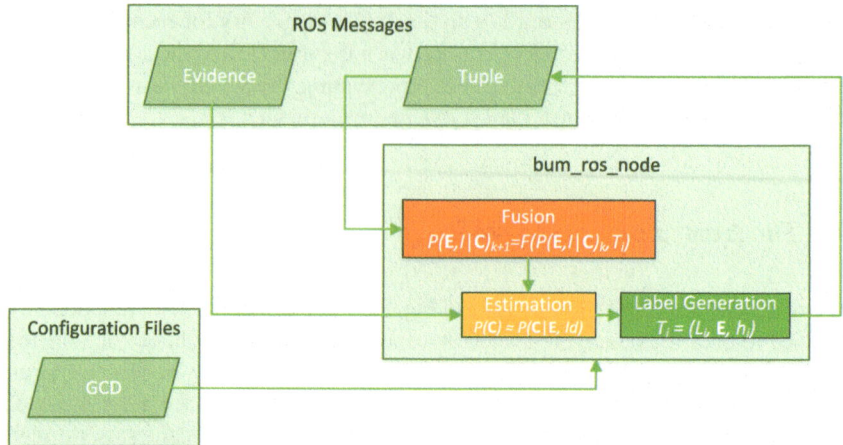

(a) An illustration of the data flow to and from the `bum_ros_node` node. It receives evidence, from which it estimates characteristics, published into the system as `Tuple` messages. These are re-used for the fusion process, if the node is configured as such. All of the process is configured by the `GCD`. This node encapsulates the functionality of Section 2, as illustrated by the colored blocks.

> **input** : `Tuple` and `Evidence` messages, `GCD`, previous system likelihood L
> **output:** `Tuple` messages
> **Callback** *tuple_ callback(Tuple)*:
>> **if** *gcd.only_ fuse_ hard and Tuple.hard or not gcd.only_ fuse_ hard* **then**
>>> | L.fuse(Tuple);
>>
>> **end**
>
> **end**
> **Callback** *evidence_ callback(Evidence)*:
>> P_c ← characteristics that can be estimated from this evidence;
>> t ← empty tuple message;
>> **for** $c \in P_c$ **do**
>>> | t.append(estimate(c));
>>
>> **end**
>> publish(t);
>
> **end**

(b) The pseudocode run by the `bum_ros_node` node. Its operation, after initialization, is entirely managed via ROS callbacks, which react to the receipt of new evidence and tuples.

Fig. 5 An illustration of the data flow and algorithm run by the `bum_ros_node` node

(a) An illustration of the data flow to and from the `data_manager` node. It receives evidence and tuples, which are written to our custom log files for later evaluation or playback.

input : Tuple and Evidence messages, GCD
output: exec_log and ev_log log files
Callback *tuple_ callback(Tuple)*:
   if *Tuple is soft evidence* **then**
     | log_ as_ execution(Tuple);
   else
     | log_ as_ evidence(Tuple);
   end
end
Callback *evidence_ callback(Evidence)*:
   | log_ as_ evidence(Evidence);
end
Function *playback(ev_ log)*:
   for *entry in ev_ log* **do**
     if *entry is hard evidence* **then**
       | $t \leftarrow$ empty tuple message;
       | t.set_ data(entry);
       | publish(t);
     else
       | $e \leftarrow$ empty evidence message;
       | e.set_ data(entry);
       | publish(e);
     end
   end
end

(b) The pseudocode run by the `data_manager` node. Its operation, after initialization, is partially managed via ROS callbacks, which react to the receipt of new evidence and tuples. The node is also able to play back log files, allowing for the reproduction of past experiments.

Fig. 6 An illustration of the data flow and algorithm run by the `data_manager` node

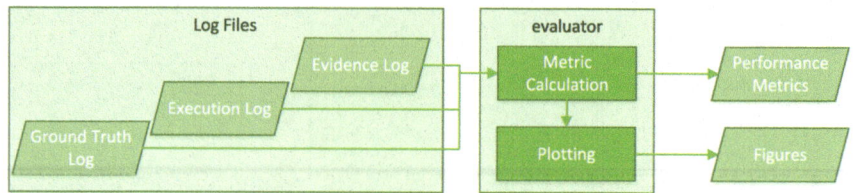

(a) An illustration of the data flow to and from the evaluator node. It reads all of the logs generated by the system, calculates performance metrics and plots figures from the data.

Fig. 7 An illustration of the data flow of the evaluator node

In the listen mode, execution is governed by a pair of callbacks, similarly to the bum_ros_node node, which are responsible for dealing the Evidence and Tuple messages.

Regardless of the execution mode, all Evidence messages captured are logged in the evidence log. Normally, Tuples are logged in the execution log as the result of estimations performed by the system. However, Tuples which contain hard evidence are logged in the evidence log, as described previously, allowing for the playback and interleave of both soft and hard evidence, and also for the differentiated treatment of system estimations and annotator-provided data.

The playback and dual modes are mainly used for experimental validation and debugging of the system. The playback mode allows for the playback of any number of evidence files, which are interleaved randomly, and does not depend on the presence of any other nodes in the system. The dual mode, on the other hand, was designed for use with a bum_ros_node node operating simultaneously. In this mode, the system plays back evidence while simultaneously executing the callbacks for both types of messages, thus simultaneously reading previously-collected evidence and collecting the results obtained by the system (Fig. 7).

3.6 The evaluator Node

The evaluator node is essentially responsible for providing visualizations during system execution, and produce offline evaluations of system performance.

If executing concurrently with the remaining system, it listens for Evidence and Tuple messages, and produces any visualizations needed, depending on the system. If executing offline, it runs iterative tests comparing the execution and ground truth logs provided and evaluates the performance of the system.

4 Usage of the bum_ros System

4.1 Set-Up

Setting up the package consists of two main steps: installing the necessary dependencies, and cloning and compiling the package. The package depends on the ProBT probabilistic computation, which needs to be installed separately, and also on depends on ROS, with ROS Indigo being the version used in our tests, and also on both Python 2 and 3. A number of extra python packages are being used, such as numpy, matplotlib, scikit-learn and scipy, which also need to be installed.

Then, it is necessary to clone the repository and build all the necessary files, which include the multiple ROS messages discussed in Sect. 3. These can then be used to run the basic non-ROS tests, or as a building block for running full-fledged system tests, as described in the remainder of this section. Listing 1.7 presents an example script to accomplish all of these steps.

```
1   # Install python dependencies
2   # This will install them to the system environment;
3   # you may prefer to use a virtualenv of pyenv environment
4   # for these tests.
5   sudo pip install numpy matplotlib scikit-learn scipy
6   sudo pip3 install numpy matplotlib scikit-learn scipy
7   # Clone the bum repository
8   git clone https://github.com/gondsm/bum
9   # cd to the root of the workspace
10  cd ..
11  # Build the message files
12  catkin_make
13  # cd into the package again
14  cd src/bum/scripts
15  # Add the ProBT package to the PYTHONPATH
16  source prep_pypl_env.sh
17  # run basic tests
18  python3 user_model_tests.py
```

Listing 1.7 Commands used to set-up the bum_ros package for basic testing.

4.2 Modes of Operation

The nodes in the bum_ros system can be combined in several manners to achieve different global operating modes. These operation modes are controlled by the parameters passed to the system, usually through a launch file similar to that illustrated in Listing 1.8. These parameters specify the paths to the files needed during execution, namely the log files and the GCD. Additionally, the `operation_mode` parameter specifies the mode of operation of the `data_manager` node, as described in Sect. 3.5. Naturally, for more complex system operation modes, these parameters can be set individually for each node via namespacing and name remapping.

```
1  <launch>
2      <!-- Four files have to be specified: the GCD, the
   ev_log, the exec_log and the gt_log -->
3      <param name="bum_ros/gcd_file" value="/home/growmeup/
   catkin_ws/src/user_model/bum_ros/config/data_gathering.gcd
   "/>
4      <param name="bum_ros/ev_log_file" value="/home/
   growmeup/catkin_ws/src/user_model/bum_ros/config/ev_log.
   yaml"/>
5      <param name="bum_ros/exec_log_file" value="/home/
   growmeup/catkin_ws/src/user_model/bum_ros/config/exec_log.
   yaml"/>
6      <param name="bum_ros/gt_log_file" value="/home/
   growmeup/catkin_ws/src/user_model/bum_ros/config/gt_log.
   yaml"/>
7      <!-- In addition, the global operation mode of the
   data_manager must be specified -->
8      <param name="bum_ros/operation_mode" value="listen"/>
9  </launch>
```

Listing 1.8 An example of a launch file for the system's parameters. In addition to the necessary file paths, the `operation_mode` parameter configures the operation of the `data_manager` node.

In addition to these parameters, the GCD can also be used to configure the operation of the `bum_ros_node` node. By combining the modes of the `data_manager`, the modes of the `bum_ros_node` node and the nodes we run, we can achieve three main operation modes for the full system:

- Regular Operation, wherein the user's characteristics are estimated and fused by the `bum_ros_node` node, with the `data_manager` logging both execution and evidence;
- Data Collection Mode, wherein only the HAL and `data_manager` nodes are run, with the goal of gathering information about the user;
- Data Fusion Mode, wherein the `bum_ros_node` is used solely for fusing information gathered by other `bum_ros` systems;
- Playback and Evaluation Mode, wherein gathered evidence is played back and fed to the `bum_ros_node`, generating log files for evaluation.

Each of the nodes can be run with the commands illustrated in Listing 1.9. If the launch and GCD files are configured correctly for each operation mode, the run command for each node is kept constant across operation modes.

```
# Run inference node
rosrun bum_ros bum_ros_node.py
# Run evaluator
rosrun bum_ros evaluator.py
# Run data manager
rosrun bum_ros data_manager.py
```

Listing 1.9 Commands for launching each of the system's nodes.

```
# [...] Characteristics and Evidence definitions
# Configuration of the node
Config:
  # Specify characteristics this node will work
on
  Active:
    # [...]
  # Will this node fuse tuples?
  Fusion: True
  # Will it only fuse hard evidence?
  Only_fuse_hard: False
  # Will tuples be published on prediction?
  Publish_on_predict: True
```

Listing 1.10 GCD configuration for the regular mode of operation.

Regular Operation This operation mode consists of the most basic way to run the bum_ros system. To achieve this mode of operation, the GCD should be configured as in Listing 1.10.

With this configuration, the node will fuse tuples whenever they are published, enriching its likelihood with previous knowledge. It also fuses both soft and hard evidence, as specified in Sect. 2. Lastly, it will publish new tuples whenever user characteristics are estimated, generating Tuple messages for fusion and, thus, closing the loop depicted on Fig. 3

To run this mode, only the system-specific HAL, the bum_ros_node node and (optionally) the data_manager node, using the commands on Listing 1.9. The data_manager is only expected to log evidence and execution and, as such, its mode should be set to listen in the launch file illustrated in Listing 1.8.

Data Fusion Mode This operation mode enables the bum_ros system to operate solely for information fusion, fusing information received from other bum_ros_node instances. To achieve this mode of operation, the GCD should be configured as in Listing 1.11.

This configuration disables the estimation abilities of the system by setting the Active parameter to an empty list. To run this mode, only the bum_ros_node node needs to be run, using the command of Listing 1.9.

Data Collection Mode This mode of operation allows for the collection of datasets into the bum_ros log files described in Sect. 3.3. As illustrated in Fig. 8, the

```
1    # [...] Characteristics and Evidence definitions
2    # Configuration of the node
3    Config:
4      # Specify characteristics this node will work
     on
5      Active: []
6      # Will this node fuse tuples?
7      Fusion: True
8      # Will it only fuse hard evidence?
9      Only_fuse_hard: False
10     # Will tuples be published on prediction?
11     Publish_on_predict: False
```

Listing 1.11 GCD configuration for the fusion mode of operation.

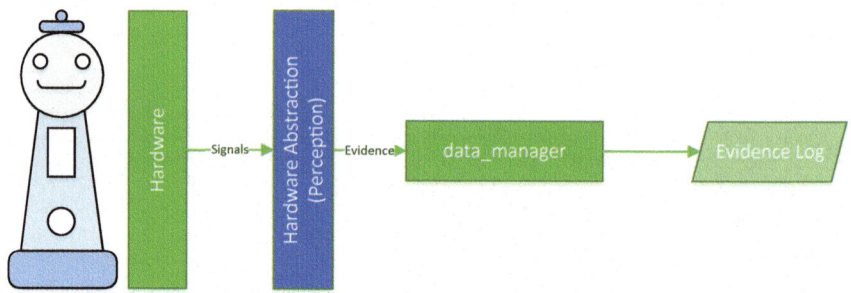

Fig. 8 An illustration of the data collection operation mode. In this mode, the system is listening to evidence produced by the HAL, which is saved to an evidence log for later playback

data_manager node replaces the bum_ros_node in receiving evidence messages, which are logged according to the structure described in Sect. 3.3. Each run of the system in this mode produces a single evidence file in the path specified in the systems parameters of Listing 1.8.

To run this mode, only the data_collector node and HAL need to be run, using the commands of Listing 1.9 and setting the operation mode to listen.

Playback and Evaluation Mode This operation mode, illustrated in Fig. 9, is meant to allow for the re-use of previously-collected data and for the repetition of experiments, sharing the goals of tools such as rosbag. It is assumed that data has been collected using the data collection mode, or that it has been converted to conform to its standards, depicted in Listing 1.2. It can be used to debug the bum_ros_node, to study the effects of randomness on system performance, or to study the system's ability to deal with multiple users by combining multiple evidence logs.

In this mode, the HAL is replaced as a source of evidence by an instance of the data_manager playing back one or more evidence logs that were collected previously. The data_manager is responsible for both playing back the data and logging the execution.

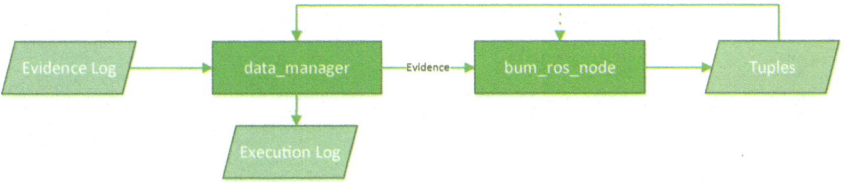

Fig. 9 An illustration of the playback operation mode. In this mode, the HAL is replaced by an evidence log collected previously in data collection mode

To run this mode, only the `bum_ros_node` and `data_manager` nodes need to be run, using the commands of Listing 1.9. The `data_manager` mode should be set to `dual`, so that the node simultaneously reads data and logs the execution of the `bum_ros_node` node. Once execution has ended, the `evaluator` node can be run on the results to produce metrics and figures.

4.3 System Topologies

One of the main goals of the `bum_ros` system is to be flexible enough to be used in several topologies and with several underlying hardware systems. These topologies can be achieved by the combination of the several nodes and modes of operation presented previously, using several instances of the nodes as building blocks for more complex systems.

In this section we discuss two types of topologies that can be achieved, as illustration of the system's flexibility:

- The single-device topology, consisting of a single interactive device, such as a robot, gaining information about its users;
- The multi-device topologies, wherein a network of distributed devices cooperates to gain information about the users, either fusing information in a centralized or distributed manner.

Single-Device Topology This topology corresponds to the system depicted in Fig. 4 In this topology, a single device gathers evidence on the user, estimates and fuses information solely for itself. This corresponds to the Regular Operation mode described in Sect. 4.2, and it can be achieved by launching the system in the manner described therein.

By itself, this topology is able to solve our essential problem: providing an underlying decision-making technique with a model of the user that it can use for adapting its actions. It can also be used to collect evidence on the user, as seen in Sect. 5, and to evaluate the system.

Multi-device Topologies The single-device topology and modes of operation can be used, through ROS, as a building block for a multi-device topology, akin to what

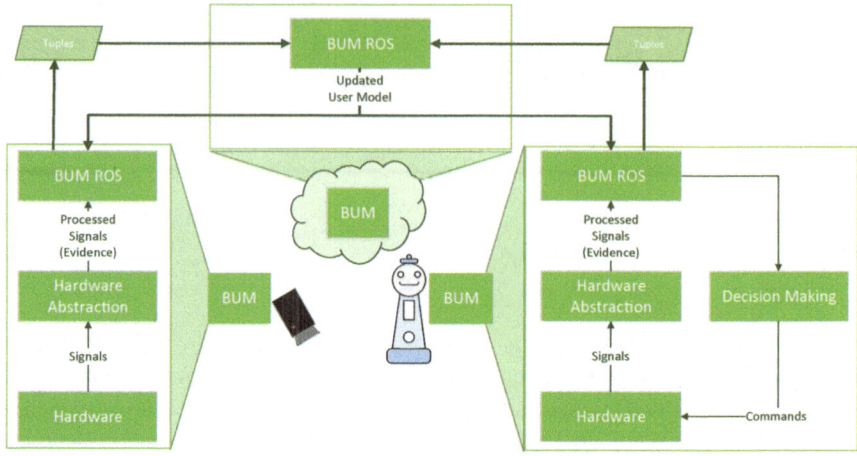

Fig. 10 An expanded illustration of the intended extensible system topology of Fig. 2. Tuples are generated by each `bum_ros_node` instance, and are fused in a centralized manner

is observed in Fig. 2. This constitutes the ultimate goal of this work, achieving a distributed, fault-tolerant network of data-gathering devices able to model a population of users seamlessly.

As described in Sect. 3, the system communicates through two main message types: `Evidence` and `Tuples`. These message types are processed by the `bum_ros_node` node with no regard to their source, i.e. any tuples and evidence received are processed.

By distributing ROS nodes across a network, we can relay the messages published in the `bum_ros/evidence` and `bum_ros/tuples` topics across several `bum_ros` systems. If each machine on the network is running an appropriate HAL, generating `Evidence` messages, and at least part of the `bum_ros` system generating `Tuple` messages, these can be relayed to other systems on the network. Thus, we can have several `bum_ros` systems cooperating and sharing information.

Thus, there are two main mechanisms for propagating information across nodes in the network: through the `Tuple` messages and through the `Evidence` messages. In turn, these can be fused either in a distributed manner, with each device fusing its own knowledge, or in a centralized manner, with a single `bum_ros_node` operating in fusion mode, as described in Sect. 4.2.

Fig. 10 illustrates the multi-device topology wherein `Tuples` are shared among systems, and fusion is centralized. This topology allows for an efficient transmission of information, since `Tuple` messages should be more compact and fewer in number than `Evidence` messages. Furthermore, this topology allows for the deployment on low-power and embedded systems, since the computational effort of information fusion is relegated to a remote machine.

At the cost of bandwidth, the whole `bum_ros` system can be moved to the cloud, leaving only the HAL running on each device. In this configuration, information

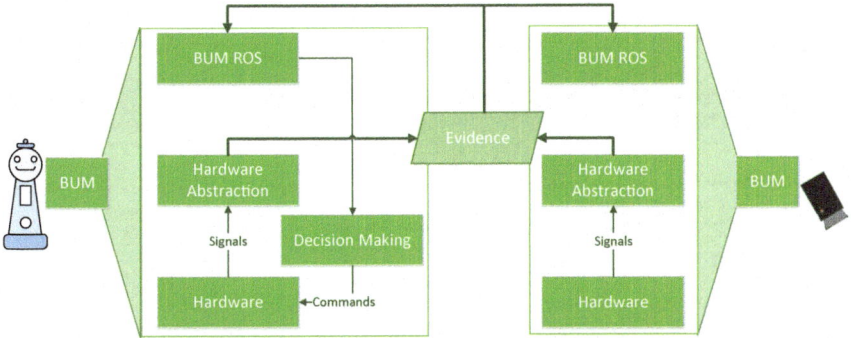

Fig. 11 An expanded illustration of the intended extensible system topology of Fig. 2. Evidence is generated by the hardware abstraction layers and is propagated to all `bum_ros_node` instances. If the same evidence is received by all nodes, their knowledge remains synchronized during execution

is propagated as `Evidence` messages, and all estimation and fusion efforts are performed remotely. This topology is, thus, ideal for embedded devices that can be used as simple data gatherers.

Fig. 11 illustrates the multi-device topology wherein `Evidence` messages are used for transmitting information across the system. In this case, fusion is distributed among the nodes, since the `Evidence` messages will trigger estimation which, in turn, will trigger fusion if the nodes are operating normally, as described in Sect. 4.2. This topology is interesting for teams of fairly powerful devices, such as Social Robots. In this case, each robot would maintain its own model of the user, potentially using it for its own decision-making purposes. When receiving exactly the same `Evidence` messages, even if in different orders, each system would maintain similar models of the user.

In turn, the distributed fusion topology could also make use of `Tuple` messages for knowledge transmission. In this case, the necessary bandwidth would be potentially reduced, also reducing the computational effort in parallel estimations. This would come, naturally, at the cost of reduced fault-tolerance in the system, since a failure in a node's estimation facilities would impede it from propagating information.

5 Experiments

Having demonstrated the mathematical workings of the model in [10], in this section we aim to better describe the experimental set-up used, and how it can be replicated,

Input : **Q** : Normal questions to be asked,
        V : Questions about volume,
        D : Questions about distance.
Output: **T** : Tuples used for training the BUM system.
asked_vol = False;
asked_dist = False;
talk_evidence = [];
while Q $\neq$ {} **do**
> // Remove random question from **Q**
> q = **Q**.pop();
> // Ask the question via robot.
> reply, time = ask(q);
> // Generate evidence from response.
> e = reply.n_words() / time;
> talk_evidence.append(e);
> send_evidence(talk_evidence);
> // Ask about volume
> **if** *asked_vol == False and rand() < 0.25* **then**
> > asked_vol = True;
> > q_v = **V**.pop();
> > reply, time = ask(q);
> > adjust_volume(reply);
> > send_tuple(reply);
>
> **end**
> // Ask about distance
> **if** *asked_dist == False and rand() < 0.25* **then**
> > asked_dist = True;
> > q_v = **D**.pop();
> > reply, time = ask(q);
> > adjust_distance(reply);
> > send_tuple(reply);
>
> **end**

end

Algorithm 1: An algorithm illustrating the operation of the user-adaptive decision-making module used for data gathering. The system gets random questions from a pool, which are prompted to the user. Randomly, the system asks about the distance or volume that it is using to communicate, and the replies used to generate hard evidence.

as well as how ROS empowers this set-up by making it easily reproducible, portable and flexible. We also provide illustrations of the results that may be obtained using our technique.

5.1 Set-Up

The modularity of the system combined with ROS's abstraction layer allow it to support different types of inputs, from physical sensor devices to data streams of simulated data. This allows the system to run on both simulated data and data collected from human users seamlessly. Data needs only to arrive correctly wrapped in `Tuple` and `Evidence` messages for the system to process it. To achieve this effect, only the HAL system depicted in Fig. 8 needs to be replaced.

For the simulated trials, the HAL was replaced with a population simulator that generated the necessary data from a simulated population of users. Simulated users were generated from a set of profiles, each profile composed essentially of a point in characteristics-space. Given a number of profiles, the simulated population was generated by applying random noise to the original profiles, thus resulting in a rich population that was segmented into clear groups. The data from the simulated users can then be published using `Evidence` and `Tuple` messages, which will be processed by the system. This process is illustrated in Figs. 12 and 13.

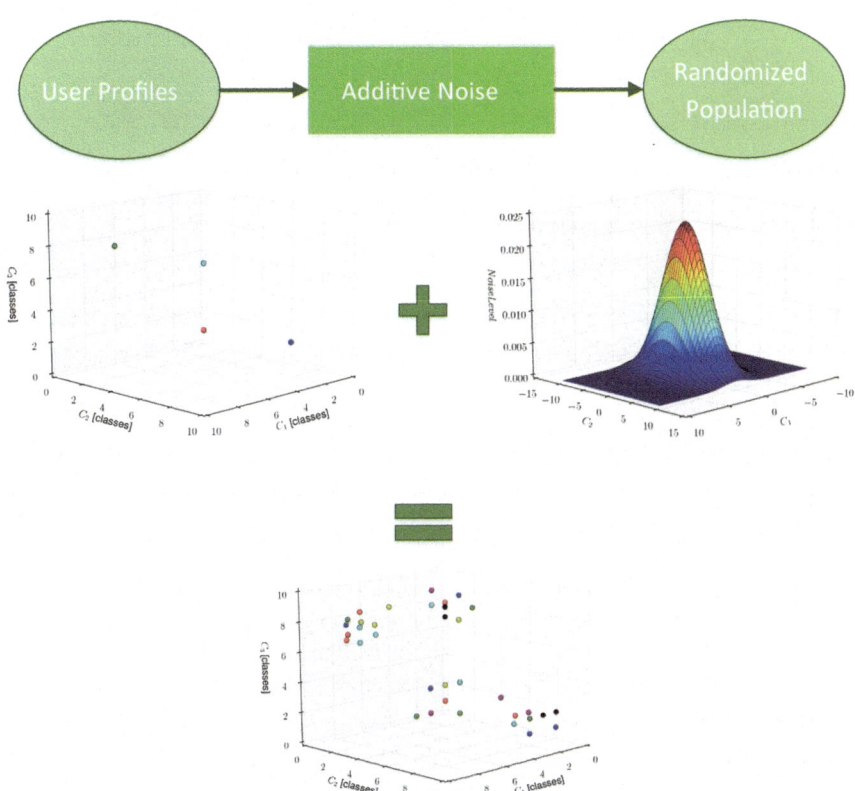

Fig. 12 An illustration of the population generation mechanism. The mechanism is fed with a small number of user profiles in characteristics space, which are perturbed with additive noise to generate a rich population that loosely follows the original profiles

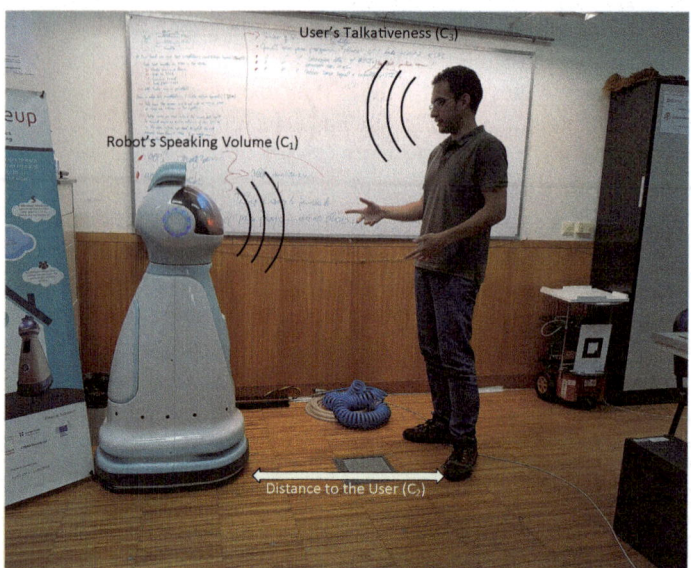

Fig. 13 A picture of the experimental environment used for data gathering. The user faced the robot, speaking naturally while answering the robot's questions. This allowed for the collection of all necessary measurements, as represented

For collecting data from real users, the HAL was replaced with an architecture that interfaced with the ROS-enabled GrowMu robot [35]. The HAL was composed of four main layers:

- The `operational` layer, composed of ROS nodes that act as drivers for single functionalities, such as speech recognition, and expose them as ROS actions;
- The `dispatcher` layer, a custom-built abstraction layer that interfaces with each individual action server, exposing all of the robot's functionality as a single ROS action;
- The `gmu_functions` layer, which interfaces with the `dispatcher` and provides a simplified Python middleware for accessing the robot's basic functions, such as speech recognition and synthesis, as well as some intermediate functionality such as asking questions, waiting for answers and ensuring that the user understood the question;
- The `conductor`, which behaved as illustrated in Algorithm 1, responsible for implementing the high-level logic as a combination of the robot's high-level functions;

This architecture is illustrated in Fig. 14. This system was used to generate the `Evidence` and `Tuple` messages that the `data_manager` in `listen` mode uses to generate its logs.

In order to mimic a natural interaction with a social robot, the GrowMu robot was used in a set-up similar to Fig. 13. This interaction consisted of a normal conversation,

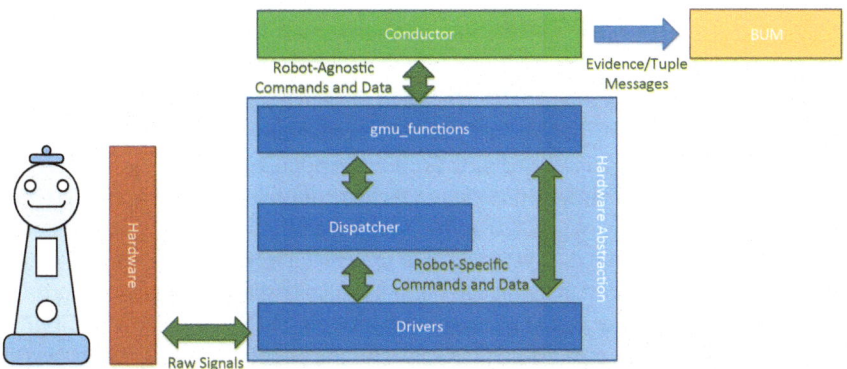

Fig. 14 An illustration of the software stack used for data gathering. A Hardware Abstraction Layer exposes the robot's functionality to the `conductor` node, which manages the interaction with the user and analyses their responses, generating tuple and evidence messages for processing at the `bum_ros` level. Basic functionality such as speaking is accessed directly, while complex functionality such as displaying webpages on the robot's screen are accessed through the Dispatcher interface

wherein the robot asked questions, and the user answered them. The goal of the interaction was to determine three characteristics of the user:

- C_1: User's preferred speech volume;
- C_2: User's preferred distance to the robot;
- C_3: User's talkativeness.

User interaction was split into sessions, with each session proceeding as follows:

1. Users were introduced to the research being performed and to what data would be gathered;
2. Users were asked about each individual characteristic, establishing a baseline "self-assessment";
3. The users interacted freely with the system;
4. The session was concluded.

Each interaction with a user resulted in a single evidence log, produced by the `data_manager`. The `data_manager` was used in data collection mode, as illustrated in Fig. 8.

The data gathered by this set-up was fed to the `bum_ros_node` node, in a set-up similar to the playback and evaluation mode of Fig. 9. The `data_manager` node was used to feed data to the `bum_ros_node` node, which generated execution logs. The `bum_ros_node` node operated normally, configured similarly to the configuration in Listing 1.1, was then responsible for, for each message received, performing estimation and fusion, gradually learning the user's preferences.

Lastly, the execution logs generated during normal execution, combined with the ground truth and evidence logs, were fed into the `evaluator` node to generate metrics and figures.

5.2 *Results and Discussion*

Fig. 15 shows the graphs produced by the evaluator node for a simulated population of 40 users. We can observe that, as the system executes and gathers data, the population that the model contains approximates the reference population. This figure represents an example can be obtained by processing the execution logs with the evaluator node, which plots the population that the system had incorporated at each iteration. This, we can observe that the system is able to achieve its main goal, to gain enough information on the user such that it can build a model that approximates the original.

Fig. 16 shows the population clusters corresponding to the populations of Fig. 15. We can observe that, as the system interacts with the population, the clusters become centered on the same coordinates as the reference clusters, which correspond to the apparent sectioning of the reference population in Fig. 15. Furthermore, as the system executes, these clusters become smaller in apparent diameter, meaning that the covariance of the underlying Gaussian mixtures decreases, indicating a refinement of the information they encode. This means that not only does the system manage to correctly estimate the characteristics of the population, it also gains further infor-

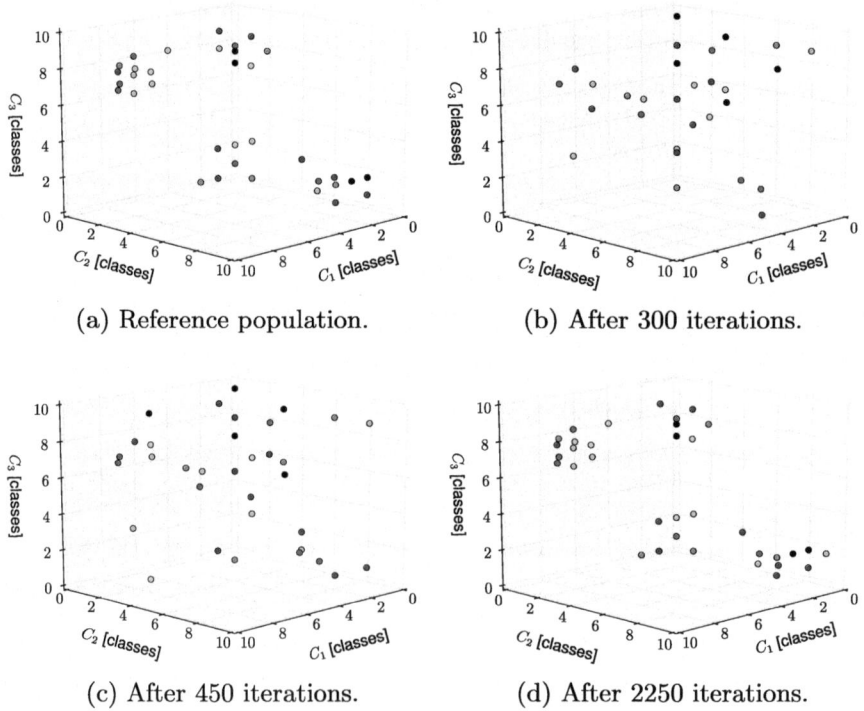

Fig. 15 An illustration of the evolution of the user population as the system executes. As more information is obtained on the users, their position in characteristics space approximates

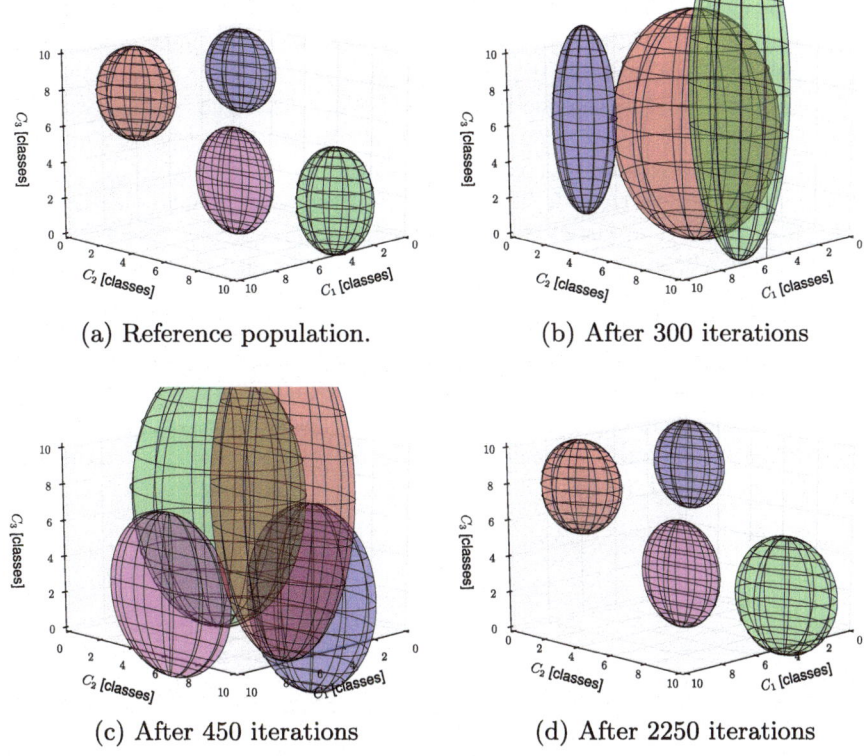

(a) Reference population.   (b) After 300 iterations

(c) After 450 iterations   (d) After 2250 iterations

Fig. 16 An illustration of the evolution of the clusters as the system executes. As the system converges, it obtains user clusters that better approximate the reference population

mation, and thus certainty in the Gaussian distributions, as the system executes. As mentioned in Sect. 2, these can be used to speed up the adaptation process or to compensate in case of a sensor failure.

6 Conclusion

In this chapter we have presented the BUM system, and its corresponding implementation in ROS.

We have presented an overview of the mathematical workings of the BUM process, including the learning, estimation and clustering sub-processes. We have then presented its implementation in ROS, providing a tutorial overview of the design choices that were made to achieve the desired functionalities, as well as of the several nodes, log and configuration files used by the system.

Several modes of operation and possible topologies of the system were also described, showing that the system, thanks to its ROS implementation, is able to achieve the level of flexibility desired. These served as basis for the implemented system that was used for experimentation using both a synthetic and real datasets. The collection and processing of these datasets was also presented, as was the evaluation process performed with the system's own facilities.

The system is now ready to be used by user-adaptive techniques, as described in Fig. 1, as a source of knowledge for generating adapted stimulus for users. Furthermore, it can be used to collect data in a distributed manner from heterogeneous devices spread over a potentially large number of environments.

In the future, it would be interesting to expand our experimental setup to make use of the largest multi-device topologies made possible by the system, using several devices and robots. It would also be interesting to explore the application of deep learning techniques to improve the learning performance of the system. Lastly, we would like to include this technique in an encompassing user-adaptive technique, as presented in Fig. 1, allowing for truly personal interaction between devices and users.

References

1. M.F. McTear, User modelling for adaptive computer systems: a survey of recent developments. Artif. Intell. Rev. **7**(3–4), 157–184 (1993)
2. A.F. Norcio, J. Stanley, Adaptive human-computer interfaces: a literature survey and perspective. IEEE Trans. Syst. Man Cybern. **19**(2), 399–408 (1989)
3. H.A. Tair, G.S. Martins, L. Santos, J. Dias, α POMDP : State-based decision making for personalized assistive robots, in *Thirty-Second AAAI Conference on Artificial Intelligence, Workshop 3: Artificial Intelligence Applied to Assistive Technologies and Smart Environments*, vol. 1 (2018)
4. N. Abdo, C. Stachniss, L. Spinello, W. Burgard, Robot, organize my shelves! tidying up objects by predicting user preferences, in *2015 IEEE International Conference on Robotics and Automation (ICRA)* (2015), pp. 1557–1564
5. K. Baraka, M. Veloso, Adaptive interaction of persistent robots to user temporal preferences, in *International Conference on Social Robotics* (Springer International Publishing, Berlin, 2015), pp. 61–71
6. M. Fiore, H. Khambhaita, An Adaptive and Proactive Human-Aware Robot Guide (2015)
7. A. Kobsa, Generic user modeling systems. User Model. User-Adapt. Interact. **11**(1–2), 49–63 (2001)
8. F. Broz, I. Nourbakhsh, R. Simmons, Planning for human-robot interaction in socially situated tasks: the impact of representing time and intention. Int. J. Soc. Robot. **5**(2), 193–214 (2013)
9. A. Tapus, A. Aly, User adaptable robot behavior, in *2011 International Conference on Collaboration Technologies and Systems (CTS)* (2011), pp. 165–167
10. G.S. Martins, L. Santos, J. Dias, BUM : bayesian user model for distributed social robots, in *26th IEEE International Symposium on Robot and Human Interactive Communication, RO-MAN* (IEEE, 2017)
11. A. Cufoglu, User profiling-a short review. Int. J. Comput. Appl. **108**(3), 1–9 (2014)
12. J.A. Konstan, J. Riedl, Recommender systems: from algorithms to user experience. User Model. User-Adapt. Interact. **22**(1–2), 101–123 (2012)

13. D. Fischinger, P. Einramhof, K. Papoutsakis, W. Wohlkinger, P. Mayer, P. Panek, S. Hofmann, T. Koertner, A. Weiss, A. Argyros, M. Vincze, Hobbit, a care robot supporting independent living at home: first prototype and lessons learned. Robot. Auton. Syst. **75**, 60–78 (2014)
14. I. Duque, K. Dautenhahn, K.L. Koay, L. Willcock, B. Christianson, A different approach of using personas in human-robot interaction: integrating Personas as computational models to modify robot companions' behaviour, in *Proceedings - IEEE International Workshop on Robot and Human Interactive Communication* (2013), pp. 424–429
15. K. Baraka, A. Paiva, M. Veloso, Expressive lights for revealing mobile service robot state. Adv. Intell. Syst. Comput. **417**, 107–119 (2016)
16. S. Nikolaidis, A. Kuznetsov, D. Hsu, S. Srinivasa, Formalizing human-robot mutual adaptation: a bounded memory model, in *Human-Robot, Interaction* (2016), pp. 75–82
17. A.B. Karami, K. Sehaba, B. Encelle, Adaptive artificial companions learning from users feedback. Adapt. Behav. **24**(2), 69–86 (2016)
18. G.S. Martins, P. Ferreira, L. Santos, J. Dias, A context-aware adaptability model for service robots, in *IJCAI-2016 Workshop on Autonomous Mobile Service Robots* (New York, 2016)
19. G.H. Lim, S.W. Hong, I. Lee, I.H. Suh, M. Beetz, Robot recommender system using affection-based episode ontology for personalization, in *Proceedings - IEEE International Workshop on Robot and Human Interactive Communication* (2013), pp. 155–160
20. A. Sekmen, P. Challa, Assessment of adaptive human-robot interactions. Knowl. Based Syst. **42**, 49–59 (2013)
21. A. Cerekovic, O. Aran, D. Gatica-Perez, Rapport with virtual agents: what do human social cues and personality explain?, in *IEEE Transactions on Affective Computing*, vol. X(X) (2016), pp. 1–1
22. A. Tapus, C. Tapus, M. Mataric, User-robot personality matching and robot behavior adaptation for post-stroke rehabilitation therapy. Intell. Serv. Robot. **1**(2), 169–183 (2008)
23. Q. Sajid, Personality-based consistent robot behavior, in *Human-Robot Interaction* (2016), pp. 635–636
24. A. Vinciarelli, G. Mohammadi, A survey of personality computing. IEEE Trans. Affect. Comput. **5**(3), 273–291 (2014)
25. F. Carmagnola, F. Cena, C. Gena, User model interoperability: a survey. User Model. User-Adapt. Interact. **21**(3), 285–331 (2011)
26. J. Fink, A. Kobsa, A review and analysis of commercial user modeling servers for personalization on the World Wide Web. User Model. User-Adapt. Interact. **10**, 209–249 (2000)
27. C.C. Evans, The official YAML web site (2001)
28. Python Software Foundation. Welcome to Python.org
29. Open Source Robotics Foundation. rospy
30. The Matplotlib Development Team. Matplotlib: Python Plotting
31. Open Source Robotics Foundation. Documentation - ROS Wiki
32. C.M. Bishop, *Pattern Recognition and Machine Learning* (2006)
33. J.F. Ferreira, J. Dias, *Probabilistic Approaches for Robotic Perception* (Springer International Publishing, Berlin, 2014)
34. C.E. Shannon, A mathematical theory of communication. Bell Syst. Tech. J. **27**(1), 379–423 (1948)
35. G.S. Martins, L. Santos, J. Dias, The GrowMeUp project and the applicability of action recognition techniques, in *Third Workshop on Recognition and Action for Scene Understanding (REACTS)*, ed. by J. Dias, F. Escolano, G. Ezzopardi, R. Marfil (Ruiz de Aloza, 2015)

ROSRemote: Using ROS on Cloud to Access Robots Remotely

Alyson Benoni Matias Pereira, Ricardo Emerson Julio and Guilherme Sousa Bastos

Abstract Cloud computing is an area that, nowadays, has been attracting a lot of researches and is expanding not only for processing data, but also for robotics. Cloud robotics is becoming a well-known subject, but it only works in a way to find a faster manner of processing data, which is almost like the idea of cloud computing. In this paper we have created a way to use cloud not only for this kind of operation but, also, to create a framework that helps users to work with ROS in a remote master, giving the possibility to create several applications that may run remotely. Using SpaceBrew, we do not have to worry about finding the robots addresses, which makes this application easier to implement because programmers only have to code as if the application is local.

Keywords Cloud computing · ROS · Robotics

1 Introduction

ROS is one of the most used frameworks in robotics which allows users to write robot software. It contains a lot of libraries that helps developers create programs to control robots or process data. Although ROS is a powerful tool, it only works offline, which means that it is not possible to create cloud applications using pure ROS.

A. B. M. Pereira (✉) · R. E. Julio · G. S. Bastos
System Engineering and Information Technology Institute, Federal University
of Itajuba, Av. BPS, 1303. UNIFEI - IESTI - Pinheirinho, Itajuba,
MG 37500-903, Brazil
e-mail: alysonmp@gmail.com
URL: http://www.unifei.edu.br

R. E. Julio
e-mail: ricardoej@unifei.edu.br

G. S. Bastos
e-mail: sousa@unifei.edu.br

© Springer International Publishing AG, part of Springer Nature 2019
A. Koubaa (ed.), *Robot Operating System (ROS)*, Studies in Computational
Intelligence 778, https://doi.org/10.1007/978-3-319-91590-6_16

Cloud applications are very common nowadays and while some companies invest their resources in creating new technologies for faster processing, others prefer to save money renting or buying big servers that solve their problems.

These companies use it to store data, process heavy calculations, or even make an application accessible for several computers at the same time. Recently, cloud computing has been used for making robots communicate among themselves and also process data needed for the robot to complete its tasks. However, the configuration of these software are hard and demand some time.

These kind of applications can be classified as Cloud Robotics [1], because they mix robotics and systems that run on cloud. Although it is an important step in robotics, it does not solve a bigger problem, which is remotely control the robots in an easy and fast way.

ROSRemote was created to fill in this blank. It is a framework that helps developers to create ROS [2] applications that can run on the cloud the same way it is used offline. With this, it is possible to create your own application or call ROSRemote to send information as well as improve the framework with other usages.

Thus, we implemented the most basic ROS commands, such as *rostopic*, *rosservice* and *rosrun*, and the users may take advantage of these tools or create their own to suit their needs.

To make the connection, ROSRemote uses a third party application that is called *SpaceBrew* [3]. It helps the communication between different ROS Masters[1] by connecting them easily by their machine IPs. With literally a mouse click, the Masters are already connected and ready to send and receive information.

The advantage of ROSRemote over other tools is that in ROSRemote the user does not need to perform any kind of configuration when using *SpaceBrew* free server, this configuration is also impossible in some cases, which make other tools less useful than ROSRemote. It only needs to create a SpaceBrew client, which is done automatically when running ROSRemote package and then connect themselves in *Spacebrew* Web Admin Page. All these steps is explained in detail more ahead in this chapter.

With ROSRemote the users can send and receive information from one ROS Master to another in the same way they would do just calling the service created in the application and passing the command as a parameter of the service.

2 Tools

2.1 Installing ROSRemote

The ROSRemote package itself does not need any installation, the user just have to clone the source code from github [4] inside the catkin workspace and normally

---

[1]ROS Master is the resource needed to run ROS on computers and robots. It provides naming and registration services, tracks publishers and subscribers to topics and services. In short, it enables individual ROS nodes to locate one another.

run it in ROS. Although ROSRemote does not need any installation, there are some dependencies necessary for the package to run, like Websockets and NodeJS.

The installation of both tools is simple. To install NodeJS the users just have to enter the following commands in the terminal.

$ sudo apt-get update

$ sudo apt-get install nodejs

After the installation it is also needed to install the npm package manager and Websocket protocol, it is possible to perform this action by inputting the following command in the terminal.

$ sudo apt-get install npm

$ sudo npm install websocket

After these commands, the user just need to run ROS and the ROSRemote package executable in the local and in the remote masters using the commands below.

$ roscore

$ rosrun cloud_ros main_cloud_ros

When rosrun is called, a *SpaceBrew* client is automatically created and the user needs to connect them using the *Web Admin Tool* as described in Sect. 2.3. After that, just follow the instructions shown in Sect. 4.1.

2.2 ROS

As described in [5] the Robot Operating System is a framework that helps writing robot software. As written in Sect. 2.3, SpaceBrew works with clients and ROS with nodes [6]. These nodes are like a Operating System process that performs calculations.

Nodes are responsible for creating topics [7]. They are buses through which all the messages are exchanged. When a node wants to send data to another one, it publishes in a Topic and those nodes which want the information, subscribe to it.

Topics are not the only feature responsible for sending and receiving data. In ROS there are also services [8], the only difference is that a Topic publishes and does not need an answer while a service publishes some information and blocks itself until an answer is received.

In ROSRemote, these topics are published locally, just like common ROS, but can be accessed remotely using this framework.

ROS is a tool that has been very used lately [boren2011exponential] and a search for some data bases proves that. A search was performed in the main data bases using the keywords "Robot Operating System" in quotes and was found 434 publications in IEEEXplore, 464 in ScienceDirect and 1004 in Springer Link, among papers, articles and conferences, this shows how ROS has been widely used for a variety of applications.

There are some different areas where ROS is used, from industrial robots to unmanned air vehicles. Among some applications can be cited the FIT IoT-LAB [9], a laboratory that helps users to conduct experiment using Internet of Things. Another interesting application is the one described by weaver2013uav that studies a way to help UAVs to perform take off, flight and landing without spending much money.

These applications does affect daily lives of common people, but they are not the only ones created using ROS. Besides them there are applications used to facilitate elderly people lives [10], which helps detect and inform when a person suffer a domestic accident. There are also some works on education areas [11], mapping and localization [12, 13], besides many other.

2.3 SpaceBrew

SpaceBrew [3] is a framework that helps to perform connections between remote objects. It works with the publisher/subscriber architecture, just like ROS. The only difference is that in ROS it is necessary to start a node in order to start working with topics, and in *SpaceBrew* a client must be created. However, they both work exactly the same way.

SpaceBrew is not a very used tool, there is just few papers that uses this framework and in all of them the authors do not get to any conclusion about it and in most of the works, they just mention that this tool is used. Among these works some of them should be highlighted. [14] work provides a context for a research of an Avatar that may communicate with others, like a video game. Another application is described by [15], in it the authors use a flying object that is capable of impose its angular momentum in a tablet. The last application that was found is one that allows a viewer interact with a video game just with the look, improving game experience [16]. Unfortunately only one author give a brief description about why *SpaceBrew* was chosen to create the applications and it is because it allow quick and easy communication among different components.

This framework easily connects two or more remote devices using their IP address and a *Web Admin Tool*, as shown in Fig. 1. To connect two clients the user only needs to click on the publisher which will send data and then click on the subscriber of the client who will receive the message.

When this is done, the framework is ready to send and receive data from remote devices and the user only needs to run ROS service in order to send the information.

For test purposes, the *SpaceBrew* developers provide a server that can be used to connect the application. On the other hand, it is also possible to create a server, for this, the code is available for free at GitHub, and may be downloaded using this link https://github.com/Spacebrew/spacebrew or clone the repository using the command:

$ git clone https://github.com/Spacebrew/spacebrew.git

After downloading or cloning the repository the user needs to install the Websocket as shown in Sect. 2.1 and the Forever Monitor, which is responsible for maintaining the server running and capable of receiving connections.

Firstly install "npm" (explained in Sect. 2.1) and use the package manager to install the Forever Monitor using the command.

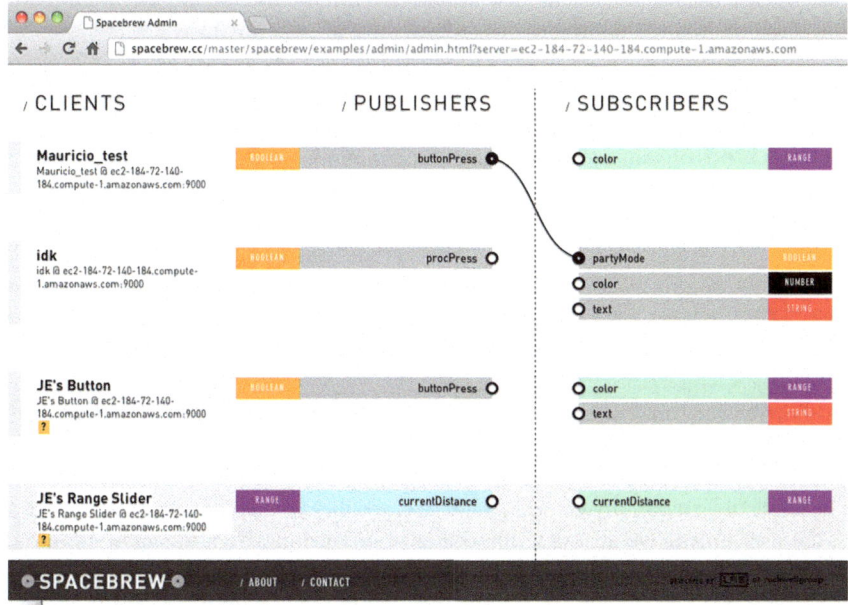

Fig. 1 *Web Admin Tool* Graphic interface implemented on SpaceBrew server. Source: http://docs. spacebrew.cc/gettingstarted/

$ npm install forever-monitor

Then use the Linux terminal to go to the downloaded *SpaceBrew* folder and run the command to start the server.

$ node node_server_forever.js

It may be necessary to install other dependencies to run the server on a local machine, but the user will get an error message and an indication of what needs to be installed and how to do it.

Use "ifconfig" or another tool to find the IP (the public IP, since ROSRemote is not necessary for running in a local network) and, in a browser, type the IP and the default port 9000, that can be changed directly in ROSRemote code.

The server is already running now and able to receive connections from ROSRemote, as long as the server is correctly pointed out in the code. More of it is explained in Sect. 4.1.

3 Related Works

In this section will be described some tools that can be used to perform a similar task than ROSRemote. Some of them were not developed specifically to work with robots but can be used for this purpose while others were develop exactly for this task.

3.1 SSH

SSH are the initials for Secure Shell and it is a data traffic [17]. Like TCP and UDP protocol The Secure Shell were not created to control with robots or work with ROS, but since it is possible to access a remote terminal with this tool, users can make use of it to control and monitor robots remotely.

To configure this tool it is necessary to create a SSH server, open a port in the firewall, if it is not already opened and, if the server is running behind a NAT (Network Address Translation) the user also needs to forward a port in order to offer external access. Then the user can connect to this SSH server and execute some tasks like access a remote terminal, execute commands in a remote computer and transfer file safely.

Although it is a tool that allows users to perform several tasks, it has some problems that may disrupt its usage and the major one is described in [18]. In this article, the author mention that in some cases, like when the user is using a 3G/4G connection it is not possible to do a port forwarding, in these cases the user can not use SSH. Besides this, the user must have access to the router in order to perform this configuration, and if not, he is not able to use this tool too.

The SSH is a very secure tool, it encrypts all the data before sending them and ensure that if someone or something intercepts this data, it will not be able do decrypt and read it. Although it is an advantage, it makes the data traffic slower than other tools that send their data without encryption.

The purpose of this tool, as described before, is not to control robots but when using it together with ROS it is possible to do all the things ROSRemote does. On the other hand ROSRemote has some advantages described in Sect. 5.5.

3.2 VPN

VPN are the initials for Virtual Private Network and it is a private network that only exists virtually [19]. It helps connecting two or more computers through a secure connection even if they have insecure connections. VPN can not be used directly with ROS, but it can be used to transfer data from one ROS Master to another one, for example.

VPN also needs some configurations before works and sometimes they are harder than SSH configurations. The user needs to create a VPN server and a client to connect them and them use hardwares, like routers and switches, that are used to make the connections. Then the user can connect to the remote server and transfer data in a safe way.

Among all VPN uses, the most common are [19]:

- Exchange confidential information;
- Access corporate networks;
- Hide activities from the internet service provider;
- Download illegal content;

VPN is a tool used to transfer data and so can not be used exactly the same way as ROSRemote, but can work the same way as SpaceBrew by transferring data. VPN also ecrypts data and like SSH this is, at the same time, an advantage and a disadvantage because it is secure but makes the traffic data exchange slower than other tools.

3.3 Rapyuta

Rapyuta is an open-source tool that helps developers to create robotic applications in the cloud [20]. It was created to remove all the heavy processing from robots, allowing them to be cheaper, with less resources and even so be able to make a decision faster than it normally would. Rapyuta allocates safe computing environments (Clones), that are strongly connected with each other [21] and to RoboEarth database [22].

RoboEarth is a database that contains experiences from robots that can be used by another robots to complete the same or similar tasks and until Rapyuta creation, it was only a database where robots could get experience. What makes Rapyuta a powerful tool is that besides it helps user to send their application to cloud in order to make all the needed processing, it also has a connection with RoboEarth. This connection helps application because the robot can automatically search for data that can help it to perform its task, make all the process in the cloud and only receive the necessary commands to reach its goal (Fig. 2).

Rapyuta is a tool that helps user to perform cloud computing, but it is not possible to control a robot using the keyboard with this tool. Figure 3 shows an overview about Rapyuta. The robots send their data to be processed in the cloud and this is done inside the clones (safe computing environment represented by the rectangles)

Fig. 2 Rapyuta overview. Source: [21]

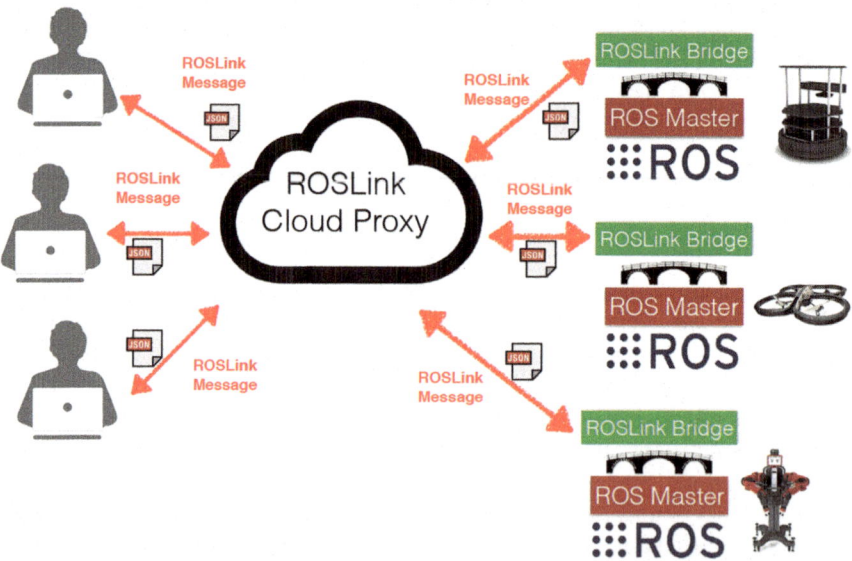

Fig. 3 ROSLink overview. Source: [18]

that are strongly connected among themselves and can exchange information. At the same time, Rapyuta cloud is connected to RoboEarth (represented by the cylinders) and it is possible, while performing the processing, consulting RoboEarth database and gather experiences from other robots.

The main difference between Rapyuta and ROSRemote is that the first helps user to perform processing outside robots, using the cloud, while ROSRemote uses cloud to transfer data among robots and assists users to control robots remotely using a common keyboard.

3.4 ROSLink

Among all the similar works that were found during the research ROSlink is the most similar one. It also has the same motivation than ROSRemote, which is the fact that ROS does not offer native support for monitoring and control robots through the internet.

In the standard ROS approach users can only connect one ROS master with one robot, which is not interesting in case it is necessary to work with a multi robot system. Trying to solve this problem, a second approach were created, called Centralized Approach. In this case several users and robots share the same ROS master. In some ways it solves the problem of multi robots, but creates other that is the lack of scalability. This means that if a lot of robots are connected to the same master, it will be overloaded and will not be able to serve all of them properly.

In the paper [18] the authors describe a new solution based on messages called ROSLink. It works with a multimaster architecture, which means that each user and robot run its own master and these masters exchange their information using a proxy in the cloud, very similar to ROSRemote. This approach can be seen on Fig. 3.

The main differences between ROSLink and ROSRemote is that the first one uses messages to send data and perform communication among devices while the second one uses a service created when the user executes ROSRemote package executable. Besides that it is possible to control a robot using keyboard command with ROSRemote, which may not be possible using ROSLink.

The messages ROSLink can send has two parts, a header and a payload, are divided in four types and can be seen here [23]:

- Presence messages;
- Motion messages;
- Sensor messages;
- Motion commands;

4 ROSRemote

4.1 ROSRemote Usage

The ROSRemote package works just like common ROS. The only difference is that instead of inserting the command directly on terminal, it is necessary to call the */send_data* service and pass as arguments what the user wants to do.

A simple example may be shown using *rostopic list*, the command that returns a list of all the topics being published at the moment. In the common ROS, the user just need to insert the words

$ rostopic list

and hit ENTER, and then all the topics are shown on the terminal. At the ROSRemote package, the user must call the service, then send the command in quotation marks

$ rosservice call /send_data "rostopic list"

then everything is done automatically and the user can see the topics being published in the remote master.

4.2 ROSRemote Package Description

This section provides an explanation about ROSRemote, showing the main classes, the data flow and the architecture of the framework. More information can be found in [24].

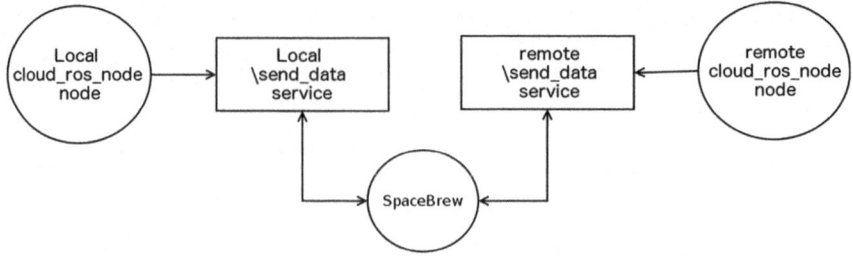

Fig. 4 Data flow of the package

Table 1 Features
implemented in ROSRemote

Tool	Commands
Rostopic	List, echo, info, type
Rosservice	List, args, call, node, type
Rosrun	Rosrun

The executable of the package just creates one node, which is responsible for creating the service that will send information from *SpaceBrew* to the remote master and, in the same way, receive the answer. The node is called "cloud_ros_node" and implements the service "/send_data", which is responsible for publishing and subscribing the information inserted by the user.

The service sends the data to *SpaceBrew* server and it re-routes the message to the remote ROS master. Then, it processes the information and returns, through *Space-Brew*, the answer to the local master. Figure 4 shows the data flow in the package.

Some main ROS features are included in the ROSRemote package, they are described in details in Sects. 4.5–4.8, but Table 1 shows all the implemented features. Besides all theses features that already exists in ROS, a new one were necessary to create, that is *roscommands*, it is used to help users to control robots and is explained in Sect. 4.8. It is also important to remember that the user can always insert their own application and improve ROSRemote.

What makes ROSRemote different from other applications, besides the facility of configuring the communication between masters, is the possibility to send a data from a second computer to a third one, without needing to receive the answer and relay it.

To exemplify it, we will assume three computers called 1, 2 and 3, that are in different networks and may be on different countries as well, showed on Fig. 5. The user 1, who is on computer 1, sends a message on broadcast to all the other computer asking the data that some topic is publishing in computer 3, for example. Computer 3 sends back the answer also on broadcast, which means that if computer 2 needs this information, it does not have to wait computer 1 receives the answer and forward the message.

A search on the main data bases available (like Scopus and ScienceDirect) have not showed any similar applications that can do this work. There are some new tools,

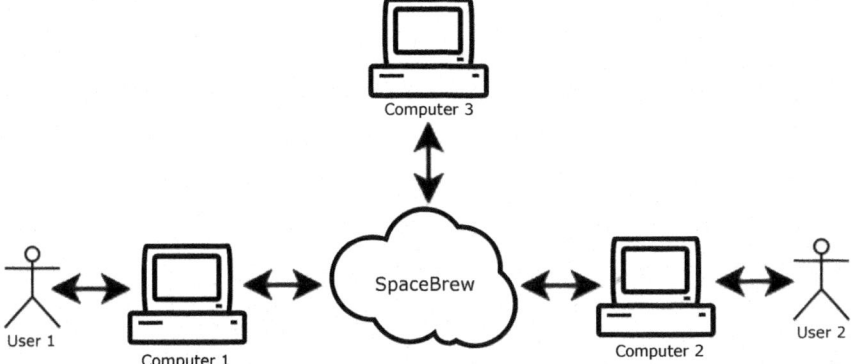

Fig. 5 Example of how ROSRemote can work

like Canopy and Rapyuta, that are starting to be used, but when this work started to be done, there were not many articles and information about them. *SpaceBrew* was the framework that was most used by the time and, even now, new applications are created every day.

4.3 ROSRemote Architecture

The data flow of the framework seems complicated, but it is really simple and was written in a way that helps developers to find where a new code should be inserted and also helps the development of new parts by showing the structure that must be followed so that the software works properly.

Figure 6 helps the understanding of the data flow in the package. Everything starts when the user types the desired command in the service *"/send_data"*. The main class receives the command and decides the correct class to send it. The subclass then creates a JSON variable that keeps all the necessary information that will be sent to the remote master.

As an example, it is shown a piece of code that sends the command *rostopic echo*.
data = 'commandRos':'rostopic', 'function':'rostopicEcho', 'action':'send', 'topic':commandSplit[2][1:]

It is possible to notice that the JSON has a few variables:

- commandRos: shows what type of command is being sent so that the main class can decide which subclass will be invoked. In this case, the command is rostopic;
- function: saves the function name that will be called in the remote master. This function is implemented in each subclass and is called when the *SpaceBrew* message, containing the data the user inserted, arrives at the remote master;
- action: shows if the message will be sent to process or if the answer is being received from the remote master;

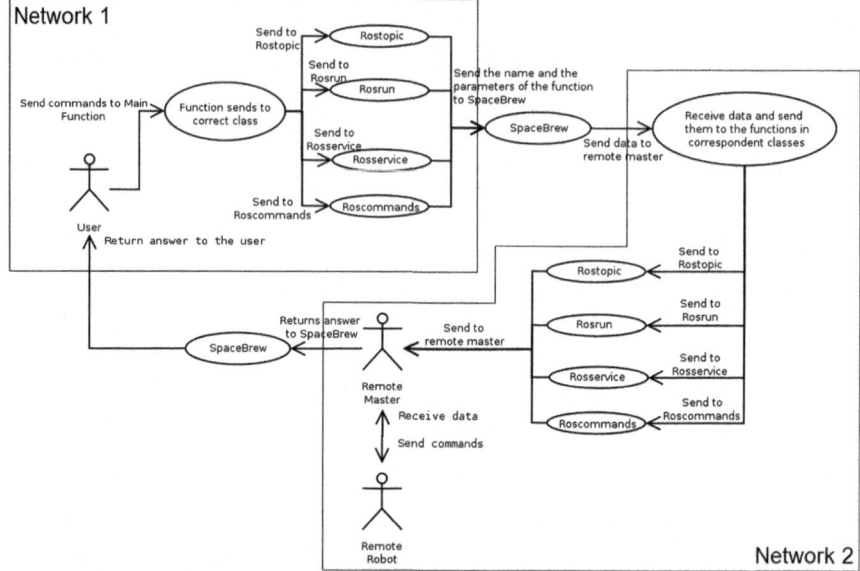

Fig. 6 Complete flow of ROSRemote application

- topic: in this example, sends the topic that will be echoed. This variable is not always needed because in some cases it is not necessary to send a topic, like on *rostopic list* command.

When the JSON variable arrives to the remote master, the latter is then able to decide to which subclass the command will be sent. In the given example, the subclass rostopic would be called.

When the *rostopic* class is requested, the *rostopicEcho* function is called, sends the command to the remote master, receives the answer and send it back, through *SpaceBrew*, to the local master, so that the user can get it.

ROSRemote implements the most common used tools in ROS, like *rostopic*, *rosservice* and *rosrun*, and, to facilitate the understanding, code implementation and maintenance, it was divided in some classes.

The main class is responsible for directing the command inserted by the user for the correct class and receive it on the remote master in order to call the matching function. Then each subclass deals with the command syntax and perform the actions needed to get the answer from the remote master. This response is sent back through *SpaceBrew* to the main class and showed to the user.

Figure 7 shows the classes implemented in the framework. The main class has some important functions, such as:

- send() - responsible for sending the command to the correct subclass, such as *rostopic*;

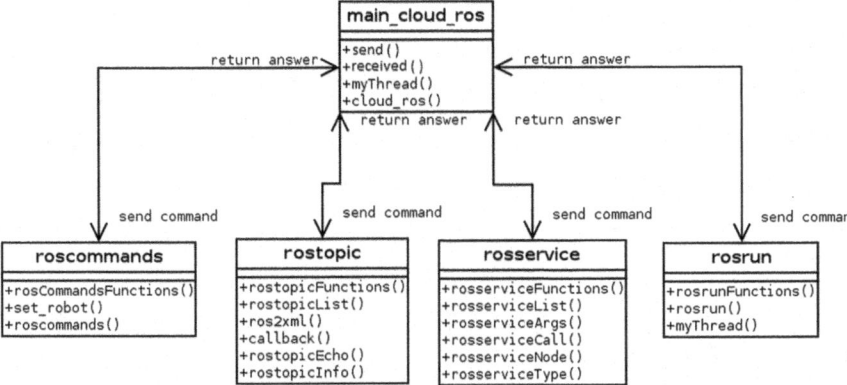

Fig. 7 UML Diagram from ROSRemote application

- received() - receives the message which is sent through *SpaceBrew*. If the message is a function, it calls the correct function. If it is an answer, it just shows it for the user;
- myThread() - responsible for keep listening teleoperational commands. At the same time the user can type a ROS command and teleoperate a robot remotely;
- cloud_ros() - the main function, responsible for creating the *SpaceBrew* client and start the /send_data service.

The code, in order to help future implementation or upgrades, is detailed in Sect. 4.4.

The sub-classes implement the code that will communicate with the remote master and all of them are very similar, differentiating only by the fact that each one deals with one ROS function.

The *rostopic* subclass is responsible for dealing with topics and is constituted by the following functions:

- **rostopicFunctions()** - responsible for verifying the command syntax and send them through *SpaceBrew*;
- **rostopicList()** - recover a list of topics running on the remote master;
- **rostopicEcho()** - shows to the user the data published by a topic in remote master;
- **callback()** - sets the publishing rate of the rostopicEcho() and keeps sending the data to the user through *SpaceBrew*;
- **ros2xml()** - converts the data received from the rostopicEcho into an XML in order to be sent by the callback() function;
- **rostopicInfo()** - shows all the information about the topic, such as which node is responsible for publishing it.

The *rosservice* subclass is responsible for dealing with services and is constituted by the following functions:

- **rosserviceFunctions**() - responsible for verify the command syntax and send it through *SpaceBrew*, just like rostopicFunctions();
- **rosserviceList**() - shows the user a list of ROS services running on the remote master;
- **rosserviceArgs**() - shows the arguments needed to call a service in the remote master;
- **rosserviceCall**() - call a service in the remote master;
- **rosserviceNode**() - shows which node is responsible for publishing a service;
- **rosserviceType**() - shows the type of the requested service;

The *rosrun* subclass is responsible for running packages and is constituted by the following functions:

- **rosrunFunctions**() - responsible for verify the command syntax and send them through *SpaceBrew*;
- **rosrun**() - runs a remote package;
- **myThread**() - create a thread responsible for dealing with the running remote package, leaving the terminal free for the user to insert another command, otherwise, the framework would stop until the package finishes running;

In order to help users to operate a robot remotely, the *roscommands* was created. It allows the user send directions with keyboard arrows for the remote robot and is constituted by the following functions:

- **rosCommandsFunctions**() - responsible for verify if the command is a keyboard arrow or another one;
- **set_robot**() - set the robot that will be controlled remotely. It is necessary because *rosCommands* publishes in the cmd_vel topic of the correct robot, and thus the framework needs to know which node is the right one. It is possible to choose one or several nodes to send the command at the same time;
- **roscommands**() - publishes the speed on cmd_vel topic, wait a fraction of a second, and publishes zeros on the speed in order to making the robot stop its movement. This is necessary to avoid robot collisions, since once published in the cmd_vel topic, the robot will follow the same speed until another publishing.

4.4   How ROSRemote Works

To send the message, the main function splits the command inserted by the user and decides to which subclass it will be sent. The Algorithm 1 shows the function *send*, responsible for this redirecting.

Algorithm 1: Code responsible for redirecting the commands.

```
1  def send(req):
2      global brew
3      command = req.command.split(" ")
4      if(command[0] == "rostopic"):
5          rostopicFunctions(req.command, brew)
6      elif(command[0] == "rosservice"):
7          rosserviceFunctions(req.command, brew)
8      elif(command[0] == "rosrun"):
9          rosrunFunctions(req.command, brew)
10     elif(command[0] == "roscommands"):
11         rosCommandsFunctions(req.command, brew)
12     else:
13         rospy.logwarn("Incorrect command syntax")
```

Whenever a message is received by any master, the function *received* is called. It checks if the action is sending a command or receiving an answer and which function should be called. These two variables are stored in the JSON message and used to dynamically call the functions.

An advantage of using Python language is that it is one of the few languages that accept dynamic function calls. It means that using a common text, it is possible to call a function and pass arguments to it. That is what is shown on Algorithm 2, lines 4 and 5, for example. It calls the function *data['function']* that is defined in *rostopic* subclass and send to it two variables: *brew* is the *SpaceBrew* client that is needed to send the message and *data['topic']* is the topic, if needed, that will be also sent.

In different commands, the JSON message will be different and it will be explained with the source code as well as a brief tutorial on how to use in Sects. 4.5.1–4.5.3.

Algorithm 2: Code for redirecting the commands received by the remote master

```
1  def received(data):
2      global brew
3      if data['action']=="send" and data['commandRos']='rostopic':
4          method = getattr(rostopic, data['function'])
5          result = method(brew, data['topic'])
6      elif data['action']=="send" and data['commandRos']=='rosservice':
7          method = getattr(rosservice, data['function'])
8          result = method(brew, data['service'], data['args'])
9      elif data['action']=="send" and data['commandRos']=='rosrun':
10         method = getattr(rosrun, data['function'])
11         result = method(brew, data['package'], data['executable'], data['parameters'])
12     elif data['action']=="send" and data['commandRos']=='roscommands':
13         method = getattr(roscommands, data['function'])
14         result = method(brew, data['commands'])
15     else:
16         rospy.logwarn(data['title']+"\n"+data['datum'])
```

To correctly connect to the server, the user needs to know the public IP. ROSRemote has as a default address *SpaceBrew* server (http://sandbox.spacebrew.cc) but it can easily be changed in the main class.

Algorithm 3: Code for redirecting the commands received by the remote master

---
1   name = "rosPy Example"
2   server = "sandbox.spacebrew.cc"
3   global brew
4   brew = Spacebrew(name=name, server=server)
5   brew.addPublisher("Publisher")
6   brew.addSubscriber("Subscriber")

Algorithm 3 shows where the developer can change the server address, in case a new one is created. Line 1 is the application name, it is possible for the user to change it and that is what will appear on *SpaceBrew* web admin page. Line 2 is the server itself, in case it does not have a name, it may changed by its public IP.

Although there is a default server and port for running all applications, the user can set it in the code if needed, and it is very simple. In line 2 just change the name of the server for the name or the IP of the server the user wants to connect to and in line 4, simply add a third variable after server, called port, and set the value. It is possible to set directly inside the brackets
brew = Spacebrew(name=name, server=server, port=9000)
or create a variable "port" that receives the number and then add it in the brackets
port = 9000
brew = Spacebrew(name=name, server=server, port=port)
The default port for *SpaceBrew* is 9000 and the server is "sandbox.spacebrew.cc".

Since *SpaceBrew* server is for public use and everybody has access to it, some care should be taken in case the application has confidential information. In this case it is really simple for another user to easily intercept the data just by changing the connection in *Web Admin Tool*. In order to not allow this happens, it is recommended that the user creates its own server or implements some kind of security, like encrypt their data, before using *SpaceBrew*.

4.5 Rostopic

Rostopic contains all the commands responsible for displaying information about existing topics, including publishers, subscribers, and, of course, the messages [7]. The ROSRemote includes some of the most used tools in *Rostopic* library, such as *list*, *echo* and *info*.

The data flow is similar to the one explained in Fig. 6. The local master sends the command to the remote master, that then knows which function to call based on the message received through JSON parameter.

When a command is typed, the ROSRemote main function splits the string by spaces and uses the first word to decide to which subclass the command will be sent. The second word is used by the subclass to send the correct JSON message to the remote master. In addition, if a command is incorrect, the software shows the user a message with the correct syntax.

The Algorithm 4 shows what the subclass does when it receives the command inserted by the user. After decide to call the subclass *rostopic*, the code decides, in this example, that the activity that needs to be done is *list* and creates a JSON message with the necessary variables. This message is sent through *SpaceBrew* in line 4.

Algorithm 4: Piece of code that shows which command was given and returns the correspondent function name.

---

1 commandSplit = command.split(" ")
2 if commandSplit[1] == "list":
3   data = { 'commandRos':'rostopic', 'function':'rostopicList', 'action':'send', 'topic':'' }
4   brew.publish("Publisher", data)
5   rospy.logwarn("command sent = "+command)

When the message is received in the remote master, it calls the function *received*, showed in Algorithm 2 and it calls the method that corresponds to the name passed in data['function'] variable, in this case, *rostopicList*.

All the functions will be described in the next sections, with its usage and source code.

4.5.1 Rostopic List

The *rostopic list* is responsible for retrieving all the topics that are being published in the remote master. In order to call it, the user must enter the following line in the terminal:

rosservice call /send_data "rostopic list"

This will make local master send a JSON message to the remote master. This message contains the command, which is *rostopic*, the function that will be called when the message is received remotely, in this case it is *rostopicList*. The action parameter simply shows if the messages are being sent by the local master or if they are received by it and the parameter topic sends a topic to be consulted. Here, an empty string is being sent, since there is no topic to be evaluated in this function.

data = 'commandRos':'rostopic', 'function':'rostopicList', 'action':'send', 'topic':''

Then the remote master will call the function *rostopicList*, showed in Algorithm 5. It gets the published topics, put them in another JSON variable and returns to the local master through *SpaceBrew*. The variable *action* receives the *receive* value, indicating that when the local master unpacks the answer, it knows that the only necessary thing to do is print the results to the user.

Algorithm 5: Code that gets all remotely published topics and send them to local master.

```
1 master = rosgraph.masterapi.Master('/rostopic')
2 answer = master.getPublishedTopics('/')
3 datum = ""
4 for i in range(0, len(answer)):
5    datum += "\n"+answer[i][0]
6 data = { 'datum':datum, 'title':"Rostopic list results", 'action':'receive' }
7 brew.publish("Publisher", data)
```

4.5.2 Rostopic Echo

The *rostopic echo* is the tool that shows the user what is being published by a topic at a certain moment. It is necessary for the user to process the data that the were sent by the master, discover if the topic is still publishing and see the messages when needed.

Just as *rostopic list*, using *rostopic echo* is simple and only requires two extra parameters: the topic and a frequency that it will be published.

To replicate this functionality in ROSRemote, it was necessary to create three functions, the *rostopicEcho*, the *callback* and the *ros2xml*, all of them communicate among themselves and play an important part on the general feature.

The *rostopicEcho* is the main one and the most difficult to implement and understand because in Python, when a user wants to subscribe to a topic, it is necessary to import the correspondent class in the code. However, the user do not know which class should be imported and can not import all existing classes in Python because this is unfeasible, since ROS has a huge amount of different classes. Because of this problem, it was necessary to create a way to dynamically import classes and Algorithm 6 shows the solution.

The first thing is that users do not need to input the type of the topic, they just type its name, so line 1 of the code shows how it was possible to get this information using *rostopic type*, which returns the type.

Algorithm 6: Code that recovers rostopic echo data and sends it for the creation of the xml that will be returned to the user.

```
1 proc = subprocess.Popen(["rostopic type "+topic], stdout=subprocess.PIPE, shell=True)
2 (data, err) = proc.communicate()
3 datum = data.split("/")
4 datum[0] += ".msg"
5 mod = __import__(datum[0], fromlist=[datum[1]])
6 klass = getattr(mod, datum[1].strip())
7 rospy.Subscriber(topic, klass, callback)
```

In Python, the subprocess module, used in line 1, allows the user to create new processes, and get their input, output and/or error pipes and obtain their return codes [25]. This means that it is possible to send the ROS command through this module and receive the answer as if it was inserted directly in the terminal.

After the software gets the type of the topic, it is possible to recreate the import as a user would normally do. The answer were split by the "/" character and a ".msg" were added to the string. Then the program is able to import, in line 5 and 6, the designated class and, in line 7, starts the callback function, responsible for publishing the topic in a certain frequency.

Algorithm 7 shows a piece of callback function where the data is published in a certain frequency. While a variable called stop is set to false, it keeps sending data in xml.

Line 2 sends the answer to the ros2xml function and saves it in the "xml" variable, then line 2 creates the JSON message, line 3 sends it to *SpaceBrew* and line 5 makes the function sleeps for the frequency asked by user.

Algorithm 7: Code that sends the answer from rostopic echo to be converted in an xml string and returns to *SpaceBrew*.

```
1 while(stop_ == False):
2    xml = ros2xml(answer, "")
3    data = { 'datum':xml, 'title':"Rostopic echo results ", 'action':'receive' }
4    brew_.publish("Publisher", data)
5    freq = freq/1
6    time.sleep(freq)
```

Finally, the method that converts the answer in a xml was found on the internet and can be seen in [26].

The syntax for this command is slightly different than the *rostopic list* because it needs two parameters, the topic and the frequency.

The topic is the name of the topic which the user wants to echo the messages, for example, cmd_vel, and the frequency is how many times per second the user wants to see the messages. It is worth mentioning that *SpaceBrew* does not allow a high frequency of data, it recognizes this as a DoS[2] attack and halt the connection.

The syntax for *rostopic echo* is

rosservice call /send_data "rostopic echo cmd_vel 1"

Where rostopic echo is the command, cmd_vel is the name of the desired topic and 1 is the frequency, in this case, once per second.

---

[2]DoS - is short for Denial of Service and happens when a server receives thousands of malicious requests per second and can not handle them, so it negates the service for new requests, leaving users without the resources they need when trying to connect to the server.

4.5.3   Rostopic Info

The *rostopic info* serves to show the user information about topics, like the node that is publishing it. Its implementation, like the rest of the features from now on, was simple. As shown in Algorithm 9, it was just necessary to send the command to the terminal through Python, get the answer and send to the user via JSON variable.

Algorithm 8: Code that sends the answer from rostopic info to *SpaceBrew*.

---

1 proc = subprocess.Popen(["rostopic info "+topic],
   stdout=subprocess.PIPE, shell=True)
2 (datum, err) = proc.communicate()
3 data = 'datum':datum, 'title':'Rostopic info results '+topic,
   'action':'receive'
4 brew.publish("Publisher", data)

In line 2 there are two variables received, but at ROSRemote only "datum" is used, there will be no errors because if the user inserted the wrong syntax, ROSRemote will send its own error message, and if the syntax is correct, ROS will send any errors that may happen and they will be stored in datum variable, to be shown to the user.

Lines 3 and 4 create the JSON variable, as in other algorithms, and send to *SpaceBrew*, to be printed to the user in the local master.

4.6   Rosrun

Rosrun [27] is a library that helps the user execute any ROS package without knowing their full path, for doing so, the only necessary thing the user needs to know is the package and executable names, for example, to run ROSRemote, as written in Sect. 2.1 it is just needed to know the package cloud_ros and the executable main_cloud_ros.

Just like all other commands, when the user calls *rosrun* the ROSRemote main class knows that the rosrun subclass needs to be initiated and, inside this subclass, all the necessary processing is done. To use this function in ROSRemote the following syntax is necessary.

rosservice call /send_data "rosrun package_name executable_name"

In this case there is an optional parameter that can be sent, obeying a syntax that had to be changed from the common ROS to ROSRemote. In ROS, to send a parameter, users have to put the parameter name followed by a ":=" and then the parameter value. In Python this is not possible because it is a reserved symbol. So in ROSRemote, to fix this problem, the ":=" were changed by "@". Therefore, the syntax is changed from

rosservice call /send_data "rosrun package_name executable_name parameter_name:=parameter_value"

to the the final syntax which is

rosservice call /send_data "rosrun package_name executable_name parameter_
name@parameter_value"

Internally, ROSRemote understands the "@" as the reserved command ":=" and
changes it before calling rosrun. It is also important to remember that the parameter
is optional and that ROSRemote works with or without it. When the user wants to
execute a package by sending *rosrun* to remote master, the main function of rosrun
subclass decides if there is or there is not a parameter, and create a JSON variable
with an empty variable "parameters" in the second case, or assign the parameter to
it in the former case, as shown in Algorithm 9.

Algorithm 9: Code that sends the answer from rostopic info to *SpaceBrew*.

---

1 data = 'commandRos':'rosrun', 'function':'rosrun', 'action':'send',
  'package':commandSplit[1], 'executable':commandSplit[2],
  'parameters':commandSplit[3]

When a package is initiated at remote master, the user can then see the topics
the respective node publishes by sending the *rostopic list* Sect. 4.5.1 command. In
order to make this application a non-blocking one, when the user starts a package
at the remote master, it creates a new thread responsible for dealing with this task.
Algorithm 10 shows how it is done.

Algorithm 10: Code that creates the thread.

---

1 class myThread (threading.Thread):
2   def __init__(self, threadID, name, counter):
3   threading.Thread.__init__(self)
4   self.threadID = threadID
5   self.name = name
6   self.counter = counter
7   def run(self):
8   global proc
9   (dados, err) = proc.communicate()

Line 3 creates a thread with a name and an ID and, and line 7 defines the function
run, responsible for starting the thread. Algorithm 11 shows how the *rosrun* function
was implemented and what happen when the thread is created and started.

Line 1 just splits the parameter from its value in order to recreate them using the
correct ROS syntax (:=) so that the package can be called in line 7, then a JSON
variable is created to show the user that the package is running and, finally, in lines
11 and 12 a thread is created so that the framework would not be blocked in the
remote master, and the answer is sent through *SpaceBrew* to the user.

Algorithm 11: Code that changes the "@" by ":=" and sends the rosrun command to the remote master.

```
1 parameters = parameters.replace("@", ":=")
2 global proc
3 if(parameters != ''):
4   proc = subprocess.Popen(["rosrun " +package +" "+ executable+" "+parameters],
    stdout=subprocess.PIPE, shell=True)
5 else:
6   proc = subprocess.Popen(["rosrun " +package +" "+ executable], stdout=subprocess.PIPE,
    shell=True)
7 data = {'datum':datum, 'title':'Package is running ', 'action':'receive'}
8 thread1 = myThread(1, "Thread-1", 1)
9 thread1.start()
10 brew.publish("Publisher", data)
```

4.7 Rosservice

The rosservice command has several commands that helps user to call the services and find out which are currently online in the local master, which nodes this service creates and gets specific information about a service, such as its type [8].

It was also implemented in the ROSRemote framework and its usage is like others and the user have to call a service and send a service as the parameter. For example purposes it is supposed that users want to start a service to make a robot go to a destiny in the remote master, and let this service be called "robot_2_destiny" so the user must input in the terminal.

rosservice call /send_data "rosservice call /robot_2_destiny"

This command will start the service on the remote master, and the user will be able to see all of its information using the commands showed in the following sections, just like it would happen in a local instance of ROS.

Just like all the rest of the framework, in ROSRemote it also has a main function responsible for creating the JSON variable with the command (rosservice), the function, depending on which was the user input, the action of sending the request or receiving the answer, the name of the service when needed and the arguments that this service requires, as shown below.

data = 'commandRos':'rosservice', 'function':'rosserviceCall', 'action':'send', 'service':commandSplit[2], 'args':argsSplit[1]

The variables service and args are not always required. For example, when the user wants only to see all the services running on the remote master, all that is needed is to send

rosservice call /send_data "rosservice list"

And the framework will return this list for the user. More of the rosservice features will be show in the next sections.

4.7.1 Rosservice List

Rosservice list helps the user to see all services that are currently running in the master, its syntax is really simple, and it is just necessary to send *rosservice list* to the remote master.

rosservice call /send_data "rosservice list"

As in other functions, ROSRemote will send the command through *SpaceBrew* and return a list of all services to the user. The Algorithm 12 shows how this function works.

Algorithm 12: Code that recovers all running services on ROS and return them through *SpaceBrew*.

---
1 proc = subprocess.Popen(["rosservice list"], stdout=subprocess.PIPE, shell=True)
2 (dados, err) = proc.communicate()
3 data = {'datum':datum, 'title':'Rosservice list results', 'action':'receive'}
4 brew.publish("Publisher", data)

The code just send to the terminal the command *rosservice list* and creates a JSON variable that is send back to the user and printed on terminal.

4.7.2 Rosservice Args

The *rosservice args* shows the arguments necessary to call a service, and takes as a parameter the name of the service that the user wants to inspect. At local master, the correct syntax for the command is:

rosservice args /service_name

And in ROSRemote, it is practically the same, the only thing the user needs is to call the function /send_data, which is responsible for sending the command to the remote master. So the syntax for *rosservice args* in the ROSRemote is:

rosservice call /send_data "rosservice args /service_name"

The Algorithm 13 shows how the command is called on remote master and returns the answer to the user.

Algorithm 13: Code that shows the necessary arguments to call a service.

---
1 proc = subprocess.Popen(["rosservice args "+service], stdout=subprocess.PIPE, shell=True)
2 (dados, err) = proc.communicate()
3 data = {'datum':datum, 'title':'Rosservice args results '+service, 'action':'receive'}
4 brew.publish("Publisher", data)

The algorithm is really simple and just get the answer from the ROS terminal, packs it in a JSON variable and sends back through *SpaceBrew*.

4.7.3 Rosservice Call

Rosservice call is responsible for calling a service passing the name of the service and the necessary arguments, when requested, to ROS, so that the service can be started. In ROS, the syntax of this command is:

rosservice call /service_name

And in ROSRemote, the user just needs to adds the service "/send_data" on the syntax, in order to send this request to the remote master. So, one must use the following command to call a remote service.

rosservice call /send_data "rosservice call /service_name"

Optionally, it is possible to send arguments when calling a service, in this case, they must be sent inside quotation marks so that ROSRemote can understand them as arguments and recreate the complete command remotely. The Algorithm 14 shows how ROSRemote creates the command *rosservice call* based on the user input.

Algorithm 14: Piece of code that indicates which command was given and returns the name of correspondent function.

```
1  elif commandSplit[1] == "call":
2    if len(commandSplit) ¡ 3:
3      rospy.logwarn("syntax = rosservice call /service")
4    elif len(commandSplit) == 3:
5      data = {'commandRos':'rosservice', 'function':'rosserviceCall',
       'action':'send', 'service':commandSplit[2], 'args':''}
6      brew.publish("Publisher", data)
7      rospy.logwarn("command sent = "+command)
8    else:
9      argsSplit = command.split("'")
10     data = {'commandRos':'rosservice', 'function':'rosserviceCall',
       'action':'send', 'service':commandSplit[2], 'args':argsSplit[1]}
11     brew.publish("Publisher", data)
12     rospy.logwarn("command sent = "+command)
```

Lines 4 and 8 decide if the user sent the command with an argument or did not, and the properly JSON variable is created in 5 and 10. In the first case, the user sends no arguments to the service (probably because it does not request any), and the variable is then created with an empty argument. In the latter one, the service needs one or more arguments in order to work, and then it is necessary to send them.

In line 9, the arguments are split by quotation marks, in order to separate them from the command itself, and then they are added as a variable called "args" in the JSON and send through *SpaceBrew*.

In this case the function called is rosserviceCall and inside it, it is necessary to verify if the variable args was filled or not, in the first case, the function, as seen in Algorithm 15 needs to create the command with the arguments inside single quotation marks, in the second case, there is no necessity of doing so, because the arguments

---

Algorithm 15: Code that starts a service with or without arguments.

---

1 if args=="":
2   proc = subprocess.Popen(["rosservice call "+service], stdout=subprocess.PIPE,
   shell=True)
3 else:
4   proc = subprocess.Popen(["rosservice call "+service+" '"+args+"' "],
   stdout=subprocess.PIPE, shell=True)
5 (dados, err) = proc.communicate()
6 data = { 'datum':datum, 'title':'Rosservice args results '+service+' '+args, 'action':'receive' }
7 brew.publish("Publisher", data)

---

are empty and sending a single quotation mark with an empty string inside would cause the program to crash.

Once the service is started, it is possible for the user to see it by sending the *rosservice list* command as it would happen if the user were working on pure ROS.

4.7.4 Rosservice Node

Rosservice node shows the name of the node that provides a certain service and, for that, is necessary the user to know the name of the service and send it as a parameter. The command input is simple, the user just have to call it using *rosservice node* and then pass the service name preceded by a slash (/service_name). The final syntax is:

rosservice node /service_name

To use this command, just like others, users have to pass it as a parameter of the /send_data service:

rosservice call /send_data "rosservice node /service_name"

Algorithm 16: Code that returns to user the node responsible for a service offered by ROS.

---

1 proc = subprocess.Popen(["rosservice node "+service],
  stdout=subprocess.PIPE, shell=True)
2 (datum, err) = proc.communicate()
3 data = { 'datum':datum, 'title':'Rosservice node results '+service,
  'action':'receive' }
4 brew.publish("Publisher", data)

The Algorithm 16 shows how this function was implemented. Line 1 sends the command to the terminal with service inserted by the user and line 3 creates the JSON variable that will be send through *SpaceBrew* in line 4.

4.7.5   Rosservice Type

The *rosservice type* command shows what is the type of a service, very useful when writing a code in any language and is necessary to call a service in it, because, as shown in the beginning of the Sect. 4.5.2, it is necessary to import the correct class in order to make the application work. So it is really important to know the service type so that it can be imported in the application.

Just like *rostopic node*, to make the syntax of *rostopic type* correct, it is necessary to type the name of the service after a slash (/service_name), so that ROS knows which service is the desired one. So, the input command would be as follows:

rosservice type /service_name

And, exactly like other inputs, to use this one in ROSRemote, the only thing needed is to call /send_data service and pass the command just like it would be in common ROS as a parameter inside quotation marks:

rosservice call /send_data "rosservice type /service_name"

It is possible to pass an extra parameter to also show the "srv" file that defines the service. In ROS there are two ways of doing it, using the rossrv command (which is not yet implemented in ROSRemote but can easily be added in the future) and passing *rossrv* as a parameter of *rosservice*. In this case, the syntax in ROS would be:

rosservice type /service_name | *rossrvshow*

And for ROSRemote the user can type the same input as a /send_data parameter, internally, ROSRemote understands the vertical slash as a separator for the arguments and recreate the command before sending to remote master. With the following syntax, the input will work just fine with this example.

rosservice call /send_data "rosservice type /service_name | *rossrvshow*"

This will trigger the function *rosserviceType*, that is shown in Algorithm 17. If the user sends any arguments with the command, then line 4 is executed, otherwise, line 2 is executed and the results are send to the user by *SpaceBrew*.

Algorithm 17: Code that returns to the user the type of the requested service.

---

1 if args=="":
2   proc = subprocess.Popen(["rosservice type "+service],
  stdout=subprocess.PIPE, shell=True)
3 else:
4   proc = subprocess.Popen(["rosservice type "+service +" | "
  +args], $stdout = subprocess.PIPE, shell = True$)
5 (datum, err) = proc.communicate()
6 data = {'datum':datum, 'title':'Rosservice type results '+service,
  'action':'receive'}
7 brew.publish("Publisher", data)

4.8 Roscommands

This feature is nonexistent in common ROS and was created in ROSRemote so that the user is able to give robots teleoperational commands. At the beginning, it was created for tests purposes, but has become a fixed feature in the application.

It works almost like ROS, but there one extra task the user has to complete before sending commands. For these teleoperational tests, RosAria was used so that it would be possible to move the robot using keyboard arrows.

For this to work, the robot must be at the same LAN than the remote master or the master must be running in the robot, then it is possible, from the local master, the reach the remote one and send the controls.

First thing the user needs to do is to connect the remote master with the robot, for this RosAria [28] teleop package were used. When using common ROS, users just need to input the following line changing robot_IP by the robots IP address and port:

rosrun rosaria RosAria _port:=robot_IP

When this is done, ROS automatically connects with the robot. To work in this with ROSRemote, the user must use the same syntax, but, instead, sending it with the service provided. So, to call *rosaria* on a remote master the user must input:

rosservice call /send_data "rosrun rosaria RosAria _port@robot_IP"

This line connects the remote master to the robot, then there is a second step that must be done in order to make the robot move, and it is call the function set_robot, that just sets the robot or robots that will be moved.

For setting the robot, the user must send the command

roscommands set_robot "robot_names"

remembering that it is possible to send more than one robot per time just by separating their names with a colon. To know the robots name, the user can send a *rostopic list* command and get the name of the nodes that are being published by the robot. In this example, the node was called "Rosaria", so the user should send:

roscommands set_robot "Rosaria"

And that is all that is needed for the user start his application. Every time a keyboard arrow is pressed, ROSRemtoe will send a JSON variable to call roscommands function, showed in Sect. 4.8.

Although this is a big code, it is really easy to understand, all it does is set the robots speed depending on which arrow the user pressed (up, down, left or right) and publish it on /cmd_vel topic of the corresponding robot.

Line 12 splits by a colon all the robots name, then, in line 14, ROS publishes this value on all the robots, wait 0.5 s, than publish a value of 0 to everyone so that it stops moving. This measure was necessary to assure that the robot will not be moving forever, then after half a second it stops and the user can send again the same command so that it keep moving to where the user wants.

Algorithm 18: Code that makes the robot moves on the remote master.

```
 1  def roscommands(brew, commands):   global robot
 2    vel = Twist()
 3    global proc
 4    if(commands == "up"):
 5      vel.linear.x = 2
 6    elif(commands == "down"):
 7      vel.linear.x = -2
 8    elif(commands == "right"):
 9      vel.angular.z = -2
10    elif(commands == "left"):
11      vel.angular.z = 2
12    robots = robot.split(":")
13    for rob in robots:
14      pub = rospy.Publisher(rob+'/cmd_vel', Twist, queue_size=10)
15      pub.publish(vel)
16      vel.linear.x = 0
17      vel.linear.y = 0
18      vel.linear.z = 0
19      vel.angular.x = 0
20      vel.angular.y = 0
21      vel.angular.z = 0
22      time.sleep(0.5)
23      pub = rospy.Publisher(rob+'/cmd_vel', Twist, queue_size=10)
24      pub.publish(vel)
```

5 Tests

To check out the viability of this framework, some tests were performed at different times of the day to guarantee that the internet oscillation would not be a major problem for the application. During all the tests, the internet speed were measured as well as the time that took for the data to be transferred, it means how much time the application took to send the command, receive the answer and show it to the user.

These tests were done using three computers and a robot in some different architectures to prove that the masters configuration does not matter for the application, as long as it is correctly connected in *SpaceBrew*.

Some tests were performed using *SpaceBrew* free server available for users all around the world to make their tests and others were done using a particular server created at Federal University of Itajuba just to check if it would make any difference in the data transfer speed.

These tests will be explained in the next sections as well as the results that were possible to get from them. In all the tests it is designated that local master means the master that sends the command and remote master is the master that receives it, but it is important to emphasize that besides the fact they were physically in the same room, they were connected in different networks, therefore it seems like they were in different places.

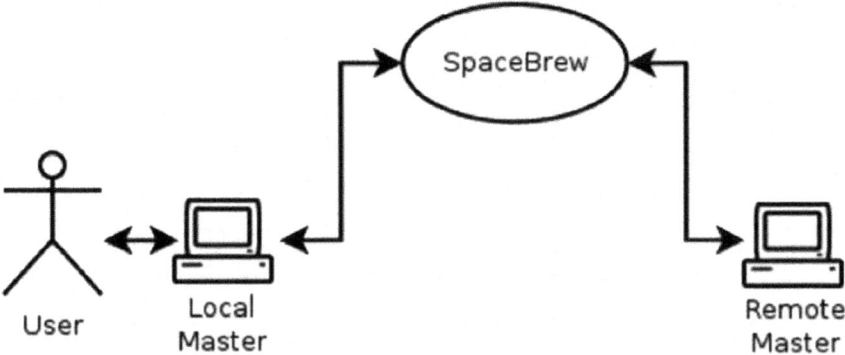

Fig. 8 Architecture of first test

5.1 Test 1 - Two Masters

The first test was the simplest one, for it was used just two masters, one local and another one remote and they communicate with each other. All *SpaceBrew* clients can send and receive information, this means that users can send data from the local and the remote master, depending on the need of the application. Figure 8 shows the architecture of the first test.

The user inputs the command in the local master and *SpaceBrew* automatically sends it to the remote one and the answer travels back the same way. This test were only made to check if the machines would communicate fast and correctly and everything werw fine during all the tests.

To check the average time and speed of the data traveling, 100 measures were taken at same circumstances but at different periods of the day, showing that, since the data traffic is not very big, it does not make any differences the period, just if the network is stable or not.

As can be noticed in Fig. 9 graphics, in 100 measurements the average time for the communication was around 0,15 s, which means that ROSRemote framework can be very helpful.

5.2 Test 2 - Two Masters and a Robot

The goal of this test was to check if it is possible to control a robot sending teleoperational commands through *SpaceBrew* and to see if the robot would respond fast to commands. Figure 10 shows the architecture of the second test.

Using roscommands it was possible to teleoperate the robot and the response time was really small, it was not possible to measure it directly in the application, like test 1, but another measurement showed that it took less than 0,1 s to make the robot moves.

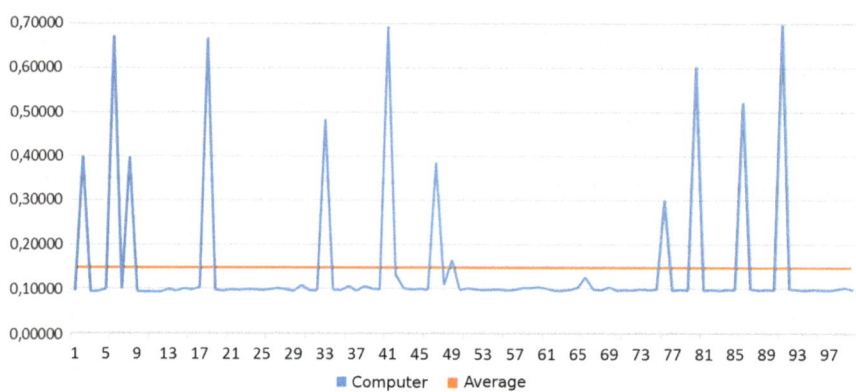

Fig. 9 Graphic for the communication time in the first test using *SpaceBrew* server

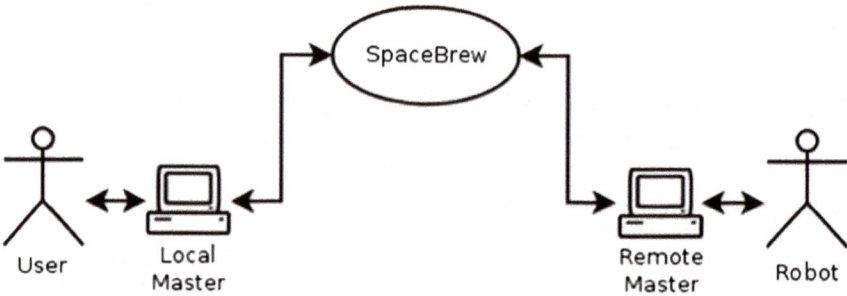

Fig. 10 Architecture of the second test

5.3 Test 3 - Three Masters

The objective of this test is to measure if the time taken to send and receive information from two different masters is more than the test made with only one master. It was also necessary to study if the messages would not arrive scrambled, it means that the messages need to reach the user in a correct order and not a piece of one answer interleaved with another.

Figure 11 shows the architecture of the third test, it is possible to notice that the local master is connected to both remote masters at the same time, and it can be configured in *SpaceBrew*. It is possible, just reconnecting *SpaceBrew* clients to make all the master communicate among themselves, it means that everybody send requests to everybody and receive the answer from all the clients.

In this case, the communication were fast and efficient, all the messages arrived in the correct order and just like the first test, the average communication time was around 0,15 s, as can be noticed in Fig. 12 graphics.

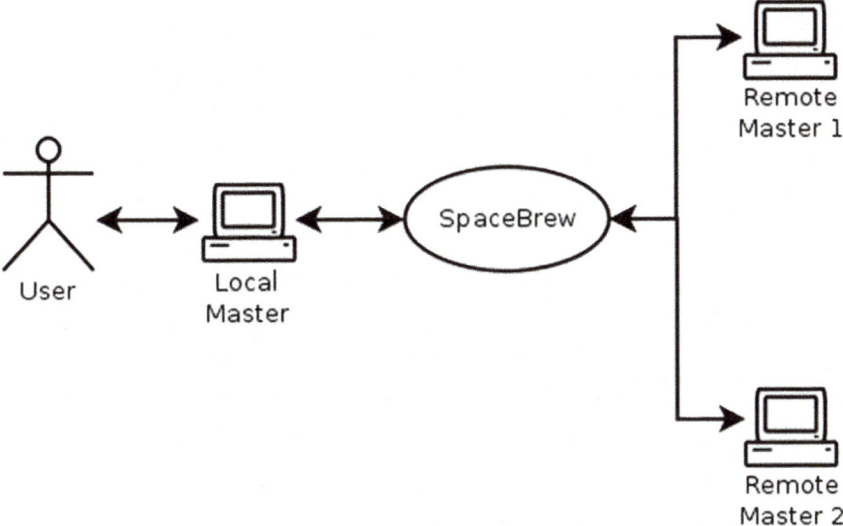

Fig. 11 Architecture of third test

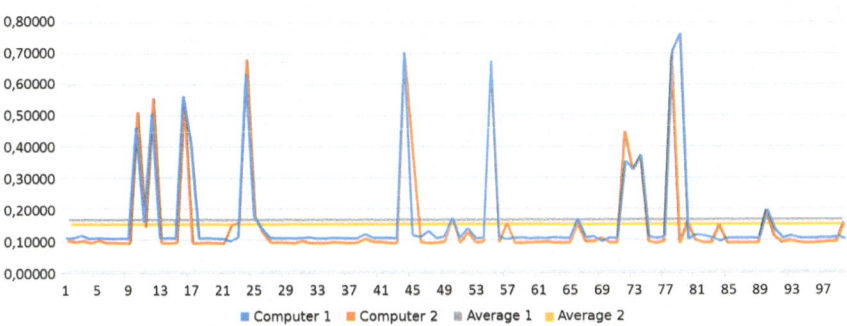

Fig. 12 Graphic for the communication time in the last test using *SpaceBrew* server

In this test the internet oscillation was a little higher than in the first one, but it does not interfere in the overall framework operation, that means that the application, even with more than one master, can be used for any application.

5.4 Test 4 - Three Masters and a Robot

The last test had as objective to determine what would happen if the user sends a command to teleoperate a robot to a master that does not have any robot connected to it. In other words, one remote master communicate with a robot and the other does not, but the local master send commands to both of them.

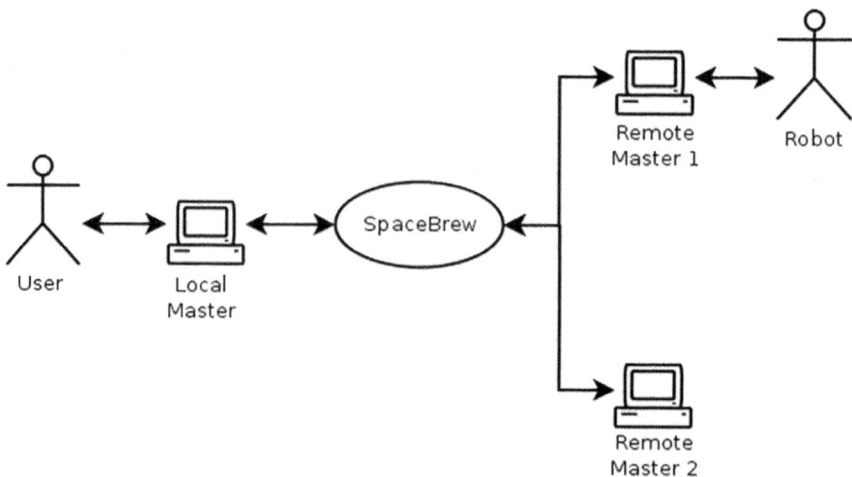

Fig. 13 Architecture of fourth test

Figure 13 shows how the test were performed. Like the third test, there were two remote masters connected to *SpaceBrew* but a robot were communicating with only one of the remote masters.

Fortunately, the results was as expected, the robot that was connected to a master moved with the command and the other machine, which had no device connected to it, returned a message to the user saying that nothing was connected to it and, therefore, was not possible to perform the action.

Even with some internet delay, the time needed for the information to return was minimal, as seen in other tests, the average time was not more than 0,16 s and the time taken for the robot to move was even less than it.

All the tests were also performed using a server created inside the Federal University of Itajuba and were also made at least 100 different communications in different periods of the day to check if the internet delay would interfere in the results.

The test done using a local server at Unifei instead of *SpaceBrew* free server proved that a local server can send data faster than the other one on the cloud. Some times is was possible to achieve really high speeds and exchange information in less than 0.001 s, but the average time was 0,0068.

Figure 14 graphics shows that the variation of time while using the server inside the University was smaller. In addition the average time was also smaller, showing that, besides for security purposes, there is much more advantages on creating a server to run these applications.

With Fig. 15 graphics it is possible to notice that the variation and traffic time using the server at Unifei was smaller than the one using *SpaceBrew* server.

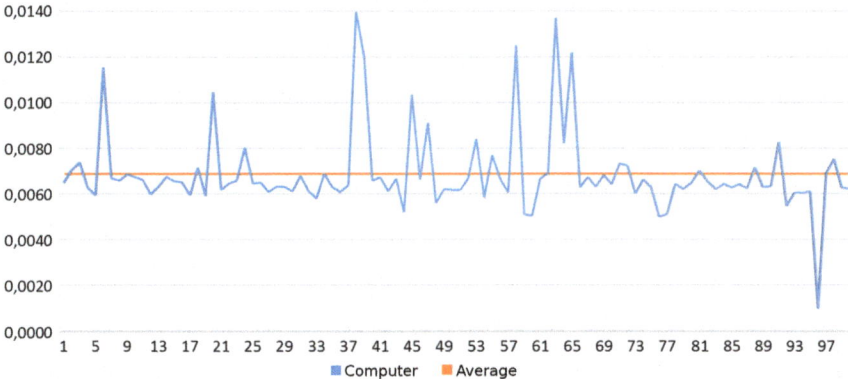

Fig. 14 Graphic of the first test using one computer and a server at Unifei

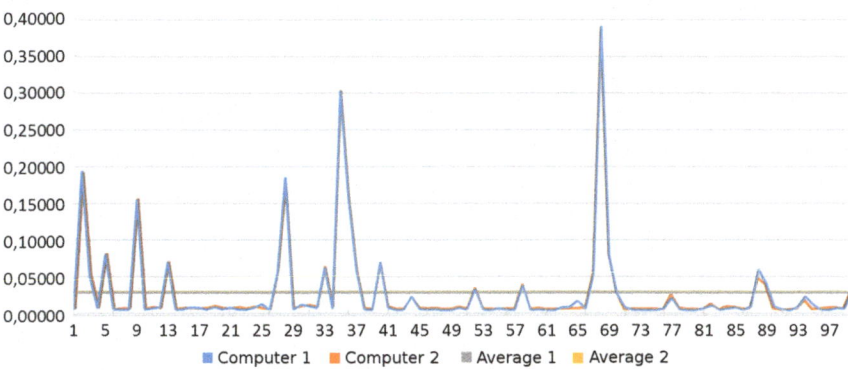

Fig. 15 Graphic of last test using two computers and a server created at Unifei

5.5 Comparison

For comparison purposes the same tests were performed using two other tools capable of transferring data: SSH and VPN. The test configuration for these two tools was the same as the one shows in Fig. 8, but it was not necessary to use *SpaceBrew* because SSH and VPN servers need to be implemented directly on the machine that the user wants to connect to.

The results of SSH were not as good as the ones from ROSRemote. The average time when using SSH was 0,33 s against 0,0068 s from ROSRemote and the internet oscillation was worse too. The traffic time for SSH can be seen in the Fig. 16 graphics.

The same way, results of VPN were worse than the ones from ROSRemote. The average time when using VPN was 0,44 s while ROSRemote obtained an average time of 0,0068 s. The traffic time for VPN can be seen in the Fig. 17 graphics.

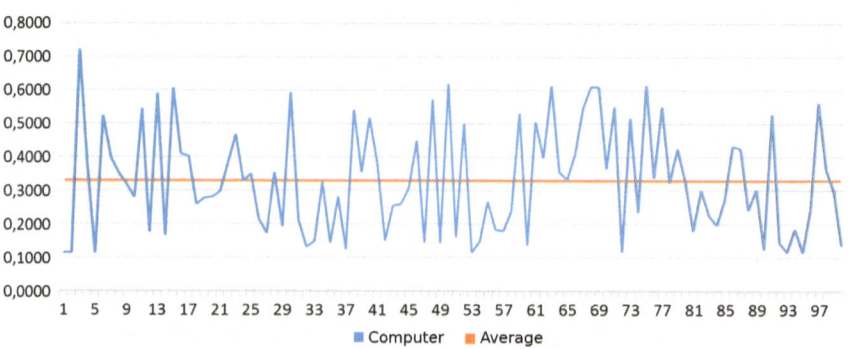

Fig. 16 Graphic of the SSH test

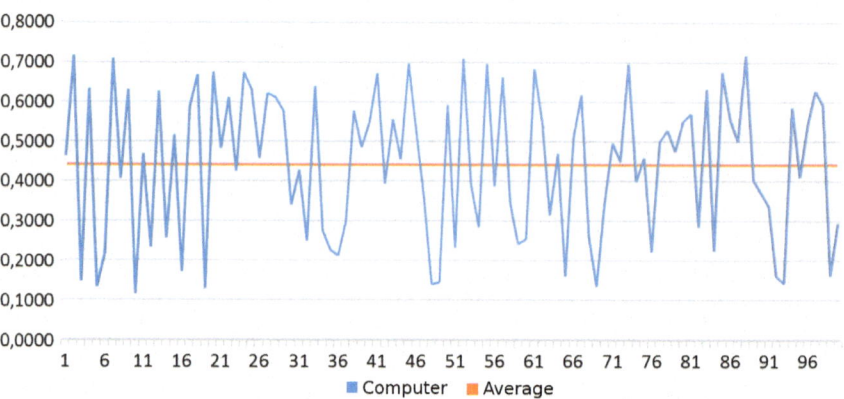

Fig. 17 Graphic of the VPN test

Besides the advantage of being faster than SSH and VPN, ROSRemote has other advantages that makes it more useful than the other two tools used for comparison. The first advantage is related to the configuration necessary to make before using SSH and VPN. After creating a server, the user needs to do a port forward in the router, in order to gain remote access. This configuration may not be possible all the time, since it needs access to routers and sometimes the user can not use this resource or does not know its password, so it is not possible to configure it. Besides, some telephony companies prohibits users to make this kind of configuration in their networks. Another problem is that, if the user is using the application in a 3G/4G network, it is just impossible to do a port forward, which prevents SSH and VPN usage.

Another advantage is that *SpaceBrew* allows user to connect and disconnect clients during execution time. It means that if the user wants to stop receiving information from one remote master, it is possible to do so without having to stop all the application, disconnecting all clients and then reconnecting only the desired ones. In addition, when user wants to connect to a client via SSH or VPN, it is necessary

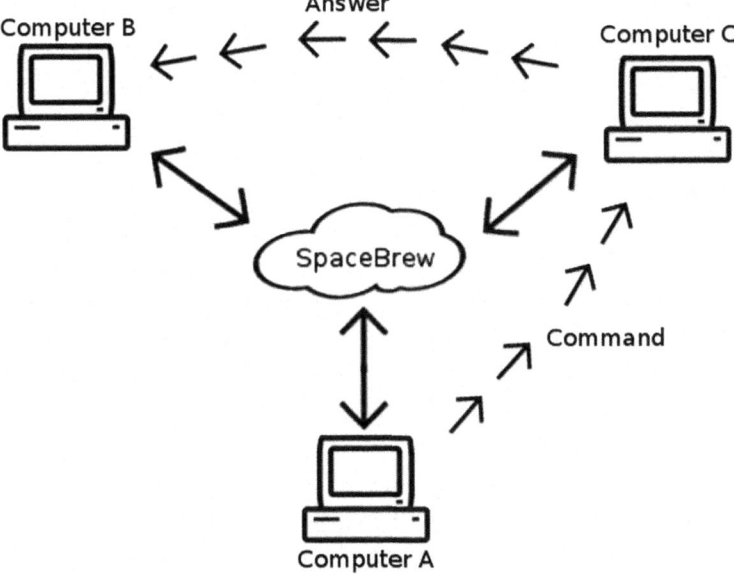

Fig. 18 Simulation of ROSRemote like users see it

to know the remote client IP address or name (in case it uses DNS). When using ROSRemote the user can give names to its applications and through Web Admin Tool, check the names to find out the remote master that is wanted to receive the information.

Besides the advantages already mentioned, ROSRemote can work with a multi-master configuration, while SSH and VPN can not. It is important because helps the user to connect different applications, devices and operating systems. Besides that, ROSRemote is independent of ROS distribution, it means that an application created in one distribution may communicate with another one that is running in a different distribution

One last advantage is the possibility of sending data to a third computer without human interaction. For example purposes is assumed three masters (A, B and C) located in different geographic locations. Computer C wants to have some data that needs to be required from computer B. Using ROSRemote it is possible to send the request from computer A and make computer B send the answer directly to computer C, as showed in Fig. 18. Internally, the answer would go to computer C through *SpaceBrew* server, but for the user, it was sent directly. On the other hand, when using SSH the user needs to send the request to computer B, wait for the answer and then forward it to computer C, as shown in Fig. 19

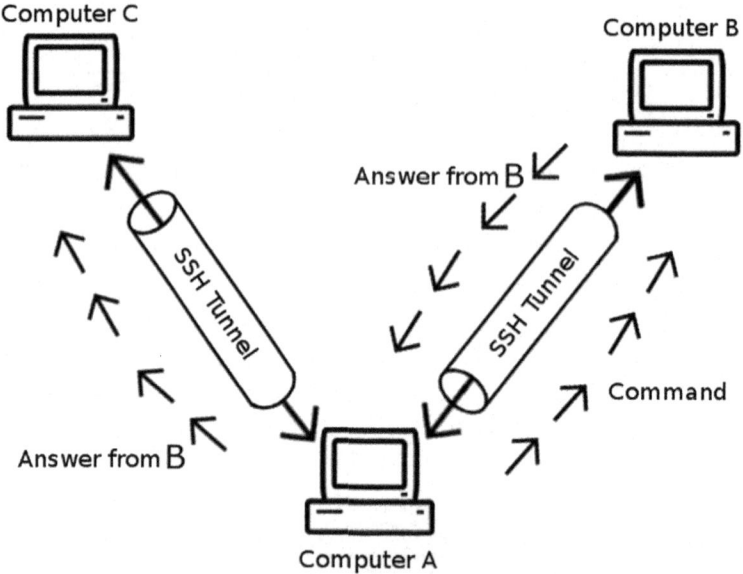

Fig. 19 Simulation of SSH like users see it

6 Conclusion

ROSRemote is considered a framework because it is possible to use it the way it has been developed and also there is the possibility to develop new features depending on the user necessity.

ROSRemote proved itself to be very efficient, since the traffic time on this tool was smaller than the other tools used in comparison. ROSRemote can send and receive information in less than a second all the time, which means that for some real time applications, ROSRemote proved to be very useful.

It is very simple to configure and use, does not need any installation and can be used with several masters at the same time. Because *SpaceBrew* has an easy-to-manage Web Admin Tool, it is possible to connect and disconnect machines with a simple click. On the other hand, SSH and VPN, used for comparison are harder to be configured and sometimes, like when users are connected to a 3G/4G, this configuration is impossible.

It is also possible to create a self-managed server, in order to improve security for the application or just for the purpose of trying to use a faster cabled connection.

References

1. H. Guoqiang, T. Wee Peng, W. Yonggang, Cloud robotics: architecture, challenges and applications. IEEE Netw. **26**, 21–28 (2012)
2. M. Quigley et al., ROS: an open-source robot operating system. in *ICRA workshop on open source software*, vol. 3.3.2 (Kobe, Japan, 2009), p. 5
3. Rockwell group, About spacebrew, 2014, http://docs.spacebrew.cc/about. Accessed 04 Aug 2017
4. P. Alyson, CloudRos (2017), https://github.com/alysonmp/ROSRemote
5. ROS, About ROS, 2007, http://www.ros.org/about-ros/. Accessed 04 Aug 2017
6. Wiki Ros, Nodes, 2017, http://wiki.ros.org/Nodes. Accessed 04 August 2017
7. Wiki Ros, Topics, 2017, http://wiki.ros.org/Topics. Accessed 04 Aug 2017
8. Wiki Ros, Services, (2017), http://wiki.ros.org/Services. Accessed 04 Aug 2017
9. C. Adjih et al., FIT IoT-LAB: a large scale open experimental IoT test bed, in *2015 IEEE 2nd World Forum on Internet of Things (WF-IoT)* (IEEE, New York, 2015), pp. 459–464
10. A. Tomoya et al., A mobile robot for following, watching and detecting falls for elderly care. Procedia Comput. Sci. **112**, 1994–2003 (2017)
11. A. Zdešar et al., Engineering education in wheeled mobile robotics, in *IFAC-Papers Online* 50.1 (2017), pp. 12173–12178
12. J. Machado Santos et al., An evaluation of 2D SLAM techniques available in robot operating system, in *2013 IEEE International Symposium on Safety, Security, and Rescue Robotics (SSRR)* (IEEE, New York, 2013), pp. 1–6
13. W.L. Audeliano, S.B. Guilherme, A hybrid self-adaptive particle filter through KLD-sampling and SAMCL, in *2017 18th International Conference on Advanced Robotics (ICAR)* (IEEE, New York, 2017), pp. 106–111
14. F. Segrera et al., Avatar medium: disembodied embodiment, fragmented communication. (2013)
15. M. Murer et al., Torquescreen: actuated ywheels for ungrounded kinaesthetic feedback in handheld devices, in *Proceedings of the Ninth International Conference on Tangible, Embedded, and Embodied Interaction* (ACM, 2015), pp. 161–164
16. F. Hemmert, G. Joost, On the other hand: embodied metaphors for interactions with mnemonic objects in live presentations, in *Proceedings of the TEI'16: Tenth International Conference on Tangible, Embedded, and Embodied Interaction* (ACM, 2016), pp. 211–217
17. T. Ylonen, C. Lonvick., The secure shell (SSH) protocol architecture, (2006)
18. A. Koubaa et al., ROSLink: bridging ROS with the internet-of-things for cloud robotics, in *Robot Operating System (ROS): The Complete Reference* vol. 2, ed. by A. Koubaa (Springer International Publishing, 2017), pp. 265–283
19. C. Scott et al., Virtual private networks, O'Reilly media, Inc. (1999)
20. Rapyuta, Rapyuta Robotics, 2015, http://www.rapyuta.org/. Accessed 20 Dec 2017
21. Gajamohan Mohanarajah et al., Rapyuta: a cloud robotics platform. IEEE Trans. Autom. Sci. Eng. **12**(2), 481–493 (2015)
22. M. Waibel et al., Roboearth. IEEE Robot. Autom. Mag. **18**(2), 69–82 (2011)
23. A. Koubaa., ROSLINK common message set, 2018, http://wiki.coins-lab.org/roslink/ROSLINKCommonMessageSet.pdf
24. A. Pereira, ROS Remote, 2017, http://wiki.ros.org/ROSRemote
25. Python, 2017, https://docs.python.org/2/library/subprocess.html. Accessed 15 Aug 2017
26. ROS answers, 2011, http://answers.ros.org/question/10330/whats-the-best-way-to-convert-a-ros-message-to-a-string-or-xml/. Accessed 15 Aug 2017
27. ROS, 2017, http://wiki.ros.org/rosbash#rosrun/. Accessed 04 Aug 2017
28. ROS, 2017, http://wiki.ros.org/ROSARIA. Accessed 04 Aug 2017

Printed by Printforce, the Netherlands